Army Officer's Guide

Army Officer's Guide

54th Edition

LTC Eric M. Hiu, USA (Ret.)

Foreword by Gen. James C. McConville, USA

STACKPOLE
BOOKS
Essex, Connecticut
Blue Ridge Summit, Pennsylvania

STACKPOLE BOOKS
An imprint of Globe Pequot, the trade division of
The Rowman & Littlefield Publishing Group, Inc.
4501 Forbes Blvd., Ste. 200
Lanham, MD 20706
www.rowman.com

Distributed by NATIONAL BOOK NETWORK

First edition published 1930. Fifty-fourth edition 2023.

This book is not an official publication of the Department of Defense or Department of the Army, nor does its publication in any way imply its endorsement by these agencies. The views presented are those of the author and do not necessarily represent the views of the Department of Defense or its Components.

British Library Cataloguing in Publication Information available

Library of Congress Cataloging-in-Publication Data

ISBN: 978-0-8117-7266-2 (paperback)
ISBN: 978-0-8117-7267-9 (electronic)

∞™ The paper used in this publication meets the minimum requirements of American National Standard for Information Sciences—Permanence of Paper for Printed Library Materials, ANSI/NISO Z39.48-1992.

Contents

PART III—SOCIAL AND FAMILY MATTERS

PART IV—QUICK REFERENCE

Foreword

The United States Army exists for one purpose: to be ready to fight and win our nation's wars as a member of the Joint Force. We win through our people—soldiers in the Regular Army, National Guard, and Army Reserve; their families; Department of the Army civilians; and Soldiers for Life, our veterans and retirees. People are the Army's greatest strength and most important weapons system. As such, they deserve the best leadership we can provide. A good leader is one who embodies the four Cs: competence, commitment, character, and caring. Good leadership is simple, but as Clausewitz observed about war and strategy, simple things can be difficult. That is why the *Army Officer's Guide* has endured for over ninety years. Since its inception in 1930, this book has ushered cadets, young officers, and all new leaders into the Profession of Arms, enabling them to become masters of their craft.

The United States military is at an inflection point, as it transitions from the post-9/11 era of prioritizing counterinsurgency and counterterrorism operations in Afghanistan and Iraq to preparing for large-scale combat operations against great powers. To remain ready to fight and win both today and tomorrow, the Army is undergoing its greatest transformation in over forty years. We have adopted Multi-Domain Operations (MDO) as the Army's capstone doctrine, the most significant change in Army doctrine since AirLand Battle. Just as our predecessors studied the 1973 Arab-Israeli War to inform AirLand Battle, we are studying lessons from Ukraine to inform MDO. MDO recognizes that we must be able to get from fort to foxhole while being contested across multiple domains—air, land, sea, space, and cyberspace. We continue to create new organizations to present multiple options to combatant commanders and multiple dilemmas to potential adversaries—organizations like our Multi-Domain Task Forces, Security Force Assistance Brigades, and the Arctic Angels of the 11th Airborne Division. We are harnessing artificial intelligence and virtual reality to augment our training. We are building our six modernization priorities, including long range precision weapons capable of penetrating enemy defenses and contributing to our own defensive abilities; helicopters that can fly farther and faster in any operating environment; and an Army network that can pass data across all systems at the speed of relevance. Perhaps, most importantly, we are taking the Army's linear, industrial-age personnel systems that managed people based on rank and MOS and replacing them with information-age talent management systems that account for knowledge, skills, behaviors, and preferences.

Like the character of war, our Army is changing. But like the nature of war, our heritage, mission, and values remain constant. This 54th edition of the *Army Officer's Guide* captures both the enduring and evolving aspects of what makes the United States Army the world's greatest fighting force. I highly recommend it as a mainstay of knowledge for all current and prospective Army officers, along with anyone seeking a better understanding of the history, culture, and operations of the United States Army.

People first! Winning matters! Army strong!

Gen. James C. McConville, U.S. Army

Disclaimer: The views presented in this foreword are my own and do not necessarily represent the views of the Department of Defense or the U.S. Army.

Acknowledgments

This edition would not have been possible without the assistance of numerous members of the U.S. Army who graciously assisted over the course of the project; their support allowed for an absolutely great product. My particular thanks go to the 40th Chief of Staff of the Army, Gen. James C. McConville, for his insightful foreword and to his team at the Army Staff. The following individuals contributed their thoughts, insights, and professional expertise to the development of much of this book: In Arlington, Virginia, Mr. Robert Dalessandro, Mr. Mike Knapp, Mr. Benjamin Brands, and Col. Chris Herrera. At the U.S. Army Cyber Center of Excellence Mr. Scott Anderson and Mr. Phillip Williams who provided the latest information on our Army's newest branch. At Fort Leavenworth, Kansas, Maj. Kyle Hatzinger who provided the latest changes to Army uniform regulations. At the U.S. Army War College in Carlisle, Pennsylvania, Col. John Bonin, USA (Ret.), who provided up-to-date information on Army organization. At Fort Eustis and Williamsburg, Virginia, Col. Larry Geddings, Maj. Cameron Albert, USA (Ret.), LTC Mike McCall, USA (Ret.). LTC Mark Young, USA (Ret.), SCPO Keith Hamlin USN (Ret.), LTC Jason Periatt, USA (Ret.), LTC Rob Taber, USA (Ret.), LTC Greg Sakimura, USA (Ret.), Col. Tom Meyer, USA (Ret.), Mr. Christopher Shearer, SCPO Charles Gum, USN (Ret.), Mr. Justin Long, Col. Doug Pietrowski, USA (Ret.), Mr. Mike Stephans and Col. Rob Marshall.

Thank you to Mr. Dave Reisch and Ms. Stephanie Otto at Stackpole Books, for allowing me the privilege of revising this edition of the *Army Officer's Guide*.

A big *mahalo* goes out to my parents Colonel (Ret.) Patrick and Bonnie Hiu for instilling in me the love of our Army and our nation.

I'd be remiss if I didn't mention my two battle buddies: Col. Theo Moore and Col. Mike Zernickow.

Special assistance was given by my daughter Capt. Jesse King and her husband Capt. Joshua King, who kept me updated on all the latest happenings in the Army at the battalion level and below.

Most of all, I would like to thank my wife, Kate, for her endless support of my "good ideas" and her keen research skills, as well as my children Jesse, Ana, Alexa and Mac for continuing to be great Army kids.

Eric M. Hiu
Williamsburg, Virginia

PART I

The Army

1

Heritage, Customs, and Courtesies of the Army

The 1942 *Officer's Guide* stated in the Foreword that the purpose of the book was to "serve as a helpful source for informative study and inspiring counsel about many of the difficult problems which face the individual as a commissioned officer of the Army of the United States." This edition is no different. You have much in common with your forebearers and the problems they faced in 1942. You will serve in an Army at war as was the case with them; you will face myriad problems in multiple venues involving your abilities as a leader, which they overcame; and you will succeed in carrying on their long tradition of service to our nation. They have passed the torch on to you; protect and guard it for future generations. In the end, soldiering is about selfless service to the nation:

THE OATHS WE TAKE

Most human cultures take oaths very seriously, and Army culture is no different. There are two reasons for this. First, lives frequently depend on leaders' given words; even in peacetime, even in garrison, the actions of leaders leave impressions or convey lessons that may have an impact on some future battlefield. Leaders' words must be unimpeachable. Oaths are the most sacred promises of all. When American soldiers enlist or reenlist, they take this oath:

> I do solemnly swear (or affirm) that I will support and defend the Constitution of the United States against all enemies, foreign and domestic; that I will bear true faith and allegiance to the same; and that I will obey the orders of the President of the United States and the orders of the officers appointed over me, according to regulations and the Uniform Code of Military Justice. So help me God.

Army officers take this oath:

> I (your name), having been appointed a (rank) in the United States Army, do solemnly swear (or affirm) that I will support and defend

> the Constitution of the United States against all enemies, foreign and domestic; that I will bear true faith and allegiance to the same; that I take this obligation freely, without any mental reservation or purpose of evasion, and that I will well and faithfully discharge the duties of the office upon which I am about to enter. So help me God.

The difference is obvious, but its meaning may not be clear. First, though, let us recognize what all American soldiers have in common. All must support and defend the Constitution against its enemies domestic and foreign and remain loyal to the American way of government and life. This requires courage, commitment, discipline, selflessness, and a regard for fellow citizens.

The key difference in the oaths—the omission of the promise to *obey* in the officer's oath—is what defines the differences between Army officers and enlisted soldiers. Officers are *legally* bound by the same Uniform Code of Military Justice (UCMJ) requirement to obey as are all other soldiers, but by their oath, they are not *morally* bound. Although it at first sounds like "doubletalk," it most assuredly is not. Officers are trusted by the American people (including enlisted soldiers), through their congressional representatives who approve commissions, with the enormous moral responsibility of knowing when to *not* obey. No system, however well thought out, however seemingly perfect, can conjure rules that are appropriate for every situation. Army regulations, orders, doctrine, and so forth are almost always right, but it is impossible to build any code of law whose logic or good sense will not at some point require a variance. That's where the officer comes in.

Officers are expected to exercise sound judgment, which sometimes—*rarely*, but sometimes—requires them to bend or break rules; to not obey those instructions, regulations, or orders that are not in the best interest of the Army at that moment. This is the single most formidable responsibility an officer can bear. This moral dispensation must never be used for personal gain or other corrupt ends. Further, although officers may be morally empowered to not obey, they are still *legally* and *morally* responsible for their actions, and it is up to the commanding officer to decide whether to prosecute an officer for exercising judgment in one of these exceptional situations.

If it sounds dicey, it is. The decision to not obey must never be taken lightly, and in the case of almost every order, directive, standard operating procedure, or regulation, swift, cheerful obedience is the appropriate response. But knowing when to do otherwise—and doing it—is the essence of what sets officers apart from enlisted soldiers. In the U.S. Army, this sort of judgment, wrapped within a full, disciplined understanding of the legal and moral impact of decisions, is *expected.*

THE ARMY IN WHICH WE SERVE

The Army in which a soldier serves is much more than his or her current unit; it is more than the sum total of all soldiers on active duty and in every component today; it is more, even, than all of them plus the Department of Defense and the civilians and family members who work for and with the Army right now.

OATH OF OFFICE - MILITARY PERSONNEL

For use of this form, see AR 135-100, the proponent agency is ODCSPER

DATA REQUIRED BY THE PRIVACY ACT OF 1974

AUTHORITY: 5 USC 3331, 552, 552a; 10 USC 10204.

PRINCIPAL PURPOSE: To create a record of the date of acceptance of appointment.

ROUTINE USES: Information is used to establish and record the date of acceptance. The SSN is used to identify the member. The date of acceptance of appointment is used in preparing statements of service and computing basic pay date.

DISCLOSURE: Completion of form is mandatory. Failure to do so will cause the appointment to be invalid.

INSTRUCTIONS

INDICATE THE APPOINTMENT FOR WHICH OATH IS BEING EXECUTED BY PLACING AN "X" IN APPROPRIATE BOX. REGULAR ARMY COMMISSIONED OFFICERS WILL ALSO SPECIFY THE BRANCH OF APPOINTMENT WHEN APPOINTED IN A SPECIAL BRANCH.

This form will be executed upon acceptance of appointment as an officer in the Army of the United States. Immediately upon receipt of notice of appointment, the appointee will, in case of acceptance of the appointment, return to the agency from which received, the oath of office *(on this form)* properly filled in, subscribed and attested. In case of non-acceptance, the notice of appointment will be returned to the agency from which received, *(by letter)* indicating the fact of non-acceptance.

COMMISSIONED OFFICERS	WARRANT OFFICERS
☐ REGULAR ARMY ____________ *(Branch, when so appointed)*	☐ REGULAR ARMY
☐ ARMY OF THE UNITED STATES, WITHOUT COMPONENT	☐ ARMY OF THE UNITED STATES, WITHOUT COMPONENT
☐ RESERVE COMMISSIONED OFFICER	☐ RESERVE WARRANT OFFICER

I, ______________________________ *(First Name, Middle Name, Last Name)* ______________________ *(Social Security Number)*

having been appointed an officer in the Army of the United States, as indicated above in the grade of ______________________________ do solemnly swear *(or affirm)* that I will support and defend the Constitution of the United States against all enemies, foreign and domestic, that I will bear true faith and allegiance to the same; that I take this obligation freely, without any mental reservation or purpose of evasion; and that I will well and faithfully discharge the duties of the office upon which I am about to enter; SO HELP ME GOD.

______________________________ *(Signature - full name as shown above)*

SWORN TO AND SUBSCRIBED BEFORE ME AT ______________________________

THIS ____________ *(Day)* DAY OF ____________ *(Month)*, ________ *(Year)*.

______________________________ *(Grade, component, or office of official administering oath)*

______________________________ *(Signature)*

FOR THE EXECUTION OF THE OATH OF OFFICE

1. Whenever any person is elected or appointed to an office of honor or trust under the Government of the United States, he/she is required before entering upon the duties of his/her office, to take and subscribe the oath prescribed by 5 USC 3331.

2. 10 USC 626 and 14309 eliminate the necessity of executing oath on promotion of officers.

3. The oath of office may be taken before any commissioned officer of any component of any Armed Force, whether or not on active duty *(10 USC 1031)*, or before any commissioned warrant officer when acting as an adjutant, assistant adjutant, acting adjutant, or personnel adjutant in any of the Armed Forces *(See UCM, Article 136; 10 USC 936)*. A commissioned warrant officer administering the oath of office will show his/her title in the block to the left of his/her signature.

4. Oath of office may also be taken before any civil officer who is authorized by the laws of the United States or by the local municipal law to administer oaths, and if so administered by a civil official, the oath must bear the official seal of the person administering the oath, or if a seal is not used by the official, the official's capacity to administer oaths must be certified to under seal by a clerk or court or other proper local official.

DA FORM 71, JUL 1999 EDITION OF DEC 1988 IS OBSOLETE USAPA V1.00

The Oath of Office

Tomorrow, new soldiers will enlist; some will depart; others will retire or go on their final leave; new officers will be commissioned; and civilians will embark on and depart for new jobs for the Army. In most civilian corporations, this would not matter, for the bottom line is almost always money; that's the way a capitalistic society works, and few Americans would want it any other way.

Thus the Army is the sum of all past and present soldiers, civilians, and loved ones who have served it. Americans may join the Army for a multitude of reasons, but they are ultimately willing to put their lives on the line *for one another*. They need not only the camaraderie of those present but also the inspiration of those who have gone before to make the ultimate sacrifice. Thus, the Army is a living, breathing entity, changing in form, scope of tasks, and operational doctrine, and changeless in mission and fundamental morality.

ARMY HERITAGE

Heritage is based on history, the truth about the past. It is more than a dry collection of dates, places, pictures, and names in books; it is more than the artifacts under glass, lights, and lock and key in museums or unit historical holdings. These are the skeleton under the frame, the touchstones that help each of us form a better comprehension of our heritage as American soldiers. Army heritage is the legacy of those who have preceded us; it is the deeds they did, the standards they set, and the expectations they left for all of us. Any comprehensive, objective look at the Army's past, however, must ultimately conclude that the 248-year heritage of the U.S. Army is one of unparalleled courage, selflessness, sacrifice, commitment, and loyalty—the very same values we must adhere to and cherish today.

In the Army, heritage is not just history, or even the inspiration and insights one can derive from its study. It is part of how we do business, day to day. It not only should be reflected in our every action but also, in fact, reflected in our customs and courtesies. No mere quaint holdovers of an irrelevant past, these are the living connections to our forebearers. To a large extent, they are what make us different from our comrades in the other uniformed services and others who put their lives on the line in service to the community or nation. Our heritage and corollary customs and courtesies help us understand our uniqueness as soldiers and to remember who we are, what we do, and why.

To learn more about Army history and heritage, visit these websites:

www.history.army.mil—The U.S. Army Center of Military History, Washington, D.C.

www.usahec.org—The U.S. Army Heritage and Education Center, Carlisle Barracks, Pennsylvania

ARMY CUSTOMS AND THEIR IMPORTANCE

A custom is an established usage. Customs include positive actions (things to do) and taboos (things to avoid doing). Like life itself, the customs that mankind observes are subject to a constant but slow process of change.

Often one of the unit social functions, called a "Hail and Farewell," is used for this purpose.

Receiving Officers of Sister Services. The officers of the host service accord a high degree of cordiality and hospitality to visiting officers from other services. This may include provisions for quarters, extension of club privileges, social invitations, introduction of the visitor to appropriate officers, and the like. But it includes above all else the hand of fellowship and comradeship to a brother or sister officer to stimulate a feeling of welcome among friends.

Military Weddings. Military weddings follow the same procedures as any other, except for additional customs that add to their color and tone. Consult your chaplain for details and arrangements that will be suitable in making wedding plans. At military weddings, all officers should wear the uniform specified in the invitation. Medals or ribbons may be worn with propriety. Badges may also be worn. Frequently, the national and unit colors are crossed just above and behind the position of the chaplain. The saber? In all likelihood, enough sabers to form the ceremonial arch of older days cannot be found. Perhaps if one is found it may be used for the first cut of the wedding cake as a polite bow to old tradition.

Reception of a Spouse. Bride and groom are usually introduced to the officers of the organization and their spouses at the next appropriate social function. Customs vary with conditions and with regard to official missions and times.

Birth of a Child. When a child is born to the family of an officer, the unit commander may send a personal letter of congratulations to the parents on behalf of the organization. By tradition, the spouses of the unit officers may purchase a silver cup from their club funds with appropriate engraving. The unit commander sends a letter of congratulations to the enlisted man and his wife or to the enlisted woman and her husband.

CUSTOMS IN CONNECTION WITH SICKNESS AND DEATH

Visiting the Sick. An officer who is sick in the hospital is visited by the officers of the unit in such numbers as may be permitted by the physician. An officer or soldier of the officer's unit visits the sick officer daily to ensure that his or her comfort or desires receive attention.

An officer's spouse who is sick in the hospital receives flowers sent in the name of the officers of the unit and their wives or husbands.

Death of an Officer or Family Member. When an officer dies, another officer is immediately designated by the commanding officer to render every possible assistance to the bereaved family. A similar courtesy may be tendered, if desired, in the case of the death of a member of an officer's family.

A letter of condolence is written by the unit commander on behalf of the brigade, regiment, group, or similar unit. Flowers are sent in the name of the officers of the unit and their spouses.

Death of an Enlisted Soldier. When an enlisted soldier dies, a letter of condolence is written to the nearest relative by the immediate commander of the deceased soldier. Flowers are sent in the name of the members of the decedent's unit for the funeral. The funeral is attended by all officers and soldiers of the deceased soldier's unit who so desire and whose duties permit.

SUPPORT OF POST AND ORGANIZATION ACTIVITIES

General. Officers are expected to support the activities of the unit to which they are assigned, as well as the activities of the entire garrison. The unit to which you are assigned consists of a closely knit group around which are entwined official duties and athletic, social, and cultural activities for the benefit of all. You are expected to support and assist, at least by your presence, many events that form a part of military life. Proper interest and pride in all activities of your unit and garrison are factors in stimulating morale and unit cohesion.

The Officers' Club. The officers' club traditionally has been the nucleus around which revolves much of the off-duty social and recreational life of officers and their spouses. At a large station, there may be branches to serve the needs of separate organizations or of distant areas. Similar establishments are provided for noncommissioned officers. The officers' club operates under the control of the station commander, often through Morale, Welfare and Recreation. Membership in the club is voluntary but is often strongly encouraged for officers assigned to the station.

In recent years, an American cultural change in attitude regarding alcohol consumption, coupled with the Army's zero-tolerance policy toward alcohol abuse, has resulted in a general decrease in club patronage. Many clubs today are struggling financially. Thus, the traditional role of the club as the center of social and recreational activity for officers and their families has changed significantly. Still, if there is a club in operation at your station, you are encouraged to become a member and use its facilities.

Attendance at Unit and Organization Parties Sponsored by Enlisted Soldiers. It is customary for officers and their spouses, when invited, to attend special social events sponsored by enlisted soldiers of a unit or organization. Conditions vary so widely throughout the service that no general customs can be identified. The best source of guidance is the commanding officer.

In general, officers, or officers and their spouses, are invited only on special occasions. When invited, officers attend in their official capacity to enhance morale and esprit. In conduct, they are mindful of the normal social amenities and are guided by the example set by the senior officer present. If the unit or organization is authorized to serve alcohol, it is accepted practice for the officer to drink in marked moderation; the nondrinker should ask for a soft drink, without excuse. Excessive drinking, exhibitionist dancing, and other ungentlemanly or unladylike behavior are frowned upon. At an appropriate time and after the customary amenities, officers depart with or immediately after the senior officer attending, leaving the party for the enjoyment of its enlisted members.

Attendance at Athletic Events. As a matter of policy, to demonstrate an interest in organization affairs as well as for personal enjoyment, officers should attend athletic events in which their unit teams participate.

Ceremonies at Holiday Dinners. On Thanksgiving, Christmas, and New Year's Day, many organizations have a tradition where the officers visit the companies during the meal or prior to the serving of the meal. The method varies widely. As an example, only the brigade, battalion, or similar commander, his or her staff, and field officers visit the dining facility just prior to the serving of dinner. Officers of the company, their families, and families of married enlisted men of the company may dine with the companies on these holidays.

THE CORRECT USE OF TITLES

Each member of the Army has a military grade, private to general, and this grade becomes his or her military title by force of regulation and custom. In official documents, a member's grade, or title, always accompanies his or her name; it is also used in conversation. Listed below are several examples. Through custom and usage, military titles are used between civilians and the military, just as custom has established the usage of "Doctor," "Professor," or "Governor."

A person who has attained a military title carries it permanently, if so choosing, including into retirement.

When addressing a subordinate, preface the soldier's last name with his or her rank. "Sergeant Jones," "Lieutenant Smith," or "Private Henderson" are appropriate means of addressing fellow soldiers of subordinate rank. Conversationally, placing a slight emphasis on the pronunciation of the rank, rather than the name, is pleasing to most ears and is better at eliciting attention than placing emphasis on the name, which may be more likely to evoke fear or alarm.

When referring to others not present, always use ranks and names. Use of last names only is disrespectful; use of pay grades (e.g. "I was talking to an O-4" or "It was that E-7") is unmilitary and degrading. All soldiers have earned their ranks, from the newest private first class to the most senior general; use them accordingly.

Titles of Commissioned Officers. Lieutenants are addressed officially as "Lieutenant." The adjectives "First" and "Second" are not used except in written communications. The same principle holds for other ranks. In conversation and in nonofficial correspondence (other than in the address itself), brigadier generals, major generals, and lieutenant generals are usually referred to and addressed as "General." Lieutenant colonels, under the same conditions, are addressed as "Colonel."

Senior officers may sometimes address subordinates as "Smith" or "Jones," but this does not give the subordinate the privilege of addressing the senior in any way other than by the senior's proper title.

"Ma'am" may be used in addressing a female officer under circumstances when the use of "Sir" would be appropriate in the case of a male officer. All chaplains are officially addressed as "Chaplain," regardless of their military grade or professional title.

Warrant Officers. The warrant officer formally ranks below second lieutenant and above cadet. He or she is extended the same privileges and respect as a commissioned officer and differs only in that there are certain regulated restrictions on command functions. Warrant officers are the Army's top-grade technicians and are addressed as "Mister" or "Miss (Mrs.)," as appropriate. Under less formal situations, warrant officers above the rank of warrant officer (WO1), are referred to as "Chief."

Titles of Cadets. Cadets of the U.S. Military Academy and the Reserve Officers Training Corps are addressed as "Cadet" officially and in written communications. Under less formal situations, they are addressed as "Mister" or "Miss."

Noncommissioned Officers. Sergeants major are addressed as "Sergeant Major." A first sergeant is addressed as "First Sergeant." Other sergeants, regardless of grade, are addressed simply as "Sergeant," and a corporal is addressed as "Corporal." Specialists are addressed as "Specialist." Privates First Class are addressed as "PFC Jones," whereas privates (in the pay grades E-1 or E-2) are addressed as "Private." The full titles of enlisted soldiers are used in official communications.

Use of Titles by Retired Personnel. Individuals retired from the armed services not on active duty are authorized to use their titles socially and in connection with commercial enterprises, subject to prescribed limitations. Official signatures will include the designated retired status after the grade, thus, "USA Retired" will be used by members on the U.S. Army Retired List (Regulars); "AUS Retired" will be used by those on the Army of the United States List.

CUSTOMS OF RANK

The Place of Honor. The place of honor is on the right. Accordingly, when a junior walks, rides, or sits with a senior, the junior takes position abreast and to the left of the senior. The deference that a young officer should pay to his or her elders pertains to this relationship. The junior should walk in step with the senior, step back and allow the senior to be the first to enter a door, and render similar acts of consideration and courtesy.

Use of the Words "Sir" and "Ma'am." The words "Sir" and "Ma'am" are used in military conversation by the junior officer in addressing a superior and by all soldiers in addressing officers. It precedes a report and a query; it follows the answer of a question. For example: "Sir, do you wish to see Sergeant Brown?" "Sir, I report as Officer of the Day." "Private Brown, Ma'am." "Thank you, Sir."

Departing Before the Commanding Officer. Officers should remain at a reception or social gathering until the commanding officer has departed.

New Year's Call on the Commanding Officer. It is Army tradition that officers and their spouses make a formal call on the commanding officer during the afternoon of New Year's Day. The pressures of the current Army mission, the desires of the commanding officer, and local or major unit custom bear upon the conduct of this event and the way it is done.

As a general guide, when the commanding officer elects to hold the event, timely information is provided regarding the time and place and the uniform to be worn. At large stations, the event usually is held at the officers' club as a reception with a receiving line; it may include a dance with light refreshments. If the senior commander does not hold the event, commanders of component units such as the brigade or battalion may choose to do so. Official funds are not provided to defray the costs. Officers whose duties of the day permit are expected to attend. In any case, think of it as a pleasant ceremonial social function, adding color to the military scene, starting the New Year in a spirit of general comradeship.

Appointments with the Commanding Officer. It is the custom to ask the adjutant, the executive officer, or an aide, as appropriate, for an appointment with the commander or other senior officer. There is no special formality about it. Just inquire, "May I see the commanding officer?" Often it is appropriate to state the reason. Take your minor administrative problems to an appropriate staff officer of your own headquarters and avoid consuming the time of your commanding officer. Save your personal requests to him or her for major matters that others cannot resolve or resolve as well.

Permission of the First Sergeant. It is the custom that enlisted personnel secure permission from the first sergeant before speaking to the company commander. It is essential to discipline that each soldier knows that he or she has the right to appeal directly to the captain for redress of wrongs.

The Open-Door Policy. The soldier's right to speak to the company commander is echoed by each commander at a higher level. It is the "open-door" policy that permits each person in the Army, regardless of rank, to appeal to the next higher commander. Indeed, this right is checked and enforced by the Inspector General. It is not uncommon for a private to talk to the battalion commander since many administrative matters are performed by the battalion staff. The officer needs to expect this and not arbitrarily bar his or her door to soldiers. Usually, if there is disagreement between a soldier and an officer, or a soldier believes that he or she has a real grievance, the soldier has the right to speak to the next senior commander and to have the matter resolved. Prior to taking a problem to the next higher commander, however, a soldier should attempt to resolve the problem using the soldier's chain of command.

Payment for Personal Services. In the past, some soldiers worked as personal servants to officers and their families in addition to their regular duties. In the Army, the custom was that such work was voluntary, with the officer compensating the soldier for the work. Historically, these soldiers were known as orderlies or strikers. Some soldiers desired such jobs to supplement their service pay, just as some soldiers take off-duty jobs, called "moonlighting." By custom and official restrictions, the use of soldiers as servants by officers and their families in garrison assignments has been terminated, except for some senior officers, and only for special reasons. For units in the field, however, in training or in combat, a senior commander may be assigned an enlisted aide for personal services so that the officer can devote maximum time to the responsibilities of command. Often the driver of the vehicle assigned to the officer performs such personal services, but generally not as a matter of assignment, and definitely not for additional pay.

TABOOS

Do Not Defame the Uniform. The officer's uniform and official or social position must not be defamed. Conduct unbecoming an officer is punishable under Article 133, Uniform Code of Military Justice. The confidence of the nation in the integrity and high standards of conduct of the officers of the Army is an asset that no individual may be permitted to lower.

Never Slink Under Cover to Avoid Retreat. As a good military person, always be proud and willing to pay homage and respect to the national flag and the national anthem. Now and then, thoughtless people in uniform are observed ducking inside a building or under other cover just to avoid a retreat ceremony and the moment of respect it includes. Or are they merely displaying their ignorance as to the purpose of the ceremony? Never slink away from an opportunity to pay respect to our flag and our anthem.

Spouses and children of Army families will wish to stand at attention and face the colors too, if the ceremony is explained to them.

Proffer No Excuses. Never volunteer excuses or explain a shortcoming unless an explanation is required. The Army demands results. More damage than good is done by proffering unsought excuses. For the most part, an officer gains respect by admitting a mistake and bearing the consequences.

Do Not Fraternize Inappropriately. It is strong Army tradition that an officer does not fraternize with enlisted soldiers as individuals in ordinary social affairs, nor gamble, nor borrow or lend money, nor drink intoxicants with them (see the discussion in chapter 3, The Army Officer).

Note, however, that in today's Army it is increasingly likely that junior officers are assigned quarters in close proximity to NCO or enlisted quarters, perhaps even occupying adjacent units in multiunit buildings. Under such circumstances, socialization by children and spouses is not only unavoidable but is acceptable and even encouraged. Socialization as families is also fine during group cookouts or similar activities. However, care is needed to avoid situations in which such socialization could be construed by others as evidence of favoritism toward the NCO or enlisted member.

Do Not Use Third Person. It is in poor taste for officers to use the third person in conversation with their seniors. For example, do not say, "Sir, does the colonel desire . . . ?" Instead, say, "Sir, is it your desire . . . ?" Most senior officers frown on the use of the third person under any condition, as it is regarded as a form of address implying servility.

Scorn Servility. Servility, "bootlicking," and deliberate courting of favor are beneath the standard of conduct expected of officers, and any who openly practice such actions earn the scorn of their associates. See chapter 3 for more details.

Avoid Praising the Commander to His or Her Face. Paying compliments directly to the commander or chief is in poor taste and suggests sycophancy, which has no place among officers. However genuine your high regard for your chief may be, to express it suggests flattery and thus may be misinterpreted. If you particularly admire your boss, you can show it by extending the standard military courtesies—and meticulously carrying out his or her policies and doing all in your power to make the organization more effective.

With respect to subordinates, however, recognition of good work on their part is an inherent part of the exercise of command; do not hesitate to commend a subordinate whose actions are praiseworthy.

Use the Phrase "Old Man" with Care. The commanding officer acquires the accolade "the Old Man" by virtue of his position and without regard to his age. When the term is used, it is more often in affection and admiration than otherwise. However, it is never used in the presence of the commanding officer; doing so would be considered disrespectful.

Avoid "Going Over an Officer's Head." The jumping of an echelon of command is called going over an officer's head (for example, a company commander making a request of the brigade commander concerning a matter that should have been presented to the commander of his or her battalion first). The act is contrary to military procedure and decidedly disrespectful. (See the earlier discussion on the open-door policy.)

Avoid Harsh Remarks. Conveying gossip, slander, harsh criticism, and faultfinding is an unofficerlike practice. In casual conversation, if you can find nothing good to say about a person, it is wiser to say nothing at all.

Avoid Vulgarity and Profanity. Foul and vulgar language larded with profanity is repulsive to most self-respecting men and women. Its use by officers is reprehensible. An officer is expected to be a lady or a gentleman, and however the traditional terms are defined, certainly they exclude the use of vulgarity and profanity in conversation.

Never Lean on a Superior Officer's Desk. Avoid leaning or lolling against a senior officer's desk. It is resented by most officers and is unmilitary. Stand erect unless invited to be seated.

Never Keep Anyone Waiting. Report at once when notified to do so. Never keep anyone waiting unnecessarily. On the drill field, when called by a senior officer, go on the double.

Avoid Having People Guess Your Name. Do not assume that an officer whom you have not seen or heard from for a considerable period will know your name when a contact is renewed. Tell him or her at once who you are, and then renew the acquaintance. If this act of courtesy is unnecessary, it will be received only as an act of thoughtfulness; if it happens to be necessary, it will save embarrassment. At official receptions, always announce your name to the aide.

Do Not Smoke. The Army officially discourages smoking, both for reasons of personal health and in deference to the wishes of nonsmokers. Smoking is not permitted in public buildings. Smoking by spectators during outdoor ceremonies, such as parades, also is considered objectionable. Indoors, in quarters or other areas where smoking may be permitted, a considerate officer who smokes should be sensitive to the wishes of others.

NCOs Do Not Work on Fatigue. A custom said to be as old as the Army is that which exempts NCOs from performing manual labor while in charge of a fatigue detail or while on fatigue.

LOCAL CUSTOMS: A WORD OF CAUTION

There is a tendency to confuse customs, traditions, and social obligations. Customs of the service that have been treated in this chapter are those that are universally observed throughout the Army. Traditions are much less formal than the recognized customs of the service. There are many more traditions than could be covered here. Many are confined to a particular unit, organization, station, or branch of service, and one is likely to become acquainted with them quickly upon reporting for a new assignment. Such traditions may catch on and become widespread because of the mobile nature of Army life, however.

There is a danger in expecting that others will accept a particular tradition once you are outside the area where it has been observed. For example, many inquiries have been received as to why previous editions of *Army Officer's Guide* offered no counsel about how an officer recognizes his or her promotion in relation to fellow officers and civilian coworkers. Research into this question reveals that, officially, there is no custom of the service to provide guidance, nor is there any well-entrenched tradition anywhere. Inasmuch as promotion is a personal thing and has differing degrees of meaning at various times and places, an officer may choose to do something or nothing, the former on any scale that befits his or her mood, position, or pocketbook at the time.

Officers should become acquainted with local traditions and customs and not unjustly criticize anyone who has no tangible understanding in the local military community.

THE SEVERAL MILITARY SALUTES

History of the Military Salute. Men of arms have used some form of the military salute as an exchange of greeting since the earliest times. It has been preserved and its use continued in all modern armies that inherit their military traditions from the Age of Chivalry. The method of rendering the salute has varied through the ages, as it still varies in form among armies today. Whatever form it has taken, it has always pertained to military personnel, and its use has been definitely restricted to those in good standing.

In the Age of Chivalry, knights were mounted and wore steel armor that covered the body completely, including the head and face. When two friendly knights met, it was the custom for each to raise the visor and expose his face to the view of the other. This was always done with the right hand, the left being used to hold the reins. It was a significant gesture of friendship and confidence, since it exposed the features and also removed the right hand—the sword hand—from the vicinity of the weapon.

The military salute is given in recognition to a comrade in the honorable profession of arms. The knightly gesture of raising the hand to the visor came to be recognized as the proper greeting between soldiers and was continued even after modern firearms had made suits of armor a thing of the past. The military salute of today is as it has always been, a unique form of greeting between military personnel.

The Different Forms of the Salute. There are several forms in which the prescribed salutes are rendered. The officer normally uses the hand salute;

however, when under arms, he or she uses the salute prescribed for the weapon with which armed. Under certain circumstances, when in civilian clothes and saluting the flag or national anthem, the member salutes by placing the right hand over the heart (see the accompanying illustration); if a male officer is wearing a hat, he salutes by removing it and holding it in his right hand, such that the hand is over the heart while the hat is over the left shoulder. In this chapter, unless stated otherwise, the hand salute is intended.

When to Use the Hand Salute and the Salute with Arms (AR 600-25). Outdoors, all soldiers in uniform are required to salute at all times when they meet and recognize persons entitled to the salute, except in public conveyances, such as trains and buses, or in public places, such as theaters, or when a salute would be manifestly inappropriate or impractical.

Salutes are exchanged between officers (commissioned and warrant) and between officers and enlisted personnel. Salutes are exchanged with personnel of the U.S. Army, Navy, Air Force, Marine Corps, and Coast Guard entitled to the salute. It is customary to salute officers of friendly foreign nations when recognized as such. The President of the United States, as the commander-in-chief, is saluted by Army personnel in uniform. Civilians may be saluted by persons in uniform when appropriate, but the uniform hat or cap is not raised as a form of salutation.

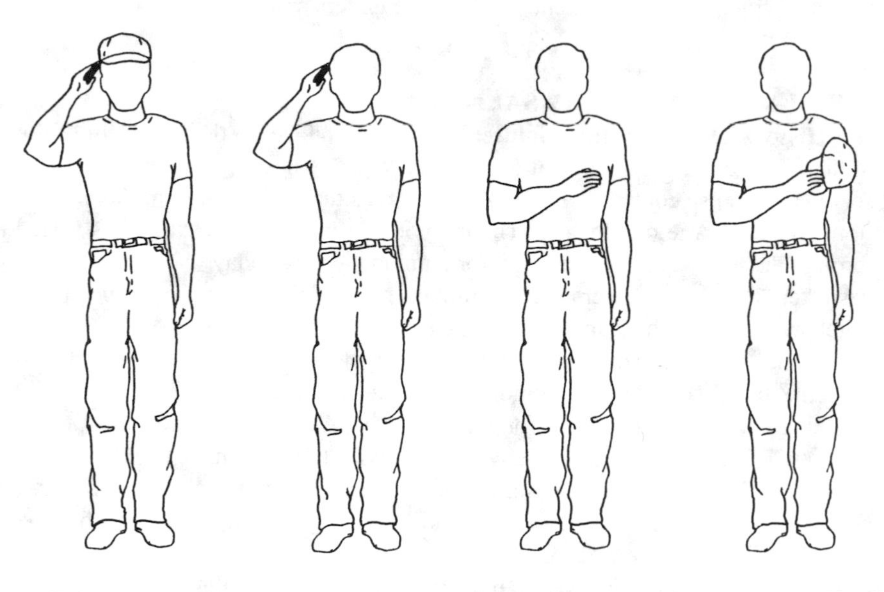

Exchanging salutes when in civilian clothes with or without hat

Saluting the flag or national anthem when in civilian clothes without hat

Saluting the flag or national anthem when in civilian clothes with hat

Forms of the Salute When in Civilian Clothes

Military personnel under arms render the salute prescribed for the weapon with which they are armed, whether or not that weapon ordinarily is prescribed as part of their equipment.

If the exchange of salutes is otherwise appropriate, it is customary, although optional, for military members in civilian clothing to exchange hand salutes upon recognition. Civilian personnel, including civilian guards, do not render the hand salute to military personnel or to other civilian personnel.

Except in formation, when a salute is prescribed, the individual either faces the person or colors saluted or turns the head so as to observe the person or colors saluted.

Covered or uncovered, salutes are exchanged in the same manner.

If running, a person comes to a walk before saluting.

The smartness with which the officer or soldier gives the salute indicates the degree of pride the member has in his or her military responsibilities. A careless or halfhearted salute is discourteous.

Methods of Saluting Used by Officers. The hand salute is the usual method. Although in most instances it is rendered while standing or marching at attention, it may be rendered while seated (e.g., an officer seated at a desk who acknowledges the salute of an officer or soldier who is making a report).

The salute by placing the *right hand over the heart* is used under three conditions. At a military funeral, all military personnel dressed in civilian clothes use this form of salute in rendering courtesies to the deceased. Male members of the services in civilian clothes and *uncovered* (without headdress) and female members in civilian clothes, *uncovered* or *covered* (with headdress), salute this way during the national anthem, "To the Color," or "Hail to the Chief." While in the same dress, this salute is used in paying homage to the national flag or color. Males in civilian clothing who are covered stand at attention, holding the hand over the heart with the headdress held in the right hand over the left shoulder as a courtesy to the national anthem or to the national flag or color.

Execution of the Hand Salute. Before the instance arrives to render the salute, stand or walk erectly with head up, chin in, and stomach muscles pulled in. Look squarely and frankly at the person to be saluted. If you are returning the salute of a soldier, execute the movements of the salute in the cadence of marching, *one*, *two*. If you are saluting a superior officer, execute the first movement and *hold* the position until the salute is acknowledged, and then complete your salute by dropping the hand smartly to your side. Do these things correctly and you will derive many rewards. Your soldiers will be quick to notice it and will vie with you in efforts to outdo their officer—a particularly healthy reaction. Thus you may set the example, which may then be extended to other matters.

The junior soldier executes the first hand salute, holds the position until it is returned by the senior, and then executes the second movement.

Accompanying the hand salute with an appropriate greeting, such as "Good morning, Sir," and its reply, "Good morning, Sergeant," is encouraged. Some units prescribe special greetings; use them when so assigned.

The salute is rendered within saluting distance, which is defined as the distance within which recognition is easy. It usually does not exceed thirty paces. The salute is begun about six paces from the person saluted or, in case the approach is outside that distance, six paces from the point of nearest approach.

Some of the more frequently observed errors in saluting are these: failure to hold the position of the salute until it is returned by the officer saluted; failure to look at the person or color saluted; failure to assume the position of attention while saluting; failure to have the thumb and fingers extended and joined, a protruding thumb being especially objectionable; a bent wrist (the hand and wrist should be in the same plane); failure to hold the upper arm horizontally. Gross errors include saluting with a cigarette in the right hand or in the mouth or saluting with the left hand in a pocket or returning a salute in a casual or perfunctory manner.

Uncovering. As a general rule, officers and enlisted personnel under arms do not remove their caps except when:

Seated as a member of or in attendance on a court or board. (Sentinels over prisoners do not uncover.)

Entering places of divine worship.

Indoors when not on duty and it is desired to remain informal.

In attendance at an official reception.

Interpretations of "Indoors" and "Outdoors." The term *outdoors* includes such buildings as drill halls, gymnasiums, and other roofed enclosures used for drill or exercise of troops. Theater marquees, covered walks, and other shelters open to the sides where a hat may be worn are also considered outdoors.

When the word *indoors* is used, it is construed to mean offices, hallways, dining halls, kitchens, orderly rooms, amusement rooms, bathrooms, libraries, dwellings, or other places of abode.

Meaning of the Term "Under Arms." The expression *under arms* is understood to mean "with arms in hand" or having attached to the person a hand arm or the equipment pertaining directly to the arm, such as cartridge belt or pistol holster.

Cannon Salute. In addition to the salutes rendered by individuals, the regulations (AR 600-25) prescribe the occasions and the procedures for rendering cannon salutes. A salute with cannon (towed, self-propelled, or tank mounted) is fired with a commissioned officer present and directing the firing. Salutes are not fired between retreat and reveille, on Sundays, or on national holidays (excluding Memorial and Independence Days) unless, in the discretion of the officer directing the honors, international courtesy or the occasion requires the exception. They are rendered at the first available opportunity thereafter, if still appropriate. The interval between rounds is normally three seconds.

The *Salute to the Union* consists of firing one gun for each state. It is fired at 1200 hours, Independence Day, at all Army installations provided with necessary equipment.

The *National Salute* consists of twenty-one guns. It is fired at 1200 hours on Memorial Day. The national flag, displayed at half-staff from reveille until noon on this day, is then hoisted to the top of the staff and so remains until retreat. In conjunction with the playing of appropriate music, this is a tribute to honored dead.

Mourning Salutes are rendered on the occasion of the death and funeral of the President or the Vice President of the United States and other high civil and military dignitaries, as prescribed in AR 600-25. The number of guns and the accompanying honors to be rendered to high dignitaries are shown in a chart later in this chapter.

The flag of the United States, national color, or national standard is always displayed at the time of firing a salute, except when firing a salute to the Union on the day of the funeral of a President, ex-President, or President-elect. On these occasions, the salute is fired at five-second intervals immediately following lowering of the flag at retreat. Personnel do not salute.

Application of Saluting Rules. The general rules for exchange of salutes are stated in an earlier paragraph. Covered or uncovered, salutes are exchanged in the same manner. The salute is rendered only once if the senior remains in the immediate vicinity and no conversation takes place.

A group of enlisted personnel or officers within the confines of military posts, camps, or stations and not in formation, on the approach of a more senior officer, is called to attention by the first person noticing the senior officer; if in formation, by the one in charge. If outdoors and not in formation, they all salute; in formation, the salute is rendered by the person in charge. If indoors, not under arms, they uncover.

Drivers of vehicles salute only when the vehicle is halted. Gate guards salute recognized officers in all vehicles. Salutes otherwise are not required by or to personnel in vehicles, although gate guards normally salute an officer's vehicle when so recognized. Military personnel in civilian attire need not exchange salutes but are encouraged to do so upon recognition. Also, military headgear need not be worn while in other than official vehicles.

Organization and detachment commanders (commissioned and noncommissioned) salute officers of higher grades by bringing the organization to attention before saluting, except when in the field.

In making reports at formations, the person making the report salutes first, regardless of rank. An example of this is the case of a battalion commander rendering a report to the adjutant at a ceremony.

All soldiers are urged to be meticulous in rendering salutes to, and in returning salutes from, fellow servicemembers. Such soldierly attitudes enhance the feeling of respect that all should feel toward comrades in arms. *The salute must never be given in a casual or perfunctory manner.*

When Not to Salute. Salutes are *not* rendered by individuals in the following cases:

- An enlisted soldier in ranks and not at attention comes to attention when addressed by an officer.
- Details (and individuals) at work. The officer or noncommissioned officer in charge, if not actively engaged at the time, salutes or acknowledges salutes for the entire detail.

- When actively engaged in games such as baseball, tennis, or golf.
- While crossing a thoroughfare, not on a military reservation, when traffic requires undivided attention.
- In churches, theaters, or places of public assemblage, or in a public conveyance.
- When carrying articles with both hands, or when otherwise so occupied as to make saluting impractical. Still, a "Good morning, Sergeant," or "Good afternoon, Sir," is manifestly appropriate.
- When on the march in combat, or under simulated combat conditions.
- While a member of the guard who is engaged in the performance of a specific duty, the proper execution of which would prevent saluting.
- On duty as a sentinel armed with a pistol. He or she stands at *raise pistol* until the challenged party has passed.
- The driver of a vehicle in motion is not required to salute.
- Indoors, except when reporting to a senior.

Reporting to a Superior Officer in His or Her Office. When reporting to a superior officer in his or her office, the subordinate (unless under arms) removes any headdress, knocks, and enters when told to do so. Upon entering, the subordinate marches up to within about two paces of the officer's desk, halts, salutes, and reports in this manner, for example: "Sir, Private Jones reports to Captain Smith," or "Sir, Lieutenant Brown reports to the Battalion Commander." After the report, conversation is carried on in the first or second person. When the business is completed, the subordinate salutes, executes about-face, and withdraws. A subordinate uncovers (unless under arms) upon entering a room where a senior officer is present.

Courtesies Exchanged When an Officer Addresses a Soldier. In general, when a conversation takes place between an officer and a soldier, the following procedure is correct: Salutes are exchanged; the conversation is completed; salutes are again exchanged. *Exceptions:* An enlisted soldier in ranks comes to attention and does not salute. Indoors, salutes are not exchanged except when reporting to an officer.

Procedures When an Officer Enters a Dining Facility. When an officer enters a dining facility, enlisted personnel seated at meals remain seated and continue eating unless the officer directs otherwise. An individual addressed by the officer ceases eating and sits at attention until completion of the conversation. In an officers' mess, although other courtesies are observed through custom, the formalities prescribed for enlisted men and women are not in effect.

Procedures When an Officer Enters a Squad Room or Tent. In a squad room or tent, individuals rise, uncover (if unarmed), and stand at attention when an officer enters. If more than one person is present, the first to perceive the officer calls, "Attention." On suitable occasions the officer commands "Rest," "As You Were," or "At Ease" when expecting to remain in the room and not desiring them to remain at attention. In officers' quarters, such courtesies are not observed.

Entering Automobiles and Small Boats. Military personnel enter automobiles and small boats in inverse order of rank; that is, the senior enters an automobile or boat last and leaves first. Subordinates, although entering the automobile first, sit to the left of the senior. The senior is always on the right.

COURTESIES TO THE NATIONAL FLAG

The Flag of the United States. There are four names in use for the flag of the United States: *flag*, *color*, *standard*, and *ensign*.

The *national color*, carried by dismounted units, measures 3 feet hoist by 4 feet fly and is trimmed on three sides with golden yellow fringe 2½ inches in width. The *standard*, identical to the *color*, is the name traditionally used by mounted, motorized, or mechanized units. The *ensign* is the naval term for the national flag (or flag indicating nationality) of any size flown from ships, small boats, and aircraft. When we speak of flags we do not mean colors, standards, or ensigns.

There are four common sizes of our national flag. The *garrison flag* is displayed on holidays and special occasions. It is 20 feet by 38 feet. The *post flag*, 10 feet by 19 feet, is for general use. The *storm flag*, 5 feet by 9 feet 6 inches, is displayed during stormy weather. The *grave decorating flag* is 7 inches hoist by 11 inches fly.

Organization Colors. Regiments and separate battalions, whose organization is fixed by tables of organization, are authorized to have organization colors symbolic of their branch and past history. Such units are "color-bearing organizations." The size is the same as the national color. The word *color*, when used alone, means the national color; the term *colors* means the national color and the organization or individual color.

Individual Colors. Individual colors, 4 feet 4 inches hoist by 5 feet 6 inches fly, are authorized by the President, Vice President, Cabinet members and their assistants, the Chairman of the Joint Chiefs of Staff, the Chief of Staff, and the Vice Chief of Staff, U.S. Army.

Pledge to the Flag. Soldiers may recite the Pledge of Allegiance as noted here. At official functions, social events, and sporting events, soldiers should, when in uniform outdoors, stand at attention, remain silent, face the flag, and render the hand salute; when in uniform indoors, stand at attention, remain silent, and face the flag; or when in civilian attire, stand at attention, face the flag with the right hand over the heart and recite the Pledge of Allegiance. Civilian headgear should be removed with the right hand and held over the left shoulder, the hand being over the heart. During military ceremonies, soldiers will not recite the Pledge of Allegiance. The words of the Pledge are:

> I pledge allegiance to the flag of the United States of America and to the republic for which it stands, one nation under God, indivisible, with liberty and justice for all.

Reveille and Retreat. The daily ceremonies of reveille and retreat constitute a dignified homage to the national flag at the beginning of the day, when it is raised, and at the end of the day, when it is lowered. Installation commanders direct the time of sounding reveille and retreat.

At every installation garrisoned by troops other than caretaking detachments, the flag is hoisted at the sound of the first note of reveille. At the last note of retreat, a gun is fired if the ceremony is on a military reservation, at which time the band or field music plays the national anthem or sound "To the Color" and the flag starts to be lowered. The lowering of the flag is regulated so as to be completed at the last note of the music. The same respect is observed by all military personnel whether the national anthem is played or "To the Color" is sounded.

The Flag at Half-Staff. The national flag is displayed at half-staff on Memorial Day until noon as a salute to the honored dead, and upon the death and funeral of military personnel and high civilian dignitaries (AR 600-25).

When the flag is displayed at half-staff it is first hoisted to the top of the staff and then lowered to the half-staff position. Before lowering the flag it is again raised to the top of the staff. For an unguyed flagstaff of one piece, the middle point of the hoist of the flag should be midway between the top and the bottom of the staff.

Memorial Day. On Memorial Day (the last Monday in May) the national flag is displayed at half-staff from reveille until noon at all Army installations. Immediately before noon the band plays an appropriate air, and at 1200 hours the national salute of twenty-one guns is fired at all installations provided with the necessary equipment for firing salutes. At the conclusion of the salute, the flag is hoisted to the top of the staff and remains so until retreat. When hoisted to the top of the staff, the flag is saluted by playing appropriate patriotic music by a band or a bugler or from a recording, depending on availability. In this manner, tribute is rendered the honored dead.

Independence Day. On Independence Day (4 July), a fifty-gun salute to the Union commemorative of the Declaration of Independence is fired at 1200 hours at all Army installations provided with the necessary equipment for firing salutes. When Independence Day occurs on a Sunday, the salute is fired the following day.

Flag Day. Flag Day is celebrated on 14 June, upon proclamation by the President. It calls upon officials of the government to display the flag on all government buildings and urges the people to observe the adoption on 14 June 1777, by the Continental Congress, of the Stars and Stripes as the official flag of the United States of America.

Salute to the President's Flag. When the President of the United States, aboard any vessel or craft flying the President's flag, passes an Army installation that is equipped to fire salutes, the installation commander causes the national salute to be fired. (See exceptions stated earlier under the discussion of cannon salutes, which would exclude firing this particular salute between retreat and reveille.)

Salute to Passing Colors. When passing or being passed by the uncased national color, military personnel render honors by executing a salute appropriate to their dress and formation as indicated previously. If indoors and not in formation, personnel assume the position of attention but do not salute. If the colors are cased, honors are not required.

Reception of an Officer on Board a Naval Vessel. The salutes to be exchanged upon boarding a naval vessel and leaving a naval vessel are prescribed in the following paragraph of United States Navy Regulations, to which all members of the Army visiting a naval vessel will conform (AR 600-25):

2108. Salutes to the National Ensign.

1. Each person in the naval service, upon coming on board a ship of the Navy, shall salute the national ensign if it is flying. He shall stop on reaching the upper platform of the accommodation ladder, or the shipboard end of the bow, face the national ensign, and render the salute, after which he shall salute the officer of the deck. On leaving the ship, he shall render the salutes in inverse order. The officer of the deck shall return both salutes in each case.
2. When passed by or passing the national ensign being carried, uncased, in a military formation, all persons in the naval service shall salute. Persons in vehicles or boats shall follow the procedure prescribed for such persons during colors.
3. The salutes prescribed in this article shall also be rendered to foreign national ensigns and aboard foreign men-of-war.

For further information on Navy courtesies and customs, see Navy Customs Army Officers Should Know, later in this chapter.

Dipping the Flag or Colors. The flag of the United States, national color, and national standard are never dipped by way of salute or compliment. The organizational color or standard is dipped in salute in all military ceremonies while the U.S. national anthem, "To the Color," or a foreign national anthem is being played, and when rendering honors to the organizational commander or an individual of higher grade, including foreign dignitaries of higher grade, but in no other case.

The U.S. Army flag is considered to be an organizational color and as such is also dipped while the U.S. national anthem, "To the Color," or a foreign national anthem is being played and when rendering honors to the Chief of Staff of the U.S. Army, his direct representative, or individual of higher grade, including foreign dignitary of equivalent or higher grade, but in no other case.

The authorized unit color salutes in all military ceremonies while the national anthem or "To the Color" is being played and when rendering honors to the organizational commander or an individual of higher rank, but in no other case.

Display and Use of the Flag. International usage forbids the display of the flag of one nation above another nation's in time of peace. When the flags of two or more nations are to be displayed, they should be flown from separate staffs, or from separate halyards, of equal size and on the same level.

The national flag, when not flown from a staff or mast, should always be hung flat, whether indoors or out. It should not be festooned over doorways or arches, tied in a bowknot, or fashioned into a rosette. When used on a rostrum, it should be displayed above and behind the speaker's desk. It should never be used to cover the speaker's desk or to drape over the front of the platform. For this latter purpose, as well as for decoration in general, bunting of the national colors should be used, and the colors should be arranged with the blue above, the white in the middle, and the red below. Under no circumstances should the flag be draped over chairs or benches, nor should any object or emblem of any kind be placed above or upon it, nor should it be hung where it can be easily contaminated or soiled. When carried with other flags, the national flag should always be on the right (as color-bearers are facing) or in front. The flag of the United States of America should be at the center and at the highest point of the group when a number of flags of states or localities or pennants of societies are grouped and displayed from staffs.

When flown at a military post, or when carried by troops, the national flag or color is never dipped by way of salute or compliment. The authorized unit color is dipped as a salute when the reviewing officer has the rank of a general officer. This is done by lowering the pike (as the staff of a color is called) to the front so that it makes an angle of about 45 degrees with the ground. The national flag is used to cover the casket at the military funeral of present or former members of the military service. It is placed lengthwise on the casket with the union at the head and over the left shoulder of the deceased. The flag is not lowered into the grave and is not allowed to touch the ground.

The display and use of the flag by civilian groups is contained in Public Law 829–77th Congress, as amended by Public Law 344–94th Congress.

Display of United Nations Flag. There are no U.S. laws or policies adopted by the United Nations that cause conflict in the display of the U.S. flag in conjunction with the United Nations flag. When the two flags are displayed together, the U.S. flag is on the right, best identified as "the marching right." This is in accordance with U.S. law. The United Nations flag code states that it can be on either side of a national flag without being subordinate to that flag. Both flags should be of the same size and displayed at the same height.

It should be noted that the United Nations flag may be displayed at military installations of the United States or carried by U.S. troops only on very specific occasions, such as the visit of high dignitaries of the United Nations, when the United Nations or high dignitaries thereof are to be honored, or as authorized by the President (AR 840-10).

COURTESIES TO THE NATIONAL ANTHEM

Whenever and wherever the national anthem, "To the Color," or "Hail to the Chief" is played outdoors, at the first note all dismounted personnel in uniform

HOW TO DISPLAY THE FLAG

1. When displayed over the middle of the street, the flag should hang vertically with the union to the north in an east-and-west street or to the east in a north-and-south street.

2. When displayed with another flag from crossed staffs, the U.S. flag should be on the right (the flag's own right), and its staff should be in front of the staff of the other flag.

3. When flying the flag at half-staff, the flag detail should first hoist the flag to the peak and then lower it to the half-staff position, but before lowering the flag for the day, they should again raise it to the peak.

4. When flags of states or cities or pennants of societies fly on the same halyard with the U.S. flag, the U.S. flag should always be at the peak.

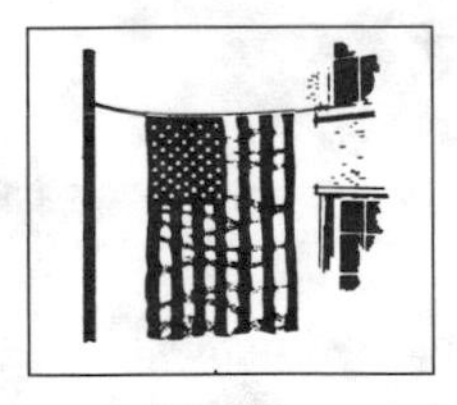

5. When the flag hangs over a sidewalk from a rope extending from house to pole at the edge of the sidewalk, the flag should go out from the building, toward the pole, union first.

6. When the flag is on display from a staff projecting horizontally or at any angle from the windowsill, balcony, or front of a building, the union of the flag should go to the peak of the staff (unless the flag is to be at half-staff).

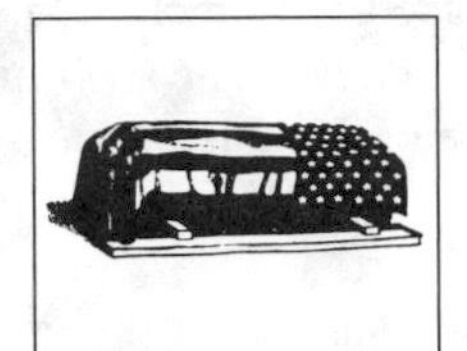

7. When the flag covers a casket, the union should be at the head and over the left shoulder of the deceased. The flag should not be lowered into the grave or allowed to touch the ground.

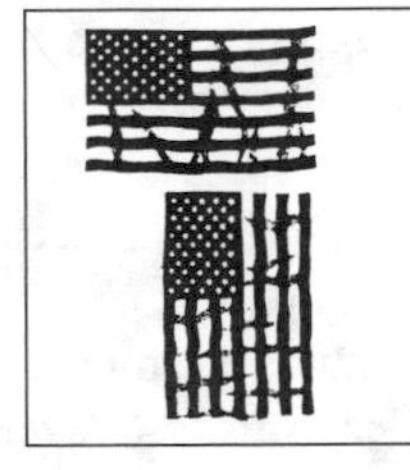

8. When the flag is on display other than by flying from a staff, it should be flat whether indoors or out. When displayed either horizontally or vertically against a wall, the union should be uppermost and to the flag's own right—that is, to the observer's left. When displayed in a window, it should appear the same way—that is, with the union or blue field to the left of the observer in the street.

9. When carried in a procession with another flag or flags, the U.S. flag should be either on the marching right or, when there is a line of other flags, in front of the center of that line.

10. When a number of flags of states or cities or pennants of societies are grouped on display from staffs with our national flag, the U.S. flag should be at the center or at the highest point of the group.

11. When the flags of two or more nations are on display, they should fly from separate staffs of the same height, and the flags should be of about equal size. International usage forbids displaying the flag of one nation above that of another nation in time of peace.

and not in formation, within saluting distance of the flag, face the flag, or the music if the flag is not in view, salute, and maintain the salute until the last note of the music is sounded. This includes personnel in athletic uniform. Men not in uniform remove the hat with the right hand and hold it at the left shoulder with the hand over the heart. If no hat is involved, they stand at attention holding the right hand over the heart. Women not in uniform should salute by placing the right hand over the heart.

Vehicles in motion are brought to a halt. Persons riding in a passenger car or on a motorcycle dismount and salute. Occupants of other types of military vehicles and buses remain seated at attention in the vehicle, the individual in charge of each vehicle dismounting and rendering the hand salute. Tank and armored vehicle commanders salute from the vehicle.

The above marks of respect are shown to the national anthem of any friendly country when it is played at official occasions.

When the national anthem is played indoors, officers and enlisted personnel stand at attention and face the music, or the flag if one is present. They do not salute unless under arms. At reveille, the procedures outlined above are followed.

The method and personnel required for raising and lowering the flag on a flagstaff are prescribed in FM 3-21.5, *Drill and Ceremonies*.

MILITARY FUNERALS

The military funeral, with its customs, precision, and courtesies, can be a source of great comfort and pride to the bereaved when executed correctly; if not executed correctly, it can add to their grief.

Officers should be thoroughly familiar with the prescribed courtesies to the military dead. This involves a knowledge of the ceremonies incident to the conduct of a military funeral, including correct procedure on the following occasions:

Officer in charge of a funeral.

Honorary pallbearer.

Command of a funeral escort.

Attendance as a mourner.

Essential references are AR 600-25 and FM 3-21.5.

For burial rights of military personnel, see chapter 19, Personal and Financial Planning.

Significance of the Military Funeral. The ceremonial customs that constitute the elements of all military funerals are rooted in ancient military usage. In many cases, these traditions are based on expedients used long ago on the battlefield in time of war. The use of a caisson as a hearse, for example, was an obvious combat improvisation. In a similar manner, the custom of covering the casket with a flag probably originated on the battlefield where caskets were not available and the flag, wrapped around the dead serviceman, served as a makeshift pall in which he could be buried. Later, these customs assumed a deeper significance than that of mere expediency. The fact that an American flag is used to cover the casket, for example, now symbolizes that the soldier served in the armed forces of the United States and that this country assumes the responsibility of burying the soldier as a solemn and sacred obligation.

Finally, the sounding of Taps over the grave has an obvious origination in military custom. Since Taps is the last bugle call the soldier hears at night, it is particularly appropriate that it be played over his grave to mark the beginning of his last, long sleep and to express hope and confidence in an ultimate reveille to come.

Courtesies at a Military Funeral. At a military funeral, all persons in the military service in uniform attending in their individual capacity face the casket and execute the hand salute at any time when the casket is being moved, while the casket is being lowered into the grave, during the firing of the volley, and while Taps is being sounded. Honorary pallbearers in uniform conform to these instructions when not in motion. Men in civilian clothes, in the above cases and during the service at the grave, stand at attention, uncovered, and hold the headdress over the left breast; if no headdress is worn, the right hand is held over the heart. Female personnel except the active pallbearers follow the example of the officiating chaplain. If he uncovers, they uncover; if he remains covered, they remain covered. When the officiating chaplain wears a biretta (clerical headpiece) during the graveside service, all personnel uncover. When the officiating chaplain wears a yarmulke (skullcap), all personnel remain covered.

The active pallbearers remain covered and do not salute while carrying the casket and while holding the flag over the casket during the service at the grave.

Women in uniform remain covered during military funerals.

Badge of Military Mourning. The badge of military mourning is a straight band of black crepe or plain black cloth 4 inches wide, worn around the left sleeve of the outer garment above the elbow. No badge of military mourning is worn with the uniform, except when prescribed by the commanding officer for funerals, or when specially ordered by the Department of the Army. For family mourning, officers are authorized to wear the sleeve band described above while at the funeral or en route thereto or therefrom.

Elements of a Military Funeral Ceremony. The military funeral ceremony that has been developed to demonstrate the nation's recognition of the debt it owes to the services and sacrifices of soldiers is based on a few simple customs and traditions. The casket of the soldier is covered with the American flag. Traditionally, the casket was transported to the cemetery on a caisson; however, since caissons are no longer used in the Army, except by the Old Guard at Arlington National Cemetery, the vehicle carrying the casket is generally a civilian hearse or sometimes a light, open Army truck or an ambulance adapted for the purpose. This should be understood wherever the word *caisson* is employed in this description.

The casket is carried from the caisson to the grave by six military body bearers. In addition to the body bearers, honorary pallbearers are usually designated who march to the cemetery alongside the caisson. At the cemetery, the casket is placed over the grave, and the body bearers hold the flag-pall waist high over the

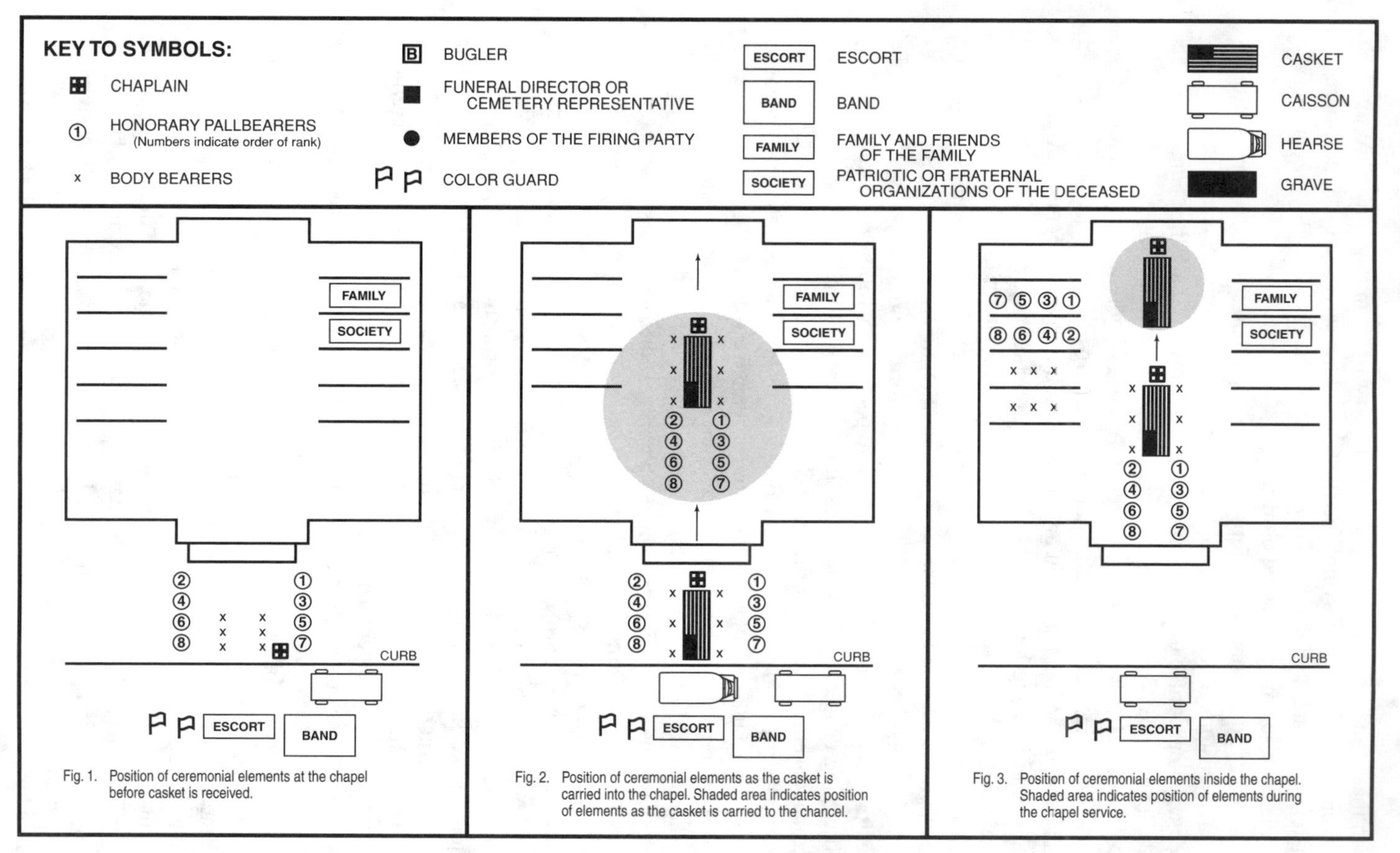

Fig. 1. Position of ceremonial elements at the chapel before casket is received.

Fig. 2. Position of ceremonial elements as the casket is carried into the chapel. Shaded area indicates position of elements as the casket is carried to the chancel.

Fig. 3. Position of ceremonial elements inside the chapel. Shaded area indicates position of elements during the chapel service.

Military Funerals, Army

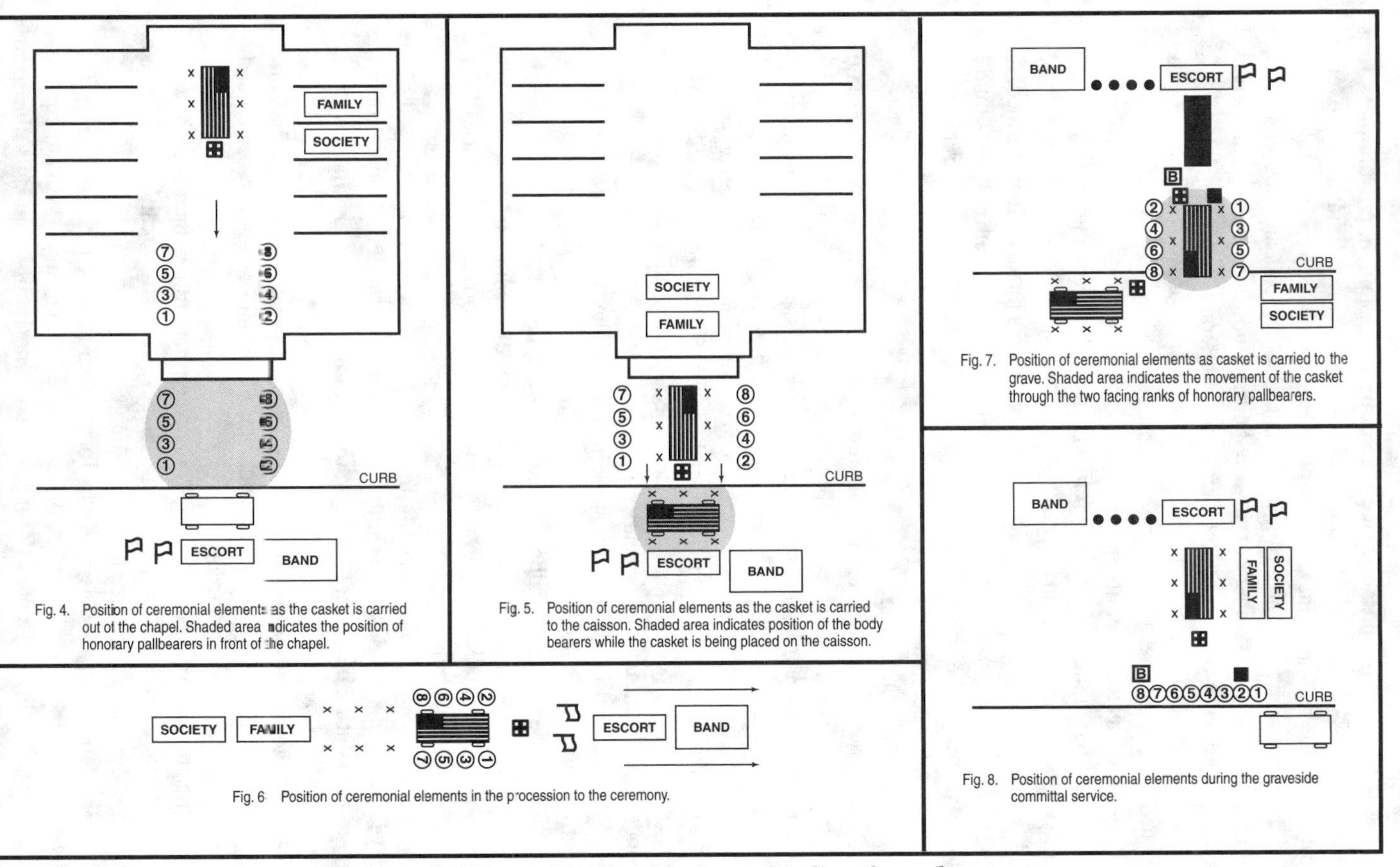

Fig. 4. Position of ceremonial elements as the casket is carried out of the chapel. Shaded area indicates the position of honorary pallbearers in front of the chapel.

Fig. 5. Position of ceremonial elements as the casket is carried to the caisson. Shaded area indicates position of the body bearers while the casket is being placed on the caisson.

Fig. 6 Position of ceremonial elements in the procession to the ceremony.

Fig. 7. Position of ceremonial elements as casket is carried to the grave. Shaded area indicates the movement of the casket through the two facing ranks of honorary pallbearers.

Fig. 8. Position of ceremonial elements during the graveside committal service.

Military Funerals, Army—*Continued*

casket. After the committal service is read by the chaplain, a firing party fires three volleys. A bugler stationed at the head of the grave sounds Taps over the casket, and the military funeral is completed. The body bearers then fold the flag, and it is presented to the next of kin. These basic elements are the foundation of all military funerals, whether last rites are being conducted over a private's casket or final honors are being paid at the grave of a general.

Honorary Pallbearers. The honorary pallbearers arrive at the chapel before the hearse arrives. They take positions in front of the entrance to the chapel in two facing ranks, as indicated in the accompanying illustration. Upon arrival of the hearse and when the body bearers remove the casket from the hearse, honorary pallbearers execute the hand salute.

When the casket is carried between the two ranks that they have formed, they come to the order, execute the appropriate facing movement, fall in behind the casket, and enter the chapel, the senior preceding the junior and marching to the right. In the chapel, they take places in the front pews to the left of the chapel as indicated in the illustration.

After the chapel service, the honorary pallbearers precede the casket in columns of two as the two active pallbearers push the church truck to the chapel entrance. The honorary pallbearers again form an aisle from the chapel entrance to the caisson or hearse and uncover or salute as prescribed. When the casket has been placed on the caisson or in the hearse, they enter their conveyances or march. When marching, the honorary pallbearers form columns of files on each side of the caisson or hearse, the leading member of each column opposite the front wheels of the caisson or hearse.

When the entrance to the burial lot is reached, the honorary pallbearers take positions on either side of the entrance. As the body bearers lift the casket from the caisson, the honorary pallbearers execute the hand salute.

When the casket has been carried past them, they come to the order and fall in behind the casket, marching to the grave site in correct precedence of rank, senior to the right and to the front.

At the grave site, they stand in line behind the chaplain at the head of the grave; the senior stands to the right and the junior to the left. They execute the hand salute during the firing of volleys, the sounding of Taps, and the lowering of the casket into the grave.

After the ceremony is over, they march off in two files behind the colors.

Family. The family arrives at the chapel before the casket is received and is seated in pews in the right front of the chapel.

When the chapel service is over, family members follow the casket down the aisle until they reach the vestibule of the chapel, where they wait until the casket is carried outside and secured to the caisson.

When the procession is ready to form, members of the family take their places in the procession immediately behind the body bearers. When the procession arrives at the grave site, the members of the family wait until the band, escort, and colors have taken their positions at the grave, and the casket is carried between the double rows of honorary pallbearers. The members of the family take their positions at the side of the grave opposite the earth mound side for the funeral service.

When the graveside ceremony is finished, a member of the family receives the interment flag from the chaplain, the cemetery representative, the officer in charge of the funeral, or the individual military escort. Upon this ending of the service, it has become customary for close friends to express regrets to the bereaved at the graveside.

NAVY CUSTOMS ARMY OFFICERS SHOULD KNOW

Courtesies. There are Navy customs applicable to shore duty and to the special situations of life on board a naval vessel. The courtesies pertaining to a naval vessel are of interest to officers of the Army when they visit or serve aboard a unit of the naval fleet.

Upon arrival at a naval vessel, it is important to alert the security staff of the rank and position of guests before boarding; this information is used to "pipe" dignitaries and guests aboard. All commanding officers, colonels, and flag officers are "piped" and announced when boarding or departing. If you are the commanding officer of the 8th Infantry Regiment, for example, you will be announced as "8th Infantry Arriving" (or "Departing"). Similarly, if the commanding general of the 1st Infantry Division or the commanding general of Fort Monroe is embarked, the pipe and announcement would be "1st Infantry Division Arriving" or "Fort Monroe Arriving."

On appropriate occasions when visiting naval vessels, officers of the armed services, except when in civilian clothes, are attended when they come aboard and when they depart by sailors known as side boys. This courtesy is also extended to commissioned officers of the armed services of foreign nations. Officers of the rank of lieutenant to major inclusive are given two side boys, from lieutenant colonel to colonel four side boys, from brigadier to major general six side boys, and lieutenant general and above eight side boys. Full guard and band are given to general officers, and for a colonel, the guard of the day, but no music.

At the order "Tend the side," the side boys fall in fore and aft of the approach to the gangway, facing each other. The boatswain's mate-of-the-watch takes station forward of them and faces aft. When the boat comes alongside, the boatswain's mate pipes and again when the visiting officer's head reaches the level of the deck. At this latter instant the side boys salute.

On departure, the ceremony is repeated in reverse; the boatswain's mate begins to pipe and side boys to salute as soon as the departing officer steps toward the gangway between the side boys. As the boat casts off, the boatswain's mate pipes again. (Shore boats and automobiles are not piped.)

You uncover when entering a space where men are at mess and in sick bay (quarters) if sick men are present. You uncover in the wardroom at all times except when under arms and passing through. All hands uncover in the captain's cabin and country, except when under arms.

You should not overtake a senior, except in emergency. In the latter case, slow, salute, and say, "By your leave, Sir." Admirals, commanding officers, and chiefs of staff when in uniform fly colors astern when embarked in boats. When officials visit, they also display their personal flags (pennants for commanding officers) in the bow. Flag officers' barges are distinguished by the appropriate number of stars on each side of the barge's hull. Captains' gigs are distinguished by the name or abbreviation of their ships surcharged by an arrow.

Use of Navy Titles. In the Navy it is customary to address officers in the grade of lieutenant commander and below, *socially*, as "Mister" or "Miss," and officers in the grade of commander and above by their titles. *Officially*, officers in both staff and line are addressed by their ranks.

Title of Commanding Officer of a Ship. Any officer in command of a ship, regardless of size or class, while exercising command is addressed as "Captain."

Visiting. Seniors come on board ship first. When reaching the deck, you face toward the colors (or aft if no colors are hoisted) and salute the colors (quarter-deck). Immediately thereafter you salute the officer of the deck (OOD) and request permission to come aboard. The usual form is, "Request permission to come aboard, Sir." The OOD is required to return both salutes.

On leaving the ship, the reverse order is observed. You salute the OOD and request permission to leave the ship. The OOD indicates when the boat is ready (if a boat is used). Each person, juniors first, salutes the OOD, then faces toward the colors (quarterdeck), salutes, and debarks.

The OOD on board ship represents the captain and as such has unquestioned authority. Only the executive and commanding officer may order him or her relieved. The authority of the OOD extends to the accommodation ladders or gangways. The OOD has the right to order any approaching boat to "lie off" and keep clear until the boat can be safely received alongside.

The OOD normally conveys orders to the embarked troops via the troop commander but in emergencies may issue orders direct to you or any person on board.

The *bridge* is the command post of the ship when under way, as the quarter-deck is at anchor. The OOD is in charge of the ship as the representative of the captain. Admittance to the bridge when under way should be at the captain's invitation or with his or her permission. You may usually obtain permission through the executive officer.

The *quarterdeck* is the seat of authority; as such it is respected. The starboard side of the quarterdeck is reserved for the captain (and admiral if a flagship). No person trespasses upon it except when necessary in the course of work or official business. All persons salute the quarterdeck when entering upon it. When pacing the deck with another officer the place of honor is outboard, and when reversing direction each turns toward the other. The port side of the quarter-deck is reserved for commissioned officers, and the crew has all the rest of the weather decks of the ship. However, every part of the deck (and the ship) is assigned to a particular division so that the crew has ample space. Not unnaturally, every division considers it has a prior though unwritten right to its own part of the ship. For gatherings such as movies, all divisions have equal privileges at the scene of assemblage. Space and chairs are reserved for officers and for CPOs, where available, and mess benches are brought up for the enlisted personnel. The seniors have the place of honor. When the captain (and admiral) arrives, those present are called to attention. The captain customarily gives "carry on" at once through the executive officer or master-at-arms.

Messes. If you take passage on board a naval vessel, you will be assigned to one of several messes on board ship, the wardroom or junior officers' mess. In off-hours a pot of coffee is usually available.

The executive officer is ex-officio the president of the wardroom mess. The wardroom officers are the division officers and the heads of departments. All officers await the arrival of the executive officer before being seated at lunch and dinner. If it is necessary for you to leave early, ask the head at your table for permission to be excused, as you would at home. The seating arrangement in the messes is by order of seniority.

Calls. Passenger officers should call on the captain of the ship. If there are many of you, you should choose a calling committee and consult the executive officer as to a convenient time to call. The latter will make arrangements with the captain.

Ceremonies. Gun salutes in the Navy are the same as in the Army, except that flag officers below the rank of fleet admiral or general of the Army are, by Navy regulations, given a gun salute upon departure only.

Saluting. By custom, Navy personnel do not salute when uncovered, although it is customary for Navy officers to return the salute of Army and Air Force personnel whether covered or not. Aboard ship, seniors are saluted only during the first greeting in the morning. The commanding officer (or any flag or general officer) is saluted whenever met.

AIR FORCE CUSTOMS ARMY OFFICERS SHOULD KNOW

Courtesies. The rules governing saluting, whether saluting other individuals or paying honor to the color or national anthem, are the same for the Air Force as for the Army.

Visiting. It is assumed that the majority of officers visiting an Air Force base are there in conjunction with air travel to or from the base. In addition to the base operations officer, who is the commander's staff officer with jurisdiction over all air traffic, the transient alert crew is charged with meeting all transient aircraft, determining the transportation requirements of transient personnel, and directing them to the various base facilities. General officers and admirals usually are met by a senior base officer. RON (remain overnight) messages may be transmitted through base operations.

Passengers from other services who desire to remain overnight at an Air Force base should make the necessary arrangements with the transient alert crew and not attach themselves to the pilot, who will be busy with his or her own responsibilities. By the same token, passengers of other services who have had a special flight arranged for them should make every effort to see that the pilot and crew are offered the same accommodations that they themselves are using, unless that base has adequate transient accommodations.

Passenger vehicles are never allowed on the ramp or flight lines unless special arrangements have been made with the base operations officer; this permission is granted only under unusual circumstances.

Travel in Military Aircraft. The aircraft commander is the final authority on the operation of any military aircraft. Passengers, regardless of rank, seniority, or service, are subject to the orders of the aircraft commander, who is solely responsible for their adherence to regulations governing conduct in and around the aircraft. In the event it is impractical for the aircraft commander to leave his or her position, orders may be transmitted through the copilot or crew chief and have the same authority as if given personally by the pilot.

The order of boarding and alighting from military aircraft—excluding the crew—varies somewhat with the nature of the mission. If a special flight is arranged for the transportation of very important persons (VIPs), official inspecting parties, or other high-ranking officers of any service, the senior member exits first, and the other members of the party follow either in order of rank or in order of seating, those nearest the hatch alighting first.

In routine transportation flights, officers are normally loaded in order of rank without regard for precedence, except that VIPs are on- and off-loaded first. In alighting, officers seated near the hatch generally debark first, and so on to those who are farthest away.

Aircraft carrying general or flag officers are usually marked with a detachable metal plate carrying stars appropriate to the highest rank aboard and are greeted on arrival by the Air Force base commander, if the destination is an Air Force base. Other aircraft are usually met by the airdrome officer, who is appointed for one day only and acts as the base commander's representative.

Since aerial flights are somewhat dependent on weather, especially when carrying passengers, the decision of the pilot to fly or not to fly or to alter the flight plan en route cannot be questioned by the passengers of whatever rank or service. Regulations governing smoking, the use of seatbelts, and the wearing of parachutes are binding on all classes of passengers.

SALUTES AND HONORS TO DISTINGUISHED OFFICERS

Certain military and civil officials in high position, including foreign officials, are accorded personal honors consisting of cannon salutes, ruffles and flourishes played by field music, and the national anthem of our country or of the foreign country, the "General's March," or a march played by the band. These honors are extended upon presentation of the escort and as part of the parade or review of troops. The accompanying chart (from AR 600-25) states the specific honors of all persons who may be accorded them. A military escort is supplied during their

ENTITLEMENT TO HONORS

Grade, Title, or Office	Number of Guns Arrival	Number of Guns Departure	Ruffles and Flourishes	Music*
President	21	21	4	National anthem or "Hail to the Chief," as appropriate
Former President or President-elect	21	21	4	National anthem
Sovereign or chief of state of a foreign country or member of reigning royal family	21	21	4	National anthem of foreign country
Vice President	19	None	4	"Hail Columbia"
Speaker of the House of Representatives; Cabinet member, President pro tempore of U.S. Senate; governor of a State, or Chief Justice of the United States	19	None	4	Honors March 1
American or foreign ambassador, or high commissioner while in country to which accredited	19	None	4	National anthem of the United States or official's country
Premier or Prime Minister	19	None	4	National anthem of official's country
Secretary of Defense, Deputy Secretary of Defense, Secretaries of the Army, Navy and Air Force, or Under Secretaries of Defense	19	19	4	Honors March 1
Chairman and Vice Chairman of the Joint Chiefs of Staff, General of the Army, Fleet Admiral, General of the Air Force; Chief and Vice Chief of Staff, Army; Chief and Vice Chief of Naval Operations; Chief and Vice Chief of Staff, Air Force; or Commandant and Assistant Commandant of the U.S. Marine Corps	19	19	4	Honors March 2, 3, or 4
Chairman of a Committee of Congress	17	None	4	Honors March 1
Assistant Secretaries of Defense and General Counsel of the Department of Defense; Under Secretary of the Army, Navy, or Air Force; Assistant Secretaries of the Army, Navy, or Air Force	17	17	4	Honors March 1
Generals or admirals (4 star)	17	17	4	Honors March 2, 3, or 4
Governors of a territory of foreign possession within the limits of their jurisdiction or American ambassadors having returned to the United States on official business	17	None	4	Honors March 1
American envoys, American ambassadors having returned to the United States but not on official business, or ministers and foreign envoys, or ministers accredited in the United States	15	None	3	Honors March 1
Lieutenant generals, vice admirals	15	None	3	Honors March 2, 3, or 4
Major general or rear admiral (upper half)	13	None	2	Honors March 2, 3, or 4
American ministers resident and ministers resident accredited to the United States	13	None	2	Honors March 1
American charges d'a-ffaires and charges d'a-ffaires accredited to the United States	11	None	1	Honors March 1
Brigadier generals or rear admirals (lower half)	11	None	1	Honors March 1 or 2
Consuls general accredited to the United States	11	None	None	Honors March 1

* Notes:
1 Honors March 1: The 32-bar medley of "The Stars and Stripes Forever."
2 Honors March 2: "General's March."
3 Honors March 3: "Admiral's March."
4 Honors March 4: "Flag Officer's March." (This march is the prescribed music for flag officers [admirals] of the U.S. Navy or the U.S. Coast Guard and general officers [generals] of the U.S. Marine Corps.)

rendition. If two or more persons entitled to honors arrive at or depart from an installation at the same time, only the senior will receive honors.

Ruffles and Flourishes. Ruffles are played on drums, flourishes on bugles. They are sounded together, once for each star of the general officer being honored and according to the accompanying table of honors for other dignitaries. Ruffles and flourishes are followed by music as prescribed in the table.

Action of the Person Receiving the Honors. It is the usual custom for the person receiving the honors to inspect the escort. The appropriate time to do this is at the conclusion of the honors rendered by the escort upon his or her reception.

During the playing of the ruffles and flourishes and music, as indicated in the chart, the person honored and those accompanying him or her, if members of the armed forces, salute at the first note of the ruffles and flourishes and remain at the salute until the last note of the music. Persons in civilian clothes salute by uncovering.

Action of Persons Witnessing the Honors. Members of the armed forces who witness the salutes and honors render the hand salute, conforming to the action of the official party. Individuals in civilian clothing uncover.

PRECEDENCE OF MILITARY ORGANIZATIONS IN FORMATION

Whenever two or more organizations of different components of the armed forces appear in the same formation, they take precedence among themselves in order as listed below. This means from right to left in line, the senior organization on the right; and from head to tail of a column, the senior organization at the head (AR 600-25).

Cadets, United States Military Academy
Midshipmen, United States Naval Academy
Cadets, United States Air Force Academy
Cadets, United States Coast Guard Academy
Midshipmen, United States Merchant Marine Academy
United States Army
United States Marine Corps
United States Navy
United States Air Force
United States Coast Guard
Army National Guard of the United States
Army Reserve
Marine Corps Reserve
Naval Reserve
Air National Guard of the United States
Air Force Reserve
Coast Guard Reserve

Other training organizations of the Army, Marine Corps, Navy, Air Force, and Coast Guard in order, respectively.

During any period when the U.S. Coast Guard operates as a part of the U.S. Navy, the cadets, U.S. Coast Guard Academy, the U.S. Coast Guard, and the Coast Guard Reserve take precedence, respectively, after the midshipmen, U.S. Naval Academy, the U.S. Navy, and the Naval Reserve.

Retirement Ceremonies

Military personnel will be given appropriate recognition at retirement, which may include reviews, ceremonies, or similar functions. Maximum publicity will be given to retirement ceremonies

The Army Song

"The Army Goes Rolling Along," the official song of the U.S. Army, is used at the conclusion of all reviews, parades, and honor guard ceremonies. Individuals will stand at attention and sing the lyrics of the Army song when it is played. Individuals will stand at attention during the playing of official songs of other services. The lyrics to the Army song are:

Intro: March along, sing our song, with the Army of the free
Count the brave, count the true, who have fought to victory
We're the Army and proud of our name
We're the Army and proudly proclaim

Verse: First to fight for the right,
And to build the Nation's might,
And the Army Goes Rolling Along
Proud of all we have done,
Fighting till the battle's won,
And the Army Goes Rolling Along.

Refrain: Then it's Hi! Hi! Hey!
The Army's on its way.
Count off the cadence loud and strong (TWO! THREE!)
For where e'er we go,
You will always know
That the Army Goes Rolling Along.

Verse: Valley Forge, Custer's ranks,
San Juan Hill and Patton's tanks,
And the Army went rolling along
Minute men, from the start,
Always fighting from the heart,
And the Army keeps rolling along.

(refrain)

Verse: Men in rags, men who froze,
Still that Army met its foes,
And the Army went rolling along.
Faith in God, then we're right,
And we'll fight with all our might,
As the Army keeps rolling along.
(refrain)

2

Missions, Purposes, and Organization

> In general, the Army includes land combat and service forces and such aviation and water transport as may be organic therein. It shall be organized, trained, and equipped primarily for prompt and sustained combat incident to operations on land. It is responsible for the preparation of land forces necessary for the effective prosecution of war except as otherwise assigned and, in accordance with integrated joint mobilization plans, for the expansion of the peacetime components of the Army to meet the needs of war.
>
> —U.S. Code 3062

For over 247 years, the Army has existed to fight and win our nation's wars as well as secure the peace. The Army contributes foundational forces to combatant commands to conduct prompt and sustained combat operations on land "against all enemies, domestic and foreign." The mission of the Army is to defeat enemy land forces, seize and control terrain, and destroy the enemy's capability and will to resist. Supported by the Air Force and Navy, and often in conjunction with the Marine Corps, the Army has forcible entry capability that allows it to conduct land operations anywhere in the world. The Army can achieve prompt and sustained land dominance across the range of military operations from support to domestic authorities, humanitarian relief, counterdrug, security assistance, peacekeeping, peace enforcement, through large-scale combat operations. The Army's strategic roles of shape, prevent, win, and consolidate through its unique, sustained landpower capabilities offer the President, Secretary of Defense, and combatant commanders' strategic options for engagement, crisis response, and warfighting. DoD Directive 5100.1 describes the primary functions of the Army.

For the Army, multidomain operations are the combined arms employment of Joint and Army capabilities that create and exploit relative advantages to achieve objectives, defeat enemy forces, and consolidate gains on behalf of joint force commanders. The following core competencies describe the Army's strengths and essential contributions to the joint force: prompt and sustained land combat; combined arms operations to include combined arms maneuver and wide area security, armored and mechanized operations, and airborne and air assault operations; and special operations to set and sustain the theater for the joint force, integrate national, multinational and joint power on land. Current and future realities require the Army to remain an agile and capable force with a joint and expeditionary mindset.

For the past two decades, the Army has met the demands of protracted overseas contingency operations around the world. The Army defeated terrorists and Saddam Hussein's forces in Iraq during Operation Enduring Freedom (OEF) and Iraqi Freedom (OIF); and facilitated the evacuation from Afghanistan during Operation Allies Refuge. Today, the Army remains a central and critical participant in ongoing Operation Atlantic Resolve assuring our NATO allies in Eastern Europe and supporting Ukraine against Russian aggression; and in Operation Inherent Resolve defeating terrorists in Iraq and Syria. The Army in 2022 has more than 140,000 soldiers deployed, forward stationed, or committed around the world in over 120 countries.

As our Army continues to engage regionally and respond globally, it is also simultaneously transforming. The Army is in the midst of one of the most significant periods of transformation and downsizing in its history. The Army is reducing from a high of 570,000 in 2014 to perhaps 450,000 by 2030 while embarking on a series of major organizational and other initiatives. The Army is internally restructuring both active and reserve component forces to better meet global commitments.

U.S. MILITARY ORGANIZATION

There are two branches of United States military chain of command. Subject to the Secretary of Defense and the provisions of Title 10 United States Code, the Department of the Army operates under the administrative command of the Secretary of the Army, with the advice of the Chief of Staff. Operationally, the Army responds to the orders of the President of the United States as commander in chief, communicated through the Secretary of Defense, with the advice and counsel of the Chairman of the Joint Chiefs of Staff, and then through the combatant commanders.

AR 10-87, *Army Commands, Army Service Component Commands, and Direct Reporting Units*, prescribes the organization and functions of the major headquarters (HQ) of both the Army's Institutional Force and Operational Forces. The institutional force ensures the readiness of all Army forces and consists of Army Commands and Direct Reporting Units whose primary mission is to generate, prepare, and sustain the operating forces of the Army and therefore, these institutional forces remain under the direction and control of the DA.

ARMY HIGHER ORGANIZATIONS

In carrying out its functions as prescribed in DoD Directive 5100.1, the U.S. Army is organized in four Army Commands (ACOMs), nine Army Service Component Commands (ASCC) of unified combatant commands (of which five are theater armies), and some thirteen direct reporting units (DRU). Army headquarters assigned as the ASCCs of a respective unified command are trained and equipped to enable that combatant command's missions. The ACOMs, ASCCs, and major DRUs are described in the following paragraphs.

ARMY COMMANDS

U.S. Army Training and Doctrine Command (TRADOC). Headquartered at Fort Eustis, Virginia, TRADOC recruits, trains, and educates the Army's soldiers; develops leaders; supports training in units; develops doctrine; and establishes standards. TRADOC's major subordinate units include U.S. Army Cadet Command, headquartered at Fort Knox, Kentucky; Combined Arms Command (CAC), headquartered at Fort Leavenworth; and the Combined Arms Support Command (CASCOM), headquartered at Fort Lee; Additionally, TRADOC operates some thirty-three schools on sixteen installations, including, among others, seventeen branch schools and centers, the Army Logistics Management College, the Army Management Staff College, and the Command and General Staff College.

U.S. Army Futures Command (AFC) headquartered in Austin, Texas, with over 26,000 personnel worldwide leads a continuous transformation of Army modernization in order to provide future warfighters with the concepts, capabilities and organizational structures they need to dominate a future battlefield. AFC integrates daily with entrepreneurs, scientists, academia, and businesses to employ an entrepreneurial spirit of accepting risks in order to design and field the future force with overmatch to deter and, when necessary, defeat current and future adversaries. AFC has three major subordinates:

- Combat Developments Command (DEVCOM) headquartered at Aberdeen Proving Ground, Maryland, ensures the dominance of Army capabilities by creating, integrating and delivering technology-enabled solutions to our soldiers. DEVCOM is the Army's organic research and development capability.
- Futures and Concepts Center (FCC) at Fort Eustis, Virginia, assesses threats and future operational environment, develops concepts, requirements and an integrated modernization path to increase lethality and overmatch to enable soldiers and units to win on future battlefields.
- Medical Research and Development Command headquartered at Fort Detrick, Maryland, manages and executes research in five basic areas: military infectious diseases, combat casualty care, military operational medicine, chemical biological defense, and clinical and rehabilitative medicine.

AFC also has several Cross-Functional Teams (CFTs) established to narrow existing capability gaps by developing capability documents, informed by experimentation and technical demonstrations, to rapidly deliver modernization priority requirements to the Army Acquisition System.

U.S. Army Materiel Command (AMC). Headquartered at Redstone Arsenal, Huntsville, Alabama, AMC provides technology, acquisition support, and logistics to the Army, the other services, and U.S. allies. "If a soldier shoots it, drives it, flies it, wears it, or eats it, AMC provides it." The command's missions range from research and development of weapon systems to maintenance and distribution of spare parts. AMC operates research, development, and engineering centers, the Army Research Laboratory, depots, arsenals, ammunition plants, and other facilities, and maintains the Army's prepositioned stocks on land and afloat. The command is also the Department of Defense executive agent for chemical weapons and conventional ammunition. AMC operates facilities in 149 locations worldwide and has over 70,000 military and civilian employees. Major subordinates include: Army Sustainment Command, Installation Management Command, Army Contracting Command, and the U.S. Army Surface Deployment and Distribution Command (USASDDC).

U.S. Army Forces Command (FORSCOM). Headquartered at Fort Bragg, North Carolina, FORSCOM is the Army's largest command responsible for mobilization planning and combat readiness of assigned active Army and USAR units in the continental United States (CONUS) and training supervision of Army National Guard during peacetime. FORSCOM commands assigned active Army operational units in CONUS, and 1st Army provides training, mobilization, and deployment support for federalized, Army National Guard and U.S. Army Reserve (USAR) units. FORSCOM commands U.S. Army Reserve Command and Army Reserve units in CONUS, Puerto Rico, and the Virgin Islands. USARC is responsible for the staffing, training, management, and deployment of USAR units to ensure their readiness for Army missions.

Major FORSCOM subordinate commands and their locations are as follows:

First Army	Rock Island Army Arsenal, IL
III Corps	Ft. Hood, TX
XVIII Corps	Ft. Bragg, NC
1st Infantry Division	Ft. Riley, KS
1st Cavalry Division	Ft. Hood, TX
1st Armored Division	Ft. Bliss, TX
3rd Infantry Division	Ft. Stewart, GA
4th Infantry Division	Ft. Carson, CO
7th Infantry Division	Joint Base Lewis-McChord, WA
10th Mountain Division	Ft. Drum, NY
82nd Airborne Division	Ft. Bragg, NC
101st Air Assault Division	Ft. Campbell, KY
National Training Center	Ft. Irwin, CA
Joint Readiness Training Center	Ft. Polk, LA
20th Support Command	Aberdeen Proving Ground, MD
32nd Army Air and Missile Defense Command	Ft. Bliss, TX
Security Forces Assistance Command	Ft. Bragg, NC

U.S. Army Reserve Command (USARC). The Chief, Army Reserve (CAR) also commands USARC. The CG of USARC is responsible for the staffing, training, and readiness of most Army Reserve units in the continental United States and Puerto Rico. In this role, the CG, USARC reports directly to Army Forces Command.

THEATER ARMIES AND ARMY SERVICE COMPONENT COMMANDS (ASCCS)

U.S. Army Pacific (USARPAC). Headquartered at Fort Shafter, Hawaii, USARPAC is the theater army of U.S. Pacific Command (PACOM). It commands and supports assigned and attached active and reserve units, installations, and activities in Alaska, Hawaii, Japan, Korea, and in possessions and trust territories administered by PACOM. Its staff oversees, evaluates, and supports the Army National Guard in Hawaii, Alaska, and Guam. USARPAC maintains a trained and ready force for employment in the Pacific theater or worldwide. USARPAC's major subordinate units include the I Corps, 25th Infantry Division; 11th Airborne Division/U.S. Army, Alaska; U.S. Army Japan; the 8th Theater Sustainment Command; 94th Army Air and Missile Defense Command; and 9th Mission Support Command (USAR). The Eighth U.S. Army in South Korea comes under the administrative control (ADCON) of USARPAC. It provides U.S. Army ground and aviation forces that come under operational control of Combined Forces Command/U.S. Forces Korea (CFC/USFK) to deter aggression against the Republic of Korea (ROK) and, should deterrence fail, to defeat that aggression. Its major subordinate units include the 2nd Infantry Division; 35th Air Defense Brigade; and the 19th Expeditionary Sustainment Command.

U.S. Army Europe and Africa; and Seventh Army (USAREUR/AF). Headquartered in Wiesbaden, Germany as a forward-based theater army, USAREUR/AF demonstrates national resolve and strategic leadership by assuring stability and security, and leading joint and combined forces in support of U.S. European Command (EUCOM). It maintains combat-ready forces to support NATO commitments as well as trained and ready forces for deployment on contingency operations in support of other commands. As the theater army of EUCOM, USAREUR commands U.S. Army units in Germany, Italy, England, the Netherlands, and elsewhere in the EUCOM area of responsibility. Major USAREUR units include Southern European Task Force, 2nd Cavalry Regiment (Stryker), the 173rd Airborne Brigade, 21st Theater Sustainment Command, 10th Army Air and Missile Defense Command, and 56th Theater Fires Command.

U.S. Army Central Command (USARCENT) and Third Army. Headquartered at Shaw AFB, South Carolina, is the theater army responsible for the support and conduct of operations throughout the U.S. Central Command (USCENTCOM) area of operations in the volatile Middle East. USARCENT maintains a continued forward presence and conducts joint and coalition exercises throughout the region. It is also prepared to respond to contingencies to deter aggression and to defeat adversaries. The 1st Theater Sustainment Command and the 160th Signal Brigade are two of its major subordinate units.

U.S. Army North and 5th Army (USARNORTH). Headquartered at Joint Base San Antonio-Fort Sam Houston, Texas, is the theater army responsible to U.S. Northern Command (USNORTHCOM) for planning the land defense of the continental United States (CONUS) and the combined Canadian–United States defense of Canada. ARNORTH provides support to civil authorities in domestic emergencies. Within the guidelines and limitations of applicable statutes, ARNORTH also provides support to federal, state, and local law enforcement agencies in the performance of their homeland security missions.

U.S. Army South and Sixth Army (USARSO). Headquartered at Joint Base San Antonio-Fort Sam Houston, Texas, USARSO commands and controls Army forces and provides theater support to U.S. Southern Command (SOUTHCOM). USARSO is the theater army of SOUTHCOM. It provides and sustains trained and ready Army forces to support full-spectrum military operations—from theater security cooperation to warfighting—in America's effort to promote regional stability. USARSO plans, programs, and provides Army support for the SOUTHCOM regional security strategy.

U.S. ARMY CYBER COMMAND (ARCYBER)

ARCYBER is the ASCC to U.S. Cyber Command. Subordinate units include Network Enterprise Technology Command (NETCOM), 1st Information Operations Command (Land), and the 780th Cyber Brigade. The command consists of more than 21,000 soldiers and civilians and is headquartered at Fort Gordon, Georgia. NETCOM/9th Signal Command. Headquartered at Fort Huachuca, Arizona, it is subordinate to ARCYBERCOM and is responsible for worldwide theater network operations and signal support. Major units include:

7th Signal Command–Ft. Gordon, GA
21st Theater Strategic Signal Brigade–Ft. Detrick, MD
93rd Theater Signal Brigade–Joint Base-Langley-Eustis, VA
106th Theater Signal Brigade–Joint Base San Antonio-Ft. Sam Houston, TX
2nd Theater Strategic Signal Brigade-Wiesbaden, Germany
1st Theater Tactical Signal Brigade–Korea
516th Theater Strategic Signal Brigade–HI
160th Theater Strategic Signal Brigade–Kuwait
Cyber Protection Brigade–Ft. Gordon, GA

U.S. Army Space and Missile Defense Command (USASMDC). Headquartered in Redstone Arsenal, Alabama. USASMDC is the ASCC to STRATCOM and to United States Space Command. It conducts space operations and provides planning, integration, control and coordination of Army forces and capabilities in support of United States Strategic Command (STRATCOM). USASMDC is the proponent for space and ground-based mid-course defense and the Army operational integrator for global missile defense. It conducts mission-related research, development, and acquisition in support of Army Title 10 responsibilities. USASMDC has subordinate elements in Germany, Okinawa, Hawaii, California, Maryland, and Virginia. The 1st Space Brigade provides Army Space Support Teams, Forward Based Mode radars, and Joint Tactical Ground Station (JTAGS) detachments to support Army and joint forces. USASMDC commands the 100th National Missile Defense Brigade (Army National Guard) in Colorado as well as the 49th Missile Defense Battalion at Fort Greely, Alaska, for command and control of the Ground Based Mid-Course Defense (GMD) System.

U.S. Army Special Operations Command (USASOC). Headquartered at Fort Bragg, North Carolina, USASOC is the ASCC of U.S. Special Operations Command (SOCOM). It provides trained and ready special forces, ranger, special operations aviation, psychological operations, and civil affairs forces to geographic combatant commanders, joint task force commanders, and U.S. embassies overseas, and is responsible for development of unique special operations doctrine, tactics, techniques, procedures, and materiel. It works closely with SOCOM, TRADOC, and Army Materiel Command (AMC). USASOC is responsible for over 28,000 Army special operations personnel, including all active, reserve, and National Guard units. Its major subordinate commands include the U.S. Army John F. Kennedy Special Warfare Center and School; 1st Special Forces Command; 95th Civil Affairs Brigade; Military Information Support Operations Command with the 4th and 8th Military Information Support Operations Groups; the Special Operations Aviation Command with the 160th Special Operations Aviation Regiment; the 75th Ranger Regiment; and the U.S. Army Special Operations Support Command.

U.S. Army Surface Deployment and Distribution Command (USASDDC). Headquartered at Scott AFB in Illinois, USASDDC (Formerly Military Traffic Management Command) is responsible for global traffic management, operation of worldwide water ports, and Department of Defense transportation engineering. Although the Army ASCC to the U.S. Transportation Command (USTRANSCOM), it is jointly staffed and subordinated to AMC.

SELECTED DIRECT REPORTING UNITS (DRUS)

U.S. Army Corps of Engineers. Headquartered in Washington, D.C., the Corps of Engineers plans, designs, builds, and operates water resources and other civil works projects such as navigation, flood control, environmental protection, and disaster response; design and management of the construction of military facilities for the Army and Air Force; and design and construction management support for other Defense and federal agencies. The Corps of Engineers consists of approximately 10,000 civilian and military men and women stationed across the United States and around the world, including Iraq, Afghanistan, Europe, Korea, the Caribbean, Latin America, and Japan. The Corps' military and civilian engineers, scientists, and other specialists work together as leaders in engineering and environmental matters.

U.S. Army Medical Command (MEDCOM). Headquartered at Joint Base San Antonio-Fort Sam Houston, Texas, it formerly provided command and control of the Army's fixed-facility medical, dental, and veterinary treatment facilities, providing preventive care, medical research and development and training institutions with over 60,000 military and civilian personnel. On 1 October 2019, operational and administrative control of all military medical facilities began transition to the Defense Health Agency.

U.S. Army Intelligence and Security Command (INSCOM). A direct reporting unit under the Army G2 with its HQ at Fort Belvoir, Virginia, INSCOM provides command of operational intelligence and security forces; conducts and synchronizes worldwide multi-discipline and all-source intelligence and security operations, and other specialized capabilities in support of Army, Joint, and Coalition Commands and the U.S. Intelligence Community. Its major units are:

National Ground Intelligence Center (2nd Intelligence Center)–Fort Belvoir, VA
66th Military Intelligence Brigade–Darmstadt, Germany
116th Aerial Intelligence Brigade–Ft. Gordon, GA
470th Military Intelligence Brigade–San Antonio, TX
500th Military Intelligence Brigade–HI
501st Military Intelligence Brigade–Seoul, ROK
513th Military Intelligence Brigade–Ft. Gordon, GA
704th Military Intelligence Brigade–Ft. Meade, MD
706th Military Intelligence Group–Ft. Gordon, GA
Army Counterintelligence Command–Ft. Meade, MD
Army Field Support Center–Ft. Meade, MD
Army Operations Group–Ft. Meade, MD
Cyber Military Intelligence Group–Ft. Gordon, GA

ARMY OPERATIONAL FORCE BASIC ORGANIZATION (IN ASCENDING ORDER)

Crews, Teams, Squads, Sections. Crews of certain weapons systems, infantry fire teams, squads, and sections are the smallest units of the Army and are led by noncommissioned officers. The smallest unit normally led by an officer is the platoon, which is led by a lieutenant and includes two or more squads. Detachments are units larger than a platoon, but smaller than a company.

Company. Companies, which usually consist of a headquarters and two or more platoons and/or sections, are usually commanded by captains, although some types are commanded by majors. In the artillery, this size unit is called a battery; in the cavalry, it is called a troop.

Battalion. Traditionally, the battalion has included its commander, his or her staff and headquarters elements, and two, three, or four companies, batteries (field artillery, air defense artillery), or troops (cavalry). The cavalry unit corresponding to the battalion is called a squadron. Battalions are normally commanded by lieutenant colonels.

Regiments. In the tactical sense, only three regiments remain, all commanded by colonels: the 2nd and 3rd Cavalry Regiments; and the 75th Infantry (Ranger) Regiment.

However, under the Combined Arms Regimental System (CARS), historic Infantry, Armor, Cavalry, Aviation, and Artillery regiments are represented by separated numbered battalions that are assigned to brigade or equivalent commands.

Regimental Affiliation Program. Under the Army's regimental affiliation program, all soldiers are assigned to a regiment. In spite of the system's original intent as an assignment factor, regimental affiliation has become largely a sentimental matter, although it is still appropriate for soldiers to respect and draw inspiration from the heritage of whichever unit they are serving in at the moment. Most soldiers belong to the regiment that corresponds to their branch. Aviation, infantry, armor, field artillery, and air defense artillery officers are generally assigned to whichever regiment of their branch they choose, usually after having served with a battalion of that regiment. Most other branches constitute a single regiment for affiliation purposes. Each regiment is allowed an honorary colonel and honorary sergeant major of the regiment. Usually retired soldiers, these honorary positions are intended to be a living link to each regiment's past. When employed properly, regimental customs and traditions, along with the honorary colonel and honorary sergeant major, can be of great value to unit morale and cohesion. Soldiers affiliated with a regiment wear regimental distinctive insignia (see appendix C, Uniforms and Appearance).

BRIGADE COMBAT TEAMS (BCTS)

As part of the Modular Force initiatives beginning in 2004, the Army transformed from a division-based to brigade-based structure from 2003-2007 with BCTs the primary building blocks for arranging landpower. All ground maneuver brigades (active and National Guard) converted to one of three designs: Armored, Infantry, or Stryker. BCTs are all composed of a brigade headquarters, two or three line infantry or tank maneuver battalions, a reconnaissance squadron, a field artillery battalion, a brigade support battalion, and brigade engineer battalion. Most BCTs remain assigned or attached to a division headquarters for administrative control and training oversight and vary in size from 3,500 to 4,500 soldiers.

Armored Brigade Combat Teams (ABCT). The Army currently has eleven ABCTs in the active force. Each ABCT has over eighty M1A1/2 main battle tanks and over one hundred Bradley Fighting Vehicles. The ARNG retains five ABCTs as well.

Infantry Brigade Combat Teams (IBCTs). The active Army currently has fourteen IBCTs of which five are airborne. These BCTs are equipped with wheeled vehicles for ground mobility and are capable of conducting air assaults when supported by Army aviation. The ARNG also has twenty IBCTs.

Stryker Brigade Combat Teams (SBCT). The active Army has six SBCTs as part of the current force. Each SBCT is be equipped with more than 300

Stryker Light Armored Vehicles in ten variants. The Army National Guard also has two SBCTs.

SUPPORT BRIGADES

As part of the Modular Army, a mix of brigades were designed to support the division and its BCTs. These brigades included a combat aviation brigade (CAB), a Maneuver Enhancement Brigade (MEB), a division artillery or field artillery brigade, and a sustainment brigade. These brigades are intended to support BCTs and carry out specific tasks in support of echelons above BCT. Unlike BCTs, multifunctional support brigades are not fixed organizations. Support brigades are designed around a base of organic elements to which a mix of additional capabilities is added, based on the campaign or major operation. To make the brigades both tailorable and effective, the brigade headquarters includes the necessary expertise to control many different capabilities. Each brigade base also includes signal and sustainment capabilities. Multifunctional brigades were normally attached to a division or corps for administrative control and training oversight.

In addition to the multifunctional support brigades, a wide variety of former non-divisional functional brigades and groups will remain in the Army structure primarily as theater-level troops. Units such as military intelligence, signal, engineers, air defense artillery, special forces, transportation, chemical, quartermaster, ordnance, and military police are organized into these brigades and groups. Most are single-arm units designed to support a particular mission but also capable of employment under corps/divisions. These brigades and groups are generally commanded by colonels.

CORPS AND DIVISIONS

Divisions. Army division headquarters (HQ) are currently capable of commanding and controlling flexible combinations of BCTs and support brigades tailored according to operational considerations. Divisions can function as operational-level headquarters, and can conduct sustained battles, engagements, and stability operations. A division can also be the Army force HQ for a joint, interagency, or multinational operational task force. Each division can serve as a JTF or JFLCC HQ (with augmentation). It is capable of rapid deployment with early entry command post capability. Under Army 2030, the Army has proposed reorganizing into divisions as units of action. The Army has proposed three types of divisions: penetration, joint forcible entry, and standard. Each penetration division will have three organic ABCTs, while each JFE will have three organic airborne or air assault IBCTs. Standard divisions will have two or three of any type BCT. Army 2030 divisions will also have organic division sustainment brigades, combat aviation brigades, protection brigades, and division artilleries, while the penetration division will also have an engineer brigade. Most divisions will also have battalion-sized MI, signal, and engineer battalions. Divisions may vary in size from 10,000 to 30,000.

Corps. Army corps HQ provide mission command over flexible combinations of divisions, BCTs, or multifunctional and single-functional brigades as needed by the operational situation. By their very nature, corps will always fight as part of a joint force, working in very close cooperation with the U.S. Air Force, Navy, and/or Marine Corps. Similarly, the nature of current world politics and U.S. treaty commitments means that corps will normally operate as part of multinational forces. Corps may also serve as an intermediate Army HQ between division and theater armies. In a contingency operation, a corps headquarters may function as the land component headquarters of a joint task force or as the joint task force headquarters. In such cases, the corps will have the responsibility for both operational (campaign) and tactical execution. Corps size may vary with the task organization. Historically, for Operation Just Cause in Panama in 1989, XVIII Corps as JTF-South had only 20,000 troops, while for Desert Shield/Desert Storm, XVIII Corps serving as an Army intermediate HQ had some 118,000 troops. In the same operation, VII Corps had over 142,000 troops with four U.S. and one British division. For Operation Iraqi Freedom, V Corps had some 130,000 troops. There are currently four corps: I, III, V, and XVIII.

ARMY SERVICE COMPONENT COMMAND/THEATER ARMY/FIELD ARMY

Field Army. Field Army HQ provide mission command over flexible combinations of corps, divisions, BCTs, or multifunctional and single-functional brigades as needed by the operational situation. Doctrinally, a field army is intended to provide an enduring HQ focused on a single peer/near peer competitor, or formed as a multi-corps HQ capable of large-scale combat operations. Eighth Army is the only active field army stationed in South Korea.

Theater Army is the historic designation for an Army Service Component Command (ASCC) assigned to a combatant commander with a surface area of responsibility and, as such, provides Army support to joint, other service, multinational, and interagency elements. These have operational as well as support functions and responsibilities. The theater army commands Army forces assigned to a unified command, and it organizes, equips, trains, maintains, and logistically sustains them. The organization of a theater army is not fixed but tailored to meet theater requirements. The theater army is primarily concerned with long-range strategic and operational planning. It prepares and coordinates the land operations plan to support the unified command's theater campaign plan. The theater army may be required to provide Army support to other services (ASOS) under executive agent and other responsibilities. For Operation Iraqi Freedom in 2003, ARCENT served as the USCENTCOM Coalition Force Land Component Command (CFLCC) with over 200,000 Army, 62,000 Marine, and 26,000 coalition forces. Of these, 56,000 were theater-level troops.

Theater Army. The organization of a theater army is dependent upon planning the mission, the stage of theater development, and the number of combat units deployed or projected to be deployed. Mission requirements will determine how many corps and divisions are needed and the size and type of enabling commands. Theater enabling commands include: theater sustainment commands, expeditionary sustainment commands, theater medical commands, air and missile defense commands, theater fires commands, engineer commands, military police commands, civil affairs commands, and signal commands-theater.

In addition to a main command post (MCP), theater armies may have either an operational command post or a contingency command post (CCP). The CCP may be used when required to conduct small scale land operations when designated as the joint land force commander or when the theater army is designated as a joint task force commander. The administrative and support activities of the CCP are much less than those of the theater army MCP, and the MCP would continue to focus on sustainment and support functions across the entire area of responsibility for the Combatant Commander. An OCP can serve as either a joint task force (JTF) headquarters or as the joint force land component commander (JFLCC) for large-scale operations where the GCC is the joint force commander (JFC) with augmentation. Each theater army headquarters and its CPs are tailored to the GCC's area of responsibility for joint and multinational operations.

TOTAL ARMY

With the advent of the Total Army approach to Army force management, the Army consists of three uniformed components: the active Army, the Army National Guard and the Army Reserve (the latter two collectively referred to as the "reserve components") as well as its civilian workforce. All are crucial to the Army's success in the execution of its extensive, worldwide missions. Due to the immense and diverse demands of the war on terrorism and the limited size of the active Army, the reserve components have been called upon to bear more of the Army's missions as an operational rather than strategic reserve more than at any time since the Second World War. At the time of writing, the Army consists of 1,000,000 soldiers, of which 473,000 are Regular Army (47 percent); the National Guard 339,000 (34 percent); and the Army Reserve 189,000 (19 percent). Of this total force, some 788,000 reside in the deployable Army forces, 32,000 are in special operating forces, over 188,000 are in general support activities, and some 63,000 active duty are in individual personnel accounts. The reserve components are planned to provide 37 percent of the BCTs; 54 percent of non-BCT artillery; 62 percent of the sustainment units above BCT; 69 percent of other units above BCT; and 88 percent of the Civil Affairs Forces.

ARMY NATIONAL GUARD

The Army National Guard has federal, state, and community functions. It is directly accessible to the President under Title 10 Authority and is responsive to state governors under Title 32 as well. Its federal function is to support national security objectives by providing trained and equipped units for prompt mobilization in the event of national emergency or war. Its state functions are to protect life and property and to preserve the peace, order, and public safety. Its community function is to participate in local, state and national programs that enhance life in America. The Army National Guard has changed from a purely strategic reserve to more of an operational reserve prepared for participation in overseas contingency operations across the range of military operations. The Army National Guard currently has eight divisional headquarters, six other theater-level commands, twenty-seven BCTs, and over ninety support brigades/groups. The ARNG BCTs currently comprise two SBCTs, twenty infantry BCTs, and five ABCTs. The National Guard Divisions are:

28th Infantry Division, Harrisburg, PA
29th Infantry Division, Ft. Belvoir, VA
34th Infantry Division, Rosemount, MN
35th Infantry Division, Ft. Leavenworth, KS
36th Infantry Division, Camp Mabry, TX
38th Infantry Division, Indianapolis, IN
40th Infantry Division, Los Alamos, CA
42nd Infantry Division, Troy, NY

ARMY RESERVE

The Army Reserve is a federal force providing trained units and qualified individuals for active duty in time of war or national emergency and at such other times as the national security requires. At the top of the U.S. Army Reserve is a three-star general with two distinct roles. The first is Chief, Army Reserve (CAR). The Army Reserve has extensive engineer, medical, training, transportation, and civil affairs assets that are well suited for domestic and humanitarian missions. The Army Reserve's capability in its primary support function is enhanced by the civilian experience and unique skills of its soldiers. The CAR reports to the Chief of Staff of the Army and represents the Army Reserve in policy and planning discussions with the Army, the Department of Defense, and Congress. The CAR is responsible for all Army Reserve soldiers, even those who report directly to the Army. Army Reserve units include twenty theater-level commands and more than seventy support brigades/groups. The Office of the Chief, Army Reserve (OCAR) is located in the Pentagon in Washington, D.C.

THE ARMY OF 2030 AND 2040

The Army of 2030 will begin to effectively employ lethal and non-lethal multidomain effects to overmatch against any adversary to shape, prevent, and win conflicts and achieve national interests. It will leverage cross-cultural and regional experts to operate among populations, promote regional security, and be interoperable with the other military services, United States government agencies and allied and partner nations. By 2040, the Army, leveraging the Total Force, will consist of a balanced versatile mix of scalable, expeditionary forces that can rapidly deploy to any place on the globe and conduct up to sustained large-scale combat operations. Composed of agile and innovative institutions, soldiers, and civilians, the United States Army of 2040 provides strategic advantage for the nation with trusted professionals who strengthen the enduring bonds between the Army and the people it serves.

3

The Army Officer

> You have to lead men in war by bringing them along to endure and display qualities of fortitude that are beyond the average man's thought of what he should be expected to do. You have to inspire them when they are hungry and exhausted and desperately uncomfortable and in great danger. Only a man of positive characteristics of leadership with the physical stamina that goes with it can function under those conditions.
>
> —General of the Army George C. Marshall

The Army officer corps is ultimately responsible for victory or defeat in war, readiness or unpreparedness in peace, success or failure in the "military operations other than war" into which the Army is increasingly being thrust as the United States assumes the mantle of responsibility conferred by its status as the world's superpower. In chapter 1, the main difference between the role of the officer and that of the enlisted soldier in our Army was identified and briefly discussed within the context of the Army, its heritage, its customs, and its courtesies. Chapter 2 provided an overview of the modern Army's organization and missions. This chapter explores, in greater depth, officers' roles and responsibilities; it also addresses the ethical and other professional requirements for becoming the kind of officer to whom General Marshall was referring in the quotation above. It is that kind of officer who wins our wars.

Field Manual 6-22, *Army Leadership, Competent, Confident, and Agile* (October 2006) lays out the Be–Know–Do model of Army leadership, as well as the values and attributes required of Army leaders. Rather than a simple reiteration of this important information, essential for all Army leaders, regardless of rank or station, this chapter expands on these themes from the perspective of Army *officers*, in consonance with the other aspects of officership as described in this guide.

SERVICE AS AN OFFICER: PROFESSION AND CALLING

The United States has been thrust into a position of world leadership, a status it did not seek. International catastrophe might occur if this responsibility is terminated. The military professionals surely will continue to be the final rampart for preserving our form of government against change by force, as well as protecting our people from outside aggression or internal insurrection. Officers must provide a living reflection of the words *Duty, Honor, Country* in their broadest, most meaningful context.

Living by this standard—at the risk of one's life—requires the expertise and rigid standards associated with a *profession*, and the dedication and focus associated with a *calling*. Service as an officer, for any length of time, requires adoption of the highest standards of technical and tactical proficiency; they are easily the equal of those of the medical or legal professions, for example. If brain surgery is a delicate and extraordinarily demanding task, so is leading a platoon of Americans in an attack on a hill. Calling for fires, placing crew-served weapons, breaching obstacles, communicating amid the din of battle, evacuating casualties, delivering vital supplies, consolidating on the objective—each of these is, in itself, a complex and challenging task, with a myriad of planned and reflexive actions. If these and counterpart activities—all the way to service at the highest levels of command—were all there is to being an officer, it would amount to a demanding and challenging profession indeed. But there is more.

The Warrior Ethos. Inspiring and motivating American soldiers to fight the battle—with every confidence that they will win that fight—is the essence of Army officership. Soldiers' willingness to fight the battle is based on confidence in their leaders, confidence in the effectiveness of their training, and in knowing that the effort of each soldier in the unit is working together toward the same end. Thoroughly trained soldiers led by competent leaders reduce the inherent risk to life and limb and motivate each soldier to put forth the maximum effort needed for victory on the battlefield.

Professional Competence. Professional competence is essential and helps to inspire confidence in subordinates, but much more is required. Totally committed, sincere, caring leadership is also required—on a "24/7" basis—and that must spring from something inside. To be effective over time, in sunshine and rain, in mud and blood, officers must hear and answer a call to service that is no less powerful than that which draws and sustains members of the clergy or other great servants of humanity. It may not come before commissioning, it may not be heard immediately thereafter, but for those who continue beyond the first few years of service as an Army officer, it will come sooner rather than later. Those who do not hear it in its many forms—mystical, practical, or otherwise—will not be able to fulfill the demands of being an officer on a sustained basis. There is no shame in failing to hear the call; there is only shame in service as an officer by "going through the motions." Our Army and our nation deserve more.

The life of the Army officer is rewarding and stimulating for those individuals who hear the call. It is an exacting life with its own code, special hazards, and rewards. Not all individuals will find the life either attractive or agreeable, which is a reason for laying the facts on the line in this chapter. The person who wants a sheltered, safe, uneventful life, with work from nine to five, a five-day week, and other niceties, is unlikely to be adaptable to the variety of missions and worldwide assignments of the Army officer. But the future security of our nation requires that a sufficient number of our fully qualified young citizens choose this career. For them, it will be a good life. There must be capable hands and minds at every level of the slender line from lieutenant to general. There is a clear need for young men and women to enter the corps of officers and to progress in grade and responsibility to meet with competence the unknown requirements of our nation in the years ahead.

THE PEOPLE WITH WHOM YOU LIVE AND WORK

In the enjoyment of any career, the quality of the people with whom you are associated is of high importance. Your success will be influenced by the manner in which they perform their duties. Among them you will find the personal friends who will enrich your life.

Much of the work of officers is with enlisted soldiers who, for the most part, are young people in their formative years, actively inclined, with good minds. They have the enthusiasms, the ambitions, and the interests of youth. There are few responsibilities equal in satisfaction to training, developing, and leading young soldiers. In these tasks you will be assisted by older soldiers who are the noncommissioned backbone of our Army and who will become individuals you will respect and whose respect you will treasure.

THE CODE OF THE ARMY OFFICER

The officer lives under a strong and inspiring code that acts as a guide in the standards of official and personal acts. It assures the officer that he or she will be associated in worthwhile missions with other officers whose loyalty to the nation, personal trustworthiness, and honor are of the highest order. Here will be found the welding of common interests that results in military teamwork, in comradeship, and in friendships with others of all ages and grades. It provides an environment that its members regard with pride and self-respect. It is the sort of life that many good men and women have loved and do love. Life under the code is good for the nation, and for most officers it is a rich and gratifying experience.

As an Army officer, you will be expected to follow the code in the performance of your official duties and in your relations with other individuals, military and civilian, on duty and off duty. It is not a harsh code, but it is demanding. Adherence to the code does not start at reveille and end with the sounding of retreat. It is a twenty-four-hours-a-day, seven-days-a-week code that will guide you throughout your military career and all the years that follow. The code sets the tone for a way of life that many honorable, dedicated men and women have found rewarding—the Army way of life.

But what is this code that is of such importance to the Army officer? It is part official and written, but it is mostly traditional and unwritten. Let us examine it more closely.

The Officer's Commission or Warrant. In addition to the oath, already discussed, the written basis for the code is expanded by the officer's commission or warrant. Examine the words of these documents and again think deeply about their meaning. In accepting your commission or warrant, you have embarked upon a life of public service to our great nation. Our nation's leaders and its citizens have every right to expect that you will discharge all assigned duties in an exemplary manner and that, by accepting your commission or warrant, you have every intention of doing your utmost to support our way of life.

Army Values. The code of the Army officer can be summarized by the Army's corporate values, which commissioned leaders will adhere to, and for which the appropriate acronym has been developed—LDRSHIP. Those letters stand for the fundamental basics of the code: Loyalty—Duty—Respect—Selfless Service—Honor—Integrity—Personal Courage. These are the recurring themes in the code that have marked the professional lives of the great military leaders who have preceded us in this profession. For more information, see www.army.mil/values.

Significance of the Code. Is this code important? All honorable professions have their own codes and precepts. The extreme importance of the code of the Army officer stems from its significance to the United States. These responsibilities, shared equally with officers of the Navy, the Marine Corps, the Air Force, and the Coast Guard, involve the security of our nation, the protection of our people, plus the support of our nation's policies in its chosen courses of action in its relations with other countries. These missions, which the officer and all members of the armed forces are obligated to accept, may lead him or her to the distant places of the world and, if combat is encountered, may involve the life or death of subordinates as well as the success or failure of the nation's mission. The duty may include placing the officer's own life on the line, for in combat, all share the hazards. These are the reasons our armed forces are maintained, and they give the code of the Army officer its vast significance.

Historically, there can be no truer foundation for the code of the Army officer than the example set by General Washington, with his own high standard of personal honor, of discipline, of personal sacrifice, of leadership, of complete devotion to his mission. Other great men in our history, whose names are household words of respect and trust, have embraced this code while adding to its strength—Generals Grant and Lee, General of the Armies Pershing, Generals of the Army Marshall, Eisenhower, MacArthur, and Bradley, and their more recent counterparts Generals Westmoreland, Schwarzkopf, and Powell.

The United States will continue to fight wars well into the future. Our nation's success will rest in great part on the strength of its officer corps. If the officer corps is professional—that is, if individual officers have prepared themselves to use their specialties to contribute to the overall military capabilities, and if they hold fast to the military virtues embodied in the code—the nation can count on victory. However, if individual officers define their duties too narrowly and if they think only of serving themselves, the nation should be prepared for defeat.

To all who shall see these presents, greeting:

Know Ye, that reposing special trust and confidence in the patriotism, valor, fidelity and abilities of **REILLY SHANE PATRICK** *, I do appoint* **HIM, MAJOR, TRANSPORTATION, U.S. ARMY RESERVES** *in the*

United States Army

to **RANK** *as such from the* **THIRD** *day of* **AUGUST** *, two thousand* **SIXTEEN** *. This officer will therefore carefully and diligently discharge the duties of the office to which appointed by doing and performing all manner of things thereunto belonging.*

And I do strictly charge and require those officers and other personnel of lesser rank to render such obedience as is due an officer of this grade and position. And, this officer is to observe and follow such orders and directions, from time to time, as may be given by the President of the United States of America or other superior officers, acting in accordance with the laws of the United States of America.

This commission is to continue in force during the pleasure of the President of the United States of America under the provisions of those public laws relating to officers of the Armed Forces of the United States of America *and the component thereof in which this appointment is made.*

Done at the city of Washington, this **THIRD** *day of* **AUGUST** *in the year of our Lord two thousand* **SIXTEEN** *and of the Independence of the United States of America the* **TWO HUNDRED AND FORTY-FIRST YEAR**

By the President:

SECRETARY OF THE ARMY

The Officer's Commission

To all who shall see these presents, greeting:
Know Ye, that reposing special trust and confidence in the patriotism, valor, fidelity and abilities of **MICHAEL K. COFFEE**

the Secretary of the Army has appointed **HIM** *a*
WARRANT OFFICER, W-1, ARMY RESERVE

in the United States Army

to rank as such from the **TWENTY-FIRST** *day of* **AUGUST** *in the year two thousand* **SIXTEEN** *. This Warrant Officer will therefore carefully and diligently discharge the duties of the office to which appointed by doing and performing all manner of things thereunto belonging. And all subordinate personnel of less rank are strictly charged and required to render such obedience as is due a Warrant Officer of this grade and position. And this Warrant Officer is to observe and follow such orders and directions, from time to time as may be given by Superior Officers and Warrant Officers acting in accordance with the laws of the United States of America.*

Done at the city of Washington, this **TWENTY-FIRST** *day of* **AUGUST** *in the year of our Lord two thousand* **SIXTEEN** *and of the Independence of the United States of America the* **TWO HUNDRED AND FORTY-FIRST YEAR**

The Adjutant General

Warrant Officer Warrant

Professionalism is the bottom line of each officer's and enlisted person's service and the way we accomplish the Army mission. This was well summarized by a very senior officer, who, to paraphrase, said:

> The Army exists to fight and win wars. That's our core expertise; it is what allows us to be called professionals. We are entrusted with the security of our nation. The tools of our trade are lethal, and we engage in operations that involve much national treasure and risk to human life. Because of what we do, our standards must be higher than those of society at large. The American public expects it of us and properly so. In the end, we earn the respect and trust of the American people because of the professionalism and integrity we demonstrate.

The military code is a standard of action. It is a firm belief that the preservation of our nation is decidedly worthwhile. It is unswerving confidence in the loyalty of our people and their sons and daughters who wear their country's uniform. It is faith. As Abraham Lincoln stated in his Cooper Union Address, 27 February 1860, "Let us have faith that right makes might; and in that faith let us to the end, dare to do our duty as we understand it."

TRADITIONAL BASIS OF THE CODE

The oath of office, the officer's commission or warrant, and the other published codes and regulations provide the official basis for the code of the Army officer. The basic standard of individual and group performance of military responsibilities is perhaps best expressed by the motto Duty, Honor, Country.

Duty. In the Army, the performance of duty to the best of one's ability is a first requirement. Missions must be accomplished to standard and on time. In thinking of duty in the Army, always remember that it includes willingness to fight, to enter areas of great personal danger, to accept the hazards of death. There is little tolerance of slipshod work or halfway measures. Your reputation depends on the successful performance of all assigned duties, regardless of their size or their perceived importance.

Most orders are of the "mission" type, which state the job to be done but leave to the officer the selection of method. As an officer, you will be expected to select methods that will accomplish the desired results, at the time required, with due regard for costs.

See work that needs doing and, without stepping into the responsibilities of others, do it. Within your own sphere of responsibility, do not wait to be told.

Stand on your own feet. Your superiors are interested in your past only insofar as it may indicate future capacity. They want to know how well you perform your duty today, so that they may estimate what you will do tomorrow. The reputation that counts most is the one you earn today.

Honor. The code requires all officers to live and conduct all their activities so that they may look all persons squarely in the eye knowing that they are honorable individuals associating themselves with other honorable individuals. Honor is the hallmark of officerlike conduct. It is the outgrowth of character. An honorable person has the knowledge to determine right from wrong and the courage to adhere to the right. It means that an officer's written or spoken word is accepted without question. Facts will be identified as facts, and opinions for what they are. Actions will be based on considerations of the good of the unit, the Army, the nation. They are all included in personal integrity and based on sound ethics.

As a part of honor, the officer is expected to lead a decent life. An officer does not lie, cheat, steal, or violate moral codes. Those who do not live honorably earn first the scorn of their associates, and if the offenses are more than trivial, they stand a fine chance of trial for conduct unbecoming an officer and a gentleman or lady.

Country. The profession of arms in which the officer is appointed a leader is a public trust. The American people maintain military forces for the preservation of their security and the sovereignty of the United States. They have the right to expect the highest standards of personal and official conduct from their officers. The responsibilities of the officer embrace all the people, and his or her oath of office prescribes the protection of the Constitution.

The Army officer holds a commission or warrant by choice. Each officer is a volunteer. He or she accepted a commission or warrant, and with it all its hazards and responsibilities as well as its rights and privileges. Each officer is a patriotic citizen who places country above self. Patriotism has this definition: "The willingness to sacrifice and endure discipline for the welfare of the community." Thomas Paine wrote, "Those who expect to reap the blessings of freedom, must, like men, undergo the fatigue of supporting it."

The Identical Standards for All. The degree to which observance of the code is expected is the same for all officers, without regard to grade, branch, sex, component, or length of service. The meaning can be illustrated by comparison with the observance of the rules of a game, such as golf; the player whose score is rarely below 100 should be as careful to avoid improving the lie of the ball in the rough as the champion.

Now let us temper this principle with reason. Inexperienced officers require instruction in what is right and wrong with regard to many military requirements, just as they must be instructed in military techniques. Some deviations will be regarded as a subject for instruction or mild rebuke. For example, it is a custom that an officer provide first for the needs of his or her soldiers before devoting time to personal needs; the officer who fails to do so may expect to be reminded of his or her duties. In contrast, an offense involving character or honor or deliberate fraud will be regarded as equally serious if committed by a newly appointed junior or by the oldest and most experienced senior.

It is true that in matters of honor the standard is the same for all. But never to be forgotten is another great truth: *The impetus for honor and honorable actions must come from the top. Officers, especially senior officers, must set a visible example.*

Integrity, the Essential Ingredient. The essential attribute of soldiers is integrity. It is the personal honor of the individual; it is the selfless devotion to duty that produces performance integrity in the discharge of individual missions; and it is the integrity of the Army as a whole in providing its share of the security of the nation in war or in peace.

The Army officer has a public vocation as an official of the government. Enlisted soldiers of the officer's unit are sworn to obey his or her lawful orders and, in so doing, perform their own public duty. In time of war, the decisions the officer makes, and the orders he or she issues, may be matters of victory or defeat, life or death. The people of our country place their trust in Army officers, and officers of the sister services, so that they can sleep peacefully at night and pursue their chosen vocations with minimum attention to military matters. The United States could not entrust its security to officers about whose integrity there is the slightest doubt.

Ethics. Much has been written in recent years regarding the need for the Army to adopt a code of ethics to guide individuals in their behavior. Particular emphasis has been placed on the need for guidelines to prescribe a proper balance between promotion of self-interest and accomplishment of the Army's mission. Development of such a set of guidelines would be difficult and, at best, would be incomplete, for no one could possibly foresee all the special situations in which some guidance might be necessary.

Is a particular set of ethical guidelines necessary? Read again the earlier words on the meaning of Duty, Honor, Country as setting the basic standard for performance. How simple it is to test one's proposed action against this standard. Is what you are proposing an honorable act? After you have taken it, will you be able to look others squarely in the eye knowing that your honor is untarnished? If so, you are on the right track. Now, examine whether your proposed action, if successful, will accomplish your mission. Will a successful outcome meet the intent as well as the letter of your assigned task, and do so within the time and resource constraints without unnecessary risk? If so, proceed. You will be doing your duty. As a final check, consider the outcome in terms of the Army and our nation. Will the results of your proposed action be in their best interest? An affirmative answer to this question, as well, means that by all three measures—Duty, Honor, Country—your proposed action is correct. You may proceed with confidence.

In most cases, the correct course of action as verified by the above checks should coincide with your own best interests, whether or not you perceive that to be the case. In any event, even if it does not, there still is no problem with a decision. Your oath of office, your commission, and your adherence to the code of the Army officer demand that you do what is right, regardless of personal outcome.

THE EXPECTATIONS OF THE U.S. CONGRESS

The Congress of the United States represents the American people. In 1997, it amended Title 10 of the U.S. Code to reinforce the code of the officer along the lines just discussed. Section 3583, "Requirement for Exemplary Conduct," reads:

All commanding officers and others in authority in the Army are required—

(1) to show in themselves a good example of virtue, honor, patriotism, and subordination;

(2) to be vigilant in inspecting the conduct of all persons who are placed under their command;

(3) to guard against and suppress all dissolute and immoral practices, and to correct, according to the laws and regulations of the Army, all persons who are guilty of them; and

(4) to take all necessary and proper measures, under the laws, regulations, and customs of the Army, to promote and safeguard the morale, the physical well-being, and the general welfare of the officers and enlisted persons under their command or charge.

This law has been on the books for the U.S. Navy for nearly 200 years. It now applies to the Army and the Air Force as well.

MILITARY TRADITIONS STEMMING FROM THE CODE

There are traditions of military service that have guided its members throughout the years of our national existence and form an important part of the code of the Army officer.

Tradition of Public Service. As an Army officer, you are a public servant. You go where you are ordered and perform the tasks that your duty requires. In peace you prepare yourself and your subordinates for the requirements of war. During war you lead American soldiers in battle against the nation's enemies. It is neither the officer nor the Army who decides when a state of war exists; that responsibility is borne by the Congress and the President. But the peace having been forfeited, it is the duty of the officer to assist in restoring it. It may cost the officer's life, and sometimes this occurs at an age at which members of other professions have retired to the garden and the front porch. The military life is devoted to public service, and the officer is a public servant.

Tradition of Accomplishing the Mission. Accomplishment of mission is recognized as the primary requirement of the military leader and indeed of all members of the service. The "Army way" of undertaking a mission is to display enthusiasm, boldness, and aggressiveness in getting any job done.

This means that big things must be done, if properly ordered or required by the mission assigned. In battle, the attack objectives must be taken and the defense objectives held. Training programs must be carried out effectively so that all are prepared for their tasks. Administrative and logistical responsibilities must be completed at the required high standards.

It means doing the little things correctly, too, as a matter of routine. Whether the task is tremendous in its scope and importance or a routine requirement, the military person is expected as a matter of course to undertake it as required and complete it up to standard and on time. In these tasks, he or she must accept the hazards of the vocation and the frequent hardships of service. This is the important military tradition that the mission must be accomplished.

Tradition of Leadership. The officer is trained to lead. From your earliest days as an officer the tradition will be impressed upon you. You will become accustomed to receiving and executing missions. This will require you to plan work, to assign missions to others, and then to see that their work is done skillfully and in cooperation with others. As you grow older, your training and experience will broaden and increase your capacities; this is generally accompanied by greater responsibilities. Thus the tradition of leadership deepens. Just as you are trained to lead others, you will be trained to be led by others, for no military person can rise so high that he or she is not responsible to another. Military leadership requires the ability to develop teamwork and at the same time be a part of the team.

Tradition of Loyalty. Loyalty is demanded of the Army officer. It extends throughout the chain of command to the President, the commander in chief. It extends to your subordinates to include the newest private. And it extends to your peers. It must be a true loyalty, and there is an essential reason for it. Members of the corps of officers have the common mission of protecting the nation and our people, which requires the coordinated best efforts of each individual. Even the suspicion of disloyalty would destroy the usefulness of any officer, for no one would trust him or her or give the officer responsibility. It must include the boss whom you may dislike; your peers with whom in a sense you are in competition; and each of your subordinates. Think about it deeply. Once trust is forfeited, it may never be regained. The loyalty of officers to the nation, to the commander in chief, to their seniors, their subordinates, their peers, has been traditional in the Army since its beginning.

Loyalty is sometimes misunderstood or, worse, misused by some in the Army. Loyalty to the Army, to the unit, to rank, and to the authority of superior positions is *required*. Similarly, loyalty to subordinates and peers is absolutely essential, until such time as they prove themselves unworthy by breaking faith with their obligations as soldiers. All this is integral to loyalty to the Constitution and the Republic, for no Army can function without these loyalties.

Sometimes, only simple things are necessary to maintain appropriate loyalties. Explaining the rationale behind orders or policies, when there is time, is one way to avoid placing a subordinate in the position of having to choose between losing credibility for passing on a seemingly ridiculous order and being disloyal to you by allowing criticism of your decision (or criticizing it himself, to save credibility). On other occasions, extraordinary measures are required, such as reporting the misdeeds of a superior to his boss or to the inspector general. Live by the ethics and requirements of the officer's code and commonsense leadership, and you will never make loyalty an issue for your peers or subordinates.

Tradition That an Officer's Word Is His or Her Bond. An officer's statement of fact, opinion, or recommendation must conform fully with his or her belief. You must take adequate care that when you make a statement of facts you can provide the evidence to support it. If you render an opinion or make a recommendation, you must have given sufficient thought to the subject to enable you to reach a reasoned conclusion. All this must be true whether the statements are oral or in writing or are just your initials extending a concurrence. The added statement "I certify" must not be interpreted as meaning "something extra" as to truth. Your word is your bond. Similarly, anything to which an officer signs his or her name means irrevocably what is said, both in letter and spirit.

Tradition of Discipline. To develop discipline within an organization, the leader must set the example of discipline. Since the unit you command is only a part of a larger organization, or of the Army, you must execute objectives or missions that reach you as orders from your own superior officers. The U.S. Army is a disciplined army. An army that is undisciplined is worth nothing. An undisciplined army is worse than useless, for it would constitute a public menace in itself. The tradition of discipline is as deeply ingrained into the mind and heart of the successful officer as the tradition of leadership.

Tradition of Readiness. One of the most striking qualities required of the Army officer is that he or she, and the troops they lead, be in a position of readiness to meet whatever tasks arise, including sudden leadership in campaign and combat.

You must be ready for an unexpected change of station or duty assignment. You must be prepared to accept and execute effectively new requirements of your mission. The tradition of readiness includes flexibility of mind and mental processes, a willingness to reach out for new ideas, and an ever-broadening capacity to undertake and do new things.

Your leadership capability and command efficiency are measured by your unit's readiness. Such readiness includes personnel, equipment, and unit training, to which must be added morale and discipline, which are evidenced when the real test—combat—comes.

Tradition of Taking Good Care of Soldiers. The officer who has the best record for accomplishment of battle missions with lowest sick and casualty rates is likely to be one who has cared best for his or her soldiers. This means that members of a command must receive thorough training for their duties; have in their possession the equipment and supplies according to their authorization and needs; and receive the best quartering and dining facilities possible under the circumstances. They must have available first-rate medical support under all conditions of service anywhere in the world. It means many other things, such as support for athletic and recreational programs, advice and help on personal problems, including legal counsel, and proper use of awards and punishment. It involves fairness and justice in all things, including spreading the workload and hazards among all eligible individuals; complete absence of favoritism; and promotions to be made on merit.

Tradition of Cooperation. Cooperation is the art of working with others to attain a common goal. This is a daily expectancy of service duties, and it is traditional that cooperation be willing and wholehearted.

Neither a commander nor a staff officer, no matter how senior, can "go it alone." An officer must cooperate with others, and others must cooperate with him or her. In any staff or any command, the problems for solution are very likely to involve two or more staff agencies or units, or a reconciliation of views between the unit commanders who execute a plan and those who plan it. It takes coordination and cooperation to accomplish a mission. The officer who neglects to cooperate with others invites failure.

Tradition of Being a Lady or Gentleman. Officers are expected to be ladies or gentlemen. This must be manifest in their moral standards, conduct, appearance, manners, and mannerisms, as well as the professional standards they establish in the performance of duty. It must be displayed in the things they avoid doing. They avoid vulgarity. They do not drink to excess. They do not avoid the payment of just bills, nor do they tender bad checks.

The general good of the officer corps demands that all individuals display the qualities of ladies or gentlemen. Great prestige attaches to officers because as a group they have generally been accepted as such. They are accepted to membership in civilian clubs and associations often because of their commission. Their credit rating is high. Their word is accepted in and out of service. Their opinions bear weight. As a group, they have the confidence and trust of the people. Officers must guard and cherish this standing, and they must realize that the unfit among them reflect on and damage the standing of all.

Tradition of Avoiding Matters of Politics. It is traditional, and also required by law, that soldiers avoid partisan politics. This is particularly important for officers. The oath of office requires each officer to serve the elected leaders of the nation without regard to their political party and without regard to the officer's own political beliefs or affiliations.

It could not and must never be otherwise. The armed forces are the final bulwark for the preservation of the Constitution and the security of the nation. We could not tolerate "Republican officers" and "Democratic officers." Loyalties go to the nation and to its form of government.

Tradition of Candor in Making Recommendations. It is a normal experience of Army officers of all grades and degrees of experience to be asked by their commanders or bosses for their opinions or recommendations. Such missions are the daily experience of staff officers, for it is always the commander who makes the decision.

The duty reaches its zenith in the recommendations made by the Chief of Staff and senior members of the Army Staff, and the commanders of Joint and Specified Commands, in their relations with the Secretary of the Army, the Secretary of Defense, and the President, and with the committees of Congress as they consider legislation governing the armed forces. How far should the military leaders go in advancing their considered views? What is their duty if their views should be challenged? What is their duty if the decision goes contrary to their convictions and their recommendations?

Here is a guide. In his first meeting with the Army Staff, Gen. Matthew B. Ridgway, former Chief of Staff of the Army, had this to say about this vital subject:

> The point I wish to make here, and to repeat it for emphasis, is that the professional military man has three primary responsibilities:
>
> *First*, to give his honest, fearless, objective, professional military opinion of what he needs to do the job the nation gives him.
>
> *Second*, if what he is given is less than the minimum he regards as essential, to give his superiors an honest, fearless, objective opinion of the consequences.
>
> *Third*, and finally, he has the duty whatever the final decision, to do the utmost with whatever is furnished.

It has never been said better.

RELATIONSHIPS AMONG OFFICERS

There is a special relationship among officers that must be understood as a part of the code. They are appointed into the service for the same general purposes, and therefore have much in common. There is trust between them because they are officers and have subscribed to the officer's oath. An understanding of the status of the officer and the relationship or responsibility of one officer to all others of the corps is a matter of primary importance.

Your Brother and Sister Officers. The officers of the Army are a cross section of the American public, drawn from all states and sections in representative numbers, and from all classes of the social order. Most officers are graduates of the United States Military Academy or of ROTC programs at colleges and universities of the nation. As a group, they are subject to the same ambitions, the same variations in viewpoint as the people whom they serve. Drawn from the whole nation by a selective process, they are fairly (no more and no less) representative of that larger group of citizens that furnishes the leaders in business, professional, and public life. The common denominator that binds Army officers together is interest in the welfare of the nation and in the national defense.

An abiding camaraderie pervades the officer corps. All worthy officers desire the success of all others, for the success of each contributes to the mission accomplishment of the unit and the Army. Most officers will go far to help fellow officers succeed; this is especially apparent when more experienced officers take well-intentioned subordinates or less experienced ones "under their wing" and give them the benefit of their wisdom and knowledge. In fact, unlike many other professions, in the Army, officers assume the competency and best intentions of others until or unless these qualities and attributes are proved otherwise.

Competition exists among Army officers, as it exists anywhere in life. Generally, however, this competition is a healthy striving for excellence measured against the Army's requirements or standards, rather than the "dog-eat-dog," petty, or ignoble infighting that characterizes some other professions. Officers who engage in such destructive behavior must be corrected immediately or eliminated from the service, as few things are more deleterious to the maintenance of good order and discipline than leaders who place their personal ambitions above those of their soldiers, their peers, their unit, and the Army.

THE ENLISTED SOLDIER

The enlisted soldier represents a cross section of American society, within appropriate age and fitness constraints. They have all volunteered to become soldiers and have been granted the privilege of military service by the Army. This means that they have met the basic medical, mental, physical, and moral fitness standards to serve as soldiers. The great majority are self-respecting, patriotic, decent, good citizens who desire to serve the Republic well. They are the inheritors of the legacy built and paid for in sweat and blood by their predecessors throughout more than 240 years of American history. With very few exceptions, they will meet the challenges of upholding that legacy with honor and great credit, whenever properly trained and led. That is your responsibility.

In return for your devotion, your loyalty, your hard work and honest effort, you will earn your soldiers' respect, their cheerful obedience, and their loyalty, even unto death. Whatever their level of education, whatever their socioeconomic background, whatever their reasons for joining the Army, this is one thing that has not changed since Valley Forge. It is the covenant between the leaders and the led, at every level, in the Army of our great Republic—an Army of free citizens who exchange much of their freedom for the privilege of service. It is the contract between *equals* serving in different capacities with different roles, responsibilities, and compensation. The Army is *not* a caste system. In the meritocracy that our Army must always be, officers are "superiors" in rank only and, in many ways, are the servants of their soldiers and units. And all are servants of the Republic.

OFFICERS' RELATIONS WITH ENLISTED PERSONNEL

Every enlisted soldier wants to feel certain of getting a fair deal from his or her commander. Indeed, each officer wants to feel that he or she, too, will get a square deal from his or her commander and the Army itself.

Basic Principles. The relationship between a commander or any superior officer and his or her personnel must be developed to stand the strains of campaign and combat, as well as the less demanding conditions at posts and stations in the United States. There is no precise parallel with other vocations against which to evaluate the need for such unusually high standards of conduct, including abstentions.

In our Army, it is strong tradition that an officer does not gamble, borrow or loan money, drink intoxicants, or participate in ordinary social association with enlisted soldiers on an individual basis. Not only may aggravated violations of the tradition be handled under the Uniform Code of Military Justice, all are matters of simple common sense. Here is why. The officer must not have favored associates, or "buddies," chosen from enlisted personnel. To do so would place in question and weaken the vital belief in the officer's impartiality, since officers have legal authority to issue orders and administer justice. Noncommissioned officers and soldiers want no favoritism in the decisions of their seniors.

There must be no justified suspicion of favoritism. The officer cannot be "one of the boys." If he or she is popular in the minds of subordinates, it should be a by-product of leadership, human fairness, knowledge, and wise decisions.

Fraternization. In today's Army, where women make up an increasingly large percentage of both the officer corps and the enlisted ranks, this matter deserves even closer attention. Since we are all human, there are bound to be occasions when officers, be they men or women, are attracted to enlisted soldiers of the opposite sex. No good can come from any relationship resulting from such attraction. Both the officer and the enlisted soldier would earn the scorn of their peers for establishing such a relationship, and the effectiveness of both members to the Army would be reduced. To any officer who may think of establishing a relationship with an enlisted soldier, we can offer only one word of advice: Don't!

ACTION UPON OUTSIDE PRESSURES

The Army officer must expect to face requests or even demands from civilian sources that we call "pressures." We are a democracy. Our people are interested in the national defense and in the citizens who are members of the armed forces. The company commander, the commander of an Army post or of a major command, or a senior official of the Department of the Army may expect to receive many requests for all sorts of information. Many Army matters are of proper interest to citizens. Included are routine requests from members of Congress or other officials who ask for proper information that they wish to supply to a constituent. Be certain that the disclosure is authorized, and if so, provide the answer with promptness and courtesy. It is routine.

But what of those pressures that seek reversal of an Army action or policy? A company commander may receive a letter from a distraught parent about his soldier son or daughter. A post commander may receive a letter from an excited mayor. Required is a consideration of the matter as represented by the outside source, balanced against the facts as known by the Army. New and important facts may be introduced from the outside that were unknown when the action under examination was taken. Army actions are made with judgment but are not infallible. *Still, Army decisions, actions, and policies must not be changed lightly or merely to accommodate an influential person, for to do so would invite endless confusion.* Even so, when the merits of the case indicate the justice of a change in a matter of some importance, the proper action may be to make the change. In all cases, choose the course of the Army's best interest, and one with which you can live as an officer worthy of trust.

REVERSING DECISIONS OF SUBORDINATES

There is an exceptional situation that deserves discussion. Assume that a negative action has been taken on a subject of importance. Assume further that an appeal to higher military authority is made, urging reversal of the negative action taken. No decision should be reversed except in those rare instances of real importance when failure to do so would injure the national interest or do a real injustice. Perhaps new facts are brought forth. Perhaps the situation was not fully known by the officer. Whatever the reason, assume that the senior officer decides it is necessary to reverse the action of a subordinate. How should it be done? *The officer must inform the subordinate of the imminent reversal, with reasons, before any other person learns of the action. Humiliation of the subordinate must be avoided. Perhaps it would be appropriate to extend an opportunity for the subordinate to change his or her own stand, although the senior must stop short of urging or seeming to force a change of viewpoint. Extend an opportunity for the subordinate to make the announcement of the reversal.* Such instances are uncommon. When reversal is considered to be necessary, the procedure must be done with judgment, lest great harm be perpetrated. The Army must be right and just, even when it is embarrassing to correct an error. Do it when necessary as an act of justice, but do it correctly.

STANDARDS OF CONDUCT FOR SOLDIERS

Standards of conduct for all military personnel are set forth in DoD 5500.7-R. These basic rules, requirements, and prohibitions are applicable to all active-duty officers and soldiers and to retired personnel. This regulation is of such importance, and so broad in scope, that officers are urged to obtain a copy and place it in their personal libraries for reference.

Code of Conduct for War. Executive Order 10631 issued in 1955, as amended by Executive Order 12017 issued in 1977, establishes a code of conduct for members of the armed forces of the United States while in combat or captivity as prisoners of war. See also AR 350-30, as amended. The order requires that specific training be given to all members liable to capture to better equip them to counter and withstand enemy efforts against them. The Code of Conduct reads as follows:

1. I am an American, fighting in the forces that guard my country and our way of life. I am prepared to give my life in their defense.
2. I will never surrender of my own free will. If in command I will never surrender the members of my command while they still have the means to resist.
3. If I am captured I will continue to resist by all means available. I will make every effort to escape and aid others to escape. I will accept neither parole nor special favors from the enemy.
4. If I become a prisoner of war, I will keep faith with my fellow prisoners. I will give no information or take part in any action that might be harmful to my comrades. If I am senior, I will take command. If not, I will obey the lawful orders of those appointed over me and will back them up in every way.
5. When questioned, should I become a prisoner of war, I am required to give name, rank, service number, and date of birth. I will evade answering further questions to the utmost of my ability. I will make no oral or

Personal Readiness. There is always the need to be prepared for entirely new conditions and missions. Operation Iraqi Freedom proved that you must be prepared for multiple deployments into an ever-changing theater of war. What started with the rapid collapse of the entire Iraqi armed forces and Saddam Hussein's Baathist regime shifted rapidly to a wide range of other operations, from humanitarian relief, to nation-building, to counterinsurgency operations, and transitioned again during surge operations and redeployment. This is a powerful testimony not only to the Army's training, but also to the personal readiness of the officers and soldiers on the ground. It appears that our country will continue its involvement in many different scales of military actions in an unpredictable pattern for the foreseeable future. We will ask our military professionals to protect the interests of our government and the security of world peace regardless of the location or the nature of the threat to be met.

As in these examples, the responsible officer must identify the new opportunities and hazards, grow in professional capacity, and rise to the requirement of his or her mission. The assignment of an important military mission provides a fertile and rewarding field for all one's wisdom, foresight, and courage. If the security of our nation and our allies seems important to you, look forward to your part in the future with anticipation and determination. There are rewards as well as trials.

The Courage to Act. You must prepare yourself to take action, often swift, sometimes hazardous, occasionally into the unknown. Your missions will be infinite in variety. As a commander in combat, your unit may be required to defeat an enemy before the enemy can defeat you. Or to seize a hill, cross a river, hold a position against strong attack. Or to unload a ship, build a road or a bridge, or clear a harbor. There is also a need for courage to act on assignments other than command, in peace as well as in war. Peacetime decisions involve courage, guts, and, in a controversial matter under development or discussion, the need to fight—with hard logic—for your own well-considered point of view.

The courage to act when action is needed is an essential requirement for an officer, whatever the nature of the mission or assignment. There must be a consideration of facts and the reaching of a decision—what is to be done? There must be planning—where and how will the action be taken, when, and by whom? Orders must be issued to subordinate leaders so that they will know their own missions, the timing, and the essential coordination with others.

These essential steps—deciding, planning, preparing, ordering—are preludes to action. They are essentials in the practice of command, or of being in charge of an administrative activity.

In battle, and even in situations other than battle, there may be a vast difference between the requirements of the military leader in comparison with leaders in other fields. *In all decisions, the military leader must be right.* He or she must be right because there will be no second chance. "Being right" must be a part of the code of the Army officer, which requires as a matter of course that you apply to vital responsibilities all your knowledge and training, and all your courage, to reach and direct a sound, workable solution at the instant needed. An 80 percent solution today is often better than a 100 percent solution tomorrow or next week.

The officer who is a commander must be the one to order "Follow me!" or "Fire!" or "Go!" *He or she must have the courage to act at the very instant action is needed.* The commander assesses the hazards and the potential losses and measures the danger of defeat as well as the fruits of victory. There is no room here for a timid soul. Lost would be the brilliant thinker with a faint heart who reaches a perfect decision tomorrow on a critical situation or a fleeting opportunity demanding action today. These are the situations that separate the best officers from those not quite up to the highest standards.

In military actions, choices may not be sharply defined. In our Army, there are two ever-present responsibilities of a commander that may seem at times to oppose each other: the requirement for accomplishment of mission and the safety and preservation of the command. Quoting Gen. James Wolfe before Quebec in 1759: "Next to valour, the best qualities in a military man are vigilance and caution."

The truth must be understood. Prompt, decisive action—so strongly advocated—is never to be confused with ill-conceived, foolhardy action with needless sacrifice of lives. Aggressiveness, which invites severe, avoidable losses, is not invited by the principle. Wisdom must accompany the swiftest action.

PERSONAL CHARACTERISTICS OF OFFICERS

Your personal integrity and professional competence and your courage to make the right decisions at the right time are the bedrock qualities on which your image is built. You cannot succeed as an officer without these, but there is more. The image you form in the minds of others is also the product of many lesser traits, attitudes, and mannerisms. Attention to these matters is important as you seek to convince others of your worthiness to be a member of the corps of officers.

Officers and soldiers of the Army of today have the same human aspirations, likes and dislikes, strengths and frailties as they had a decade ago or ten decades ago. All are responsive to human appeals, and some are prone to succumb to human temptations. Listed below are a number of human characteristics, or qualities, that are especially worthy of consideration. It is people who make the Army, and people who win its victories. Human relationships are important, always.

Guidelines

Be Strong on Principle. This lets people know where you stand. It helps create a feeling of confidence among associates.

Be Coolheaded. Learn the facts, establish the truth, then stand up and be heard. Those who do are sought out for preferred assignments.

Be Willing to Adjust. Without compromising mission or basic issues, there is a need to hear and consider the views or the interests of others. This often involves negotiation or adjustment in order to gain understanding and agreement.

Be Certain to Preserve Unity. Disrespect and distrust develop disunity, and disunity invites defeat. Recognize and treat with respect members of the sister services. Recognize that branches of the Army other than your own are important, and treat their members with the respect that is their due. The good officer builds unity.

Don't Be a Diehard. Don't be the last to identify a truth or adopt the new. Although a military leader must often follow the hard right instead of the easy wrong, there must be an avoidance of sheer bullheadedness. (A bullheaded person has been defined as a person with very determined ideas, many of them unsound, to which he or she clings with more than usual tenacity.)

Don't Be a Timid Soul. Don't fear to advance your considered views or to take issue or to point out a fault, just for fear of causing annoyance. Don't go along with the crowd for fear of expressing a contrary view. Don't hold back to let another grasp the lead. The retiring person on the fringe of discussion, who may have the skill to reason and arrive at sound conclusions but lacks the fortitude to express or defend them, will have little influence on history or the outcome of battles. The new idea or the as-yet unheard solution may be the one that delivers success.

Don't Be a Know-it-all. As ineffective as the timid soul is the know-it-all, or wise guy, who has a glib solution to all problems with or without knowledge of the subject or time for consideration. In a meeting, this person is the one to advance an immediate solution without waiting to hear the facts bearing on the problem. "Off with his head," before hearing the evidence. Or, for example, a platoon leader may direct an unwise action because he or she failed to explain the task and listen to the suggestions of the NCOs. Neither the timid soul nor the wise guy will have much, if any, influence on history, and any influence of the latter is likely to be negative. The timid soul will supply nothing, either good or bad; but the wise guy may do much harm until his or her associates learn better. Aggressive advancement of hasty conclusions may sound plausible and be adopted, only to develop bugs later.

Earn the Loyalty of Subordinates. It is not enough to seek this loyalty; you must earn it. Fair treatment and the human touch will help. If there is something good to which your soldiers are entitled and are not getting, go after it and get it. Stand up for your troops or for your employees. Although you should not conceal their transgressions or let their offenses go unpunished, you can certainly identify and encourage their strengths. Be the first to get the full share of improved facilities for living, for work, or for recreation, if you can, and seek opportunities for your subordinates to advance their own goals. When the "boss" does these things, his or her rewards may be abundant. And when he or she fails to do so, the penalties may be severe. When subordinates have thoughtful reason to respect and trust their commander or chief, the results can be important.

Avoid the Political. Be somewhat guarded in your political comments. Officers are entitled to have opinions on matters of politics and national policy, as are any citizens. They have the same right and duty to vote and record their convictions. But it is unwise for officers to become too outspoken in approval or disapproval of any political party or of the leading members of any party. Indeed, officers are prohibited from using disrespectful language about government officials. Army officers serve both Democratic and Republican (or other party) administrations. Soldiers must be nonpolitical and serve each with equal zeal. The higher the grade they achieve, the more important this caution becomes.

LEADERSHIP ADVICE FOR THE TRENCHES

As a young lieutenant, I'll never forget meeting the iconic Gen. Bruce C. Clarke, renowned for his service from World War I through the Korean War, and particularly illustrious as one of the Army's foremost leadership experts. Suddenly, I had the opportunity to ask his advice on how to be successful in the Army! Without missing a beat, he replied to my question, "Lieutenant, it only takes two of three things to be successful in the Army. You have to be really good at what you do, you have to be really lucky, or you have to know someone." He continued, "Any two of the three will do; I was really lucky and knew someone."

The general had a good laugh at my expense, and then shared some sage advice: Trust your noncommissioned officers, care for your soldiers, and always be part of the solution, not part of the problem.

Clarke personified the World War II leader; he was tough, but deeply cared for his soldiers. Much has changed since he was a lieutenant, but being a problem solver and caring for your soldiers will go a long way toward being an effective leader.

Particularly good advice for a young officer appeared on the Platoon Leader forum, http://platoonleader.army.mil, as part of a post titled: "Congrats . . . you're a platoon leader. Now what?" by Nate Palisca. Following are excerpts from his advice. Although this post addressed assignment as a platoon leader, the tips apply equally to staff officer assignments.

TAKING CHARGE AS A PLATOON LEADER

This is it. After countless hours spent earning your commission, you finally have it. A platoon. *Your* platoon. Sure, the goal at West Point, Officer Candidate School, and Reserve Officer Training Corps was always to get your gold bar, but now you can truthfully call yourself a platoon leader. So what happens now? Here is a list of ten things to focus on in your first ninety days as a platoon leader.

You and Your Platoon Sergeant. Without question, your platoon sergeant should be the first person in the platoon you talk to. Sit down with him or her and ask for a rundown on the platoon. Some ideas for questions: How trained are the soldiers on critical tasks? How do we stand on PT? How is morale in the platoon? What tracking systems do you have in place, and can I get a copy? How is maintenance? What are our SOPs? What training is planned? How do you visualize our relationship as platoon leader/platoon sergeant? Are there any major personnel issues/legal cases? Who are the problem children? Who are the studs? It is important that you form your own opinions about your people, but the extremes (good and bad) are easy, and your platoon sergeant should know.

You and Your Platoon. You can meet with your NCOs before talking to the soldiers if you want to, but it's not necessary. Definitely schedule a time to talk to your platoon as a whole soon after taking charge. They will be curious about you, and they need to see your face as soon as possible. You're not just another new guy; you're the new lieutenant. Let them know who you are and where you're coming from. You don't have to give them a hardcore Patton-style speech; just be yourself. Your soldiers will be able to see through any smoke and mirrors. Do not, under any circumstances, criticize the last platoon leader. That shows insecurity, weakness, and unprofessionalism on your part, and it's simply in bad taste. Let your soldiers know that you've heard good things about the platoon and that you are excited to be there—it helps if you actually mean it.

Counseling. Counseling a sergeant first class who has twelve or more years of time in service can be an intimidating proposition, especially when you have all of about six months under your belt. How are you supposed to tell him or her anything? It all comes down to how you approach it. Given that each person is unique, there is no set method that guarantees success. Take a week or two to get to know your platoon sergeant before initiating counseling, and then proceed with a well-thought-out plan. The bottom

continued on next page

LEADERSHIP ADVICE FOR THE TRENCHES *continued*

line on counseling is that you owe it to your platoon sergeant to let him know what your expectations are and to give him some direction. Don't let this one slide. It could come back to haunt you when it's time to write your platoon sergeant's Noncommissioned Officer Evaluation Report (NCOER).

Playing Well with Others. Cultivate good working relationships with the key players in your world. In general, people are much more willing to help you if they see you as a friendly face. Examples include your commander, executive officer, first sergeant, other platoon leaders, supply sergeant, maintenance team chief, headquarters platoon sergeant, and anyone else you will be working with on a regular basis.

Being with Your Soldiers. Be visible to your soldiers without being nosy or overbearing. This is easier said than done, but it is worth your effort to get it right. Yes, you will make a first impression when you initially talk to your platoon, but it will take a while for soldiers to "feel you out" and see what kind of platoon leader you are.

Standard Operating Procedures (SOP). If your platoon has a written SOP, read, learn, and internalize it. If not, get with your NCOs and start drafting one. Everyone in your platoon needs to be on the same sheet of music if you are going to be successful.

OER Support Forms. If your commander doesn't give you a copy of his Officer Evaluation Report (OER) Support Form (DA Form 67-9-1), ask for it. You should have a copy of your rater's and senior rater's support forms. Both of those forms will not only give you ideas on how to write yours, but also show you how your bosses think yours should be formatted. After you have your bosses' support forms, start drafting your junior officer developmental support form (JODSF) and your support form. You might think that your commander will remember all of the great and wonderful things you did—and he will, after you remind him. A good support form is a tool that helps you get a fair shake on your OER.

Layouts and Property Accountability. You should (and likely will) do a layout of all of the equipment assigned to your platoon. You will inventory and sign for all of it, thus taking responsibility for your platoon's equipment. You can't afford to screw this up. Here's some good advice: While you are counting widgets and gizmos, ensure you are looking for serviceability in addition to just how many are there. If you have a piece of equipment that is unserviceable, it's not doing you any good, so you may as well be missing it. After your layout, you will need to ensure your hand receipt is correct (so you don't sign for things that aren't there) and begin the process of filling your shortages. This is where your XO and supply sergeant can be very helpful. Both of them can guide you through the process of ordering replacement equipment.

Systems Improvement. Look around at how your platoon conducts business. Start thinking about what I call "systems improvement." Look for things that could make your soldiers' lives easier. Sometimes that may be a tool or piece of equipment that they never had; sometimes it's streamlining a system. Whatever it may be, your guys will appreciate your looking out for their welfare. One important caveat to that is don't make changes just for the sake of change or to be able to say, "Look what I did." That won't help anybody. It will only serve to make you look ego-driven and alienate you from your platoon. Just observe, ask your soldiers/NCOs, and give it some thought.

You and Your Ego. You don't have to tell your NCOs that you are in charge. They know how you fit into the platoon. That being said, listen to your NCOs and pick their brains for the knowledge that only their years of experience can bring. The bottom line is that you are in charge, but your opinion isn't the only one that matters.

Avoid the "First Name" Problem. There has developed in the Army an excessive use of first names while on duty by seniors to juniors, officer to officer, even officer to soldier. The problem deserves consideration.

As prescribed by military courtesy, members of the service are addressed by their proper titles on all official occasions. On social occasions, the practice should be governed by good taste. In athletics, hunting and fishing trips, or other activities wholly unconnected with official relationships, it is merely two or more kindred spirits out for the enjoyment of a hobby. This discussion pertains to the use of first names on official duty.

This is not to advocate the "stuffed shirt" attitude or the martinet. As leaders, officers have stern duties involving discipline, reward and punishment, selection for promotion, and selection for favored assignment or hazardous duty. *Officers must never play favorites or by their actions permit the suspicion that they are doing so.* Officers simply cannot be on terms of personal familiarity today, and then tomorrow, when duty requires that they admonish an individual, revert to the official title. There is just one rule to follow on this first-name business in official relationships: *Don't use it.*

Consider the Influence of Spouse and Family. A spouse's attitude and conduct within the military or civilian community in which the officer performs his or her duties may have a profound bearing on the officer's effectiveness. If the spouse exerts a wholesome, normal, pleasant influence, he or she will meet the basic requirements for a happy, relaxing home life, resulting in a happy and confident officer on duty. Some assignments require semiofficial social appearances or duties on the part of spouses, but these are mostly limited to joint service or diplomatic assignments or to actual command positions. An act of social misconduct on the part of the spouse will probably not directly damage the officer's career, but it could be an indirect factor both by its effect on the surrounding community and by its effect on the personal security and confidence of the officer. Guidance can be found in chapter 17, The Social Side of Army Life.

Avoid Tarnishing your Reputation. It is not sufficient to strive through job performance to construct a good image. You must also strive to avoid any tarnish to the image you have fashioned in the minds of others. Officers may be accorded preferment on the intangibles of faith, confidence, and the like, but they are punished, held back, or eliminated as a result of definite acts of transgression or shortcomings. Conduct yourself in such a manner as to avoid the likelihood of any black marks appearing on your evaluation reports or in your other military records, or even causing raised eyebrows among your associates.

Shortcomings in Personal Conduct

Intemperance in the Use of Liquor. Excessive drinking, or the results thereof, has probably ruined the careers of more officers than any other cause. Drinking to any degree of intoxication while on duty is punishable and can have a severe impact on an officer's record. Similarly, excessive drinking when off duty, particularly when your resultant behavior reflects on the service or your intemperance lowers your performance of duty, can result in a marred record. The single beer at noon can develop into a martini. Remember that it is far easier not to start a bad habit than to stop one, and that it is easier to avoid a bad episode than to undo one. An alcohol-related incident may result in immediate dismissal and certainly will result in a shortened career. Few commanders or seniors of any grade insist on their officers being teetotalers. Rather, they expect that there will be about the same percentage of users and abstainers as in the general population. An officer who wishes to remain in the Army and enjoy a successful career, however, must avoid excesses in the use of liquor and avoid permitting its immoderate use to reduce the effectiveness of his or her work or to warrant criticism for personal conduct.

Use of Illegal Drugs. Whereas there is little tolerance regarding intemperance in the use of alcohol among the corps of officers, there is *zero* tolerance regarding the use of drugs. Any drug-related incident will surely result in the immediate separation of an officer from the service. Any officer or officer candidate who feels that he or she needs or desires to use drugs has chosen the wrong vocation.

Financial Difficulties. Officers have the same family and personal financial responsibilities and difficulties as other citizens. Debt is credit and is a standard practice in our economy. Trouble arises when the officer assumes greater obligations than he or she can discharge within the time limit of the agreement to pay. When debts have not been paid when due, merchants often write to the commander or to the Adjutant General and ask for assistance in collecting the debt. The receipt of such a letter is very harmful to the reputation of an officer. It indicates a deficiency in conducting personal affairs and implies a weakness in the ability to administer affairs of the Army. Army policy provides for garnishment of pay to settle just debts.

Avoid financial difficulties! Live within your income—with some to spare, if possible. Put some funds aside for emergencies and as a start toward retirement. Should you become submerged in debt, make an exact list of the indebtedness, go to your creditors, and seek to arrange a plan of payment that is agreeable to them and within your means. If complaints are likely to be filed, go to your commanding officer and lay the cards on the table. He or she may be able to suggest a solution. At least your commander will know that your intentions are honorable.

The Army cannot permit its officers to acquire a general reputation of being poor credit risks, for such a condition would reflect on all who wear the uniform. Such action uncurbed in a few officers harms the general credit standing of all. The officer in debt must have a definite understanding with creditors as to when and how much he or she can pay on overdue accounts. Having reached such an agreement, the officer must meet these obligations or negotiate a new understanding. The worst action an officer can take is to permit the creditor to reach a conclusion that the debtor seeks to evade an honest and just debt. More than a few officers have been considered for elimination because they have failed to observe these simple practices.

Transgressions of Moral Codes. Officers are expected to be gentlemen or ladies in their personal as well as their official conduct. Tarnished images and low evaluation ratings inevitably follow the transgression of moral codes.

Violations of Honor. An officer's word, signature, or initials of concurrence must be backed by facts, accuracy, and honor. There must be no doubt as to the truth of any statement that bears his or her authentication. The officer's oral or written word is his or her bond. If your statements, either written or oral, contain opinions as well as facts, take care to identify the opinions for what they are.

The acceptance of an officer as an honorable individual by all who meet or know him or her is traditional in the Army. Thus it has always been, and thus it must always be if the Army is to retain its status as a reliable instrument to implement national policy. It goes without saying that acts of deliberate cheating, lying, or quibbling cannot be condoned. Violations of honor will surely be filed in the minds of others as well as noted on evaluation reports or other records.

Misconduct. Misconduct or misuse of official position will affect the evaluation report and tarnish the reputation of an officer; serious frauds, minor acts for personal gain, personal use of government property, abuse of privileges, and rowdy or disorderly conduct are examples of these weaknesses.

RELATIONSHIP WITH THE "BOSS"

Let us turn briefly now to your relationship with your commander, the "boss," for it is your image in his or her eyes that determines your evaluation report ratings, on which all else hinges.

There are some very human and commonsense elements involved in earning good reports, but one fact must be clearly understood: *The boss will be pleased only when there has been effective execution of duty throughout the period.* This has nothing to do with laughing at poor jokes, dating an unattractive daughter or son, performing subservient actions, or making unworthy efforts to avoid the results of poor work.

Much has been written about the way a senior officer should treat junior officers and soldiers. Now we put the shoe on the other foot. Beyond the conventional prescribed actions and attitudes, how should the subordinate treat the superior?

Treat your commander or chief, your boss, as a man or woman. At the same time, do your own part of acting as an adult. You are both mature, trained officials of government. The junior officer is not a child standing in the presence of a stern parent, nor an ignorant tyro facing a person of unique skill. You are both capable officers associated in a joint undertaking. In most cases, a senior is older in years. Treat your chief honorably and respectfully behind his or her back as well as face to face. Be businesslike. Stand on your own feet.

Respect the authority of your commander. Understand his or her mission and responsibilities. Your commander's success depends, in part, on the results of your work. Teamwork means harmonious effort with all members of the organization—not just with the boss.

Keep your boss informed. Give him or her the information needed as to your own progress, and give it straight, clear, and on time. Don't conceal bad news or try to slip it past. Don't embellish the good news. Your chief needs essential facts for his or her own planning and for coordination with other subordinates. There are few furies equal to that of a military leader who makes a wrong decision that was based on incorrect or misleading reports from a subordinate.

Apply standard military courtesies and customs of the service, neither overdrawn nor underdrawn. Know these things and apply them sensibly. Let the chief set the pace in the degree of formality or informality of routine official contacts. Especially let this be the case in outside-of-duty relationships. In these latter contacts, a chicf may call you "Pat," but be certain to address him or her as "Captain." Extending compliments to your chief is in poor taste and will be so regarded. Knowing how to work for and with your commander is an important goal, and it deserves careful consideration.

Know your chief's military and personal background. Identify his or her human likes and dislikes. Be attuned and responsive to these human characteristics of your chief, as you also avoid acts or mannerisms that serve no real purpose and that are upsetting to him or her. An understanding of the person for whom you work is helpful in developing proper teamwork.

Some commanders issue broad, mission-type orders and encourage subordinates to exercise considerable latitude as to the method or procedure for execution. Others give detailed instructions that they wish followed meticulously. Do not take advantage of the one with unnecessary delay or laziness, and don't rebel at the latter because you prefer your chief to work with a looser rein. In fact, don't fight this problem at all. Learn what your chief wants and conform to it.

Engrave this in your mind: You are a part of your commander's team, not he or she of yours. You are the head of your own team, with your own subordinates. Develop your team wisely.

SYCOPHANCY—ENEMY OF MERITOCRACY

When an organization is downsized, it places unusual pressures on its members. This is particularly true of an organization such as the Army, to which there is no corresponding parallel group. Let's face it: Service to the Republic can be pursued along a variety of avenues, but you can't be an officer in some other army if this one decides it doesn't want you anymore. There is no other organization in which you can command an air defense artillery battalion or coordinate the plans and training of an armored brigade. The pressures to "make the cut" are therefore tremendous—especially in an age in which some thirty-eight-year-old majors who have spent their entire adult lives in the Army are not selected for retention until retirement four years hence.

Of course, the great majority of officers do not buckle to this pressure. They refuse to be seduced by the "dark side" and continue their performance of assigned duties to the best of their abilities. However, one of the most distressing, creeping tendencies among some officers who cannot bear the strain has been the inclination toward sycophancy, the attempt to curry personal favor with superiors outside the line of duty.

Sycophancy has many other names—bootlicking, apple-polishing, and several more colorful but less printable synonyms. We all know what it is, and most of us abhor it. It is alien to American culture; even sixth-graders detest the "teacher's pet" who always seems to be laughing hardest at the teacher's jokes, making a show of knowing the answer, or—worst of all—snitching on the kid who threw the paper airplane across the room. In a sophomoric way, almost all of us despise this sort of behavior (and often those who display it), but sycophancy in the Army is much more serious and is intolerable under any circumstances.

The reason sycophancy is alien to mainstream American culture is that our democratic traditions emphasize *merit* as the basis for advancement. As mentioned before, the Army—and especially the officer corps—must be a strict meritocracy. Sycophants attempt to elicit personal favor with the boss (or the boss's boss) to compensate for their own lack of merit. Such actions are deliberately designed to thwart those who might otherwise advance based on their competence and record of genuine achievement; thus, sycophancy is the enemy of meritocracy and is antithetical to the good of the Army. If any officers other than the best are promoted or selected for positions of great responsibility, soldiers may die. At the highest levels, in wartime, battles or even wars may be lost if we do not have the most competent and forthright officers in charge. In peacetime, sycophancy at any level may lead to poor decisions being made. Even good leaders are often only as good as the information they get, and if they are being advised by subordinates who are telling them only what they think the boss wants to hear . . . well, you know the rest. Many insightful historians agree, for instance, that pervasive sycophancy in the German general officer corps before and during World War II not only allowed Hitler's rise to power, but tolerated his bungling of much of the German war effort. The cost of sycophancy, in peace or war, is too high.

Officers must be careful to differentiate between being appropriately social, or properly respectful, and being sycophantic or obsequious. They must also discern the difference between those who do their duty to the best of their abilities and those who are licking the boss's boots. To the experienced eye, this distinction is usually fairly obvious, but to the novice, it may be confusing. Here are some good rules:

- Officers who work hard, all the time, are probably not sycophants. Officers who work hard only when the boss is watching probably are.
- Officers who are avid golfers and who accept invitations to occasionally play with the boss are probably not sycophants. Officers who suddenly take up golf when they find out that the boss likes to play probably are.
- Officers who consistently, respectfully, tactfully express their honest opinions and recommendations to the boss are probably not sycophants.
- Officers whose only expressed opinions always coincide exactly with those of the boss—and who often vigorously express them in public—probably are.

Finally, sycophancy is unacceptable not only because it undermines meritocratic advancement but also because its close cousins—backstabbing, character assassination, and, worst of all, pure corruption are usually lurking right around the corner. Individuals who will stoop to sycophancy often do not stop there. All are destructive and intolerable among U.S. Army officers and must be eliminated.

RIGHTS, PRIVILEGES, RESTRICTIONS, AND BENEFITS

Since 1973, when conscription ended, joining the U.S. Army has been a privilege extended only to those who desire to serve and who are qualified by regulations. Service in the Army is not a *right.* It is denied to the physically or medically unfit, those judged too old or too young, the mentally incapable, criminals, and the morally or socially unsuitable. In short, military service is denied to anyone whose presence is deleterious to the accomplishment of the Army's mission or to the maintenance of good order and discipline.

In exchange for the privilege of service, those who join the Army must give up many other privileges—and even a few rights—normally afforded to civilians. The restrictions on officers are in some ways different from those on enlisted soldiers, but generally, they are similar. Essentially, when you become an officer, you lose those rights and privileges that are inimical to the proper regulation of the Army of a democratic republic. At the same time, you are granted certain privileges associated historically or legally with being an American soldier, and even more that are implicit to service as an officer. There are also benefits granted in compensation for and consideration of military service.

RIGHTS

American soldiers are citizens, and their rights are abridged only to the extent required for the maintenance of an effective army. Greater deprivation of rights not only is unnecessary but also could separate the military too far from the civilian sector of society, an inherently dangerous situation in a republic such as ours. The retention of too many rights deprives commanders of the flexibility and security essential for building and maintaining an effective military force.

Taking the oath of office as an officer entitles the individual at once to certain rights of military service. Other rights accrue only by completing specified requirements, such as the right to retirement, which accrues only to those who have honorably completed a stipulated period of service.

The Right to Wear the Uniform of the U.S. Army. This is a right granted singularly to American soldiers. Donning the uniform links you—in a material way—to all those who have gone before. Do not unduly take advantage of this right: Wear the uniform proudly and smartly, and do not tolerate it being worn by others in an improper way. In its varying forms, the uniform of the U.S. Army has been worn since 1775 by millions of soldiers who have made uncounted sacrifices to create, sustain, and defend the Republic. Other than American citizenship itself, few more meaningful rights could ever be granted simply for taking an oath.

The Right of Officers to Command. In the commission granted an officer by the President are these words: "And I do strictly charge and require those officers and other personnel of lesser rank to render such obedience as is due an officer of this grade and position." The commission itself may be regarded as the basic document that gives military officers the right to exercise command and to exact obedience to proper orders. Warrant officers, when assigned duties as station, unit, or detachment commanders, are vested with all powers usually exercised by commissioned officers except as indicated in AR 600-20, *Army Command Policy*. Commissioned warrant officers are vested with the same powers as other commissioned officers.

The Right to Draw Pay and Allowances. Pay scales for grade and length of service are established by law. The rights to pay and allowances may be suspended, in part, by action of a court-martial or forfeited, in part, by absence without leave.

The Right to Receive Medical Attention. In addition to routine medical and physical maintenance, soldiers are entitled to appropriate medical or dental care for the treatment of wounds, injuries, or disease. In fact, refusal to accept treatment ruled to be necessary may be punishable by court-martial.

The Right to Individual Protection Afforded by the Uniform Code of Military Justice. All members of the military service are under the jurisdiction of the Uniform Code of Military Justice (UCMJ). Many persons regard the *Manual for Courts-Martial, United States*, as merely the authorization for courts-martial and the implementation of their procedures as a means of maintaining discipline or punishing crime. This is a shallow view. Except for the punitive articles (Articles 77 to 134), the code pertains in considerable measure to the protection of individual rights. In fact, even Melvin Belli, one of the most liberal commentators on American jurisprudence, has declared the UCMJ and its corollary processes to be the fairest system of justice in America, providing the most complete protection of the rights granted by the Constitution.

On 19 December 2003, the President signed into law H.R. 100, the Servicemembers' Civil Relief Act (SCRA). The SCRA expanded and improved the former Soldiers' and Sailors' Civil Relief Act (SSCRA), passed in 1940. The SCRA provides a wide range of protections for persons joining military service, reservists and national guardsmen called to active duty in the military and deployed servicemembers. The purpose of the SCRA is to postpone or suspend certain civil obligations to enable servicemembers to focus on duty and to relieve stress on the family members of deployed servicemembers. Some of the types of protections the SCRA afford servicemembers are excessive interest on outstanding credit card debt, delay or extensions on pending civil trials, prevention of double taxation, and terminations of lease.

The Right of Equal Opportunity. It is a policy of the Army to conduct all its activities in a manner that is free from racial discrimination and that provides equal opportunity and treatment of all uniformed members irrespective of their sex, race, color, religion, or national origin. This applies to on-duty matters and off-duty situations, including on-post housing, transportation, facilities, and schooling. AR 600-20 prescribes the regulations that enact the principles of Title 11 of the Civil Rights Act of 1964. Strict compliance with these regulations is more than a matter of law; it is a matter of the maintenance of the good order and discipline essential in the army of a democratic republic.

Redress of Just Grievances. Article 138, UCMJ, prescribes procedures by which any member of the military service may seek redress of grievance. It is probably the single greatest advantage of military service over civilian life, wherein most grievances must be settled by laborious and expensive judicial procedures—if they are settled at all. Each officer should become fully acquainted with this process, for the maintenance of justice—and *fairness*—within the Army is of the utmost importance for the success of meritocracy.

The Right to File Claims for Losses Incident to Service. The Military Personnel Claims Act (AR 27-20) establishes the right of military personnel on active duty to file claims for losses of personal property incident to military service. A claim for loss may be submitted for consideration and in proper cases will be approved for payment. Examples of claims in covered areas include:

- Damage to property located at quarters or other authorized places from fire, flood, hurricane, or other serious occurrence.
- Transportation losses.
- Marine or aircraft disaster.
- Enemy action or incidental to public service.
- Money held in trust for others and personal funds, under some conditions.
- Motor vehicles lost when used in mandatory performance of military duty and during authorized shipment overseas.

To secure reimbursement for losses, it is necessary to establish the facts. Immediately upon an occurrence that may justify a claim, the officer should begin collecting essential documents, statements of witnesses, or other material to support the claim.

The Right to Vote. Legislation enacted by Congress in 1955 establishes the right of voting by members of the armed forces, and commanding officers are required to establish facilities for absentee voting for members of their commands. See AR 608-20. Uniformed personnel are subject to Department of Defense Directive (DoDD) 1344.10, Political Activities by Members of the Armed Force, www.dtic.mil/whs/directives/corres/pdf/134410p.pdf. This directive enforces the traditional concept that members on active duty should not engage in partisan political activity.

The Right to Retire. After satisfying specific requirements of honorable service or having endured physical disability beyond a fixed degree, officers of the armed forces have the right to retire.

The Right to Be Buried in a National Cemetery. The rights of a deceased soldier to be buried in a national military cemetery are discussed in chapter 19, Personal and Financial Planning.

PRIVILEGES

The greatest privilege afforded an Army officer is the opportunity to lead American soldiers. Even when not assigned as a unit commander, officers are looked to for leadership in the U.S. Army. All other privileges afforded officers derive mainly from this singular, sacred opportunity. In other words, officer privileges are not perquisites of rank but rather concessions to the flexibility essential for effective leadership and duty performance. The old expression "Rank hath its privileges" rightly had a lesser-known rejoinder, namely, "and responsibilities." In the army of a democratic republic, privileges do not accrue to persons of higher rank by virtue of some sort of assumption or recognition of social or professional superiority. Instead, officers are afforded privileges so that they may better perform the duties of increasing responsibility that are entrusted to and expected of them.

Time Management and Accountability. Generally, officers are assumed to be doing their duty and pursuing their unit's or soldiers' best interests—within the intent of the commander and using their best judgment. As officers ascend in rank, they generally also incur increased responsibilities. For these reasons, officers are often allowed privileges regarding their use of time. When not in the field or certain school environments, officers are not subject to the same accountability requirements as enlisted soldiers. For example, officers are usually not required to be present at "first call" or to attend the several daily mandatory formations at which their enlisted subordinates must be present. In garrison, officers are generally required only to attend scheduled meetings and to check in with the boss at specified times of the day; otherwise, they are assumed to be where their duties require them to be.

Bachelor Officer Housing. Except in some school environments, bachelor (including "geographic" bachelor) officers are allowed to live in bachelor officers' quarters, which are generally significantly more roomy and better appointed than the barracks accommodations afforded bachelor enlisted soldiers. Of course, bachelor officers are usually also allowed to live off post. This is not because they are more deserving of luxury than their soldiers but because their duties require a more independent lifestyle. Generally, officers are afforded living arrangements similar to those of senior NCOs, who need the same flexibility for the performance of their duties.

Rank and Position. Certain privileges accrue to certain duty positions, by tradition and by practicality. As long as they are used within the boundaries of Army tradition and common sense, few will begrudge them. However, when taken to extremes or used to subordinates' distinct and purposeless disadvantage, they become counterproductive. For example, a few officers in key positions (such as commanders) are generally afforded the privilege of exclusive use of reserved parking spaces near their offices, due to the need for rapid departure and return during their long duty days. No one begrudges this privilege. However, when nearly *all* the spaces near the barracks are reserved for officers—few of whom need them before reveille or much after retreat—the soldiers who are deprived of the opportunity to park near the barracks resent it, and rightfully so.

As a rule of thumb, if you need it to get the job done, if it's legal, supported by tradition, and it doesn't unduly inconvenience your subordinates, take the privilege. If not, play by the same rules as everyone else. This Army is a meritocracy, not an aristocracy.

RESTRICTIONS

Generally, the restrictions imposed on Army officers can be divided into two categories: *temporal* and *professional.*

Temporal Restrictions. The Army is a complex organization with multiple missions across a broad spectrum, from support of civil authorities to the prosecution of high-intensity war. Increasingly, officers are spending more time either deployed on tactical missions or subject to short-notice recall for rapid deployment. The restrictions inherent to these situations have an impact on family life, on personal time, and sometimes, perversely, on professional self-development. Limitations on the use of "off-duty" time are part of being an Army officer and cannot be ignored. See part III for some ideas on how to make optimal use of the reduced time available with your family or on your own.

Professional Restrictions. With our democratic form of government, the existence of a strong standing Army is possible only by strong restrictions on its use and the activities of its soldiers, especially its officers. Some restrictions, such as those imposed on gifts to superiors, are corollary to the need to maintain a strict meritocracy in the Army. Others, such as those restricting commercial enterprise or gain, are designed to ensure the trust and confidence of the American people in their Army and its officer corps. Whenever you have questions about restrictions, consult with the staff judge advocate (SJA) ethics counselor, and seriously consider his or her advice before acting.

Posse Comitatus. The *Posse Comitatus* Act severely restricts the participation of Army forces in most law-enforcement activities. Although it is a complicated law with many complex corollaries, essentially, this act restricts the use of the Army as a national police force, except in cases expressly approved by Congress. Officers participating in or assigned to the various joint task forces and other organizations called on to support domestic law-enforcement agencies should carefully study the obligations and prohibitions impending on their duties. When in doubt, consult your post or unit SJA ethics counselor.

Political Activity and Comment. The short answer is, stay out of politics. As you ascend in rank, this may become more difficult, but to the greatest extent possible, Army officers should remain apart from political controversy. As servants of the Republic and defenders of the Constitution, officers have no legitimate reason to officially take sides, apart from their legally affirmed right to vote and to contact their representatives in Congress. Officers must serve the constitutionally legitimate civilian government and, given this duty, should refrain from making public pronouncements of political opinion. In addition to speech, political bumper stickers on privately owned vehicles, campaign signs in front of quarters, and the like are public statements, and as such, are inappropriate and, in some cases, prohibited.

The UCMJ clearly delineates certain restrictions of this sort. *Soldiers on active duty may not use their official authority or influence to interfere with an election or affect the course or outcome thereof.* They are not permitted to participate in any way in political management or political campaigns. This includes the making of political speeches, activity at political conventions or on political committees, the publication of articles, or any other activity that might influence an election or the solicitation of votes for themselves or others.

This restriction extends to political activity beyond election campaigns. Participation by soldiers in public demonstrations not sanctioned by competent authority, including those pertaining to civil rights, is *prohibited* under the following circumstances:

- During the hours they are required to be present for duty.
- When they are in uniform.
- When they are on a military reservation.
- When they are in a foreign country.
- When their activities constitute a breach of law and order.
- When violence is reasonably likely to result.

Clearly the principles are twofold: The demonstration is not to be (mistakenly) interpreted as being officially sanctioned by the Army, nor is the Army to be discredited by the presence of a member (see AR 600-20).

Finally, just as all soldiers are forbidden to publicly speak disrespectfully of their commissioned superiors—even if they are not present at the time—Article 88, UCMJ, also requires that *"Any commissioned officer who uses contemptuous words against the President, the Vice President, Congress, the Secretary of Defense, the Secretary of a military department, the Secretary of Homeland Security, or the Governor or legislature of any State, Territory, or other possession of the United States in which he is on duty or present shall be punished as a court-martial may direct."* It is therefore professionally appropriate to refrain from public criticism of civilian officials, no matter how odious their conduct or loathsome their attitudes. Public criticism of an official may be interpreted as a statement with political intent; this applies to public pronouncements, speeches, published letters or editorials, or any other means of expression in a public forum. As always, if you have any questions about what constitutes allowable activity, contact your SJA ethics counselor.

Classified Information. In the course of performing their duties, officers are often granted access to classified information or information that has been officially deemed by competent authorities to be injurious to the security of the United States if openly divulged. Officers must be mindful of the restrictions placed on the use and communication of such information and the punitive action that may be taken against them for its improper handling or use. Classified documents, whatever their classification, are serious business. No matter how frequently or commonly you handle them, treat them with the same respect you show weapons and ammunition. *Never* allow yourself to become complacent or lackadaisical in their storage, transmission, transportation, or destruction.

Commercial Use of Military Titles. Military titles may not be used in connection with a commercial enterprise by individuals on active duty. This applies specifically to Regular Army personnel on the active list, and retired and reserve component personnel on extended active duty. Authorship of material for publication is exempted from this provision, but such material may be subject to review and clearance by the Department of Defense. See your SJA ethics counselor for details.

Acting as Attorney or Agent. No military member or civilian employee of the Army or of the Department of the Army whose official duties are concerned with patent activities can act as agent or attorney in connection with the inventions or patent rights of others, except when such action is part of the official duties of the person so acting (see AR 27-60).

Representing Clients. An officer previously assigned to military duty is disqualified for life from representing in any manner or capacity any interest opposed to the United States or with which he or she was directly connected during government service. There is also a two-year ban on representing anyone on a matter that was previously under the officer's official responsibility.

Acting as Consultant for a Private Enterprise. No member of the military establishment on the active list or on active duty, and no civilian employee of the Army or the Department of the Army, can act as a consultant for a private enterprise with regard to any matter in which the government is interested.

Contributions or Presents to Superiors. In keeping with the imperative of maintaining a meritocracy, military and civilian personnel of the Department of the Army cannot solicit contributions from other government employees, military or civilian, for a gift for an official superior; cannot make a donation as a gift to an official superior; and cannot accept a gift from other government personnel subordinate to themselves. However, *voluntary* gifts or contributions of nominal value are permitted on special occasions such as marriage, transfer, illness, or retirement, provided any gifts acquired with such contributions do not exceed a nominal value.

Acceptance of Gratuities. The acceptance of gratuities by either military or civilian personnel of the Department of the Army, *or members of their families*, from those who have or seek business with the Department of Defense or from those whose business interests are affected by DoD functions is strictly forbidden. Such acceptance, no matter how innocently tendered or received, may be a source of embarrassment to the Army, may affect the objective judgment of the personnel involved, and may impair the public confidence in the integrity of the government. With certain limited exceptions, DA personnel cannot solicit, accept, or agree to accept any gratuity for themselves, members of their families, or others, either directly or indirectly from any source with business interests of the type noted above. The exceptions are detailed in DoD 5500.7-R. Officers are urged to study the regulation to be fully aware of both the details and the philosophy involved. In case of doubt, consult your SJA ethics counselor.

General Prohibition. All Department of the Army personnel, military and civilian, must avoid any action, whether or not specifically prohibited by regulations, that might result in or reasonably be expected to create even the *appearance* of the following:

- Using public office for private gain.
- Giving preferential treatment to any person or entity.
- Impeding government efficiency or economy.
- Losing independence or impartiality.
- Making a government decision outside official channels.
- Affecting adversely the confidence of the public in the integrity of the government.

Honorary Titles. Conferring honorary titles of military rank upon civilians is prohibited.

Refusal of Medical Treatment. Any soldier may be investigated by a board of medical officers and subsequently be separated from the Army or be subject to trial by court-martial for refusing to submit to a surgical or dental operation or to medical or dental treatment, at the hands of military authorities, if it is designed to restore or increase his or her fitness for service.

Restrictions on Writing for Publication. Officers with ideas to share with other professionals or the public at large should grasp the opportunity to write for service magazines or journals, magazines of general circulation, or books. It is the principal means of communicating the point of view of the military professional, just as it is also the principal means for expanding sources of information and reference on military subjects.

Active-duty officers desiring to write about current or recent military subjects must obtain the approval of the Department of the Army before furnishing a manuscript to a publisher, except for articles for service journals and official Army publications. Material prepared for service journals may be completed as an official duty using military facilities and clerical assistance. Otherwise, officers should write only on "off-duty" time, using their own resources. Writing must not interfere with the performance of duty. If profit or the payment of royalties is involved, check with your SJA ethics counselor; the rules about this are complex and subject to frequent change. One way to simplify the situation is to donate all proceeds to charity; this is often done by military authors whose writings clearly derive from their immediate military experience.

As with all professional conduct, your overwhelming concern must be with doing what is best for the Army. Be careful to avoid the appearance of self-congratulation, of self-aggrandizement, or of criticism of specific units or serving or recently retired Army officers. Even veiled references can cause a great deal of hard feeling and can easily redound to the author's disadvantage. For officers who are ladies or gentlemen, open fora are not the places for ad hominem criticism.

This is not to say that officers should not engage in well-thought-out, crisply written comment on Army policies, programs, or larger trends. There are, however, proper media for such writing, military service journals being the most appropriate. Similarly, alternative views on issues requiring more comprehensive treatment may be expressed in books, but remember that when you express opinions in public, your ideas will be subject to a variety of interpretations. It's like sports—the bigger the audience, the wider the spectrum of opinions (including negative ones) about your talents and abilities. In any case, when writing on any military subject, submit your manuscripts to the local public affairs officer (PAO) before sending them to a commercial publisher. This may save all concerned a great deal of anguish. There is no requirement to do this with a submission to a service journal.

One final note: Rightly or wrongly, there exists among some officers a clear prejudice against those who publish extensively. In some cases, this negative attitude has been fully justified by a small number of high-profile officers who have neglected their duties in favor of their writing activities. Keep in mind that when you write for publication, you may need to redouble your efforts to validly demonstrate that your writing is not interfering with your all-important, overarching performance of your *assigned* duties. Conversely, there are few more satisfying surprises for a superior than to find out that a quietly professional, dedicated, hardworking, productive subordinate has produced an excellent written product that is going to be published. Such instances of "above and beyond" professional endeavor can be one mark of a truly outstanding officer.

Public Speaking Engagements. The restrictions on public speaking stem from the need to ensure that the Army "speaks with one voice" on important issues, as well as security concerns. *Never* make political speeches, and *never* publicly disagree with current Army policies. Such speech, no matter how well intended, will ultimately reflect poorly on you and the Army, which may appear fractious and rife with dissent.

In contrast, eloquent, thoughtful comments by Army officers in public fora reflect well on the Army and can help the public better understand the activities for which their tax dollars are paying. Veterans' groups, in particular, appreciate an excellent presentation by a qualified, knowledgeable officer, who can make them feel appreciated and still "in the loop" as creators of the Army heritage.

Before accepting an invitation to speak publicly, consult your local PAO. He or she will be able to apprise you of the advisability of acceptance and give you some tips about how to graciously avoid inappropriate subjects, such as classified information or politics. Basically, if you stick to talking about subjects about which you are sincere and have solid, personal knowledge, you will not go wrong; the experience will be rewarding and meaningful for all concerned.

As always, before accepting any royalties or gratuities, consult your SJA ethics counselor. If the subject is brought up in advance of your engagement, apprise your hosts of restrictions, to avoid embarrassment on the part of all concerned.

Restrictions on Personal Conduct. The conduct of officers must always be above reproach. Because officers are always on duty, there is no such thing as "off-duty conduct." This is not just a hoary old platitude or an anachronistic tradition but a real, practical requirement for the maintenance of good order and discipline in the Army. Officers quickly lose the respect and trust of their subordinates when they compromise their integrity or adherence to the law. As discussed before, there are some exceptional occasions when officers rightly choose not to comply with some order, policy, or regulation, when doing so is in the best interest of mission accomplishment or the ultimate welfare of the Army and the Republic. Even then, they may be subject to adverse actions for disobedience. This moral and legal dilemma is part of what sets officers' duties and authorities apart from those of NCOs.

On no account, however, may officers *ever* fail to comply with orders or regulations out of a desire for personal benefit, sloth, indecency, dishonesty, injustice, or cruelty. Not everyone can be expected to meet ideal standards or to always demonstrate all the attributes demanded of the "perfect" officer, but there is a limit below which the acts of officers or cadets cannot be tolerated. While many of these unacceptable actions may be dealt with administratively—that is, by counseling statements, letters of admonishment or reprimand, or adverse comments on efficiency reports—the most extreme examples are punishable under the UCMJ.

Article 133, UCMJ, prohibits "conduct unbecoming an officer" and can be used to justify punishment by general court-martial offenses, including but not limited to:

- Knowingly making false official statements.
- Dishonorably neglecting to pay debts.
- Opening and reading another's letters or email without authority.
- Drafting a check on a bank where the officer knows or reasonably should know that there are not funds to meet it, and without intending that there should be.
- Using insulting or defamatory language to another officer in his or her presence, or about the officer to other military personnel.
- Being grossly drunk and conspicuously disorderly in a public place.
- Associating publicly with notorious prostitutes.
- Failing without good cause to support one's family.

Adultery. There are few instances in which the restrictions on military personnel differ so drastically from those on other American citizens. Adultery is punishable under the UCMJ, and Army policy shows no evidence of changing.

Restrictions of an Officer Under Arrest. An officer under arrest is subject to the following restrictions:

- Cannot exercise command of any kind.
- Must restrict himself or herself as directed.
- Cannot bear arms.
- Cannot visit his or her commanding officer or other superior officer, unless directed to do so.
- Must make requests of every nature in writing, unless otherwise directed.
- Must, unless otherwise directed, fall in and follow in the rear of his or her organization at formations and on the march.

SOCIAL MEDIA CONSIDERATIONS FOR LEADERS

One of the most significant changes we have seen in the last decade is the overarching influence of social media on our world, particularly on the way we communicate through an ever-growing host of platforms. The Army has largely embraced social media as a means to communicate its strategic messaging and has encouraged its use by both soldiers and their families.

The Internet has proved to be one of the most powerful communication tools available to the leader; however, the negative side of this power has been demonstrated by the employment of cyber warfare against the government and the armed forces by both state and non-state actors. Cyber warfare, such as the attack against the Office of Personnel Management that affected thousands of servicemembers and DoD civilian employees, often targets personal identifiable information (PII). One of the DoD's principal concerns is employing the communications power of social media while safeguarding the PII of its members and their families. If your soldiers or civilians are affected by this breach, helpful information on both the attack and on monitoring your PII is located at www.opm.gov/news/latest-news/announcements.

Soldiers naturally seek out involvement in social media platforms, in part to connect and interact with individuals who share similar interests. Soldiers are authorized—and often encouraged—to use and belong to a variety of social media platforms, as long as their involvement does not violate unit policy and the basic guidelines of the Uniform Code of Military Justice (UCMJ).

All leaders should communicate social media expectations with their soldiers, as it is important to outline unit policy and make sure all soldiers know what they can and cannot do when using these various platforms. The following tips and guidelines come from the Army's social media policy, which can be found at www.army.mil/socialmedia.

Soldiers using social media must abide by the UCMJ at all times. Commenting, posting, or linking to material that violates the UCMJ or the basic rules of soldier conduct is prohibited. Social media enables soldiers to speak freely about their activities and interests; however, soldiers are subject to UCMJ even when off-duty, and talking negatively about supervisors or releasing sensitive information is a punishable offense. It is important that all soldiers know that they still represent the Army even when logged onto a social media platform.

MIA and KIA

Social media can play an important role (good or bad) in managing trauma in a unit. This is particularly crucial in the handling of MIA and KIA situations. If you are captured, members of the media may look at your Facebook profile or those of your family to find out more about you; captors may also turn to Facebook to pull information for interrogation purposes. As such, it's vitally important that you and your family restrict privacy settings as much as possible.

Bear in mind that details concerning soldiers killed in action cannot be released until twenty-four hours after the next of kin has been notified and after the information has been released by the DoD at www.defensegov/releases. In our social media culture, this has become more difficult to enforce. It's important that all friends, family, and fellow soldiers know that information about individuals killed in action must not be released before the next of kin is notified. Always follow unit and Army protocol when it comes to MIA and KIA situations.

In addition, when a soldier is injured, be sure to avoid posting any medical information when providing updates on his or her condition.

Online Relationships

It is natural that Army leaders may interact on the same social media platforms as their subordinates. How leaders connect and interact with their subordinates online is left to their discretion, but it is advised that the online relationship function in the same manner as the professional relationship.

Leader Conduct Online

When you are in a position of leadership, your online conduct should be professional. By using social media, you are essentially providing a permanent record of what you say. If you would not say it in front of a formation, do not say it online. It is also your responsibility to monitor your soldiers' conduct on social media platforms. If you find evidence of a soldier violating command policy or the UCMJ online, you should respond in the same manner as you would if you witnessed the infraction in any other environment.

It is not appropriate to use your rank, job, and/or responsibilities to promote yourself online for personal or financial gain. Such actions can damage the image of the Army and your individual command. Similarly, treat any requests from nongovernmental blogs for guest posts as a media request and coordinate them with your public affairs officer. Army regulations forbid accepting compensation for such posts.

Keep in mind that the Hatch Act and DoD directives limit partisan political activities. Avoid any political postings, such as publishing partisan political articles, letters, or endorsements, or soliciting votes for or against a party, candidate, or cause.

SHOULD LEADERS "FOLLOW" THOSE IN THEIR COMMAND?

This depends solely on how a leader plans to use social media. If you are using social media as a way to receive command and unit information along with installation updates, then following subordinate members of your command is appropriate. However, if you are using it to keep in touch with family and friends, it is rarely appropriate to follow your subordinates. Leaders cannot require members of their unit to accept a friend request from their personal account.

SOCIAL MEDIA TIPS FOR OPSEC

In addition to safeguarding PII, the Army's primary concern surrounding social media platforms focuses on operational security (OPSEC). Here are a few tips that will help you steer clear of OPSEC-related issues:

- Do not reveal sensitive information about yourself, such as schedules, locations, or any other details that could be dangerous to you or your unit if released.
- Before posting, ask yourself, "What could the wrong person do with this information? Could it compromise the safety of myself, my family or my unit?"
- Consider turning off the GPS function of your smartphone and digital camera so that photos and videos are not geotagged, a feature that reveals your location to other people within your network.
- Photos and videos can go viral quickly. Review them closely for any sensitive information before posting.
- Talk to your family about OPSEC and be sure they know what can and cannot be posted.
- Look closely at all privacy settings. Set security options to allow visibility to "friends only."

Following are some ways to make potentially dangerous social media posts safer.

Dangerous Post: I'm assigned to XYZ unit at ABC camp in ABC city, Afghanistan.
Safer Post: I'm deployed to Afghanistan.

Dangerous Post: My soldier will be leaving Kuwait and heading to Afghanistan in three days.
Safer Post: My soldier deployed this week.

Dangerous Post: I'm coming back at XYZ time on XYZ day.
Safer Post: I'll be home this summer.

Dangerous Post: My family is back in Edwardsville, Illinois.
Safer Post: My family is in the Midwest.

Social Media and Army Families

Social media helps keep families and soldiers connected, which is vitally important to unit well-being. Family Readiness Groups (FRGs) have turned to social media as a way to provide information on what is happening at an installation. Many also now host online discussion sections where the FRG, soldiers, and their families can post information and photos about installation news and activities.

For more information on DoD policies, which drive the Army policy, visit www.defense.gov/socialmedia. For specific Army information, visit www.army.mil/socialmedia.

BENEFITS

The Army and a grateful citizenry afford many benefits to military personnel. Keep in mind that these benefits may change, however, as they depend on the continued financial support of the people as determined through their elected representatives and the President of the United States.

Use of Exchange, Commissary, Theater, and Medical Facilities. These sales tax-free facilities are operated on an other than for-profit basis to pass on the lowest possible prices to their patrons. They are conveniently located on post to allow easy access for soldiers and their families, and generally are staffed by personnel who cater to the needs and desires of the Army community. Authorized patrons and their family members must display appropriate identification to take advantage of certain benefits, such as obtaining medical service by family members, patronage of the post exchange or commissary, attendance at theaters of the Army and Air Force Exchange Service, and others. Corresponding facilities of other services (Air Force, Navy, Marines) are also open to Army personnel and their families.

Leave. Under current laws and regulations, military personnel are entitled to accumulate leave and to take it when their superiors deem that their duties so permit. Readiness, training, or tactical considerations govern the boss's decision. If soldiers absent themselves without this permission they are subject to disciplinary action. Leave is accrued at the rate of 2.5 days per month. Soldiers may accumulate up to 90 days of leave but may carry forward only 60 days of leave into the new fiscal year.

Official Army policy is that short leaves, taken frequently, are the preferred usage of this benefit. In terms of what is most conducive to the maintenance of a good individual attitude and least stressful on unit leadership, this is certainly valid. However, given the operational tempo of the current Army, it is unlikely that many officers will be able to take advantage of this concept. Take leave when you can, and remember, no one should be indispensable. If you are, you have a weak unit, and you have probably failed to properly develop your subordinates.

Discounts on Travel, Recreation, and Other Off-Duty Activities. Many transportation companies, theme parks, museums, and the like offer discounts to active-duty soldiers. If your post has a travel agency, it can usually apprise you of the availability and the scope of the discounts; your local recreational services office can do the same for activities within its purview. If circumstances prevent consulting these professionals, it is appropriate to ask the airlines and other ticket vendors about the availability of active-duty discounts.

Recreation Services. Although many services were cut back significantly the last few decades, there are still many little-known "good deals" available through your post recreation services office. From tours to full suites of camping equipment, significant opportunities for low-cost, quality recreational opportunities are often available.

Billeting, Guest Houses, and Travel Camps. Although the variety and quantity of these facilities are shrinking with the closure of many installations, there are still a significant number of accommodations available to the traveling soldier or family. In recent years, most of these accommodations have been brought to a standard that is comparable—or superior—to average commercial hostelries and represent not only great convenience but also tremendous value. A chosen few are truly spectacular and represent some of the last "best-kept secrets" of military benefits. Naturally, personnel on leave have lowest priority, but before you travel, check with the billeting office at any U.S. military installation near your intended destination (or along the way), and look into the possibility of obtaining quarters. See chapter 9 for more details.

LEADERSHIP IN THE FUTURE

At the time of writing, the United States has been at war for more than two decades. The demands of the Global War on Terrorism were enormous. Those demands have not changed as we face threats from near peer competitors. Frequent rotations to combat zones, either as part of deploying units or individually, place great stress on officers' physical and emotional endurance. Rotations also induce pressure on families and other loved ones. These factors only increase the importance of Army customs and heritage, as they provide touchstones for dealing with the challenges of protracted war conditions. American soldiers have done all this before. In the last 100 years alone, there have been even tougher challenges, such as those posed by world wars from which practically no one came back until it was over, over there.

While certainly the main focus of Army officers' efforts, combat is not the only effort that officers may anticipate in the future. Increasing involvement in stability and support operations, antiterrorism operations, and the challenges of sustaining combat readiness in a multidomain environment all mitigate in favor of both new emphases and time-proven fundamentals.

The speed and distance of operations enabled by integrated capabilities will also have many corollary impacts and requirements for leader development. Technological developments will change the way leader actions are applied; however, leadership will remain the most essential dynamic of combat power for the U.S. Army. Leaders must have a keen awareness of the world around them and recognize how political, cultural, economic, and ethical factors impact on military operations. Future leaders will require a broader understanding of the interrelationships of peace and conflict and the demands they place on leadership.

The Army will require in its leaders the capabilities and traits essential for success in the twenty-first century. These build on inherent American cultural values, as well as enduring Army values and traditions, and must be consciously cultivated for our leaders to optimize their effectiveness and that of their soldiers and their organizations. High-velocity, lethal operations executed by multidomain-capable formations and equipment will place a premium on the values, attributes, skills, and actions required of Army leaders. These qualities must be developed through mentoring, advanced training, and experience. Early and continuous efforts to identify leaders whose performance clearly indicates possession of these qualities must be aggressively conducted to build the kind of Army that will prevail on current and future battlefields.

Values. Leaders of the future must be morally, ethically, and legally prepared for the nature of future operations. Excellence and depth of character contribute significantly to how leaders act; they form the foundation of the moral framework essential for leader credibility and effectiveness. Army values include the code of the officer discussed earlier.

To this end, the Army seeks to cultivate a leadership culture that mitigates against the incorporation of fleeting trends or passing moral fashions into its values set. Equally important, a thorough familiarity with military law, appropriate aspects of local and international law, and, at the higher levels of Army leadership, constitutional law will be indispensable. With a commonly understood and embraced set of values and practical understanding of the law, leaders can derive the full measure of freedom to make the morally and ethically demanding judgments essential for success in fast-moving, complex future operations.

Attributes. Leaders' mental, physical, and emotional attributes are the fundamental qualities and personal characteristics that, along with values, constitute their character. Attributes influence actions and thereby influence the Army, the unit, and their subordinates.

Flexibility. Mental attributes take on a new importance in an era in which operations must be planned, coordinated, and executed with unprecedented speed. Leaders must develop the physical and mental agility, in thought and deed, that translates organizational flexibility into operational success. Versatility, cognitive resilience, and rapid decision making characterize this flexibility, as well as intuition and the ability to recognize and synthesize information to exploit the opportunities provided by the information dominance sought and maintained by future operations. In an operational environment in which the conditions—and therefore the missions—may change with unpredictable rapidity, combinations of offensive, defensive, stability, and support missions will vary quickly and unpredictably. Leaders must possess a highly developed understanding of the operational requirements for each of these missions and be able to simultaneously plan and execute various combinations of these mission types within a single overall campaign, shifting focus and priority based on operational requirements. To develop the flexibility to accomplish this, leaders must repetitively train under conditions that approximate the complex and rapidly changing nature of the environments and conditions they will encounter in the operational environment.

Discipline. An increasingly unsupervised and ambiguous operational environment requires leaders who possess superb discipline, develop it in their subordinates and units, and implicitly encourage it in their peers. Assuming a continued all-volunteer force, such discipline can best be instilled through thorough, holistic understanding of the mission, the commander's intent, and the enforcement of clearly articulated individual and unit standards of performance. The discipline that springs from a focus on the unit, devotion to fellow soldiers, and a genuine dedication to the mission is most likely to achieve the level of performance necessary in the decentralized, fast-paced operations of the future.

Judgment. The cross-domain operations essential for the Army of the twenty-first century to succeed will place leaders at all levels in situations in which their on-the-spot judgments will make the difference between victory or defeat, success or failure, life or death. These decisions will be made under the constant view of international media. Sober, thoughtful, considered, but equally agile and bold decisions in stressful, dynamic conditions will be absolutely essential to success. To make these critical judgments on a sustained basis, leaders will require exceptional degrees of physical fitness, mental agility, and psychological resilience.

The shared situational awareness that is a cornerstone of multidomain operations must be tempered by leaders' candor and appreciation of that quality in their subordinates. The essential complement to the shared vision that will accelerate operations to a war-winning velocity is a culture in which leaders possess a keen sense of individual judgment. To avoid the potentially disastrous pitfalls of groupthink that could easily arise in an information-enabled force, leaders must value forthrightness and frankness and encourage it in their organizations.

Initiative. The lethality and dispersion of the battlespace in the twenty-first century will require increased initiative on the part of leaders at all levels. Leader initiative should be seen as the partner attribute to all those discussed so far. Without discipline, independent action can lead to chaos; without judgment, it can lead to disaster; without flexibility, it is impossible. From the opposite perspective, without initiative, discipline is merely stifling; judgment can become purely dogmatic; and flexibility actually becomes pointless meandering. Decentralized, distributed, fast-paced operations require leaders at every level to routinely exercise initiative; they must not only practice it regularly themselves but also encourage and foster it in their subordinates. Information Age technologies will go a long way in improving the information available to leaders; it will not eliminate the "fog of war." Although the extent and size of the battlespace will change, the requirement for leaders to act with bold, audacious initiative with less than perfect information will be even more important in the future to maintain the rapid pace so essential for operational success.

Leaders must be willing to provide the latitude their subordinates need to develop a capability to exercise initiative. Initiative must be emphasized from the earliest leader training. In an Army in which technological developments will make it possible for leaders to monitor subordinate unit and even individual soldier actions from great distances and much higher echelons, leaders must be conditioned to trust their subordinates and allow them greater latitude in the commission of their duties. Effective development of a cultural emphasis on demonstrating confidence in subordinates and willingness to accept responsibility for subordinates' actions will be absolutely critical for deriving the maximum benefits from emerging information technologies.

Commitment. In a professional environment that will undoubtedly become more demanding of leaders' time and effort, leaders must be completely committed to the nation, the Army, their unit's mission, and their subordinates. Although this is a long-standing Army tradition, the challenges of the future place new importance on this critical attribute.

A deep and cherished knowledge of Army and unit traditions—including personal contact with unit veterans—is one way to elicit commitment and discipline. Once acquainted with the historical and human sources of their unit's traditions, most soldiers will go quite far to avoid letting the unit down.

Skills. The multidomain-capable organizations emerging from Army transformation initiatives and contingency orientation of the future force require leaders who possess the highest levels of competence in more technical, tactical, interpersonal, and sociological skills than ever before. Leaders, active and reserve, will manage technologically advanced communications and weapon systems and operate independently in complex, rapidly changing operational environments with inherently different cultures and languages. They must be capable of rapidly building mission-tailored teams from forces that can operate effectively in joint, interagency, and multinational environments. The sheer quantities of knowledge and skills to be mastered in the development of such competence will have to be taught through a balanced, integrated combination of institutional, unit, and self-development programs. Formal and informal instruction will be important elements of leader development, but leaders must be able to apply what they have learned and further develop their skills in the hands-on environments of operational assignments. Given the greater lethality of the future battlespace and the potential consequences of mission failure in an increasingly visually connected world, an extremely high level of competence will become an indispensable quality in Army leaders. Hard, practical assessments of leader abilities will be essential for building a culture in which military skills and tactical and technical capability define "competence," and in which competence becomes the foremost criterion for professional advancement.

Conceptualization is another fundamental leader skill. Rapid decision making, sound reasoning, and intellectual synthesis will be essential companions to the possession of technical knowledge and discrete operational skills. Such critical, adaptive thinking facilitates the creativity that is essential to exercise the initiative that will fully enable the potential of digitized operations. The possession of appropriate operational knowledge and adroit conceptualization skills will combine to enable the sort of bold, audacious actions that will be the hallmark of twenty-first-century Army leaders and the ability to effectively link tactical and operational successes to achieve strategy-level objectives.

Actions. Even leaders possessing all the necessary values, attributes, and skills are ineffective if they do not apply them through their actions. Due to the requirement to be prepared for success across a broad range of military operations, the volume of tasks that must be efficiently and effectively executed will require intensive management efforts by all concerned. Hence, leaders will have to impart their knowledge and their intent for the conduct of training with precision. There will be little time and fewer resources to expend on repeating improperly coordinated or ill-conceived training events. Communications will have to be precise about the letter of what is to be accomplished in training, but more importantly, leaders will have to clearly communicate the intent of training to be conducted.

The increased size of the battlespace and the stresses concomitant to frequent organizational mission tailoring will impose significant challenges to leader credibility and referent power. The increased distances between leaders and their subordinates and the longer intervals between face-to-face interactions will require leaders to optimize the time they do have in the company of their soldiers. To this end, whenever possible, leaders will have to take special pains to demonstrate their skills to inspire confidence in their decisions among their subordinates. Also absolutely essential will be actions to enhance leader referent power: sharing hardship; setting the example; and looking after their soldiers' professional needs, health, and welfare. Only through such actions—the opportunities for which will become fewer due to the wider dispersion and faster tempo of future operations—can leaders inspire their soldiers to perform with the discipline and initiative required for victory in the twenty-first-century battlespace.

4

The Army in the Joint and Interagency Team

As war has become increasingly complex through the ages, the need for the knowledge of how to fight and win it has also increased. Before gunpowder was introduced into military operations during the Middle Ages, military science at the tactical level could be divided into three branches of knowledge: how to fight on foot (infantry), how to fight on horseback (cavalry), and how to support lengthy sieges and protracted operations (engineers and logisticians). Early writing on the art of war focused on how to conduct combined arms operations.

Gunpowder and subsequent developments in military science forced armies to diversify and create new branches of knowledge, which had led to new combat capabilities. As a result, new military specialties were institutionalized, including ordnance (to build and maintain guns); coast artillery (to operate against navies); armor (to provide mobile protection from the effects of gunfire); engineers (to provide construction support and better protection from the effects of gunfire, and more); air defense artillery (to operate new types of guns and missiles against aircraft); Army aviation (to operate organic Army aircraft); and Special Forces.

The history of the artilleryman illustrates how one military function evolved into a branch of service. The earliest artillerymen were considered technicians, because of the complexity of aiming and firing the ancient cannon. They fought mainly from long distances, usually during relatively bloodless sieges. Because artillerymen generally did not engage in close or personal combat, they were not considered to be soldiers. Only when King Gustavus Adolphus of Sweden applied field artillery's deadly effects against military formations on the battlefields of the Thirty Years War (1618–48) did artillerymen finally begin to be accepted as legitimate soldiers. They shared the same risks and dangers as soldiers, and therefore had to adopt the same code of honor and ethics as the infantrymen and the cavalrymen.

This brief history illustrates important themes about the organizational structure of the U.S. Army. Each branch has come about as a result of an identified need for military units with specialized knowledge and abilities. All branches today are necessary for the successful conduct of military operations. All officers are equal in their status as soldiers, and all are expected to follow the same standards of conduct and the same dedication to duty, honor, and country. Therefore, it does not matter in which branch an officer serves, as long as he or she fully demonstrates the professional ethic that is characteristic of the U.S. Army. To be sure, stereotypes exist for members of all branches, but in each branch, it is one's individual character and professional expertise that gains universal acceptance and respect by all.

Just as in the past, today's changing environment forces the Army to adapt to conduct operations across multiple domains and contested spaces. Branches may evolve or even mutate as combined arms operations change. Most recently, the Cyber Branch was created to address the acknowledgment that cyber operations will play a central role in any future conflict. The current development efforts related to the Army changing to meet increasing demands has resulted in divisions and brigades becoming increasingly more lethal as new capabilities are added to their formations. If anything, an officer's knowledge of other branches' skills and competencies will only increase over time, as units with different capabilities are integrated at lower tactical levels.

Beyond the trend toward lower-level integration of different types of army units is the increasingly *joint* nature of military operations at lower levels of command. Over the last two decades, semi-permanent joint organizations, such as Joint Task Force–Bravo in Central America, have become more common. Multi-service or joint operations have become common, and almost all deployments for operations outside of the United States are now conducted as joint task forces where the resources of Army, Navy, Marine Corps, and Air Force units operate together. The past joint operations in Afghanistan involving U.S. Army, U.S. Marines, Air Force, Navy, and Special Forces elements is a good example of this trend.

THE OTHER ARMED FORCES

Congress has organized the national defense and defined the function of each armed service through Title 10, U.S. Code. Within the executive branch, these are found in DoD Directive 5100.1, *Functions of the Department of Defense and Its Components*. All the U.S. Armed Forces—Army, Air Force, Navy, Marine Corps, Coast Guard, and Space Force—are required to respond globally. The capabilities of each service complement those of the other services in the execution of national and theater strategies. Since the Defense Reorganization Act of 1947, all United States military operations are commanded by a joint force commander (JFC).

The United States Air Force (USAF) and Space Force (USSF)

Air Force and Space Force capabilities—intelligence, logistical airlift, and combat—are invaluable for creating the conditions for success before and during land operations. The Army works closely with the Air Force as air support of ground operations is an integral part of each phase of a joint military operation. The Space Force conducts operations that allow us the freedom of action and access to our essential space-based systems. USAF strategic and intratheater airlift delivers Army forces into the area of operations. Fires from USAF and USSF systems are integrated into air and land operations. In addition, the USAF provides computer network defense intelligence, surveillance, and reconnaissance functions as part of joint plans.

Army aviation, air defense, military intelligence, and field artillery capabilities are often integrated with Air Force counterair, counterland, theater reconnaissance, and surveillance missions. Planning and executing successful, integrated land and air operations as part of a synchronized joint campaign is a principal task of a joint force commander.

The U.S. Navy (USN) and Marine Corps (USMC)

The U.S. Navy and Marine Corps conduct operations in oceans and littoral (coastal) regions. The USN's two basic functions are sea control operations and maritime power projection. Sea control is the uninhibited use of selected sea areas and the associated underwater and air spaces. Sea control is essential for the movement of Army and air forces, and their follow-on supplies, to overseas trouble spots. Maritime power projection covers a broad spectrum of joint operations, including naval bombardment, employment of carrier-based aircraft, creation of lodgments by amphibious assault, use of maritime prepositioned equipment, employment of naval expeditionary capabilities ashore, and strategic sealift, all of which are important warfighting capabilities used by the Army.

The Marine Corps, with its expeditionary character and potent forcible entry capabilities, affords the joint commander the ability to respond rapidly and seize lodgments suitable for force projection, and conduct operations on land and in the air for a relatively short duration. The flexible air and ground capabilities of Marine Air-Ground Task Forces (MAGTFs) complement those of the Army for a highly flexible force capable of decisive land operations in any environment. Once ashore, MAGTF operations are normally integrated with and supported by the Army under a Joint Force Land Component Commander.

The U.S. Coast Guard

The Coast Guard is an armed service that operates in peacetime under the Department of Homeland Security. In domestic support operations, the Coast Guard does not have the same restrictions on its employment as the Army, and it can sometimes be employed by federal authorities in ways the Army is legally barred from acting. It has a statutory civil law enforcement mission and authority. Army forces support Coast Guard forces, especially during drug interdiction and border control operations. When directed by the President or upon a formal declaration of war, the Coast Guard becomes a specialized service under the Navy. The Coast Guard and Navy cooperate in naval coastal warfare missions during peace, conflict, and war. During deployment and redeployment operations, the Coast Guard supports force projection. It protects military shipping at seaports of embarkation and debarkation in the U.S. and overseas. The Coast Guard supports joint commanders with port security and maritime patrol operations.

THE ARMY IN INTERAGENCY OPERATIONS

Stability and civil support operations, previously called military operations other than war, are a reality for many soldiers today who now find themselves operating as part of an interagency team conducting unified action. These teams may include United States governmental agencies, such as the Department of State, the U.S. Agency for International Development, and the Federal Emergency Management Agency; various law enforcement agencies such as the FBI or U.S. Border Patrol; and perhaps also nongovernmental organizations (NGOs), private voluntary organizations (PVOs), or international organizations (IOs). In many cases, a military service may not even be the lead organization. This is frequently the case with Army support to counterdrug or arms control missions, foreign humanitarian assistance, and domestic or overseas support missions.

Army involvement in interagency operations such as service on Provincial Reconstruction Teams requires officers to possess exceptional knowledge of the culture of the foreign nations in which some interagency operations take place *and* the culture of the domestic and international agencies with which the Army is working.

In some operations there are also legal considerations of which the officer must be aware. This is especially true in U.S. domestic support operations, where laws may restrict what Army units can do. When in doubt or uncertain, consult an Army judge advocate officer or the joint force legal advisor.

THE ARMY IN MULTINATIONAL OPERATIONS

The Army has conducted multinational operations since the successful operations with the French Army and Navy during the American Revolution. Beginning with the World Wars, the U.S. Army has participated in numerous coalition operations and has maintained significant involvement in NATO and Korea. Modern multinational coalition operations are extremely complicated and require soldiers to develop sophisticated knowledge of the environment and its people.

SOLDIERS, ALWAYS

The need for an officer's constant development and improvement should not lead to a neglect of his or her basic soldierly skills and values. Soldiers must remain focused on warfighting, and officers on leading soldiers to victory. The trend toward engagement in combined, joint, and interagency operations notwithstanding, officers must always demonstrate the values, attributes, and actions required to inspire and effectively lead their soldiers in combat. These are all learned and internalized during an officer's earliest branch assignments. But as the officer advances in grade and experience, he or she needs to know the roles, capabilities, and assets of the other Army branches and the other services and agencies so that a full and accurate Army contribution can be made to each joint, interagency, and multinational operation.

5

Combat Arms Branches

The combat arms branches principally participate in direct tactical and operational land combat with the enemy. In general, they include units that carry or employ small or large weapons systems. They are charged with engaging and destroying enemy forces.

AIR DEFENSE ARTILLERY

Air Defense Artillery (ADA) is a combat arms branch of the Army, providing air and missile defenses for a theater commander. The mission of the ADA is to protect the force and selected physical assets from aerial attack, missile attack, and surveillance. This ensures our force decisive victory with minimum casualties by offering freedom of maneuver and force protection. ADA units also protect command and control centers that manage the battle and enable our forces to sustain the war by protecting logistic centers and other vital theater geopolitical and military assets. Air Defense Artillery units maintain a high state of readiness for immediate worldwide deployment, and many are forward deployed, from Europe, to the Persian Gulf, and to Korea.

ADA officers perform duties dealing with the employment of air defense missile units and the directing of tactical operations and targeting techniques used by the various weapon systems. They also evaluate tactical situations and system capabilities to direct the engagement of hostile aircraft and missiles according to doctrine. Besides possessing leadership skills, an ADA officer must have extensive tactical and technical expertise to handle the highly specialized air defense weapon systems in a combined arms environment.

Guided missiles, in the sense that they can be controlled and guided to specific targets, are a recent phenomenon in warfare, although military use of self-propelled projectiles dates back at least to thirteenth-century China. British use of rockets during the War of 1812 is immortalized in our own national anthem.

The first surface-to-air missile, later known as the Nike-Ajax, was fired at Fort Bliss on 20 October 1953. This missile evolved over the years into the successful Nike-Hercules and Hawk missiles. Both of these were replaced by the Patriot missile system. With its multicapable, phased-array radar and long-range, high-altitude missile, the Patriot is the most formidable air defense system in the world today. It has the capability to intercept incoming tactical ballistic missiles (TBMs), a role it performed with success in the Gulf War. Upgrades to the Patriot system continue to provide greater capabilities against TBMs, and the antimissile system, Terminal High Altitude Area Defense (THAAD), gives the Army tremendously increased capabilities against TBMs.

The early, short-range gun systems, such as Vulcan, have now been replaced with systems based on the Stinger missile to protect the division and corps maneuver forces. The missile system is mounted on a HUMMV or the new Stryker-based M-SHORAD platform and found in divisions and corps areas. The systems have a twenty-four-hour, all-weather, shoot-on-the-move capability that makes it a formidable weapon against low-flying aircraft and unmanned aerial vehicles.

Commanding troops who control such highly technical equipment requires unique leadership abilities. The U.S. Army Air Defense Artillery School at Fort Sill, Oklahoma, which is the home of air defense leadership and technical training, traces its lineage to the Artillery School for Instruction. This school, the oldest service school in the Army, laid the foundation for the present system of military education in the Army. Over the years, as a result of various budget constraints or world situations, the school opened and closed and was known under several names, including the Artillery School of Practice and the Artillery School of the United States Army. Finally in 1907 the War Department separated artillery into the Field Artillery and Coast Artillery branches. Coast Artillerymen prided themselves on their ability to not just hit a target, but to hit ships moving simultaneously in three dimensions, bobbing on huge ocean swells. In 1917, the Army created the first antiaircraft training center in France, manned by Coast Artillery officers and intended to train soldiers to counter the rapidly improving enemy air capability. In February 1918, a five-week antiaircraft course was added to the Coast Artillery curriculum at Fort Monroe for those officers en route to France during World War I, where they would engage even more challenging targets.

In the interwar years, Coast Artillery remained a proponent of antiaircraft development, but interest waned until World War II events. The German onslaught in Poland and the West combined with the disasters in the Pacific to

Design of the Air Defense Artillery insignia was first approved in 1957 for the then single artillery branch. The crossed field guns indicate branch ties to the field artillery, the superimposed missile symbolizing modern developments. This insignia became the identification of the ADA when it was authorized as a separate branch in 1968.

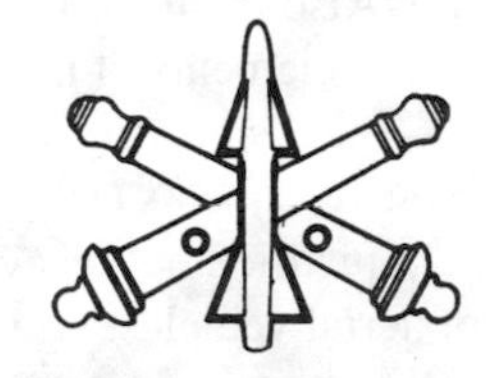

brought recognition of the need for effective air defense. In March 1942 the Army Antiaircraft Artillery was organized into a separate command under Army Ground Forces at Camp Davis, North Carolina. In October 1944 the antiaircraft school was moved to Fort Bliss, Texas, its present home. In July 1957 the school was officially renamed the U.S. Army Air Defense School, and in 1983 the title was changed to U.S. Army Air Defense Artillery School to recognize its lineage. The ADA School is located at Fort Sill, Oklahoma, as part of the Fires Center.

Currently officers commissioned in the ADA start their careers with attendance at the ADA Basic Officer Leader Course at the Air Defense Artillery School. During this eighteen-week course they are prepared for their first duty assignments and become familiar with the family of air defense weapons and their tactical employment. Based on needs of the Army and their personal preference, student officers are either trained in short-range air defense (SHORAD), which maneuvers with brigade combat teams or divisions; or specialize in high- to medium-range (HIMAD) missile systems or Long Range Precision Fires (LRPF), which defend division, corps, theater, and geopolitical assets.

After this basic schooling, new officers are normally assigned as Air and Missile Defense platoon leaders. As SHORAD platoon leaders, their troops will be deployed to provide close-in air defense to combat elements, high priority maneuver units, and high priority critical assets. Patriot leaders are employed to protect forces in all types of operations. Patriot platoon leaders will be deployed with their battery or as part of a composite air and missile defense (AMD) battalion to protect entering forces, airfields, seaports, transportation centers, population centers, and command, control, communication and intelligence activities and even geopolitical assets. The officers' responsibilities include the operational training and tactical employment of the platoon, the maintenance of equipment, and the welfare and morale of the soldiers.

Through this training and experience, the ADA officer normally qualifies for positions of increasing responsibility as an Air Defense Artillery unit executive officer or staff officer in the various Air Defense organizations deployed around the world. These positions provide for both new challenges and professional development opportunities in a branch with a vital mission for preserving peace.

The next phase of formal training occurs between the third and fifth year of commissioned service and is known as the ADA Captains Career Course, a training course conducted at the Air Defense Artillery School. The course prepares officers to command at the battery level, perform as battalion and brigade staff officers, and serve as air defense airspace management cell (ADAM) officers-in-charge. Also included is instruction in the tactical employment of ADA batteries and other elements of the combined arms team.

A quick-reacting air defense, provided by a family of complementary weapon systems, is essential for success on the modern battlefield. This requirement is justified by the ever-increasing destructive power of today's modern aircraft, aerial vehicles, cruise missiles, and ballistic missiles, especially those that are designed to carry weapons of mass destruction. The ADA provides numerous deterrent weapons in support of land warfare operations. Research into ballistic missile defense technology for the defense of CONUS is an ADA responsibility. The burden of these numerous, vital, and diverse responsibilities rests with the dedicated, capable officers of the Air Defense Artillery.

The ADA continues to be one of the major combat arms of the Army, and proper employment of ADA units makes the difference in force protection. Ever prepared and vigilant, the branch is ready to live up to its motto of "First to Fire" anywhere in defense of the free world. As technology continues to advance, the ADA will require dedicated officers who possess the leadership and tactical and technical skills that will enable them to meet the many diverse challenges within the branch and the Army in the future. For the most current branch information, visit http://sill-www.army.mil/ADAschool.

ARMOR

The responsibility for the development and conduct of mounted maneuver, originally the province of the horse cavalry, rests with the U.S. Army's Armor branch. Although modern technology has produced weaponry and mobility systems that are far more efficient than the horse, the incomparable spirit of the old cavalry and the impetuous character of its leaders are instantly recognizable in its modern counterpart. The impulse for devastating attack that has governed the tactics of mounted warfare from antiquity continues to dominate the doctrines and combat operations of the Armor branch. Its three subcomponents of armored cavalry, air cavalry, and armor provide the Army with its most powerful reconnaissance and offensive forces, all of which are trained to maneuver and fight with flexibility and shock.

The concept of combat systems augmented by armor protection and increased mobility is not a new one. Military leaders of the most ancient cultures constantly sought means to increase the individual's lethality on the battlefield while rendering him impervious to harm.

The distinguished history of the U.S. Cavalry dates from the Revolutionary War.

From 1868 until the turn of the century, those cavalry units retained on active status after the Civil War were sent to the western frontier to combat and subdue the Plains Indians. Conventional infantry proved to be ineffective against the mounted raiding parties of the Sioux and the Comanches, and ten cavalry regiments dispersed among fifty-five posts throughout the West finally brought peace

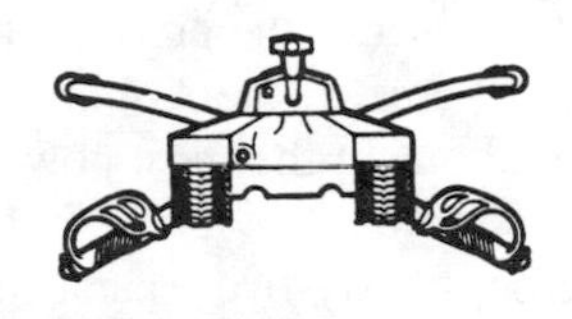

Approved in 1950, the current Armor insignia blends the past and present. Its base is formed by the traditional crossed sabers adopted for the Cavalry in 1851, on which a front view of the M26 tank is superimposed. It symbolizes Armor's heritage from the horse cavalry, as well as its role today as a mechanized force.

to the new territories. In every major ground action until World War I the U.S. Cavalry led the way. General Pershing's pursuit of the Mexican bandit Pancho Villa in 1916 was the last major action of the Army's horse cavalry. From that point on, cavalry would undergo a gradual mechanization that would lead to its present incorporation into the Armor branch's combined arms team.

World War I heralded the birth of the team's second component, the tank. First developed by the British Royal Navy, and disguised for intelligence purposes as "tanks for water in Russia" or "tanks for water in Mesopotamia," tanks saw their first engagement in September 1916, when they participated in the Battle of the Somme. The American Tank Corps, initially part of the infantry branch, fought its first battle in 1918, using French tanks due to the absence of American-built machines. During the Meuse-Argonne attack, General Pershing offered "anything in the A.E.F." for 500 additional tanks, but they were simply not available. By the end of the war, tanks had been employed by the British, French, Germans, and Americans in more than ninety engagements.

Although the value of tanks in modern warfare had proved substantial, few major military leaders were prepared to recognize this. The U.S. Army—together with the principal European armies—believed that tanks should be relegated to an infantry-support role. Consequently, through the National Defense Act of 1920, the Tank Corps was assigned to the Chief of Infantry. The first Tank School was organized at Fort Meade, Maryland, but the program was subsequently moved to Fort Benning, Georgia, and renamed the Tank Section of the Infantry School. During this same period, the Chief of Cavalry was authorized to develop mechanized weapons, and Fort Knox, Kentucky, was designated as the new home of the mechanized cavalry.

Between wars, tanks, and more importantly, the development of highly mobile combined arms units and a doctrine that would facilitate their revolutionary impact were advocated by many in the Army, from the lowest to the highest levels. In the 1920s, the euphoria of the victory in World War I and the booming economy prevented many outside the Army from investing the funds essential for developing such machines and organizations; in the 1930s, the Great Depression accomplished the same.

Although Germany was also profoundly affected by the same forces—to an even greater degree—the development of a command economy after 1933 enabled swift rearmament and the creation of organizations about which the other armies of the world could only dream. The success of these combined arms panzer divisions in 1939 and especially in the spring of 1940 (in France and Belgium) awakened the long-moribund idea of armored divisions among those with the power of the purse in the United States. A War Department order of 10 July 1940 created the Armored Force, and within four years, the U.S. Army would field one of the most powerful armored forces the world had ever seen, second only to the Soviets'. Sixteen armored divisions, thirteen mechanized cavalry groups, sixty-five separate tank battalions, seventy-eight tank destroyer battalions, and three separate mechanized cavalry squadrons constituted the Army's armored strength by the time of Operation Overlord. Deployed exclusively to North Africa, Sicily, Italy, and the European Theater, U.S. Army armored divisions were the most fully integrated combined arms forces in the world during that war. Not even the vaunted *Wehrmacht* could field divisions with three brigades (or "combat commands," as they were known in those days), each with a battalion of tanks, a battalion of armored infantry, and a battalion of armored, self-propelled artillery. The separate tank and tank destroyer battalions—which generally supported infantry divisions at the ratio of one of each per division—gave U.S. Army infantry divisions, with their towed artillery and ability to move battalions of infantry in trucks, a de facto combined arms capability greater than any other army on earth.

When the Korean conflict erupted in June 1950, North Korean–manned Soviet T-34 tanks accompanied the communists in their southward drive. The U.S. Army, which had not originally considered the terrain in that part of the world "tankable," fought without them until mid-July, when the first American tanks reached the peninsula. By August there were over 500 tanks in action within the Pusan Perimeter, outnumbering the enemy's by more than five to one. For the remainder of the war, tank units of battalion size and smaller participated in most combat actions.

After the Korean truce, the Army launched a serious investigation into the possibilities of using rotary-wing aircraft in combat operations. The first sky cavalry unit, equipped with unarmed helicopters for reconnaissance purposes only, underwent testing during exercise Sagebrush in 1955. Shortly after the Sagebrush exercise, the Continental Army Command directed the Army Aviation School to establish a project entitled Armed Helicopter Mobile Task Force. In 1958, the Armor School was first charged with the responsibility of preparing the doctrine for tactical employment of Air Cavalry. By 1962, three Air Cavalry troops had been organized: one at Fort Knox, Kentucky, one at Fort Carson, Colorado, and another at Fort Hood, Texas. Also in 1962 the Defense Department directed Gen. Hamilton Howze to establish an Air Mobility Requirements Review Board. The Howze Board established the requirement for organizing an Air Assault Division (the 11th) and Air Transport Bridge (the 10th). The experiments conducted were highly successful and resulted in the deployment of the division to Vietnam in 1965, just days after being redesignated as the 1st Cavalry Division. Its air cavalry squadron proved so successful that additional air cavalry units were organized and deployed to Vietnam. The earliest helicopter gunships were the UH-1 Hueys—originally designed as troop carriers. A later arrival was the AH-1G Cobra—the first aircraft in the Army inventory designed specifically as a weapons platform. Although the terrain in South Vietnam initially was considered unsuitable for armored vehicles, as had been the case earlier in Korea, the air cavalry soon was complemented on the ground by the employment, between 1965 and 1973, of three medium tank battalions and seventeen armored cavalry squadrons and separate troops. These units employed the M48A2 and M48A3 tanks, the M551 Sheridan Armored Reconnaissance/Airborne Assault Vehicle, and the versatile M113 Armored Personnel Carrier. These three dimensions of armor, working in conjunction with both mechanized and dismounted infantry forces, achieved substantial tactical success during the war in the Republic of Vietnam. Near the end of U.S. troop participation in Vietnam, armored units, both ground and air, constituted over 54 percent of the total combat maneuver forces and were among the last units to redeploy to the United States.

During the 1980s, the durable M60 series of tanks, which had been the mainstay of Army armor units since the early 1960s, were phased out in favor of the faster, harder-hitting M1 series. The M1-series tank offered increased armor protection, better integrated fire control, and sustained higher speed and mobility. The M1A1 tank and its successor, the M1A2, mount a 120mm main gun (replacing the 105mm gun on the earlier version). The M1A2, with its integrated vehicle information system (IVIS), will be the main battle tank well into the twenty-first century. The M3 Bradley cavalry vehicle, which entered production in 1982, is well suited to the cavalry missions of reconnaissance and security. Like the M1-series tanks, the Bradley furnishes greater speed and mobility as well as increased armor protection and firepower.

Contemporary armor and cavalry tactics emphasize mobility, firepower, and shock action to overcome an enemy force. The combined arms team concept includes tanks, armored and air cavalry, mechanized infantry/artillery/engineers, and Army aviation, all supported by a flexible communications network and a highly mobile and responsive combat service support system. Though the tank continues to be the principal armor-defeating weapon in the combined arms team, it is intended for general application against the entire enemy force. Armor is continually evolving to meet worldwide challenges and potential threats. With the inception of the Aviation branch in 1983, the Armor branch transferred the advocacy of scout and attack helicopters to the Aviation center at Fort Rucker, Alabama.

Armor officers have educational patterns and career programs similar to those found in other branches. The Armor School at Fort Benning, Georgia, offers the Armor Officer Basic and Career Courses, as well as specialized instructional programs for noncommissioned officers and senior-grade commissioned officers. Notable among the special courses for officers is the Junior Officers' Maintenance Course. In the eight weeks of the course, students receive meticulous instruction on maintenance management, supervision and inspection of vehicular maintenance and repair procedures, and the complex materiel requirements of armor/cavalry field operations.

Upon completion of the career course, armor officers usually gain experience as company or troop commanders. Then they can expect assignments in battalions, brigades, or divisions as staff officers. Armor officers are also in demand as planners, directors, and special staff officers at major Army headquarters, on joint staffs, at the Department of the Army, and in research and development projects related to combined arms concepts.

Armor is rich in tradition and exciting in potential. Officers who apply and are selected for assignment to the Armor branch are assured of unusually dynamic and rewarding service. For the most current information on the Armor branch, visit www.benning.army.mil/armor.

AVIATION

The origins of the Aviation branch go back to the formation of the Balloon Corps in the Army of the Potomac during the Civil War. The Aviation branch was created by the approval of the Secretary of the Army on 12 April 1983, and implementation of the new branch began on 6 June 1983. Its warfighting mission is to find, fix, and destroy any enemy through fire and maneuver and to provide both combat and combat service support in coordinated operations as an integrated member of the combined arms team. In conjunction with other combat arms units, aviation units provide a quickly deployable contingency force for the projection of power and the protection of national interests.

In addition to aviation brigades at corps and higher echelons, there is an aviation brigade in every Army division as one of its maneuver brigades. The aviation resources of the divisions are located in these highly mobile and flexible aviation brigades, which are tailored to fight as members of the Army combined arms team (as well as joint or coalition combined arms teams). Aviation brigades provide unprecedented tactical flexibility to division and corps commanders, since aviation is best suited to conduct attack, air assault, reconnaissance, show-of-force, intelligence, and logistical operations. The aviation brigade commander also controls tactical ground forces, if missions so require.

Army aviation, as distinguished from the Army Air Corps, had its beginning during World War II. On 6 June 1942, the War Department approved organic Army aviation as an adjunct of the field artillery. The Department of Air Training was established at Fort Sill, Oklahoma, and the first class of aviators and mechanics began training on 15 January 1942. Army aviation spotter planes were used in all theaters of operation during World War II. The missions directed from them for field artillery and naval gunfire wreaked havoc on major enemy units, disrupted communications and supply lines, and made life miserable for the enemy. In addition, the spotter planes served as liaison aircraft, flew medical evacuation missions, made supply drops to beleaguered units, and transported staff personnel and commanders throughout the war zones.

After World War II, the National Security Act of 1947 separated the Army Air Corps from the Army and established it as the U.S. Air Force. Specific tactical missions were delineated for the new Air Force and for the remaining aviation

Approved in 1983, the Aviation insignia is similar but not identical to the branch insignia of the old Army Air Corps. It consists of a silver propeller superimposed on gold wings.

resources organic to the Army. One of the most significant events during this period happened in 1945, when Capt. Robert J. Ely became the first Army aviator to pilot a helicopter. In 1946 the Army bought the Bell Model YR-13 helicopter, later designating it the H-13 in 1948. The first class of Army helicopter pilots began on 1 September 1947 at San Marcos, Texas.

With the beginning of the Korean War on 25 June 1950, the mission of Army aviation took a new direction. The Korean War was the first war in which the helicopter was used on a regular basis and played an important role, particularly in the area of medical evacuation. The Army flew its first helicopter medical evacuation mission on 3 January 1951. By the end of the war, Army helicopter pilots had evacuated 21,212 wounded personnel. By that time, the Army had fielded two transportation helicopter companies to Korea, the 6th and 13th, each having twenty-one H-19 (12-place) helicopters. They were used to transport soldiers and supplies and were involved in medical evacuation and prisoner repatriation. The Korean War proved the value of the helicopter.

During the 1950s and early 1960s, the Army's inventory of aircraft increased dramatically, as both fixed-wing and rotary-wing machines were added. The CV-2 Caribou light tactical transport (later transferred to the Air Force and redesigned C7A), the OV-1 Mohawk reconnaissance aircraft, and the U-21 Ute and C-12 Huron light transports all increased the Army's mobility. (The C-12 still exists today in highly sophisticated electronic reconnaissance versions.) Even more types of helicopters joined the inventory, including three that are still in the inventory today: the UH-1 Iroquois (more commonly known as the "Huey"), the CH-47 Chinook, and the AH-1 Cobra.

In 1962, the Secretary of Defense directed that an in-depth study be made of tactical mobility of the Army ground forces, particularly with regard to the potential for air mobility. Gen. Hamilton H. Howze, who earlier had been the first director of Army aviation, was tasked to establish a board to study air mobility capabilities for the Army. Recommendations of the Howze Board resulted in the formation of the 11th Air Assault Division at Fort Benning, Georgia, to undergo tests. The tests proved the tactical mobility of the helicopter. On 1 July 1965, the 11th Air Assault Division became the 1st Cavalry Division (Airmobile).

Two airmobile divisions served in Vietnam: the 1st Cavalry and, after it was converted from an airborne division, the 101st Airborne Division (Airmobile). With initially large numbers of airborne volunteers and daring and aggressive leadership, these first combat airmobile units established the air assault elements of the U.S. Army as elite organizations. The 1st Aviation Brigade, activated in Vietnam in 1966, assisted many other units in the conduct of airmobile operations with the more than 4,000 aircraft under its administrative control at the peak of fighting. In the jungles, swamps, and hills of Vietnam, the helicopter became the U.S. Army's vehicle of choice for a variety of tasks as diverse as the many missions the Army conducted in its attempt to win that war. Tactical infantry assaults and extractions, reconnaissance and surveillance, direct fire support with rockets and guns, medical evacuation, personnel and cargo transportation, direction of artillery fires, and recovery of downed aircraft and crews were all common missions for Army aviation units in that conflict. Although it was not enough to sway the strategic outcome of the war, Army aviation came of age during the Vietnam War and established the U.S. Army as the most tactically mobile and flexible in the world.

"Uncommon valor was a common virtue" was said of Army aviation personnel in Vietnam. Seven Army aviators received the Congressional Medal of Honor (two of them posthumously) for their acts of heroism. Many times Army aviators flew into areas from which they received withering ground fire, yet they still fulfilled their missions. One of the most well-known missions was that of "dustoff" (aeromedical evacuation) units that went into many hot LZs to extract wounded troops.

In the 1970s and 1980s, Army aviation underwent further changes in doctrine, weaponry, and mission. The Army aviation community used the 1970s to develop new tactics, such as nap-of-the-earth and adverse weather flying with and without night vision devices. Doctrine developed during the 1970s emphasized the use of helicopters in a European combat scenario and included high-intensity combat and deep penetration capabilities. This new Army aviation doctrine and training provided soldiers and units that were ready for threats and situations in places other than the European theater and that were deployable, agile, and lethal.

Army aviation supported forces in Grenada in October 1983 to protect American citizens during that island nation's instability. Army aviation deployed to Panama in December 1989 to take part in Operation Just Cause. In both operations, Army aviators conducted special operations, aeroscout, attack, air assault, and resupply missions in concert with ground maneuver elements. They demonstrated their capability to fly and fight at night, delivering accurate fire with pinpoint precision.

When Iraqi forces rolled into Kuwait in early August 1990, Army aviation units were among the first deployed to Saudi Arabia to draw "a line in the sand" as part of Operation Desert Shield. As Army aviation units arrived in Saudi Arabia, they provided a security screen around ports and airfields for the arrival of coalition forces, and unhampered by the rugged desert terrain and lack of roads, they conducted force-penetration missions.

Army aviators flying AH-64 Apaches fired the first shots of Operation Desert Storm, blasting Iraqi early-warning radar sites, creating a gap in the Iraqi air defense coverage, and allowing coalition air forces to initiate the air campaign. When the ground war began, Army aviation conducted a wide range of fast-paced missions alongside coalition forces and went deep into Iraqi territory, covering vast distances with speed and agility surpassing any seen before. When ground combat ended after one hundred hours, Army aviation remained active in support of Operation Provide Comfort, flying humanitarian relief missions in southern Iraq and Turkey.

The versatility of Army aviation units makes them indispensable. Most recently, Army aviation units supported combat operations in Afghanistan and Iraq, along with deployments in Europe and Korea. They fly in support of counterdrug and humanitarian relief operations in Central America, transport political dignitaries to the ever-dangerous Korean DMZ, and even support civil authorities within the continental United States with medevac and search-and-rescue missions. At the extreme ends of the Army aviation spectrum, Army aviators fly special operations with state-of-the-art covert aircraft and also fly jets (Gulfstream IIIs and IVs) in support of nontactical VIP transportation worldwide. With state-of-the-art equipment such as the Apache Longbow, the Aviation branch offers one of the most diversified and challenging opportunities for service in the Army.

Planned changes in the Army school system have not yet been finalized, but currently newly commissioned aviation officers attend the Aviation Officer Basic Course (AOBC) and the Initial Entry Rotary Wing Course (IERW) at Fort Rucker, Alabama. The course is conducted in consecutive phases totaling forty-five to forty-nine weeks. Phase I starts with four weeks devoted to teaching basic officer skills and provides an introduction to the Aviation branch. The officer next attends IERW, which consists of thirty-four to thirty-six weeks of basic flight instruction in the TH-67 Creek, the UH-1 Huey, or the OH-58C Kiowa, followed by two weeks of intensive aviation logistics study. Phase II, the final phase of the basic course, consists of five weeks and two days of aviation tactical training. Many officers can then expect to immediately attend a follow-on transition course into advanced aircraft, such as the UH-60 Blackhawk, the AH-64 Apache, or the CH-47 Chinook.

Assignments are governed by Army requirements. Commissioned officers are normally initially assigned as platoon leaders in a variety of units, depending on the type of aircraft they fly.

Between the third and sixth years of service, aviation commissioned officers attend the Aviation Officer Career Course. This course prepares aviation captains for command and other leadership positions and provides a sound training background in combined arms tactics. Following graduation, the aviation officer continues to build experience through assignments to staff positions at different levels and command at company or troop level.

All aviation officers receive continuing aviation education. As they rotate between assignments, their aviation skills and aircraft qualifications are upgraded to allow them to operate whatever system the new assignment requires. Examples are advanced aircraft qualification courses, the fixed-wing qualification course, and the maintenance manager/maintenance test pilot course.

Because of the highly technical nature of aviation and some of the functional areas (97, 53, 52, 51, 49), there are opportunities for aviation officers to pursue graduate studies that support an area of concentration or a functional area requirement. Some of these study areas are logistics, research and development, and engineering sciences. A graduate degree is not a requirement for promotion, but the Army will, in some cases, support advanced civil schooling that is a job or position prerequisite.

To become an aviation warrant officer, training begins with six weeks of intensive military development at the Warrant Officer Candidate School (WOCS). Upon successful completion of WOCS, candidates are appointed to the rank of WO1. The warrant officer then attends thirty-six to forty weeks of IERW, which includes training in the TH-67, UH-1, or OH-58C. This is followed by the Warrant Officer Basic Course (WOBC), an intensive four-week preparation for the warrant officer's first assignment as a combat aviator. Many students can then expect an immediate transition course to an advanced aircraft, such as the UH-60, AH-64, or CH-47D.

Aviation warrant officers remain at unit level in flying assignments throughout their careers. Warrant officers may advance as high as MW-5 and serve on an aviation brigade staff as a safety officer, maintenance officer, standardization officer, or flight operations officer. For the most current information on the Army Aviation branch, see www.rucker.army.mil.

CORPS OF ENGINEERS

The Corps of Engineers is one of the most diversified branches of the Army. In addition to its traditional military role, the corps has a history of engineering achievements that have benefited the nation as a whole. Engineer officers such as Lewis and Clark figured prominently in the development of our nation from the Atlantic states to the Pacific Ocean and beyond. Today's corps is a key member of the combined arms team, and its activities encompass both military engineering and civil works and all related planning, organization, training, operation, supply, and maintenance.

Engineer officers are responsible for training and leading troops in combat, topographic engineering, and construction operations essential to the Army in the field. They direct the operation and maintenance of Army facilities worldwide. They develop and manage the Army's extensive military construction and civil works programs, and they provide engineering and topographic expertise for other federal agencies and even foreign countries when program size and national interest dictate.

Today, engineers stand prepared to support the battlefield commander by executing their threefold tactical mission of mobility, countermobility, and survivability, as well as by providing topographical support to the Army. In addition, they continue to maintain the nation's waterways and contribute to other special national projects, such as America's space program. Engineers participate in combat operations as the terrain shapers for the maneuver commander. They

The triple-turreted castle identifying members of the Corps of Engineers was adopted in 1840 and symbolizes two major functions: construction and fortification.

assault fortifications as well as construct them; they reduce obstacles as well as create them; they provide topographic and terrain analysis support; and they participate in assault river crossings and amphibious operations in addition to other combat engineering tasks.

Types of engineer organizations to which an officer will likely be assigned include the various combat engineer units supporting or organic to infantry or armored divisions, serving at company, battalion, and brigade level; construction companies; float bridge companies; medium girder bridge companies; port construction companies; engineer equipment companies; topographic companies; and utilities maintenance units. In addition to the diversified unit commander positions, engineer officer assignments include special jobs such as observer/controller at a combined arms training center, division engineer, district engineer, facilities engineer, topographic engineer, project engineer, staff engineer, and civil engineer. Officers serve as engineer staff officers at all levels, coordinating, planning, and providing staff supervision of engineer operations, including support of the Air Force.

Corps of Engineer officers have outstanding opportunities for service that blend sound professional development programs with the personal satisfaction of serving with many of the Army's talented officers. Most junior officers are assigned to combat engineer units, so they will have opportunities to develop and polish their leadership skills while working with outstanding soldiers.

The Engineer branch offers numerous educational opportunities, ranging from professional engineer status to graduate school. For technically qualified engineer officers, civil engineer duty or an engineer troop assignment can be counted as engineering experience required for the professional engineer examination. The Engineer School maintains an active program to assist engineers in preparing for the examination. Engineer officers also are encouraged to affiliate with professional organizations and submit papers and articles for publication. The Engineer School has a graduate degree program established with the University of Missouri that gives credit for the valuable training received in the Officer Basic and Career Courses. This allows most officers the opportunity to earn a master's degree in conjunction with their required military schooling.

The home of the Corps of Engineers is Fort Leonard Wood, Missouri, which serves as the center of Army engineer activities. All engineer officers begin their careers there and return periodically for specialized training. This professional training, which takes place at Fort Leonard Wood and other locations, includes conduct of combat engineer operations as part of the combined arms team; supervising the design, construction, and contract administration of military construction and government civil works projects; developing, producing, and reproducing maps; surveying and mapping projects; bridge classification and construction; terrain studies; and natural resource and environmental studies. Engineer officers are also trained to test and evaluate all military engineering hardware and software, and they plan, construct, repair, and rehabilitate posts, camps, stations, airfields, structures, ports, harbors, roads, inland waterways, railways, pipelines, and utility plants and systems. These and myriad related responsibilities involving real estate are reflected in extensive and diversified engineer units and duty positions.

Officers commissioned into the Corps of Engineers share a rich and glorious legacy of honorable service to the nation and the Army for more than two centuries. As an integral part of the U.S. Army, the Corps of Engineers traces its beginning to 16 June 1775, when the Continental Congress provided for a chief engineer with two assistants "at The Grand Army." On 3 July 1775, General Washington appointed Richard Gridley to this post, and one of his first efforts was to lay out the defenses of Breed's Hill. He later directed the fortifications that forced the British to evacuate Boston in March 1776. Two years later, Congress authorized three companies of sappers (the British term for engineer soldiers) and miners. These first three companies of engineers built the siege works and fortifications that brought a final victory for the colonies at Yorktown. In 1779 Congress formally established the first Commandant of the Corps of Engineers. In 1802 Congress provided for the present corps and "constituted" it a military academy at West Point, New York. Thereafter, for nearly sixty-four years, the U.S. Military Academy was almost exclusively an engineering school. Army engineer officers in charge of operations were educated at West Point and commissioned as topographical engineers. They battled nature to survey and map the Great Plains, the Rockies, and the Columbia River. Their mission was to obtain the scientific data necessary for opening up the frontier for settlement.

Much of the credit for American victories in the Mexican War goes to engineers. The siege operations ending in the surrender of Veracruz, Cerro Gordo, Chapultepec, and Mexico City were all directed by Army engineers. The roll call of engineers in the Mexican War reads like a roster of famous Union and Confederate generals: George B. McClellan, Robert E. Lee, George G. Meade, Joseph E. Johnston, P.G.T. Beauregard, and Henry Halleck. Military engineering became more complex during the Civil War, with extensive pontoon bridging, planning of defensive positions and railroad supply lines, and siege contributions among the impressive list of engineering achievements.

World War I was a time of significant accomplishment for the Corps of Engineers. The corps developed many of the military functions it has today: construction of ports, docks, roads, bridges, transportation facilities, camps, hospitals, and depots; and responsibility for mapmaking, camouflage, mine warfare, obstacle emplacement and reduction, and terrain analysis. Engineer battalions became organic to all types of divisions at that time.

In 1940, as the Army burgeoned from its peacetime strength of 190,000 to several million thanks to the institution of the nation's first peacetime draft, the Corps of Engineers improved and repaired the facilities built in the 1930s by the Civil Conservation Corps (CCC) and built entirely new facilities as well. During World War II, engineers conducted a multitude of missions: building roads through the mountains and jungles of Burma, China, and India; clearing deadly beach obstacles on myriad Pacific islands and atolls; crossing desperately defended rivers in Italy; and finally blasting a way through the imposing fortifications of Fortress Europe.

In Vietnam, Grenada, and Panama, engineers were called on to perform countless tasks so American forces could complete their missions successfully. During Desert Shield/Desert Storm, engineers made possible the rapid buildup of allied forces in Saudi Arabia by constructing base camps, field hospitals, airfields, helipads, and main supply routes. During Desert Storm, combat engineers led the way by punching through Iraqi defensive berms. Engineers rolled with, and sometimes ahead of, the armor and infantry, breaching and clearing obstacles, building combat trails, and destroying Iraqi equipment on the strategically executed record-breaking drive toward Baghdad.

As a combat arms branch, a secondary mission of every combat engineer unit is to serve as infantry. In fact, most Army engineers who have earned the Medal of Honor were awarded it for actions while their units served in this capacity. U.S. Army engineers have served with distinction as infantry in every American war from the War of Independence through Vietnam.

Engineers also supported construction of the highly successful space program, including NASA headquarters in Houston and the John F. Kennedy Space Center launching facilities at Cape Canaveral. Before the mapping, charting, and geodesy functions of all the services were consolidated in a single DoD agency (the Defense Mapping Agency) in 1972, the U.S. Army Topographic Command, which had responsibility for the corps' geodesy and mapping mission, had begun mapping the moon. This imaginative and aggressive thinking is also being applied on earth, as the corps works hand in glove with the Environmental Protection Agency to prevent further pollution of waterways and to restore them to their former purity and beauty. For the most current information on the Engineer Regiment, visit www.wood.army.mil/usaes.

FIELD ARTILLERY

The American field artillery was born on 17 November 1775, when Henry Knox, a twenty-six-year-old self-taught artilleryman, was appointed Chief of the Continental Artillery. From its beginning, the field artillery has been unique as a combat arm. Having the unbeatable combination of daring soldiers and deadly weapons, the field artillery has participated in every military conflict in which our country has been involved and has amply earned the title "King of Battle."

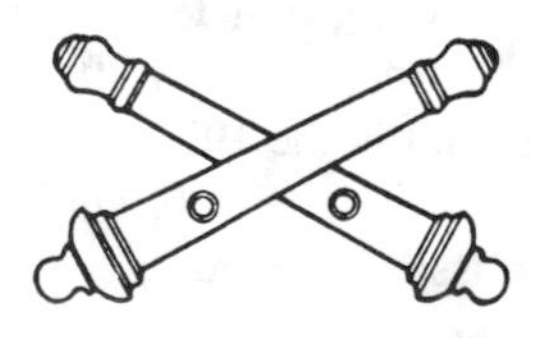

Field Artillery officers have been identified by the insignia of crossed cannons since 1834, and although the design went through several variations with the establishment of the Coast Artillery and the advent of missiles, the current design is basically that of the original.

During the Revolutionary War, the continental artillery, under the command of Alexander Hamilton, performed magnificently at the Battle of Trenton, and the skill of American gunners forced the British into siege trenches at Yorktown.

In 1784, when all of the Army was abolished except for a single detachment of eighty men to guard government stores, those men were artillerymen. Today's 1st Battalion, 5th Field Artillery, carries on this unit's heritage and lineage. Thus the Artillery is the only part of the Army that has been in continuous service since the Revolution. The many uses and value of artillery in nineteenth-century military campaigns, including the Civil War, are far too comprehensive to detail here.

It was not until 1907 that two separate artillery corps were established. The Field Artillery and the Coast Artillery were organized with specific missions obvious from the names, and during World War I the Coast Artillery was given the additional job of developing railroad-mounted and antiaircraft artillery pieces.

World War II brought about an enormous leap of capability in field artillery, which became an integral and indispensable part of the victorious U.S. Army combined arms team. The integration of radios—used by forward observers to adjust fires from foxholes, armored vehicles, or light aircraft—at once vastly increased the responsiveness, accuracy, and survivability of the field artillery. Artillery also fought the war with full mechanization, as all pieces were either towed by trucks or mounted on tracked, self-propelled chassis. From nimble 75mm pack howitzers to mammoth 203mm guns and 240mm howitzers, U.S. Army field artillery dominated the battlefields of the Pacific and Europe and was the single biggest killer of enemy troops.

In the wake of the Germans' successful use of intermediate-range ballistic missiles, the Field Artillery branch began development of a series of its own rockets. Corporal, Sergeant, Lacrosse, Little John, Honest John, Pershing, and Lance were some of the most well known, and although never used in combat, they helped keep the tenuous balance of forces that allowed the Western allies to eventually win the Cold War. In the early 1950s, the Army also fielded the first artillery piece capable of firing a nuclear warhead; with the introduction of the 280mm gun and subsequent development of nuclear warheads for other projectiles and rockets, the field artillery became the combat arm of the Army most deeply involved with nuclear weapons.

Conventional artillery has also developed significantly, with the greatest strides coming in digitization of the fire control system, direct connection of sensors (such as target acquisition radars) to firing batteries, and the ability to fire artillery rounds indirectly at tanks with great precision. The 1980s' introduction of the Multiple Launch Rocket System (MLRS) and its several devastating munitions represented a new and extremely effective use of rocket technology, even as the Lances and Pershings were retired. By greatly enhancing the precision, mobility, survivability, and firepower of field artillery systems, these developments have maintained field artillery units as the most lethal and most survivable combat units in the Army today.

The mission of the field artillery is more challenging than ever before—to destroy, neutralize, or suppress the enemy by cannon, rocket, and missile fire and to integrate all supporting fires into combined arms operations. The key to the field artilleryman's job is to focus combat power; everyone looks to him to integrate all fire support. That includes coordinating the employment of field artillery, tactical air forces, naval guns, Army aviation, mortars, and electronic warfare. As a key player in the combined arms team, the field artillery supports the ground-gaining elements of the team—infantry, armor, and cavalry—by placing massive and accurate firepower in the right places at the right time and in the right proportions.

The field artillery team integrates people, weapons, and support systems. Its leaders work together to plan and coordinate fire support, acquire targets, compute fire direction data, and deliver fires from all organic and supporting weapons. Providing the optimum level of fire support requires a synchronized effort with dedicated, competent leaders; well-trained soldiers; sophisticated equipment; effective command and control; and a mix of modern weapon systems capable of delivering accurate and devastating fires on target. There are four major components of the field artillery team: the firing battery, the fire support team, the fire direction center, and target acquisition.

Leadership is gained by company-grade field artillery officers in the firing batteries, providing a sound stepping-stone to leadership challenges at higher levels of command and staff. The focal point of the field artillery is the line of metal—the firing batteries of field artillery battalions. Firing platoons, commanded by field artillery lieutenants, and firing batteries, commanded by field artillery captains, are the delivery units for the impressive array of artillery weapons.

All other efforts of the field artillery team, as provided by fire support, target acquisition, and fire direction elements, serve but one purpose: to help the firing units place responsive, accurate, and lethal fires on target. Such fires can be accurately placed anywhere on the battlefield.

Howitzers, rockets, and missiles are the muscle of field artillery. Currently, towed field artillery systems include the M119A1 105mm howitzer and the M777 155mm howitzer. Self-propelled systems include the M109-series 155mm howitzer, M142 HIMARS, and the MLRS. The artillery's current missile system is the Army Tactical Missile System (ATACMS), Blocks 1 and 2, which provides long-range fires against personnel and materiel targets. ATACMS is fired from the MLRS launcher.

The company fire support officer (FSO), a field artillery lieutenant, leads the fire support team (FIST). He and his team are responsible for planning and coordinating the fires of the infantry or armor company or cavalry troop the FIST is supporting.

The company FSO works with the maneuver company or troop commander to develop a fire support plan for the unit's scheme of maneuver. The plan integrates all available artillery fires as well as those of the mortars organic to the maneuver units. When these systems are employed, the FIST calls for and adjusts fires on enemy targets.

An important tool of the company FSO is the ground/vehicular laser locator designator (G/VLLD); it can be ground-mounted or mounted in the fire support vehicle (FSV). The G/VLLD determines range to targets and designates targets for laser-guided munitions such as the tank-killing Copperhead round or the Hellfire missile.

When you consider the total firepower available to the company FSO, you can easily see how much responsibility the field artillery places on its junior officers. It is genuinely a position of trust. Further, the fire support available to the FSO is not limited just to artillery and mortars. When the tactical situation warrants, the company FSO may employ naval gunfire and joint close air support as well.

The fire direction center (FDC) is the nerve center of the field artillery. The fire direction officer (FDO) and his team translate the FIST's calls for fire into firing data for the guns by digital means using a computer network consisting of the FIST digital message device (DMD), the battalion FDC's automated fire direction computer system, the battery FDC's computer system (BCS) or backup computer system (BUCS), and a gun display unit (GDU) on each firing howitzer. Using this automated network, the FDC can now "place steel on target" seconds after the FIST requests fire.

The target acquisition element of the field artillery team is another vital link in the fire support system. While the FIST acquires targets visible to frontline troops, the target acquisition assets of the artillery locate more distant targets not visible to forward observers. This task is accomplished by using highly sophisticated and effective target-locating radar systems.

After completing the Field Artillery Officer Basic Course, where a field artilleryman learns fundamental officer skills, a field artillery lieutenant can expect assignment to a variety of duties, usually at battery or battalion level. Assignments include platoon leader, company fire support officer, battery fire direction officer, or battalion staff officer in cannon, rocket, and missile units. Officers assigned to missile and target acquisition units perform different but comparable junior officer leadership duties. All officers receive effective on-the-job training and valuable troop-leading experience.

After three or four years of troop experience, officers attend the Field Artillery Officer Career Course as part of their professional development. Successful completion of this course marks the officer as ready for command. Following the career course, the officer can expect to command an artillery battery and fill key staff assignments that require integrating field artillery knowledge and experience with the overall scheme of combined arms operations. Typical assignments for captains include battery commander, battalion or brigade fire support officer, or staff officer at brigade or division artillery.

As officers gain more experience and years of service, they become competitive for advanced schooling and assignments of increased responsibility, including command at battalion and brigade levels.

The dynamic nature of field artillery requires competent, well-trained leadership at all levels to maintain the established standard of excellence and the proud "Redleg" tradition. As a field artillery officer, you can expect progressive education, training, and duty assignments to prepare you for higher rank and greater responsibility. The field artillery will continue to reign as the King of Battle. "The future belongs to the field artillery." For the most current information on the Field Artillery, visit sill-www.army.mil/usafas.

INFANTRY

The infantry is the oldest of the combat arms. The infantry branch is the basic ground-gaining and retaining arm of the Army. Its mission is "to close with the enemy by fire and maneuver to destroy or capture him, and to repel his assault by fire, close combat, and counterattack."

Today's infantryman can move by land, sea, or air. The modern infantryman may fight on foot or go into action by parachute, helicopter, assault boat, or the Bradley Fighting Vehicle. The infantry can operate at night, under any climatic conditions, and can overcome natural and manmade obstacles that would stop other forces.

Infantry officers can serve in a wide variety of units. Light infantry units are rapidly deployable and especially useful in missions conducted in complex and urban terrain, due to their streamlined nature and lack of dependence on masses of heavy equipment for transportation. In the active component today, light infantry units are the closest to the traditional "grunts" of earlier days. Air assault infantry units are more heavily equipped and, although fully capable of dismounted operations, typically incorporate the large numbers of helicopters available in the air assault division's aviation groups into their tactical plans. Airborne infantry are uniquely qualified and equipped for arrival in combat by parachute but are fully capable of dismounted or air assault operations as well. More heavily equipped than air assault infantry, they are nevertheless highly strategically mobile due to their ability to be delivered from aircraft while in flight. Mechanized infantry are the most heavily armed and protected infantry, riding into battle in Bradley Infantry Fighting Vehicles, which are armed with 25mm automatic cannon, antitank missiles, and machine guns. Although also trained for dismounted operations, the great strength of mechanized infantry is the mobility afforded by the Bradleys and the Stryker combat vehicle, as well as their awe-

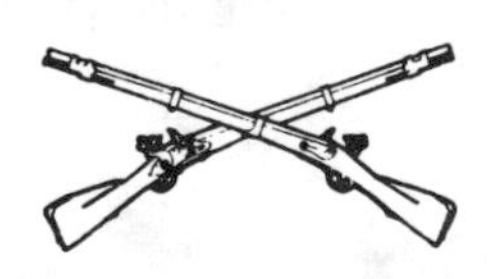

In 1875, a design of crossed muskets was authorized for wear by infantry officers. Several subsequent changes reflecting newer models ended in 1924 when the present design, based on the first type of musket used by U.S. Army troops, was approved.

some firepower. Ranger infantry are trained to go into battle by parachute, special operations helicopter, small boat, or dismounted. Ranger reconnaissance units are also trained in scuba and military free-fall parachuting as additional means of reaching the battlefield. Although trained as infantry, ranger infantry units are increasingly being used as special operations forces, with a variety of associated missions. Officers are accepted into ranger infantry only after graduating from the Army ranger and airborne schools, establishing an outstanding record in other infantry units, and passing a demanding assessment and selection program conducted by the 75th Ranger Regiment. That program must be passed before each subsequent ranger infantry assignment as well.

Although several National Guard infantry units can trace their lineage to colonial militia organizations from the seventeenth century, the oldest infantry unit in the active component is the 3rd Infantry, or Old Guard, which was authorized by Congress in June 1781. Hundreds of infantry units have been authorized since then, and many reorganizations have been made, but the purpose of the infantry remains the same.

Perhaps two key words in describing an infantry officer's role are *leadership* and *adaptability.* The satisfaction and personal development the officer experiences from the leadership of soldiers and the management of challenging staff assignments prove invaluable both within the military and in later civilian pursuits. Normally, all infantry officers have the opportunity to lead a platoon during their initial tour of duty.

Leadership in the infantry is an especially visceral process. Motivating infantry soldiers to accomplish their missions—whether they are conducting the night live-fire assaults in training, patrolling a zone of separation as part of a peacekeeping mission, or defending a hill against an enemy attack in the decisive operation of war—requires supremely *human* as well as tactical and technical skills. Infantry officers can expect to develop an especially close bond with their men and to be challenged to the limits of their abilities to lead them with the excellence they deserve and demand. In return, infantry officers are rewarded with the sure knowledge that they have done the hardest job in the Army and that they have earned the respect of the world's toughest audience—the American infantryman.

Duty as an infantry platoon leader or company commander is an opportunity to lead and practice leadership in a demanding, complex job. It is the management of priorities among collective as well as individual training and provides the young infantry lieutenant the opportunity to make tough decisions, accept responsibilities, hone leadership skills, and develop proficiency in the most demanding tactical and technical techniques associated with the infantry branch in both combat and peacetime roles. During the first two years of service, infantry lieutenants can expect to experience at least one staff assignment. Generally these staff assignments are at the troop level in positions such as assistant operating officer or motor officer in battalion- or brigade-size units.

A second major area in which infantry officers can expect to serve in staff roles is the Army's special commands, where training, administration, and management are the principal functions of the organization. For example, infantry lieutenants serve at Fort Benning at the training center. Duty positions include company training officer and instructor. Newly commissioned infantry officers receive training in basic military techniques to enable them to serve competently and confidently in any initial assignments they might face. Infantry officers must first receive a solid background at the small-unit level. Official policy holds that it is desirable for young infantry officers to serve in both light infantry and mechanized infantry assignments early to develop broad-based tactical knowledge for later field-grade development. Later in their service, they may be eligible for additional schooling opportunities, both military and civilian.

The first assignment for newly commissioned infantry officers is attendance at the Basic Officer Leader Course (IBOLC) at Fort Benning, Georgia. The course is nineteen weeks in duration and is composed of approximately one-third classroom work and two-thirds field practical exercises. It is designed to prepare each newly commissioned infantry officer to train and lead an infantry platoon and to appreciate fully his or her role in the organization and functioning of the infantry rifle company. The officers study tactics, leadership, management, and administration. The course of instruction, presented in a learning environment enhanced by a thoroughly professional and experienced faculty, is constantly revised to present only the most current concepts and techniques in an interesting as well as informative manner.

To enhance an officer's background between major schooling periods, some temporary duty (TDY) courses of instruction are offered. Infantry lieutenants of all components should volunteer for Ranger School to achieve the professional development goal of successful service in all types of infantry units. Two-thirds of all infantry platoon leader positions are coded for ranger-qualified lieutenants, and infantry branch policy is to provide ranger training for all IOBC graduates who volunteer and meet the prerequisites. Participation in ranger training develops in-depth unit combat skills and stresses night tactical operations and leader skills and endurance in various geographical settings. This training provides the Army with tactically competent, aggressive, self-disciplined, and confident officers who are prepared to train and lead infantry units in combat.

The Airborne Course is a three-week program designed to qualify volunteers in military parachuting. The course is divided into three phases (ground week, tower week, and jump week), during which an individual progresses from physical conditioning and mastery of parachute landing falls to control of an opened parachute and proper exiting from an aircraft. The Bradley Commander's Course is designed to train leaders on the employment and operation of the most advanced infantry fighting vehicle in the world.

Airborne and Ranger Schools are also important for infantry officers because, in the eyes of many fellow infantrymen, they are rites of passage. Although all infantry officers should be evaluated based on their demonstrated performance of their assigned duties, remember that they could be required to go into combat in their *first week* in an infantry unit. Although it is true that the U.S. Army has won most of its wars without *any* airborne rangers, it is also true that very few of those wars were fought on the no-notice, "come as you are" basis of recent and future conflicts. Ranger and airborne training provides additional self-confidence and, in the case of Ranger School, hones some of the battlefield skills that might otherwise be developed only over long experience. Further, it is indisputable that the Americans who joined the Army, fought as infantry, and won our major wars came from a significantly different society; without mincing words, they were generally much physically and emotionally tougher than most Americans are today. Infantry combat is still a dirty, exhausting, terrifying affair, requiring some very tough men to prevail in its practice. The ranger tab and jump wings do not prove that you are a highly competent super trooper, nor do they mean that you are better than those who do not have them; they are simply mute evidence that you have demonstrated the courage, tenacity, and professional commitment to attend and complete rigorous and demanding training designed to make you the best combat leader possible. Completing the Ranger and Airborne Courses will make an implicit statement about your abilities and attitude that will inspire confidence in your subordinates, peers, and superiors alike. If you are an infantryman, go to Ranger School and jump school, and do your damnedest. For the most current information on the infantry, visit www.benning.army.mil/infantry/infantry.htm

SPECIAL FORCES

Special Forces (SF) branch was authorized by the Secretary of the Army on 9 April 1987 and made official by General Order 35 on 19 June 1987. The decision to make SF a branch was based on an analysis of the current and future threat, the integration of SF into the Army's warfighting doctrine and force structure, and the various SF missions. Creation of the branch also solved a number of long-standing personnel management and professional development inadequacies.

Special forces missions—special reconnaissance, direct action, foreign internal defense, unconventional warfare, and counterterrorism—are an important part of the Army's doctrine. Special Forces Mobile Training Teams (MTTs) travel around the world to teach soldier skills to the military and paramilitary forces of friendly foreign countries. In fact, more than one-third of the security assistance teams sent from the United States to other countries come from SF units, which make up only one percent of the Army. Special forces assignments are world-

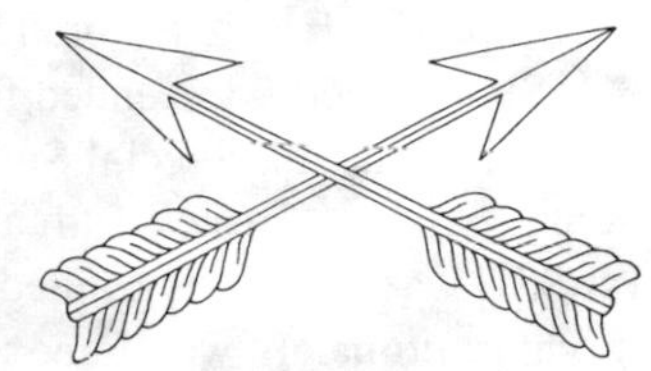

Approved in 1987, the crossed arrows of the Special Forces insignia were worn during the late 1800s by American Indian scouts and during World War II by officers of the First Special Service Force.

wide, and expansion of SF has created a number of new assignments, offering more opportunity for responsibility and especially meaningful service.

On 16 April 1987, the Department of Defense activated the U.S. Special Operations Command (USSOCOM) as a unified command reporting to the Joint Chiefs of Staff. USSOCOM is headquartered at McDill Air Force Base, Florida. It has operational command of all Army, Navy, and Air Force special operations forces. The U.S. Army Special Operations Command (USASOC) was created in December 1989 at Fort Bragg, North Carolina, to enhance the readiness of Army special operations forces. USASOC is the Army component USSOCOM.

USASOC has three major subordinate commands: the U.S. Army Special Forces Command (Airborne) (USASFC(A)), the U.S. Army Civil Affairs and Psychological Operations Command (Airborne) (USACAPOC(A)), and the U.S. Army John F. Kennedy Special Warfare Center and School. In addition, two major subordinate units—the 75th Ranger Regiment, headquartered at Fort Benning, Georgia, and the 160th Special Operations Aviation Regiment (Airborne) at Fort Campbell, Kentucky—report directly to USASOC. USASFC(A) currently has five active-duty groups: the 1st SFG(A) headquartered at Fort Lewis, Washington, with a battalion in Okinawa; the 3rd SFG(A) headquartered at Fort Bragg, North Carolina; the 5th SFG(A) at Fort Campbell, Kentucky; the 7th SFG(A) headquartered at Elgin AFB, Florida; and the 10th SFG(A) headquartered at Fort Carson, Colorado, with a battalion in Germany. The special forces command also provides oversight of the organization and training of two groups in the Army National Guard—the 19th SFG(A) headquartered at Draper, Utah, and the 20th SFG(A) headquartered at Birmingham, Alabama.

The 4th Psychological Operations Group (Airborne) and the 95th Civil Affairs BDE (Airborne) are colocated at Fort Bragg with USACAPOC(A) and constitute the Army's active-duty resources in these disciplines. The bulk of the Army's Civil Affairs (CA) and Psychological Operations (PSYOP) capability is found in the Army Reserve. Reporting to USACAPOC(A) from the Army Reserve are the 350th, 351st, 352nd, and 353rd Civil Affairs Commands, the 358th Civil Affairs Brigade, the 151st Theater Information Operations Group, and the 2nd and 7th PSYOP Groups.

The U.S. Army John F. Kennedy Special Warfare Center and School, originally activated in 1952 as the U.S. Army Psychological Warfare Center, conducts training for SF, Psychological Operations, and Civil Affairs personnel, as well as survival, evasion, resistance, and escape (SERE) training and other courses for specialized operations. The center is also responsible for SF doctrine and research, the development of new SF equipment and force modernization, and, as the proponent for the SF branch, the direction of personnel policy development.

SF is directly descended from the World War II Office of Strategic Services (OSS) and the 1st Special Service Force (1st SSF). Though separate organizations, each was formed with a mission to conduct unconventional warfare. The 1st SSF was originally trained for sabotage missions in German-occupied Norway but ultimately won fame in the Aleutian Islands, Italy, and southern France.

OSS Operational Groups (OGs) and Jedburgh teams parachuted behind enemy lines to organize resistance fighters—Detachment 101 in Burma, and the OGs and Jedburghs in France and Italy. Late in the war, the 1st SSF was deactivated; shortly after the war, the OSS was also dissolved. With the exception of limited employment of partisans and intelligence-gathering operations during the Korean War by the United Nations Partisan Force–Korea (UNPFK), unconventional warfare was quiescent after 1945.

It wasn't until 20 June 1952 that a new military unit, the 10th Special Forces Group, was formed at Fort Bragg to meet the need for a force able to wage guerrilla warfare in the event of a Russian invasion of western Europe. The new unit was commanded by Col. Aaron Bank, a former OSS Jedburgh team member. Bank's new group included many former OSS, 1st SSF, ranger, and airborne veterans, trained in jungle and underwater operations, demolitions, airborne techniques, and foreign weapons.

The new special forces group grew quickly. In November 1953, the group split. Half deployed to Flint Kaserne in Bad Tölz, Germany, retaining the name of the 10th Group. The half remaining at Fort Bragg became the 77th Special Forces Group. In June 1957, a cadre from the 77th moved to Okinawa to form the 1st SFG. In June 1960, the 1st Special Forces was formed under the Combat Arms Regimental System to be the parent regiment of all Special Forces Groups. The 77th Group was then renamed the 7th SFG.

By 1957, SF soldiers from the 1st and 7th Groups were serving temporary duty as advisers in South Vietnam, where government troops were fighting Communist guerrillas. In September 1961, the 5th Special Forces Group was formed, orienting on the Republic of Vietnam.

On 12 October 1961, special forces met one of its staunchest supporters—President John F. Kennedy. After reviewing the unit, President Kennedy authorized the green beret as a "symbol of courage, a badge of distinction"; it is the only uniform headgear in the Army with presidential authorization.

Soon, the new 5th SFG was sending increasing numbers of MTTs to South Vietnam. The first elements of the group moved to Vietnam in 1962, and the deployment was completed by October 1964, when the group headquarters moved into its permanent base at Nha Trang. Special forces missions in Vietnam included training in counterinsurgency and strike operations and conducting civic action programs. They also trained various mountain tribes, collectively called Montagnards, to fight Viet Cong and North Vietnamese forces in the central highlands.

During the same period, SF MTTs traveled to various Latin American countries, Laos, and Liberia. To meet the demands of the growing mission, SF expanded by forming the 3rd, 6th, and 8th SF Groups.

Toward the close of the Vietnam War, troop reductions began. The 3rd SFG was deactivated in December 1960, the 6th in March 1971, the 8th in June 1972, and the 1st in June 1974. With cutbacks of almost 70 percent, even the strength of remaining SF units was considerably reduced.

Increasing awareness of the need for special forces in the early 1980s resulted in a revitalization and expansion of the branch. The year 1983 saw three major milestones in the development of special forces: special forces NCOs received the separate Career Management Field (CMF) of 18; the Special Forces Warrant Officer program, military occupational specialty (MOS) 180A, was created; and special forces was recognized for the first time as a career field for commissioned officers with the creation of a secondary specialty, specialty code (SC) 18 (which became functional area 18 in 1984). Also in 1984, the 1st SFG was reactivated.

The commander of the Special Forces Operational Detachment "A" (SFODA) is the leader who must direct and employ the other SF experts. He is trained in the principles, strategies, and tactics of all SF missions as well as other special operations such as psychological operations and civil affairs. He must meet the same standards as the other members of his team and be thoroughly familiar with their skills.

Special forces is a nonaccession branch, open to any officer in the Officer Personnel Management Directorate (OPMD)–managed branches. Officers are eligible to request branch transfer to special forces during the same year as their consideration for promotion to captain, normally the third to fourth year of active commissioned service. Officers who meet the basic requirements contained in chapter 18, DA Pamphlet 600-3, are then screened by an annual board and nominated to attend the Special Forces Assessment and Selection (SFAS) program. Conducted at Camp Mackall, North Carolina, the three-week SFAS is designed to select officers and NCOs who demonstrate the greatest potential for success in the demanding Special Forces Qualification Course (SFQC) and Special Forces Detachment Officer Qualification Course (SFDOQC). Officers normally attend SFAS in a temporary duty status while assigned to their home unit, while en route to their branch Officer Advanced Course (OAC), or after OAC.

Once an officer is both an SFAS selectee and an OAC graduate, he is programmed for attendance at the twenty-six-week SFDOQC, which includes a permanent-change-of-station move to Fort Bragg, North Carolina. The training is divided into two phases. The MOS phase is an eighteen-week period covering common skills training, including land navigation and survival, evasion, resistance, and escape (SERE). Officers receive training in each SF specialty (weapons, engineering, medical, and communications), underground and resistance movements, guerrilla warfare, and counterinsurgency operations. In preparation for the field phase, the MOS phase has three isolation practical exercises. Training is conducted at Fort Bragg and nearby Camp Mackall. The field phase is the final qualification period of training. Training is six weeks long and integrates and tests both common skills and specialty training. Students are organized into operational detachments to execute all their previous training, including the organization of simulated guerrilla forces. Students are inserted by static-line parachute into North Carolina's Uwharrie National Forest for a Special Operations warfare exercise known as "Robin Sage."

Beyond their basic qualifications, SF officers may need additional training either before or after assignment to an operational SF group. This training may include language training for a particular geographic area; training for operations in arctic, jungle, desert, or mountain environments; and advanced skills, such as military free-fall parachuting and underwater operations.

Special forces captains are branch qualified for promotion once they complete a branch OAC, the SFDOQC, and eighteen months (plus or minus six months) of successful SFODA command.

Promotion opportunities are comparable to those of other combat arms branches, according to the Army's OPMD. Command opportunities for SF captains and majors are the best in the Army, and command opportunities for lieutenant colonels are comparable to those of infantry lieutenant colonels.

6

Combat Support Branches

The combat support branches are focused on providing force enhancement and operational support to combat arms through expertise in their specific assigned specialties.

CHEMICAL CORPS

The Chemical branch is a combat support branch aligned under the Maneuver, Fires and Effects (MFE) functional category and is focused primarily on war fighting operations and training that supports all aspects of Combating Weapons of Mass Destruction (WMD): nonproliferation, counter proliferation, and consequence management. The Chemical Corps is focused on operations and training in support of chemical, biological, radiological and nuclear (CBRN) defense; sensitive site exploitation and assessment; multi-spectral obscuration and flame employment; CBRN vulnerability assessment; biological and chemical arms control verification; smoke and flame munitions technology and management; chemical weapons storage and demilitarization; WMD force protection programs; CBRN foreign and domestic consequence management; and CBRN military support to civil authorities. The branch provides the Army with a highly trained corps of CBRN experts in both the Active and Reserve components that advise commanders and staffs at all levels in the Department of Defense. The Chemical Corps has a wide variety of scientific research, development, and material management functions relating to CBRN systems and combat development. These involve management of CBRN defensive equipment from their conception through development and employment. All CBRN officers will have the opportunity to serve in these challenging activities.

The Chemical Corps insignia consists of crossed retorts and a benzene ring; first adopted in 1917, the insignia alludes to the chemistry-related functions of the corps.

To be effective, CBRN officers must know how their expertise can best support the Army, joint, interagency, intergovernmental, multinational (JIIM), and combined arms operations. For this reason CBRN lieutenants begin their service as the assistant S-3/CBRN officer in a combat arms battalion or as a platoon leader in a chemical unit. These positions involve direct interface with combat units and participation in combined arms operations. Lieutenants can also be platoon leaders for Stryker or FOX recon elements in Brigade Combat Teams; Biological Integrated Detection System platoons; Reconnaissance platoons (Stryker or Fox); HAZMAT capable Recon platoons; or Smoke (Mechanized or Wheel) platoons. CBRN lieutenants in these jobs gain the tactical experience that will prepare them for higher-level command and staff positions or service school instructor jobs. The Chemical Corps service progression in both the Active and Reserve components continues to offer diverse command opportunity at company through brigade level and positions in major operationally oriented staffs.

Many CBRN officers will have the opportunity to serve in one of the unique duty positions listed below:

(a) Chemical Reconnaissance Detachment Cdr. (SF Group) (CPT).
(b) National Guard Civil Support Teams (WMD-CST) (LT-LTC).
(c) Instructor (USMA Chemistry/Life Science Dept.) (CPT-MAJ).
(d) Technical Escort Battalion (LT-LTC).
(e) Defense Threat Reduction Agency (DTRA) (MAJ-COL).
(f) U.S. Nuclear and Chemical Agency (USANCA) (LTC-COL).
(g) 20th Support Command (CBRNE) (CPT-COL).
(h) 75th Ranger Regiment (LT and MAJ).

Modern chemical warfare began with World War I. The first large-scale attack was made by the Germans on 22 April 1915 at Ypres, Belgium. Chlorine gas was used in that attack that opened up a five-mile break in the French lines and caused thousands of casualties; later in the war phosgene and mustard gases were also used. Shortly after the United States entered the war, the Gas Service was created in France to coordinate all uses of gas by the American Expeditionary Force. This service also had the major responsibility of defense against chemical warfare.

On 18 June 1918 the War Department created the Chemical Warfare Service as the agency responsible for all matters related to chemical warfare. This was because Gen. John J. Pershing believed that a future war might involve chemical agents and the U.S. Army had to be ready. In 1920 this service became a permanent part of the Army, which recognized the enduring threat from the chemical weapons that had caused one-fourth of our casualties during the Great War.

World War II did not see a repeat of the use of chemical weapons, largely because Germany and Japan realized that the training of the Allies made us much less susceptible to chemical attack and that the certainty of retaliation in kind would result in greater casualties than they were willing to accept. The Allies were not, however, aware of the development of nerve gas by the Germans, a weapon that, if used, could have influenced the war in Europe in a dramatic way. Although chemical weapons were not used to a significant extent during the war, the Chemical Corps played a critical role throughout the war in all theaters. Smoke and flame missions contributed greatly to the success of many combat operations. The first waves at Omaha Beach included small units of a chemical decontamination company that were there specifically to monitor the beach for nuclear and/or chemical contamination. Chemical mortar battalions became a valuable asset; their 4.2-inch mortars delivered quick, accurate, and devastating fire.

The Chemical Corps continued to provide critical support to the Army in Korea with its flame, smoke, and mortars, and in Vietnam with its "tunnel rats," "people sniffers," flame, and smoke. By the late 1960s and early 1970s, however, the political climate and declining interest in CBRN warfare led to a move to abolish the Chemical Corps. Then the Yom Kippur War of 1973 provided dramatic evidence that chemical warfare was still a real threat. Soviet equipment captured by the Israelis revealed a startling secret. This captured equipment clearly demonstrated a significant capability for offensive and defensive chemical operations. Clearly, the Soviets did not intend to forgo the use of a weapon that could significantly affect the outcome of battle. As a result of this discovery, the Army reversed itself and, beginning in 1975, embarked upon a major expansion of the Chemical Corps that is still in progress today.

The Chemical Corps played a vital role during Operations Desert Shield and Desert Storm, when we faced an enemy with a significant chemical and biological capability. Through a combination of aggressive offensive actions and outstanding CBRN defense readiness, Iraq was deterred from using weapons of mass destruction.

In the twenty-first century the threat of WMD warfare looms even larger. Dozens of countries are known, or suspected, to have WMD capabilities. Countries with ties to international terrorism are developing WMD warfare capabilities and have actually used them with considerable success. To meet this continuing threat, the Chemical Corps units have the capability to provide each Brigade Combat Team (BCT), division, and corps with CBRN reconnaissance, decontamination, and large-area smoke and obscuration. Moreover, there are CBRN Defense Staff Organizations to ensure that commanders have access to expert CBRN advice. This infrastructure consists of Chemical Corps soldiers in every combat unit from the company to Army level. To meet the threat of WMD used against the United States, the National Guard's full-time Civil Support Teams (CST), Weapons of Mass Destruction were created in 1998 for immediate response to a WMD incident or disaster anywhere in the United States. The CSTs represent both the DoD's and the Chemical Corps' first response to WMD and have been employed to respond to hazards such as anthrax, chemical hazards, radiological isotopes, and terrorist events such as the attacks on the World Trade Center. Additionally the teams have responded to the 2003 Columbia Space Shuttle disaster and the 2005 hurricanes that affected the Gulf states. Chemical officers from the ranks of lieutenant to lieutenant colonel now serve in the fifty-seven CSTs across our nation and territories.

In recent combat operations, CBRN soldiers deployed with every unit going to Afghanistan or Iraq. While Chemical units deployed to theater did not perform their traditional CBRN defense role, they demonstrated the Chemical Corps' versatility and relevancy to the Army by performing a multitude of force protection missions including site and convoy security. Commanders in the field uniformly offer praise for their CBRN soldiers and units for their contributions to their mission accomplishment.

The threat of WMDs being used against the United States is greater than at any time in recent history. With the increasing proliferation of WMDs and delivery system technology, this threat continues to grow as the Global War on Terrorism continues. The spread of this technology around the globe creates new challenges for the Chemical Corps. The corps standard is *"Chemical Corps—Capable Now."*

As the Chemical Corps evolves during the twenty-first century, it continues its role as steward of the Army's CBRN defense readiness. The Chemical Corps is the only service in the DoD

- With a dedicated CBRN force structure.
- That provides large-scale decontamination capability to include fixed site, terrain, and personnel.
- That includes a dual mission Corps which supports contingency operations as well as Homeland Security missions.
- Which trains and commissions full-time officers with a primary CBRN mission in all components of the Army, both Active and Reserve.
- Which provides expertise and capabilities in CBRN for Civil Affairs.

The Chemical Corps will continue to offer great challenges and opportunities. Given the unique skills and capabilities of the Chemical Corps, CBRN officers will be involved in supporting peacetime requirements at both the national and international levels. Organized since 1986 under the Army Regimental System, the Chemical Corps offers an exciting and rewarding career. Chemical Corps officers share a proud tradition that traces its roots to the First Gas Regiment of World War I and provides a firm foundation for a bright future.

The Chemical Corps home and school is located at Fort Leonard Wood, Missouri, as part of the Maneuver Support Center with the Engineer and Military Police schools. For more information about the Chemical Corps and school, see https://home.army.mil/wood.

CIVIL AFFAIRS

Never before in the history of the U.S. Army have Civil Affairs (CA) units or personnel been as crucial to the success of American military operations or as widely employed as they are today. Civil affairs officers today face the widest imaginable array of challenges and are among the most likely to be deployed to "real world" situations.

The Department of the Army approved the design for the Civil Affairs insignia on 1 June 1956. The globe indicates the worldwide areas of Civil Affairs operations. The torch, which is from the Statue of Liberty, a symbol associated with the spirit of the United States, also represents enlightened performance of duty. The scroll and sword depict the civil and military aspects of the organization's mission.

The Civil Affairs/Military Government branch was established as an Army Reserve branch on 17 August 1955. Modern-era Army civil affairs operations began with the allied military government of parts of the German Rhineland following World War I. Greater challenges faced the Army's Civil Affairs Division (formed in late 1943) in post–World War II Europe, where 80 million people, including the citizens of former enemy nations, had to be cared for and governed effectively while their society was being reformed. Simultaneously, millions of refugees—displaced persons who could not or did not want to go back to the lands of their origins—had to be resettled or repatriated.

Redesignated as the Civil Affairs branch on 2 October 1959, Army Civil Affairs personnel have continued to provide guidance to commanders in a broad spectrum of activities ranging from host nation relations to the assumption of executive, legislative, and judicial duties in occupied or liberated areas. The Vietnam War demanded a great deal from Army CA personnel, who operated programs to win the "hearts and minds" of the Vietnamese. Often acting in conjunction with special forces and psychological operations units, CA personnel organized and supported medical and veterinary assistance programs in remote or underdeveloped regions.

Following Vietnam, CA personnel played key roles in replacing the communist government in Grenada after Operation Urgent Fury by building a working governmental infrastructure. After Operation Just Cause, they were instrumental in establishing a demilitarized police force and democratic government in Panama. Throughout the 1980s, as the United States attempted to contain, and ultimately roll back, communism in El Salvador and Honduras, CA personnel provided key links to regional governments and operated humanitarian and civil affairs programs designed to counter the appeal of communist propaganda.

Beginning with Operations Desert Shield and Desert Storm, and continuing throughout the 1990s, CA personnel participated in a growing number of stability and support operations at home and abroad that have become a major activity of the Army in the post–Cold War era.

The Civil Affairs' mission is to support the commander's relationship with the civil authorities and civilian populace, promote mission legitimacy, and enhance military effectiveness. The CA branch is a nonaccession combat support branch in the U.S. Army Reserve (USAR). The active duty counterpart for officers is Civil Affairs (FA 38), one of three functional areas of the Operations career field.

The only area of concentration (AOC) within the CA branch is the Civil Affairs Officer, General (38A). CA officers command or serve in CA units or in S-5/G-5 and joint staff positions, as well as in other positions requiring general military expertise and knowledge to work with other special, general, and joint staffs; the ability to plan, direct, and participate in the conduct of both civil-military operations (CMOs) and support to civil administrations; the ability to provide the interface between the U.S. military, foreign governments, civilian relief agencies, and other U.S. governmental agencies; diplomacy and skills to advise and interact with senior officials of foreign governments; the ability to analyze economic, social, cultural, psychological, or political aspects of an area; the ability to conduct coordination or liaison with foreign civil or military personnel; the ability to prepare economic, cultural, governmental, special functional studies, assessments, and estimates of a regional area; the ability to coordinate with civil authorities and enhance, develop, establish, or control civil infrastructures in operational areas in support of friendly military operations; the knowledge to provide advice and assistance to civil, paramilitary, and military leaders of U.S. and foreign nations involving CMO matters; and the knowledge and ability to conduct cross-cultural communications to facilitate interaction with foreign governmental officials, soldiers, and civilians.

The breadth and scope of these requirements often require advanced civil schooling and foreign language training, and offer opportunities for expanding officers' professional horizons.

Civil affairs units and personnel can also execute and support missions such as foreign nation support; support and resources identification; populace and resources control; dislocated civilian resettlement; ration control; curfew and travel restrictions; licensing; humanitarian assistance; foreign disaster relief; noncombatant evacuation operations (NEO); foreign displaced civilian support; and military civic action.

Since most CA forces are in the reserve component, these soldiers bring to the Army skills they practice in the civilian sector. It is not unusual to find judges, physicians, bankers, educators, health inspectors, fire chiefs, and others with public safety, public administration, public health, legal affairs, commercial and financial management, public finance, education, public works and utilities, mass media communications, transportation management and operations, food and agricultural services, and cultural affairs civilian skills in CA units.

Most civil affairs units provide support to conventional units by providing teams to augment staff at battalion level or higher, or to staff civil military operations centers (CMOCs). Active or reserve CA officers not assigned to civil affairs units can expect to be assigned as principal or assistant S-5s, G-5s, or to CMOC/civil affairs offices on joint staffs at all levels. Culturally trained, linguistically capable CA soldiers may also provide functional expertise for foreign internal defense operations and unconventional warfare operations in support of special operations.

The Army Reserve includes three CA Foreign Internal Defense/ Unconventional Warfare (FID/UW) battalions that assist special forces with medical and engineer support resources during selected missions.

Reserve component CA officers are among the most often activated and deployed, both individually and as part of CA units. Active component CA officers similarly find themselves frequently deployed on many important and interesting deployments. The operational environment of today and of the foreseeable future guarantees that CA officers will continue to be needed.

CYBER

With the development of computer networks in the 1950s and the first permanent ARPANET (original name for Internet) links in the late 1960s, the DoD began sharing and storing data electronically. It did not take long for military officials to advocate for caution with the realization that information stored on these networks could be targeted by insiders and spies. While the military made efforts to secure its networks, DoD officials realized as more nations became connected to these global networks, an opportunity arose to use this to our advantage. By 1979, the National Security Agency (NSA) recognized that any computer system connected to a network could be penetrated by a knowledgeable user. In the 1980s, the DoD engaged in early forms of computer network exploitation against Cold War adversaries while these same adversaries attempted to do the same to the U.S. as evidenced by the 1986 discovery of the Soviet KGB paying West German hackers to steal U.S. military secrets. A 1990 article described how the Army desired to hire contracted hackers to develop computer viruses to unleash on adversaries.

As the 1990s arrived with the advent of the World Wide Web (WWW) portion of the Internet, global use of this worldwide network exploded in both the public and private sectors. The intelligence community's exploration of offensive cyberspace operations (OCO) emerged in response to the growing importance of digital communications to Signals Intelligence (SIGINT) collection. The Army's Military Intelligence (MI) branch began to view the collection of data via the penetration of adversary networks as another form of SIGINT. In 1995, the 704th MI BDE at Fort Meade—the Army's SIGINT brigade—was tasked to explore the development of a computer network operations force. Over the next several years, this unit evolved several times, becoming the Army Network Warfare Battalion in 2008, and redesignated the 744th MI BN the following year to conduct tool development, expeditionary, and remote cyberspace operations. The 744th helped develop several capabilities and programs that directly supported joint war fighters in their efforts to conduct kill/capture operations during the Global War on Terror.

While the MI branch became involved in OCO, the Signal Corps, based on the branch's history of applying new technologies, began using computers upon their arrival. Since the mid-1940s, the Army had partnered with private industry to build computers as more people and organizations began using computers for communication. The Signal Corps, as the Army's "Communicators," viewed these digital communications as information in need of secure processing. The growing complexity of security threats to the Army's portion of the WWW, compelled the Army to centralize global command and control under one Army command—U.S. Army Network Enterprise Technology Command (NETCOM)—which activated in 2002. NETCOM forces engineer, operate, sustain, and defend the Army's portion of the DoD's Global Information Grid, otherwise known as LandWarNet. With most Army communications taking place on e-mail, NETCOM was tasked to ensure its security and integrity. All these efforts by the Signal Corps placed it predominantly on the side of defensive cyberspace operations (DCO).

Following the establishment of United States Cyber Command (USCYBERCOM) in 2009, the Secretary of Defense directed all the services to create their own subordinate cyber units to USCYBERCOM. For the Army, this meant the formation of Army Cyber Command (ARCYBER) in 2010 at Fort Belvoir, Virginia (later moving to Fort Gordon, Georgia, in 2020). After the creation of ARCYBER, the aforementioned 744th MI BN transitioned into the 780th MI BDE in 2011 and continued to support Army OCO while the Signal Corps continued to support DCO.

However, by 2012 it became clear to Army senior leadership that the existing split branch solution of MI and Signal was inadequate. The main problem arose from the arrangement of OCO and DCO personnel being assigned outside the normal career paths of their respective branches, where they had to think in a maneuver mindset, not to mention receive extensive and expensive cyber training, but then go back to the Signal or MI corps at large.

On 20 February 2013, U.S. Army Training and Doctrine Command (TRADOC) Commander, Gen. Robert Cone, gave an important briefing at an Association of the U.S. Army Symposium where he called for the formal creation of a Cyber School and career field within the U.S. Army. He stated the Army needed to "start developing career paths for cyber warriors as we move to the future." Army Chief of Staff, Gen. Raymond Odierno, quickly approved the establishment of a consolidated Army Cyber School at Fort Gordon, Georgia. The purpose of the Cyber School would be to unify and integrate training, including a new cyber career field for officers, warrant officers, and enlisted Soldiers.

TRADOC announced in January 2014 that the Signal Center of Excellence at Fort Gordon would thereafter be known as the Cyber Center of Excellence (CCOE) with both the Signal and Cyber Schools operating under this new two-star command. During this period, the Army also directed the Cyber School to oversee Electromagnetic Warfare (EW) training. The Cyber Protection Brigade also activated in 2014 at Fort Gordon to serve as the Army's primary DCO unit.

On 4 August 2014, the U.S. Army Cyber School headquarters was unveiled by CCOE commander Maj. Gen. LaWarren Patterson and Col. Jennifer Buckner, the first dual-hatted Chief of the Cyber Branch and Cyber School Commandant. The Army officially established the Cyber Branch on 1 September 2014, pursuant to the authority of Section 3063(a)(13), Title 10, U.S. Code. This was the first new Army branch and career field for all cohorts since the creation of Army Special Forces in 1987.

The Army began accepting applicants to the new 17-series career fields for officers (17A), warrant officers (170A), and enlisted soldiers (17C). The duties, functions, and positions to be associated with Army cyber operations were carved out roles within the Signal and MI Corps. As one leader noted, they were "recoloring these roles from MI Blue and Signal Orange to Cyber Gray." The Army chose the 17-series MOS for cyberspace personnel to reflect the branch's designation as a maneuver space. After a year of hard work mapping the courses and building the branch, the first U.S. Army Cyber Basic Officer Leader Course (BOLC) began in August 2015.

A BRIEF BACKGROUND ON HOW ARMY ELECTROMAGNETIC WARFARE BECAME PART OF ARMY CYBER

During the Cold War, Electromagnetic Warfare (EW) units and their highly classified capabilities were segregated from mainstream conventional Army forces. Over time, this separation caused EW to evolve into a highly specialized and much underappreciated sub-component of the U.S. Army Military Intelligence community.

After the collapse of the Soviet Union in 1991, the Army slowly divested itself of EW. The Army used EW in Operation Desert Storm in 1991 within aircraft and land operations, but this was the last main utilization for some time and the Army's EW capability slowly atrophied and became ineffective.

In the years leading up to the terrorist attacks on 11 September 2001, the Army conducted very little EW within its forces. After 9/11 and the deployment of U.S. forces supporting Operation Enduring Freedom (OEF) in Afghanistan and Operation Iraqi Freedom (OIF), U.S. ground forces encountered radio controlled improvised explosive devices (RCIED) for the first time, and the Army had to rely on Air Force and Navy EW personnel to conduct EW for the Army. As a result, Army leadership looked to revive its own EW capability.

In 2004, the Combined Arms College (CAC) designated the Fires Center of Excellence (FCoE) at Fort Sill, Oklahoma, as the location to house a new EW School. The Army established Career Field (CF) 29 in January 2009 with new career field designations: 29A (EW Officer), 290A (EW Technician), and 29E (EW NCO). Career Field 29 established EW personnel capability at all echelons in response to RCIED threats, focusing primarily on maintaining IED jammers and on planning and synchronizing Joint EW capabilities.

In January 2014, Army leadership announced that all Army EW personnel serving under CF 29 would transition to the new Cyber Branch in the next several years. This became official on 1 October 2018, as the Army formally merged EW professionals into the Cyber Branch, recoding 29As to 17B, 290As to 170B, and

The Cyber Branch insignia as authorized in 2015 consists of two crossed lightning bolts, surmounted by a vertical dagger, point up, all gold. The lightning bolts symbolize the intelligence, security, and communications originators of the modern Cyber Branch. The dagger denotes readiness to prevent global cyber occurrences.

29Es to 17E. All EW training moved from Fort Sill to Fort Gordon in 2020-2021. Now as part of the Cyber Branch, EW officers, warrant officers, and enlisted personnel continue to serve at every tactical echelon in every theater of operations.

Newest Cyber MOS

The establishment of the 17D Cyber Capabilities Development Officer and 170D Cyber Capabilities Developer Technician military occupational specialties on 1 October 2021, and their requisite training courses, became the most recent Cyber School and Branch milestone.

MILITARY INTELLIGENCE CORPS

Military intelligence has a history that dates back to the beginning of human conflict. It has been evident that foreknowledge of the capabilities and probable courses of action of an enemy, or a potential enemy, is of great value to a government and its military commanders in making sound decisions for the conduct of state affairs and military operations.

Military Intelligence is a basic branch and a combat support arm of the Army. Its officers are primarily concerned with the intelligence aspects of the Army's mission. This field of activity encompasses intelligence, counterintelligence, cryptologic and signals intelligence, electronic warfare, operations security, order of battle, interrogation, aerial surveillance, imagery interpretation, and all related planning, organization, training, and operations. Intelligence officers are assigned to both branch material and branch immaterial positions within all Army, joint, and combined commands and staffs.

Up through World War I, the United States found its Army ill-prepared in all fields of military intelligence, due to the lack of a consistent policy of planning and coordination in this field. From a very small information division under the Adjutant General, the Military Intelligence Division (MID) appeared in 1918 under the General Staff of the War Department. The Corps of Intelligence Police (CIP) was formed in 1917 as the counterintelligence agency of the Army. Following the end of World War I, these intelligence agencies were reduced drastically during the cutbacks and reductions of the 1920s.

It was not until World War II, then, that military intelligence, due to excellent staff planning and coordination, began to take on the broad range and professional nature that characterize this field today. The Military Intelligence Service was organized early in the war and began to gather specialists in intelligence and intelligence-related areas. Among these were linguists, language and area studies students, professional investigators, geographers, economic and technological experts, world travelers, and editors. Counterintelligence training was reinstituted in February 1941, and the official designation of the organization became the Counter Intelligence Corps in 1942.

The Signal Security Agency was created in 1943 under the Chief Signal Officer and assumed the responsibilities and performed the functions formerly carried on by the Signal Intelligence Service.

In September 1945, the U.S. Army Security Agency (USASA) was created and placed under the direction of the Assistant Chief of Staff, G2, Intelligence, Department of the Army. It was redesignated as a major field command of the Department of the Army in 1964.

In June 1962, the Military Intelligence branch, composed of ASA, Intelligence Corps, and strategic and combat intelligence officers, was formally created to meet the growing requirements for control and career guidance of the increasing numbers of officers in the intelligence field. It was designated as Army Intelligence and Security branch.

In July 1967, the Army Intelligence and Security branch was redesignated the Military Intelligence branch, and its mission was changed from combat service support to combat support. In 1971, the United States Army Intelligence Center and School, the home of Military Intelligence, was established at Fort Huachuca, Arizona.

On 1 July 1987, the Military Intelligence Corps was activated as a regiment under the U.S. Army Regimental System.

The primary function of military intelligence (MI) officers is the collection, analysis, production, and dissemination of intelligence. To accomplish this function, it is essential that they possess comprehensive knowledge of military strategy and tactics.

MI officers lead, manage, and direct intelligence planning and operations at the strategic, operational, and tactical levels of war. At the strategic level, the MI officer assesses the capabilities and limitations of actual or potential adversaries for the national and departmental-level decision makers who develop national plans and strategy. At the operational level, MI officers assist in the development

A symbolic sun, patterned after that of the mythical Helios, god of the sun, who could see and hear everything, provides the base for the Military Intelligence design. The sun's rays indicate the worldwide mission of the branch; the superimposed rose revives the ancient symbol of secrecy; and the partially concealed dagger refers to the aggressiveness, protection, and element of physical danger inherent to branch operations.

and execution of campaign plans and major operations within a theater of operations. They direct, supervise, and employ theater-level intelligence assets to assess the capabilities and limitations of enemy and adversary alliances; geography, weather, and climate; and risks associated with enemy and friendly courses of action. At the tactical level, MI officers command, direct, supervise, and employ organic intelligence assets. They also plan for the optimum use of nonorganic intelligence assets to reduce the commander's uncertainty concerning the enemy, the terrain, and the weather; to assess risks associated with friendly and enemy courses of action; and to counter or neutralize the multidiscipline hostile intelligence threat.

At all levels, MI officers plan for, supervise, and perform collection and analysis of raw intelligence information and produce and disseminate finished all-source intelligence products for commanders and other intelligence consumers. MI officers also plan, coordinate, and participate in deception operations, operations security (OPSEC), electronic warfare operations, and counterintelligence, including countersignals intelligence (Counter-SIGINT), counterimagery intelligence (Counter-IMINT), and counterhuman intelligence (Counter-HUMINT).

MI officers serve as staff officers and command MI units at all levels throughout the Army, the Department of Defense, and the intelligence community, which includes assignments with such national agencies as the National Security Agency and the Defense Intelligence Agency. MI officers are initially trained and assigned as all-source intelligence officers (35D). After attendance at the MI Officer Advanced Course, some officers may receive additional training in a second MI area of concentration, depending on their follow-up assignments.

All newly commissioned military intelligence officers attend the Military Intelligence Basic Officers Leader Course conducted at the U.S. Army Intelligence Center and School. These officers receive training in common Army skills and in all-source intelligence. The all-source intelligence training covers the entire spectrum of intelligence and accents the application to the tactical environment. Upon graduation from this course, most officers go to tactical assignments.

Branch detail officers constitute a significant portion of total Military Intelligence accessions. Officers selected for branch detail attend the basic course of their detail branch and serve a tour of up to four years with that branch. MI officers may be detailed to the following branches: Infantry, Armor, Field Artillery, Air Defense Artillery, and Chemical. Branch detail officers are encouraged to lead troops at every opportunity. Additionally, they should maintain contact with MI officers and soldiers in the unit and seek duty in an S-2 section if given the opportunity. Following the completion of their branch detail and attendance at the MI Officer Transition Course and the MI Officer Career Course (MIOCC), branch detail officers follow the same career progression patterns as other MI officers.

Officers who desire military intelligence, even though they are participating in the branch detail program, should request military intelligence as their first choice for branch selection. Officers who desire to serve in the branch detail program should still identify MI as their first choice but should indicate in the narrative portion of their requests which branch they desire for detailing.

The growing complexity of international events, the rapidly increasing technological sophistication of multidiscipline intelligence collection and analysis systems, and the constraints implicit in a smaller, vastly streamlined Army are but a few of the challenges facing the military intelligence officer corps. All MI officers are expected to know, understand, and be able to function in all intelligence disciplines (IMINT, Counterintelligence, HUMINT, and SIGINT/Electronic Warfare) at all levels—tactical, operational, and strategic. As a consequence, officers attend the MI career course between their third and sixth years of active commissioned service. The course develops highly skilled, multidiscipline intelligence officer leaders who can perform extremely well in a wide variety of intelligence duties.

Some officers receive additional training in a second MI area of concentration, depending on their follow-up assignments. The areas of concentration skills with the MI Corps include strategic intelligence analyst (35B), imagery intelligence officer (35C), all-source intelligence officer (35D), counterintelligence officer (35E), human intelligence officer (35F), and signals intelligence/electronic warfare officer (35G).

MI officers can apply for a number of skill identifier (SI)-producing programs. As a general rule, selected MI officers may participate in only one of the following programs: Military Operations Training Course (MOTC), Junior Officer Cryptologic Career Program (JOCCP), or Defense Sensor Interpretation and Application Training Program (DSIATP).

The Military Intelligence Corps offers outstanding service opportunities through a wide variety of assignments and education. In military intelligence, an officer may contribute at all echelons in many different ways. For the most current information on Military Intelligence, visit https://www.army.mil//USAICoE.

MILITARY POLICE CORPS

The Military Police Corps is a basic branch of the Army. Its members perform combat, combat support, and combat service support missions. Military police (MPs) contribute to battlefield success by conducting combat operations against opposing forces in U.S. rear areas. They provide combat support by expediting the movement of critical combat resources and by collecting and processing enemy prisoners of war and evacuating them from the battle area. Additionally, military police provide security for critical Army facilities and resources, such as command posts and special ammunition, and conduct law-enforcement operations to ensure a secure environment for the Army community. Military police units provide support on a flexible mission basis, keyed to the commander's priorities.

A provost marshal was appointed to General Washington's Army of the United Colonies in January 1776, and two years later Congress passed a resolution establishing a "Provost Corps, to be . . . mounted on horseback and armed and accoutered as Light Dragoons." At the same time, General Washington directed the corps to apprehend "deserters, marauders, drunkards, rioters, and stragglers" and to perform other military police duties.

Between the American Revolution and the Civil War, there were no military police units in the Army. Instead, temporary duty personnel performed military police functions. The Civil War witnessed the establishment of the position of Provost Marshal General of the United States. That official's prime responsibility was to enforce the North's draft laws. To assist him in that endeavor, Congress in 1863 created the Veterans Reserve Corps. Besides aiding in the enforcement of conscription, the members of this organization also served as Home Guards, prisoner of war escorts, railroad security guards, and garrison troops. In addition, each field army had its own provost marshal. He commanded details of provost guards drawn from the individual army's regiments. Their functions were to maintain march discipline, discourage desertion, and return stragglers to their parent commands. All these positions and units disappeared following the end of the war.

It was not until World War I that the Office of Provost Marshal General was recreated. Once again its primary function was to enforce the draft system. It was also during that conflict that the Army took a first tentative step toward the creation of a permanent Military Police Corps. A month before the armistice, the War Department approved the creation of such an organization. This proved to be a premature development, however. Except for a handful of active and reserve companies, the corps was disbanded following the conclusion of the war.

Finally, on 26 September 1941, the Secretary of War approved the creation of the Military Police Corps, and it became a branch of the Army. During World War II, it grew to include 200,000 enlisted men and 9,250 officers. Among other duties, they protected war plants and supplies, escorted and guarded prisoners of war, controlled traffic, and fought enemy infiltrators. During the Korean War, military police were responsible for controlling large numbers of refugees in addition to their normal police functions. It was here that the helicopter was first used by military police for battlefield circulation control and area security.

The Military Police Corps won widespread praise for its performance of duty during operations in the Republic of Vietnam. In 1965, the 198th Military Police Brigade was activated at Fort Bragg, North Carolina. It deployed to Vietnam, becoming operational there on 26 September 1966, the first of its kind to be deployed in combat. Brigade missions were expanded to include port and harbor security and infantry-type tactical operations. During the Tet Offensive of 1968, military police distinguished themselves in the defense of the U.S. Embassy and

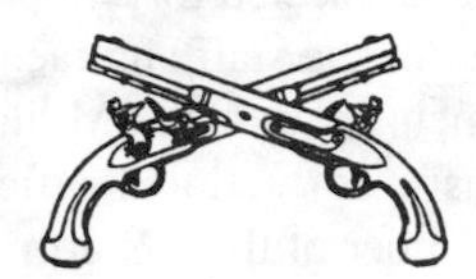

The crossed pistols insignia of the Military Police Corps was officially adopted in 1922. The model for the insignia was the 1806 Harpers Ferry pistol—the first official U.S. Army handgun.

other critical installations in Saigon, while keeping vital roads and waterways open throughout the Republic of Vietnam. The activities of the Military Police Corps in its combat support role led to the designation of the corps as an arm as well as a service on 14 October 1968. This designation reflects formal recognition by the Department of the Army of the combat role that the Military Police Corps has always performed. This combat role was further reinforced by military police in all major Army operations since 1983. The Military Police Corps' versatility as both a combat force and a law-enforcement agency has made it a "force of choice."

Today, military police personnel and units perform a wide range of combat, combat support, and combat service support operations. These include battlefield circulation control, area security, enemy prisoner of war/civilian internee operations, and law and order operations. In the garrison environment, military police are responsible for law enforcement, physical security, criminal investigations, and the confinement and correctional treatment of U.S. military prisoners. The fundamental objective for the Military Police Corps in garrison-related functions is to protect and assist fellow soldiers and their families.

Service in the Military Police Corps is one of wide variety, different assignments, and educational opportunities. The Military Police Corps continually provides opportunities for the education of its officers. From their first comprehensive training in military police operations until shortly before they retire, officers of the Military Police Corps receive formal professional training at progressively advanced levels. At the Military Police School, new officers are taught the principles and techniques of small-unit leadership, tactical operations, management, and military police operations. Emphasis is placed on developing platoon leaders who are fully capable of immediately deploying in support of war or operations other than war. Subsequent military schooling includes the MP Officer Career Course and, in all likelihood, other specialty courses as the opportunities arise. In past years, military police officers have taken part in graduate study programs in the fields of education, personnel management, area studies, comptrollership, operations research, ADP systems, criminology, correctional administration, and police science and administration.

The Military Police Corps' combination of battlefield and peacetime functions provides its officers with real opportunities not only to achieve a unique blend of skills but also to serve in a variety of challenging assignments. Since the Military Police Corps serves the entire Army, an officer may be assigned to almost any type of organization stationed wherever Army units are throughout the world. The initial assignment of a junior officer will probably be to an MP company or battalion but could also be to a specific functional law-enforcement position on any Army installation. Later, an officer could be selected for company command, for duty as a staff MP officer, or as a member of the staff and faculty at the U.S. Army Military Police School, an ROTC element, or the U.S. Military Academy. Other possible assignments include duty as a correctional officer, physical security officer, reserve component adviser, member of the Department of the Army Staff, with the U.S. Army Criminal Investigation Command, or a variety of other branch-related and branch-immaterial positions. Officers naturally take on more responsibility in their assignments as they progress in rank and experience.

The most exciting chapter in the history of the MP Corps is now taking shape. A greatly expanded role on the battlefield has brought with it significant increases in firepower and combat effectiveness. At the same time, the motto "Of the Troops and for the Troops" is taking on renewed meaning for military police serving Army communities throughout the world. The pace is fast and the esprit is high as the Military Police Corps continues to welcome all professional challenges that the future may hold. For the most current information on the MP Corps, visit http://home.army.mil/wood.

SIGNAL CORPS

In the entire history of the U.S. Army since the invention of the semaphore, it would be difficult to find a commander who would not agree that the commander's span of control depends on the success of communications. Signaling was, of course, the first effective means of fast communication by elements in the field—hence the name of this branch. Today's Signal Corps uses the latest communications and automation technology to support diverse requirements extending from the White House to the foxhole.

Within the classification of combat, combat support, and combat service support, the Signal Corps is a combat support branch that also performs combat service support missions. The overall mission includes the collective, integrated, and synchronized use of information technology in the form of systems, networks, services, and resources supporting command, control, communications, and computer (C4) requirements in organizations at all operational levels during peacetime, war, and operations other than war.

The Signal Corps traces its beginning from 21 June 1860, when Maj. Albert J. Myer, who had developed the signaling system we know as semaphore as a result of his work with the deaf, was appointed Signal Officer of the Army. The Signal Corps was officially established as a branch in March 1863.

Subsequently, signal officers and enlisted soldiers were deeply involved in every aspect of the nation's growth and exploration. The use of the telegraph became a tactical necessity during the Civil War. Later, signal soldiers played important roles in developing the National Weather Service, linking the West to the East as the country expanded, exploring the Arctic, and opening Alaska to the Gold Rush. Thousands of miles of telegraph wire were installed and maintained during this period.

War with Spain in 1898 found only eight officers and fifty-two enlisted soldiers assigned to the Signal Corps. Two legislative acts, however, quickly authorized the creation of a voluntary Signal Corps, which resulted in the creation of seventeen companies, each composed of four officers and fifty-five enlisted soldiers. Signal soldiers, using the telephone for the first time in combat—as well as

The present design of crossed flags and torch was adopted in 1884, although enlisted men of the acting Signal Corps had worn crossed flags since 1868. The insignia represents the signaling system, invented by the first signal officer, which used flags during the day and torches at night.

the telegraph, heliograph, and observation balloons—served in Cuba, Puerto Rico, and the Philippines.

Ballooning, which became a signal responsibility in 1885, led to the corps' development and control of aviation during World War I. The Signal Corps had been busily engaged in the development of military aircraft since the first successful flight by the Wright brothers and, in fact, purchased the Army's first aircraft from the Wright brothers in 1908.

An aviation section was authorized in 1914. By 1918, 16,000 officers and 147,000 enlisted soldiers were engaged in air operations, and in May of that year they became the branch known as the Air Corps.

During the same period, many innovations in sound-producing communications systems were developed, and the first permanent Signal Corps post and training center was activated at Fort Monmouth, New Jersey. Among the new kinds of equipment designed by corps personnel were vacuum-tube radios and detection gear to indicate approaching aircraft. In the field, the Signal Corps provided the required meteorological service for artillery and aviation, and photography operations were expanded to include motion pictures.

World War II saw the next major contribution of communications and sound to military operations. Radio had been perfected to the highest levels of performance, and the development of radar, sonar, and radio-controlled weapons systems was only the beginning of technological advances not even dreamed of a few years before.

Conventional as well as unconventional warfare intensified the need for highly mobile forces during the Korean conflict. The Signal Corps responded to new demands and met the challenges with the help of technological advances, which included VHF radio, improved radar, and the development of the Army Area Communications System.

The first major communications system employed in Vietnam was developed in response to a need for high-quality telephone and message circuits between key locations. Code-named "Backporch," it used tropospheric scatter radio links capable of providing multiple circuits between locations more than 200 miles apart. Funded by the Air Force and deployed by the Signal Corps, the system had vans containing troposcatter terminals capable of transmitting and receiving up to seventy-two voice channels simultaneously. Recognizable by the sixty-foot-tall "billboard antennae," the system was operated by the 39th Signal Battalion.

The inadequacy and unreliability of the radio circuits linking Vietnam, Hawaii, and Washington led to the Syncom satellite communications service, which provided a relay link between Hawaii and Saigon. Also, it marked the first use of satellite communications in a combat zone.

Further demands for a communications system that could meet the needs of a vastly expanding U.S. presence in Vietnam led ultimately to a commercial fixed-station system. Known as the integrated wideband communications system, this was the Southeast Asian link in the global defense communications system.

Support for Operation Just Cause (20 December 1989 to 31 January 1990) in the Republic of Panama required real-time command, control, and communications (C3) systems. Employing an integrated network that included frequency-modulated and high-frequency radio, tactical satellite communications, facsimile, and radio teletype, the Signal Corps kept the lines of communication open from the moment of notification until the operation was over.

The Signal Corps' experience in Operation Desert Shield/Desert Storm demonstrated the Army's reliance on sophisticated communications-electronics systems. Global positioning system (GPS) receivers, using signals from navigation satellites, allowed thousands of soldiers to safely navigate the featureless desert terrain. Satellite communication systems were used extensively for critical command and control. The Signal Corps also fielded the largest automatic switched network in history, connecting the Army with its sister services and coalition forces, spanning Saudi Arabia, Iraq, and Kuwait with seamless connectivity to the continental United States. Operation Desert Shield/Desert Storm created a template to structure the Signal Corps for the twenty-first century.

Today, the responsibilities of the Signal Corps are more varied than ever. Installing, operating, maintaining, and reconfiguring networks of information systems for theater/tactical, strategic, and sustaining base operations; operating the Army portion of the global defense communications system; training officer and enlisted signal specialists; developing signal warfighting doctrine; carrying out research and development projects; managing the acquisition, distribution, and fielding of information systems and networks; and leading the Army's efforts in information management and the digitization of the battlefield offer unsurpassed technological opportunities for signal officers of all ranks.

The complexities of today's Army require a flexible Signal Corps organization. In addition to TOE Signal organizations, signal personnel are employed in practically every organizational structure throughout the Army. Major roles for Signal Corps personnel are found within the U.S. Army Information Systems Command, U.S. Army Materiel Command, U.S. Army Training and Doctrine Command, U.S. Army Forces Command, and Department of the Army and Department of Defense agencies.

The U.S. Army Information Systems Command (USAISC), with its headquarters at Fort Huachuca, Arizona, is the Army's signal organization charged with the global mission of operating and maintaining assigned information systems at echelons above corps and supporting theater/tactical, strategic, and sustaining base operations. The command also is responsible for planning, developing, engineering, acquiring, and installing information systems; developing and implementing information systems; and advising, assisting, and providing tactical support. USAISC provides the interconnection of theater armies with activities of the Department of Defense, whether locally or over intercontinental distances.

A corps signal brigade is the signal organization expressly formed to provide the planning, engineering, installation, supervision, and control of the automation and communications networks within the corps area. Each corps signal brigade is structured to meet the needs of the corps it supports. The brigade assigns its various organic signal battalions to plan, engineer, install, operate, and maintain the integrated network of command, control, communications, and computer systems and acts as the integrator of information systems serving the corps headquarters down to each division and separate combat brigade. Each division has its own organic signal battalion. In addition, at each combat brigade within the division, there is a Signal Corps officer who serves as the combat brigade signal officer. Every maneuver battalion assigned to the combat brigade—whether an infantry battalion, an armored battalion, or a mechanized battalion—also has its own signal officer, who is responsible for organic command, control, communications, and computer operations. Signal officers also serve in combat support and combat service support brigades and battalions.

Service in the Signal Corps offers a wide range of training and assignment possibilities, punctuated at appropriate intervals with educational opportunities paralleling branch career programs.

The U.S. Army Cyber Center of Excellence at Fort Gordon, Georgia, is the Army's principal training facility for the Signal Corps. The Signal School provides military education and practical training to prepare officers for positions throughout the Signal Corps in theater/tactical, strategic, and sustaining base operations.

Fort Gordon is also the home of the Signal Corps Regiment, designated on 1 June 1986. All signal soldiers are affiliated with the regiment, but those who typically serve in a unit affiliated with a combat arms regiment may also elect to associate with that unit's regiment.

The majority of all signal officers are commissioned as signal operations officers. During the first eight years, officers attend the basic and career courses, with assignments designed to provide opportunities for the development of leadership skills. After the Signal Basic Officer Leader Course at Fort Gordon, a challenging career lies ahead. Graduates of the Signal Basic Officer Leader Course normally begin their careers as platoon leaders in a theater, corps, or division signal battalion. Subsequent assignments to signal battalion operational and staff positions provide the lieutenants with a broad base of experience in signal missions, functions, tactics, and techniques. Graduates of the Signal Officer Career Course may command signal companies, be assigned as battalion signal officers in maneuver battalions, or serve in staff officer positions in signal battalions and brigades. Signal officers typically are involved in engineering, configuring, installing, integrating, and managing local area, wide area, and global command, control, communications, and computer (C4) networks in support of Army operations.

From about the eighth through the fifteenth year of service, officers continue developing expertise in branch duties but may concentrate their efforts in an increasing managerial capacity for handling the most complex information systems activities of the Army and the Department of Defense.

In summary, the Signal Corps provides the expertise and facilities that support information systems activities at every level in the Army: theater/tactical, strategic, and sustaining base. It continually researches, develops, and improves the equipment required to provide that support. The Signal Corps trains its officers and enlisted personnel as professionals, equipped with the most sophisticated equipment available. The Signal Corps provides the media for command and control of all Army elements during both peace and war. For the most current information on the Signal Corps, visit https://cybercoe.army.mil/SIGNALSCH.

7

Force Sustainment Branches

Force sustainment branches provide logistical support and services focused principally on supply, maintenance, transportation, health, and other services required by combat unit soldiers to continue their missions. They are force enabler branches.

ADJUTANT GENERAL'S CORPS

The Adjutant General's (AG) Corps is the Army proponent for Human Resources (HR) support. AG Corps doctrine, FM 1-0, *Human Resources Support*, identifies ten core competencies designed to provide timely and dependable HR support to combatant commanders engaged in military operations across the full spectrum of conflict. They include:

- Personnel Readiness Management (PRM)
- Personnel Accountability and Strength Reporting (PASR)
- Personnel Information Management (PIM)
- Reception, Replacement, Return to Duty (RTD), Rest and Recuperation, Redeployment Operations (R5)
- Casualty Operations
- Essential Personnel Services (EPS)
- Postal Operations
- Morale, Welfare, and Recreation (MWR) Operations
- Human Resource Planning and Staff Operations
- Band Operations.

AG Corps officers are trained, equipped, and prepared to deploy to all parts of the globe to perform HR support in operational environments that vary in complexity and scope, and which serve to protect the American people and U.S. national security interests abroad.

The origins of the Adjutant General's Corps date back to the establishment of the U.S. Army during the American Revolution. On 16 June 1775 the Continental Congress appointed Horatio Gates the first Adjutant General of the Army and commissioned him in the grade of brigadier general. Gates's mission derived from the old Latin word "adjutare," meaning to aid or assist. In Gates's case, it meant assisting Gen. George Washington in all aspects of Continental Army operations. As the second officer commissioned behind Washington, Gates's appointment established the Adjutant General's Corps as the second oldest branch in the Army. Accordingly, he is honored as "father of the AG Corps." As adviser and principal assistant to General Washington, Gates succeeded in organizing the Continental Army into regiments and brigades and implemented the system that generated the first "strength return" of the U.S. Army. He also proved himself to be an able field commander leading the Continental Army to victory over the British at the Battle of Saratoga (September–October 1777), considered by many to be a major turning point in the Revolutionary War.

The Adjutant General's Department, along with other general staff offices in the War Department, was established in March 1813 in response to wartime emergencies created by the War of 1812. For the remainder of the nineteenth century and into the period following the Spanish-American War in 1898, the Adjutant General of the Army served principally as the chief administrator and coordinator of all War Department business, a position similar in stature to today's Chief of Staff of the Army. Maj. Gen. Roger Jones, Adjutant General of the Army for twenty-seven years from 1825 to 1852, is generally credited with promoting his office and the AG Department above other staff offices and assuming a primary leadership role in the War Department that would last into the twentieth century and until the U.S. Army adopted the general staff system in 1903.

Lessons learned from the U.S. Army's experience in World War I reshaped the Army's understanding of "personnel" work and the mission of the Adjutant General's Department. The mobilization of the nation for war and the drafting of millions of American citizens into military service required the development of an efficient and effective "personnel system" to procure, induct, classify, and assign individuals to a host of military specialties and units. Managers of this system, the first generation of military "personnel specialists," were found in all Army units down to the regimental level, assisting unit Adjutant Generals to manage unit strength; track casualties; identify the need for replacements, procure

The unique branch insignia of the Adjutant General Corps became its official symbol on 14 December 1872. The shield was previously used by topographic engineers as an authentication device on military maps. Thirteen embossed stars replaced the letters T. E. to create the branch insignia worn by Adjutant General Corps officers today. The insignia symbolizes the trust placed in the branch by the Army as representing the values of the nation and the authority to speak "for the commander."

them by specialty and assign them within the unit; and account for all transfers, promotions, and discharges of individual soldiers.

Under Brig. Gen. Robert C. Davis, Theater Adjutant General of the American Expeditionary Force, the theater Central Records Office grew to become the largest administrative operation of the Allied armies, employing more than six thousand officers, enlisted soldiers, and civilians to track unit strength, casualties, and the location, assignment, and other key personnel information on 4 million American soldiers who had deployed to the Western Front in France. In addition to responsibilities already mentioned, the Theater Adjutant General managed all headquarters correspondence, the war prison, muster rolls, records, postal affairs, the YMCA, and printing for the deployed Army. World War I became the defining experience for the modern Adjutant General's Corps and the soldiers who perform the Army's human resources support mission in the twenty-first century.

Today, officers of the Adjutant General's Corps serve in a wide variety of challenging staff and command positions. They are skilled and innovative professionals who support commanders and soldiers from company level to the Theater Army. Following completion of the Basic Officer Leadership Course (BOLC), AG Corps second lieutenants attend Phase III training at the U.S. Army Adjutant General School, Fort Jackson, South Carolina. During Phase III, officers learn the basic skills and knowledge expected of Adjutant General's Corps personnel. Students follow a course outline involving classes in field training exercises (FTX), command post exercises (CPX), military skills, professional development, and personnel management systems. Students also receive instruction on automated personnel systems in today's "brigade-centric" operating environment. Lieutenants can expect to serve as battalion or possibly brigade S1 (HR) officers during field training and command post exercises. Students also serve in various class leadership positions such as class leader, class S3, and physical fitness officer. Overall, the course is designed to provide a rigorous and relevant training experience for the HR professional called to support commanders and soldiers in today's complex operating environment. Graduating AG lieutenants serve as brigade S1 strength managers, battalion S1s, company executive officers, or platoon leaders for postal, casualty, or R5 (Reception, Replacement, Return to Duty, Rest and Recuperation, Redeployment) platoons, positions found in combat and non-combat Army units.

After three to five years of service, lieutenants return to Fort Jackson as captains to attend the Adjutant General Captains Career Course (AGCCC) designed to produce technically and tactically proficient officers able to command or supervise HR personnel. AGCCC technical core instruction provides officers with the knowledge, skill, and confidence to function effectively in various branch assignments during war or peacetime. Common core instruction encourages officers to think creatively and critically and provides them with a working familiarity of staff operations, military intelligence, and logistics under garrison, field, and tactical conditions. After graduation from the Career Course, AG officers may be assigned to high-level staffs, attend graduate school, or serve as battalion or brigade S1 staff officers. AG captains may also be selected to command a company, serve as a Military Entrance Processing Station (MEPS) operations officer, personnel officer, HR plans and operations officer, postal officer, strength management officer, military transition team officer, service school instructor, or associate bandmaster of a major command or special band. For more information, please go to the U.S. Army Adjutant General School website at www.agsssi-www.army.mil.

For more than 225 years, from the battlegrounds of Lexington and Concord to Operation Iraqi Freedom, the United States Army continues to require soldiers, like the adjutants of old, to assist commanders in the successful accomplishment of the Army mission. Today, the Adjutant General's Corps is the Army's primary provider of Human Resource Management. Through the dedication and commitment of its officers, the Adjutant General's Corps continues to improve upon both the means and methods of support that Horatio Gates and his able assistants provided to our nation's first Army in 1775. For more information, visit www.ags.army.mil.

FINANCE CORPS

The U.S. Army Finance Corps was born amid the tumult of the American Revolution, 1775–1781. On 27 July 1775, fully one year before the United States declared its independence from Great Britain, the Continental Congress elected James Warren as Paymaster General of the Continental Army. Warren's appointment followed the 16 June 1775 establishment of several staff officer positions intended to provide administrative and logistical support to Gen. George Washington, Commander in Chief of the Continental Army. Modeled after the British Army's Paymaster General, Congress intended the holder of the office to centralize control over the disbursement of funds for soldier pay and other purposes that were expected to flow from both the Congress and the separate colonial governments. Warren's service to our nation during its first war became the foundation upon which the mission of today's Finance Corps was built.

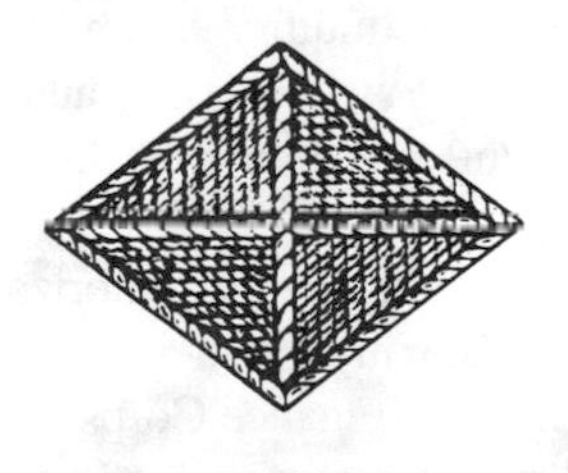

The diamond insignia worn by members of the Finance Corps is not without historical significance. Prior to the minting of coins, goldsmiths used the orle (diamond) as a mark of identification and to stamp the weight on gold bullion used in trade. Originally chosen by General Washington for his paymaster general, it was approved by the Secretary of War in 1896 to be the insignia of the Pay Department, and later the Finance Corps. The finance diamond represents the four basic functions of the Finance Corps (accounting, disbursing, administration, and auditing), with the inner lines representing the coordination between the pay and procurement agencies of the Army.

In the period immediately following the War of 1812, Congress on 24 April 1816 established the Pay Department as a permanent part of the military establishment. At the time, it consisted of one paymaster-general of the army, paymasters for each regiment of the Army, and one paymaster for each battalion of the corps of artillery. All paymasters were to render "regular and punctual payment of their respective regiments or corps," and discharge the duties of a district paymaster when directed to do so by the Paymaster-General. The Pay Department functioned as part of the War Department staff until 1912 when it was absorbed by the Quartermaster Department. However, the experience of World War I quickly proved the former arrangement costly and impractical. Altogether, the lessons of war pointed toward the reestablishment of an independent financial arm of the Army staff. In 1920, Congress established the Finance Department headed by Brig. Gen. Herbert M. Lord, the first Chief of Finance. Gen. Peyton C. March, Chief of Staff of the Army, underscored the historic significance of the event when he pointed to the fact that the Army now required "professional soldiers who were also experts familiar with all sides of Army financial operations just as the infantryman was familiar with all sides of infantry operations." To produce the financial experts of which March spoke, Brigadier General Lord established the Army Finance School at Fort Washington, Maryland, in 1920. With Lord's strong commitment to the conscientious management of the funds and resources entrusted to the Army, the modern Finance Corps was born.

Following World War II, the Army Organizational Act of 1950 established the "Finance Corps" as a permanent specialty branch of the U.S. Army. As a specialty branch, the Finance Corps assumed responsibility for managing the careers and professional progression of finance officers and enlisted personnel, soldiers who were now permanently assigned to the duty rather than detailed into it. The branch also became responsible for developing the doctrine, organization, equipment, and training programs that would ready it and its soldiers for wartime service. Since 1950, the Finance Corps has mastered certain automated systems that have improved immeasurably the service and support available to commanders and soldiers stationed all over the world. Centralized automated pay systems generate active duty and retired military payrolls with real-time inquiry capability. Other programs include automated accounting and budget systems, and automated financial systems for travel, commercial accounts, civilian pay, Army Reserve pay, and disbursing.

In October 2006 the Commandant of the Finance School received authorization to change the name of the Finance School to the U.S. Army Financial Management School. The name change signified the fact that the Finance Corps had become the Army proponent for military finance and resource management, the latter previously a separate function belonging to military comptrollers. In the merger of finance and comptroller competencies, the Finance Corps significantly broadened its responsibilities by adding budgeting and financial planning to its list of proponent functions.

The Financial Management School annually trains thousands of military and civilian personnel from all parts of the Department of Defense in subjects ranging from disbursing to financial management. Among those students are newly commissioned finance officers, both active duty and reserve component, who come to Fort Jackson to attend the twelve-week Army Basic Officer Leadership Course (BOLC). During Phase B training, officers learn the basic skills and knowledge expected of Finance Corps personnel. Graduating lieutenants often receive additional training in subjects such as military accounting or take advantage of extra-branch opportunities like airborne training at the Army's Airborne School, Fort Benning, Georgia. A few finance lieutenants volunteer or are selected for participation in the branch detail program where they attend a combat arms officer basic course and serve two years in a combat arms unit. Immediately following the two-year branch detail, lieutenants then begin their careers as finance officers by attending the four-week Finance Officer Branch Qualification Course at the Financial Management School.

Finance lieutenants will likely become cash-control officers or disbursing officers during their first duty assignment. In these positions, the officer becomes an agent of the U.S. Treasury responsible for managing large sums of cash and negotiable instruments. The lieutenant supervises both military and civilian employees and maintains a network of automated systems to facilitate the kind of reliable mission support to which commanders and soldiers have become accustomed. Another possibility is assignment as chief of a military pay section that maintains military pay accounts for personnel in the area serviced by the unit. Modern computer technology links finance offices around the world to the Department of Defense Finance and Accounting Service. The officer is responsible for the smooth and uninterrupted flow of transactions that ensure timely and accurate pay support for soldiers served by the section. The lieutenant could also become the chief or assistant chief of a pay and examination branch that is responsible for not only military pay, but other financial transactions that include civilian pay, travel, and commercial accounts. Other opportunities might include assignment to a finance unit staff to serve as a plans officer developing training and contingency plans for subordinate units, conducting unit inspections, and arranging logistical support and other staff-related programs. Assignment to a non-tactical finance unit or major command staff is also a possibility. In this scenario, the lieutenant would likely work with and perhaps supervise a predominantly civilian work force.

Between the fifth and eighth year of service, finance officers return to Fort Jackson to attend the Captains Career Course (CCC). The CCC is taught in small groups to encourage the development of leadership skills for company-grade officers. Tactical training is balanced with technical training to prepare officers for both the peace and wartime missions of the Finance Corps. The CCC culminates with a command post and field training exercise that measures student knowledge and understanding of finance doctrine and operations, and common tactical skills that all Army officers are expected to possess. In 2008, the Army merged Branch

Code 44 (Finance) and Functional Area 45 (Comptroller) into Branch Code 36 (Financial Management). Multi-functional Finance and Comptroller Professionals who can operate in both the Financial Operations and the Comptroller domains are the future. For more information, please go to the U.S. Army Financial Management School website at www.finance.army.mil.

Although the Army's smallest branch, the Finance Corps presents challenging job opportunities for young soldiers in all corners of the globe. Strengthened by the establishment of the Finance Corps Regiment on 7 May 1987, finance soldiers are bound together in a mutual understanding of the significant history, customs, and traditions of the Finance Corps. Members of the Regiment remain deeply committed to the finance mission, the Army, and the nation it has served for more than 225 years. For more information, visit http://www.ssi.army.mil.

ORDNANCE CORPS

"Armament for Peace" is the motto of the Ordnance Corps. The ordnance mission is to support the development, production, acquisition, and sustainment of weapons systems, ammunition, missiles, electronics, and ground mobility materiel during peace and war to provide combat power to the U.S. Army.

The explosive growth of science and technology in the years since World War II has contributed significantly to the development of military materiel, as evidenced during Operations Desert Shield/Desert Storm. The major developments of World War II—nuclear weapons, missiles, and electronics—have been exploited, refined, and applied in a variety of ways. As the pace and complexity of technological advancements have grown, the interaction between military materiel and operations has become increasingly important and has also had a significant impact on the manner of accomplishing the support mission. Ordnance officers can be found worldwide, both in the field and in garrison, performing the enormous task of maintaining all types of equipment used by both our armed forces and those of our allies.

Ordnance officers command companies, battalions, plants and arsenals, depots, groups, and division support commands (DISCOMs), and routinely occupy command and senior staff positions at the one-, two-, and three-star general rank (such as support commands, corps support commands (COSCOMs), and Army staff). There are dozens of separate areas of concentration and military occupational skills, and ordnance personnel are assigned to virtually every unit in the Army.

The "shell and flame" insignia of the Ordnance Corps is patterned after a similar design used by British troops. It became official in 1832. Considered the oldest of the branch insignia, it represents the early explosive devices and properly symbolizes that particular branch function today.

Ordnance officers may serve in two challenging areas of concentration: maintenance and munitions management and explosive ordnance disposal.

Materiel maintenance management officers are responsible for integrated maintenance and repair parts supply support of Army conventional weapons systems, small arms, artillery, and fire control equipment; missile systems and their associated ground support equipment; electronics; tracked and wheeled vehicles; and engineer and power generation equipment. Maintenance functions include metalworking, fabrication, welding, inspection, test, service, calibration, repair, overhaul, and reclamation. These officers are also involved in the Army's test, measurement, and diagnostic equipment (TMDE) program. The role of every officer in materiel maintenance management is to effectively and efficiently ensure that the maximum number of weapons systems are operational, ready, and available to combat commanders. Management of the Army's maintenance is an increasingly sophisticated challenge. Officers serving in this area of concentration can expect to command and serve in organizations throughout the Army structure. Duty positions filled by materiel maintenance management officers require a comprehensive knowledge of maintenance management techniques and integrated logistics support as it applies to multiple commodity areas.

When working in munitions management, officers are trained for assignments around the world involving conventional and chemical warheads.

Support functions include supply, maintenance, surveillance, inspection, stock control, and security, as well as maintenance of associated testing and materiel handling equipment. The overall and increasing technical sophistication of various munitions in the Army's inventory and the use of robotics in the ammunition field will require officers serving in this AOC to develop expertise in several engineering technologies and management techniques.

Explosive ordnance disposal (EOD) officers provide a unique and critical service to the Army. EOD officers are responsible for identifying, locating, rendering safe, handling, removing, salvaging, and disposing of U.S. and foreign unexploded conventional, chemical, and nuclear munitions. Officers serving in this AOC can expect to command and serve in EOD units throughout the world and to serve on major command staffs as advisers on EOD matters. EOD officers advise and assist law-enforcement agencies in the removal and disarming of explosive devices, provide support and protection to the President of the United States and other high-ranking American officials, and support intelligence activities through analysis of foreign munitions.

The organization of ordnance support today still follows the concept and philosophy upon which it was founded—direct support to the combat arms. The Ordnance Corps had its beginning with an act of the Continental Congress on 27 May 1775, which established a committee to study and plan for the supply of weapons and other war materiel to the Continental Army. Soon after the act was passed, General Washington appointed Ezekiel Cheever to be the Commissary General of Artillery Stores. Cheever functioned in essence as a civilian Ordnance Chief, with Maj. Gen. Henry Knox, Chief of Artillery, in charge of all military components.

From this beginning up to the present, the Ordnance Corps has changed as the Army changed. Early in the Revolutionary War, design of weapons was not considered an ordnance function, primarily because U.S. Army weapons were either purchased abroad or captured from the enemy. Individual weapons were usually the property of the individual soldier. However, in 1777 the first arsenal for the manufacture of weapons was established at Springfield, Massachusetts. During the same year a storage facility was set up at Carlisle, Pennsylvania.

In 1801, Eli Whitney demonstrated a primitive method of mass-producing weapons, a process that revolutionized the manufacture of weapons. It was quickly adopted by our military armories, leading to formal authorization of the Ordnance Department on 14 May 1812. The first formally designated chief was Col. Decius Wadsworth. Full-scale mass production was not possible, however, until the eve of the Civil War.

After the War of 1812, the Chief of Ordnance was responsible for contracting for arms and ammunition, for supervising the government armories and storage depots, and for recruiting and training officers to be attached to regiments, corps, and garrisons. By 1816, five federal arsenals were in operation: Springfield and Harpers Ferry, making small arms; Watervliet, producing artillery equipment and ammunition; Watertown, producing small arms ammunition and gun carriages; and Frankford, making ammunition. Two others, Rock Island and Picatinny, were added before or during the Civil War. Early in the Civil War, Harpers Ferry was destroyed.

As a result of the large-scale, widely dispersed operations during the Civil War, ordnance concepts and activities were greatly improved.

The war with Spain brought new and difficult problems to the Ordnance Department. Procurement methods had to be revamped and methods of supply expanded to meet the requirements of our first overseas conflict. Most war supplies had been stored at established arsenals and depots; in most cases they were delivered directly to military units in field locations. When forces were being prepared for shipment by sea, a large depot was set up in Tampa, Florida, to complete the equipping of units moving through the port. This port depot and those field depots set up in Cuba were the beginning of Ordnance field service. World Wars I and II and the Korean conflict did not change the mission of the Ordnance Department, renamed the Ordnance Corps in 1950, but the amounts of items procured, distributed, and maintained stagger the imagination. During this period, small arms procurement mounted into the millions, rounds of ammunition into the billions, and other items produced increased proportionally. The Ordnance Department became the proponent for wheeled vehicles after 1942. Ordnance field service was expanded, along with the training of personnel and the development of techniques, in order to be responsive to the needs of the fighting forces.

The Atomic Age began an entirely new era of weapons and weapons systems and the continuing improvement of more conventional equipment. To meet these changes and the more complex requirements of a modern army, and to plan for the future, a new agency was created. The Army Materiel Command was given the responsibility for procurement and distribution of materiel that had formerly been divided among the various technical services. The Ordnance Corps, which in the past had been represented in organizational structures by name and was somewhat restricted in areas of employment, now furnished ordnance-trained personnel to staff all logistic elements of the Army. Since 1983, the reconstituted Office, Chief of Ordnance, has reassumed responsibility for the career management of ordnance personnel, working with the assignment officers of the ordnance branch of the Officer Personnel Management Directorate.

Ordnance officers have always played a major role in the research and development (R&D) of Army weapons systems. Until the late 1930s, the chiefs of other branches gave requirements for equipment to the Ordnance Corps, and ordnance engineers would produce them. From 1942 on, the Ordnance Board became the R&D arm of the Chief of Ordnance, with the mission of designing equipment, developing training for its use, studying improvements in operational employment, and developing ordnance unit structures.

In 1936, all ordnance training was centralized at Aberdeen Proving Ground, Maryland. As a result, the Ordnance Field School for enlisted men was moved from Raritan Arsenal in July 1940 and combined with the officer school to form the Ordnance School. This consolidated school for a time handled most of the school training for the corps.

In July 1973, the U.S. Army Chemical School was merged with the Ordnance School, giving that school the added mission of providing the defense establishment with trained NBC specialists and continuously updated doctrine in the chemical field. However, the Chemical Corps was reestablished as a basic branch of the Army in 1976, and in 1979 the Chemical School was moved to Fort McClellan, Alabama, under the control of the Chemical Corps.

On 1 November 1983, the Office Chief of Ordnance (OCO) was established to assist the Commanding General of the Ordnance Center and School in overseeing the operations and management of the two separate schools and focusing on ordnance-related matters worldwide. The Commanding General of the Ordnance Center and School formally became the Chief of Ordnance on 18 October 1985, when the Ordnance Corps became the Army's first combat service support branch to be recognized under the U.S. Army Regimental System. This office concentrates in a collective sense on personnel management. The OCO further assists the chief by developing Ordnance Corps policy and providing for the welfare of corps personnel. Materiel systems integration, training, doctrine, and force structure issues that cross organizational lines and/or have a broad impact on the Ordnance Corps are coordinated on behalf of the Chief of Ordnance by ordnance "cells" at the U.S. Army Combined Arms Support Command, Fort Lee, Virginia.

In addition to these Army training establishments, use is made of civilian universities and technical training establishments for selected officers and enlisted personnel.

The ordnance officer of today's Army is a materiel-oriented leader. He or she functions and operates in an atmosphere of electronics, mobility, and high-technology weaponry—rifles, machine guns, artillery, trucks, tanks, rockets, missiles, and conventional firepower. The ordnance officer must have a basic qualification in military organization, operations, and tactics in addition to technical, managerial, and leadership abilities in a variety of duties. Ordnance officers are maintenance or ammunition managers at the Company, Battalion, Squadron, Group, Regiment, Brigade, Division, Corps HQ, Theater Sustainment Command, or Expeditionary Sustainment Commands. There are dozens of military occupational skills for enlisted soldiers and ordnance personnel are assigned to virtually every unit in the Army.

In staff positions, ordnance officers provide advice and assistance to all commanders and other staff officers at all levels on ordnance matters. They are also assigned to instruct and train military and civilian personnel in maintenance, repair parts, ammunition, missile, and general logistical support doctrine and procedures. In assignments to planning staffs, the ordnance officer helps to develop concepts, doctrines, policies, and procedures for furnishing ammunition and maintenance support to the Army.

Performing integrated commodity (life cycle) management of ground vehicles, missile systems, weapons systems, and ammunition; force protection and explosives safety duties related to EOD; formulating plans, programs, and policies for industrial mobilization; and providing technical supervision and inspection for designated commodities are some of the fascinating assignments awaiting those who qualify, choose, and are selected for the Ordnance Corps.

Because of its widely diversified field mission, many tactical units are authorized ordnance officers. From the ammunition and maintenance companies, and EOD detachments, to the depots, arsenals, and ordnance plants, assignments are varied and diversified. Ordnance jobs for company-grade officers include platoon leader, maintenance officer, technical supply officer, production control officer, ammunition stock control officer, EOD officer, EOD company commander, company commander, service school instructor, maintenance test officer, and range officer.

Further schooling at the Ordnance Center and School, as well as at selected civilian institutions, is usually the future of Army ordnance officers in the grade of captain and above.

Changes in weaponry and the need to maintain weapons in fighting condition will be a primary activity so long as armies exist. Therefore, the young officer selecting ordnance will quickly encounter exciting challenges and responsibilities in the ordnance area. Initial assignments are usually as maintenance or munitions platoon leaders. Such assignments involve leading and managing supply and maintenance personnel and maintenance of sophisticated Army equipment, such as the M-1 Abrams tank or high-technology missile systems.

The current range of ordnance service possibilities is extensive. With weapons systems and equipment ranging from conventional to laser (computer and Space Age technology), the Ordnance Corps of today and the future presents a tremendous opportunity for those who qualify. For more information, please visit www.goordnance.army.mil.

One of the more complex branch insignias is that of the Quartermaster Corps. A key, symbolic of storekeeping, is crossed with a sword indicating "military" and superimposed on a wagon wheel, pertaining to the delivery of supplies. The stars and spokes of the wheel represent the original thirteen colonies and the origin of the corps during the American Revolution. The eagle is used as a national symbol.

QUARTERMASTER CORPS

"Supporting Victory" is the motto of the Quartermaster Corps. Advanced data processing equipment, sophisticated communications networks, and modern transportation techniques are the tools of the Quartermaster Corps officers in performing their logistical support missions around the world.

The corps has come a long way since 16 June 1775, when the first Quartermaster General was appointed to provide some items of camp equipment and the transportation for the Army. Having virtually no money and no authority and dependent on the several states for supplies, the early Quartermaster Department was hard put to accomplish its mission. Yet the quartermaster played an important role in the successful defense of liberty made by the young nation. Military supply was largely under civilian control in the post–Revolutionary War period. At the outbreak of the war with England in 1812, Congress appointed a brigadier general to supply the Army. In 1814, quartermaster sergeants first appeared and were assigned to each of the three regiments of riflemen.

From 1818 to 1860, the Quartermaster General was Brig. Gen. Thomas S. Jesup, a remarkably able administrator. During his tenure, the Quartermaster Department made great strides, emerging as the integrated, permanent supply agency of the Army. In this period the Quartermaster Department took over the procurement and distribution of clothing and other items of supply.

The Civil War brought the development of an effective depot system, and railroads were used extensively in establishing supply lines. The supply system and procedures developed during this period formed the basis for supply doctrine until World War I. In 1962, the department assumed responsibility for the burial of the war dead and the maintenance of national cemeteries.

In 1912, Congress consolidated the former Subsistence, Pay, and Quartermaster Departments to create the Quartermaster Corps with its own officers and troops. The First World War showed the increasing importance of supply. The United States participated long enough to give the Quartermaster Corps its first set of huge figures: nearly 4 billion pounds of food valued at $727 million, $1 billion spent for clothing, and 3,606,000 tons of supplies procured and used during the Great War.

To fill an increasing need for specialists in Army supply problems, the Quartermaster School was begun in a small way in 1910 at the Philadelphia Quartermaster Depot. The school remained in Philadelphia, except for a short period at

Camp Johnston, Florida, during World War I, until it moved to Camp Lee (now Fort Lee), Virginia, in 1941. The school continues today to train the officers and enlisted specialists assigned to the Army's sophisticated logistical system.

During World War II, the Quartermaster Corps sent more and a greater variety of supplies to more men in more places in the world than any other quartermaster activity in the history of the world. It was during this conflict that transportation and construction were transferred to the Transportation Corps and the Corps of Engineers, respectively. With the loss of these two original functions, the Quartermaster Corps concentrated entirely on its supply and service missions.

The Quartermaster Corps pioneered in the field of air supply of ground troops, using both free-fall and parachute delivery extensively in Korea as a regular means of supply for the first time in military history.

When the Department of the Army was reorganized in 1961–62, the Office of the Quartermaster General was abolished and its functions and responsibilities reassigned to the Department of the Army staff agencies and commands established under the functional concept. The Quartermaster Corps, however, remained as one of the Army's important technical branches, and quartermaster personnel are still performing the logistical functions within the new functional framework. Beginning in 1983, the Commandant of the Quartermaster School assumed the traditional role of Quartermaster General of the Army, and the school was designated the home of the Quartermaster Corps.

Officers assigned to the Quartermaster Corps, though they function as members of a team within the current complex of logistical concepts, are still identified with the basics of supply—supply management, procurement, cataloging, inventory, storage, distribution, salvage, and disposal of all material except medical and cryptological items. A host of supply activities and service support responsibilities include mortuary affairs, laundries and showers, issue points at reception centers, petroleum and water distribution, water purification, and petroleum product testing.

Quartermaster officers are part of numerous TDA organizations to provide staff advice and counsel on supply and service operations. Within the TOE authorizations, officers can expect assignments to units such as the expeditionary sustainment command, sustainment brigade, brigade support battalion, combat service support battalion, petroleum operation and supply company, airdrop support company, airdrop equipment repair company, field service company, and mortuary affairs company. Among the positions offered by this branch are petroleum management officer, supply and services officer, and aerial delivery officer, all of which provide unusual opportunities for the individual with a logistics-related background and interests. Materiel systems integration, training, doctrine, and force structure issues that cross organizational lines and/or have a broad impact on the Quartermaster Corps are coordinated on behalf of the Quartermaster General by quartermaster "cells" at the U.S. Army Combined Arms Support Command, Fort Lee, Virginia.

Formal schooling, including the Logistics Basic Officer Leadership Course and the Combined Logistics Captains Career Course at Fort Lee, is in the future of all quartermaster officers. Each junior officer can expect training and early duty assignment in one of the supply fields of petroleum management, aerial delivery, or supply and services management. During career development, many special educational opportunities at the Quartermaster School and selected civilian institutions are open to quartermaster officers.

Today's highly mobile Army requires professionalism in all stages of combat service support. Whether computing requirements for repair parts or operating a petroleum pipeline to keep helicopters in the air, the quartermaster officer provides the technical knowledge to get the job done right, and now. For the dedicated officer, the Quartermaster Corps offers unlimited opportunities for personal and professional development. For more information on the Quartermaster Corps, please see the website at www.quartermaster.army.mil.

TRANSPORTATION CORPS

Firepower and mobility are the two fundamental, inseparable capabilities that ensure success in tactical operations, and the basic ingredient of mobility is transportation—modern equipment to do the job in the most effective, fastest, and safest way, with quality people who are technically and tactically proficient. That's the Transportation Corps (TC)—the Spearhead of Logistics.

The Transportation Corps, established 31 July 1942, grew out of the increased necessity for centralized control and operations as a result of the Army's expansion during the mobilization and war years of 1940–42. Prior to that time, transportation responsibilities were split between the Quartermaster Corps and Corps of Engineers. The establishment of the Transportation Corps came at a critical time. The problems of moving millions of soldiers and uncountable tons of supplies to virtually every corner of the world had no precedent in history. Fledgling officers, many drafted from civilian counterpart jobs, became the nucleus of the new branch and admirably accomplished many of the most successful transport operations ever recorded. Every type of carrier was used. Ships of all sizes and description (the Army maintains its own watercraft fleet), aircraft, railroads, trucks, buses, and military vehicles were put to work to get the job done. Volumes have been published on the statistics and accomplishments of the Transportation Corps during

Transportation by rail (represented by a flanged, winged wheel on a rail), air (represented by wings on the wheel), land (symbolized by the shield used as standard U.S. highway markers), and water (indicated by the ship's steering wheel) make up the Transportation Corps insignia. The current design was approved in 1942 and is based on a similar one in use since 1919.

World War II. Suffice it to say that the corps earned a place in the heart of every fighting man who reached for ammunition and found it.

The vital role of the Transportation Corps has been repeatedly demonstrated in all of the Army's operations since 1942. Active, reserve, and National Guard transportation units were deployed to provide theater-wide transportation services and operate air, motor, sea, and rail terminals and networks. These established networks, in conjunction with commercial airlift and sealift and host nation support, allowed the Transportation Corps to rapidly deploy, sustain, and maneuver an overwhelming force.

Fort Eustis, Virginia, is the home of the Transportation Corps and the U.S. Army Transportation School. TC officers attend the Logistics Basic Officer Leadership Course and Combined Logistics Officer Career Course at Fort Lee, Virginia. They see the full range of responsibilities, from the systems operating within the continental United States—air, rail, highway, and inland waterways—to the many foreign ports and air bases used to funnel supplies and equipment to Army troops around the world. TC officers also learn where and how military, civilian, and host nation transportation fits into overall logistical planning for the fighting force and review the need for research and development. Materiel systems integration, training, doctrine, and force structure issues that cross organizational lines and/or have a broad impact on the Transportation Corps are coordinated on behalf of the Chief of Transportation by transportation "cells" at the U.S. Army Combined Arms Support Command, Fort Lee, Virginia.

Transportation officers must gain a thorough knowledge of all modes of transportation and transportation systems, develop management skills to control use of these systems, and be competent leaders to guide and motivate the people who operate the systems. The organizational structure of the Army includes highway transport units, which use various types of trucks; marine terminal units, which operate ports, discharge and load cargo, and use various types of boats or lighters ranging from the 2,000-ton-capacity Logistics Support Vessel (LSV) to the Lighter Air Cushion Vehicle, 30-ton (LACV 30) hovercraft; cargo transfer units, which manage and conduct loading and off-loading of cargo; movement control units, which control the flow of transportation assets for all transportation modes; and, in the reserve component, railroad operating units. This diversity of missions provides TC officers with a range of assignments and professional possibilities.

Opportunity takes many forms for the TC officer. First, duty assignments are available throughout the world. Some of the more exotic include Australia, Hawaii, Japan, Korea, Turkey, and all the places where the Army is deployed. Of course, assignments are also available throughout the continental United States (CONUS) and Europe. Command opportunity is outstanding at company, battalion, and brigade levels. TC officers compete for transportation commands as well as multifunctional logistics commands. Educational opportunities abound.

The Transportation Corps' future looks to the increased use of robotics and automation and missions in space for officers and enlisted personnel with specialties in cargo handling and documentation. The Transportation Corps will provide the leadership and the means to enable the Army to move into space. Also, as the Army draws down the force and becomes more CONUS-based, the focus will turn toward improved strategic deployment capability. The Transportation Corps is at the forefront of that effort and remains the keystone to getting forces, equipment, and supplies to the fight. It is the Spearhead of Logistics. For more information on the Transportation Corps, please see the website at https://transportation.army.mil.

LOGISTICS CORPS

The Logistics Corps consists of officers in the historical Ordnance, Quartermaster, and Transportation regiments, as well as the multifunctional Logistics branch, and civilians who work in logistics career fields. This section explains how logistics officers are accessed, assigned, and professionally developed, as well as describes the skills, knowledge, and attributes needed for such officers to succeed in today's Army.

Army logistics dates back to the early days of the American Revolution with the establishment of the Quartermaster Department in June 1775. The Ordnance Department followed during the War of 1812. World War II saw the creation of the Transportation Corps in July 1942. These three branches, and their supporting civilians, have long, distinguished records of superior service and are vital components of the total Army force structure. In 1993, FA 90 was created within the Operations Career Field in order to support the development of multifunctional logisticians. Since then, the FA 90 designation has been used to signify officers skilled across the functional logistics branches. In 2005, as part of an Officer Personnel Management System review, an effort was undertaken to examine how to further advance the notion of multifunctional logistics leaders. The result was the creation of a Logistics branch for officers in the grade of captain through colonel and the formal recognition of a Logistics Officer Corps as approved by the Army Chief of Staff in May 2006. The Logistics branch official establishment date was 1 January 2008.

The Logistics Branch insignia, a ship's steering wheel crossed by a key and cannon and a stylized star in the center, represents the union of these functional areas. The key represents the Quartermaster Corps' responsibility to provide supplies and services. The ship's wheel denotes the Transportation Corps' movement of troops, supplies, and equipment. The cannon symbolizes the Ordnance Corps' maintenance and munitions responsibilities. Finally, the stylized star in the center represents the unity and integration of all these functions.

The Logistics Officer Corps includes all commissioned officers holding a branch or military occupational specialty (MOS) within the Logistics Corps. Commissioned officers accessed as lieutenants into one of the three Logistics Corps functional branches (Transportation, Ordnance, and Quartermaster) develop their functional branch skills for the first four years of their careers. Upon promotion to captain and successful completion of the Captains Career Course, officers are inducted into the Logistics branch. Their original functional branch, Ordnance, Quartermaster, or Transportation, becomes their secondary area of concentration (AOC) and qualifies them for functional assignments.

Logistics branch officers plan, integrate, and direct all types of sustainment activities in order to operate effectively on the modern battlefield, enabling Army forces to initiate and sustain unified land operations. The nature of twenty-first-century warfare mandates Logistics branch officers maintain competence in all facets of logistics; therefore, the Logistics branch merges Transportation, Ordnance, and Quartermaster basic branch officers into one unified branch at the rank of captain.

The Commanding General, U.S. Army Combined Arms Support Command is the proponent for the Logistics branch and the contact office is the Logistics Branch Proponency Office at the Combined Arms Command, Fort Lee, Virginia, 23801; 804-734-1188 or 804-734-0312.

Logistics branch officers serve in both operating and generating forces and require extensive knowledge and experience in planning, preparing, executing, and assessing the sustainment warfighting function subcomponent of logistics. Logistics tasks include ammunition management, supply, field services, transportation, maintenance, distribution, operational contract support, and life-cycle logistics. Logistics branch officers also serve as the Army's Explosive Ordnance Disposal (EOD) experts. EOD is a highly technical area and is a sub-function of the protection warfighting function. Logistics branch officers must be familiar with the other sustainment warfighting function subcomponents, including Personnel Services (human resource support, financial management operations, legal support, religious support, and band support) and Health Services Support. Logistics officers support Special Operations Forces, Joint Forces, and Defense Support to Civil Operations, and handle more functions as further changes are made to the Army's mission in Multidomain Operations.

A Logistics officer is multifunctionally developed and an expert in a function or specified skill set in order to support unified land operations. The ultimate goal for Logistics branch officers is to never let a mission fail due to lack of quality logistics support. Mission success ultimately requires the proper balance between technical skills and the ability to understand and apply the appropriate tactical skills at the right moment. Success also requires honed conceptual skills, enabling officers to handle changing situations and ideas. They must be experts in integrating the various aspects of logistics into the commander's plan and must be heavily experienced in multifunctional logistics. They must have a basic competence in the skills, knowledge, and attributes of supply, maintenance, and transportation.

Leaders must be grounded in Army Values and the Warrior Ethos, competent in their core proficiencies, and broadly experienced enough to operate across the levels of war—tactical, operational, and strategic. Officers' knowledge must include the ability to function at the strategic level of logistics. Understanding the Industrial Enterprise of the Army, leveraging its capabilities, and operating as a part of it are critical to our supported warriors.

Operating in support of unified action partners in joint, interagency, intergovernmental, and multinational (JIIM) environments while leveraging capabilities beyond the Army in achieving objectives is critical to success. Logistics leaders place the welfare of their soldiers above their own. These officers are self-reliant, agile, and proactive leaders who work in asymmetric and unpredictable environments where time available for mission analysis is constrained, but where sound, timely decisions are urgent. They are responsive to rapidly evolving operational environments and they improvise ways and means to accomplish the mission when doctrinal approaches do not apply. From the Basic Course forward, tactics are an essential skill set and are incorporated into the education of logistics officers from inception to the end of their careers.

The Logistics branch incorporates five AOCs and four skill identifiers (SI). All AOCs and SIs are open to all Logistics branch officers. Logistics branch officers all have a common, primary AOC of 90A00. Logistics branch officers also hold a secondary AOC indicating their functional specialty (90A88, 90A91, 90A92) within force sustainment. Officers continue their regimental affiliation with the Transportation, Ordnance, or Quartermaster regiments based on their secondary AOC. Officers should refer to the branch sections of this chapter for detailed description and criteria for each of the AOCs.

8

Special Branches

Special branches are those in which officers must have particular civilian education or professional certification. They include the Chaplain Corps, Judge Advocate General Corps, and the six corps within the Army Medical Department. Generally, officers have already received their professional educations when they join these corps, although a significant number of officers served in the Army previously in other branches—as either officers or enlisted soldiers—and returned to the colors after gaining their professional certifications elsewhere. In a few cases, such as physician's assistants (members of the Army Medical Specialist Corps), the Army provides the educational opportunity to members without a break in service.

CHAPLAIN CORPS

Clergy serving in the U.S. Army are called chaplains. To qualify for commissioning as an Army chaplain, a member of the clergy must have satisfactorily completed college and theological or equivalent graduate training acceptable to the Department of the Army. The chaplain must be endorsed by a particular faith or denomination for ministry to Army personnel. A direct commission as an Army officer is the usual procedure. Theological students may be commissioned as chaplain candidates and begin training before graduation and ordination.

As a member of the clergy, the military chaplain is a representative of a particular religious faith. The primary mission of a chaplain is to perform ministry by conducting religious services and by providing a complete program of religious education for American soldiers, family members, and authorized civilians. Chaplains conduct sacraments, rites, and ordinances consistent with their endorsing faith group. Counseling is one of the primary religious functions of the Army chaplain. Chaplains counsel on religious and quasi-religious subjects in chapels, hospitals, and quarters and in combat, training, and recreational areas.

The chaplain is also a staff officer and serves on the special staff of the commander. In this capacity, the chaplain advises the commander on matters of religion, morals, and morale as affected by religion. The chaplain maintains

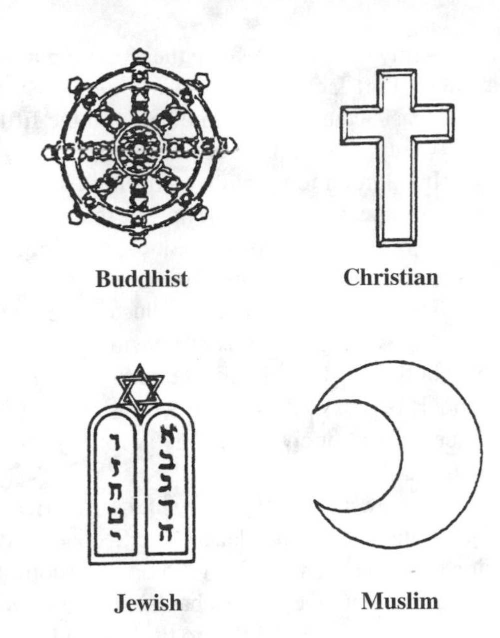

The Chaplain Corps is authorized four distinctive insignia: the cross, the tablets, the Buddhist wheel, and the crescent. The cross is a symbol of Christianity and is worn by all Christian chaplains with no distinction between Protestant and Catholic. The cross was approved as an insignia in 1898. Jewish chaplains wear insignia symbolizing the tablets of Moses, which have Hebrew characters representing the Ten Commandments. Above the tablets of Moses is the six-pointed Star of David. The Jewish insignia was approved in 1918 and was changed to incorporate the Hebrew characters in 1981. Buddhist chaplains wear the Buddhist wheel, symbolizing the perfectness and everlastingness of Buddhist teachings. The eight spokes of the wheel indicate the eightfold Right Path, the most fundamental teachings of Buddha. The Buddhist insignia was approved in 1991. Muslim chaplains wear the silver color crescent symbolizing the Muslim faith. The Muslim insignia was approved in 1993.

liaison with civilian religious groups and welfare agencies to facilitate cooperative programs. The chaplain has responsibility for participating in civic action projects and for advising the commander on matters of religion in the culture of the local inhabitants in overseas areas of operations.

Chaplains become deeply involved in the spiritual welfare of the command and accompany units in training to remain sensitive to their morale and familiar with unit missions. By sharing the hardships associated with field duty, chaplains build trust and confidence among soldiers and set an example of patience and good cheer. By being available to the soldiers and families of the command, listening to their problems, and assisting them or directing them to the appropriate agencies for help, chaplains at once help maintain high morale and help units discharge their responsibilities. Few things are more highly prized by commanders, in peace or in war, than an excellent chaplain.

In combat, the chaplain performs the functions of religious ministry, including the spiritual care of the wounded and dying, as well as prisoners of war. The Army Forward Thrust program assigns chaplains to maneuver battalions, as far forward as possible. When captured and imprisoned, special status is accorded under the Geneva Convention, which permits the chaplain to continue ministry among fellow prisoners of war.

Army chaplains receive their training at the U.S. Army Chaplain Center and School, Fort Jackson, South Carolina. Early in their service, there is an eleven-week basic course; sometime between the fifth and seventh year of service, there is a twenty-one-week career course.

In addition to specialized military schooling, the Army chaplain is considered for special civil schooling in one or more disciplines, including religious education, clinical pastoral education, drug and alcohol abuse, and preaching. Other continuing education programs, some available from the Chaplain Center and School, supplement these educational opportunities.

The Army chaplain is responsible for providing religious support to all members of the unit of assignment and to a designated area of responsibility. Religious support is provided as permitted by the chaplain's endorsing faith group and by other military or civilian clergy according to the religious needs of the command and/or area.

The organized chaplaincy in the American Army was established prior to the Declaration of Independence. The Second Continental Congress created the position of chaplain on 29 July 1775 on the recommendation of Gen. George Washington. Since that time, Army chaplains have served in all areas of the world, from the battlefields of the Civil War to the Bataan Death March and in the jungles of South Vietnam to the sands of Southwest Asia. Since the founding of the Chaplain Corps, more than 270 chaplains have died as a result of hostilities. They continue to minister to the spiritual needs of our soldiers in theater, often at the risk of their own personal safety.

In recognizing the professional character of the chaplains, the Department of the Army provides chaplains with enlisted assistants. These enlisted volunteers, known as chaplain assistants, provide administrative support, perform as vehicle drivers, and guard the chaplain's life in combat. (Chaplains are defined as noncombatants by the Geneva Convention of 1949.) The enlisted assistant is an important link between chaplains and service personnel in the religious program within the U.S. Army. Together, the chaplain and the assistant constitute the Unit Ministry Team. For more information on the Chaplain Corps, visit https://isairl.tradoc.army.mil.

JUDGE ADVOCATE GENERAL CORPS

The legal affairs of the Army and its soldiers are entrusted to this special group of officers, all of whom are graduates of accredited law schools, members of their state's bar, and commissioned in the Judge Advocate General Corps. This corps traces its beginning from 29 July 1775, when William Tudor—a leading Boston lawyer—became the first Judge Advocate of the Army. A year later the designation of Judge Advocate General and the rank of lieutenant colonel were prescribed for this office.

Until 1802, other individuals were designated as Judge Advocate of the Army, but after that date and until the corps was officially established, the term *Judge Advocate* was used rather freely in designating officers whose primary function was prosecuting courts-martial and advising military commanders on matters pertaining to military justice and enforcing discipline in the Army.

By an act of Congress in 1849, the office of Judge Advocate of the Army was established on a permanent basis, but a corps of officers was not authorized until 1862. In 1864, the Judge Advocate was granted the rank and pay of a brigadier general, and the Bureau of Military Justice was created as a forerunner of the Office of the Judge Advocate General. A merger of the bureau and the corps of judge advocates in the field took place in 1884 and resulted in the creation of the Judge Advocate General Department, by which it was known until 1948, when "corps" was substituted for "department."

Military law grew in scope and intensity from the time of the Civil War so that today the Judge Advocate General Corps performs many duties and services not envisioned by those who created it. From the Judge Advocate General, who acts as legal adviser to the Secretary of the Army and all other agencies within Headquarters, Department of the Army, to a legal officer at a small installation, members of the corps face day-to-day situations far more diversified than in most general civilian law practices. Today's judge advocate is skilled in the law of nations, the environment, labor relations, and contracts, as well as in the Army criminal justice system. He or she also advises the commander on installation problems, personnel administration, patent and tax law, and claims by and against the government.

A primary responsibility of the Judge Advocate General Corps is the complete administration of the Uniform Code of Military Justice. To fulfill its ever-increasing military justice responsibility, the role of the corps has been greatly increased to provide legal advice and counsel for accused soldiers in areas where previously no such right was guaranteed. Moreover, constant refinements in the Uniform Code of Military Justice and the implementing regulations foster an atmosphere of professionalism, free from the command restraints and influence of the past. For example, the judge advocates who act as military judges in courts-martial are organized in a separate command, the U.S. Army Legal Services Agency, and are assigned and rated independently.

Other services to and protections for individual soldiers have been introduced. Like the military judge, the military defense counsel has been placed in a separate organization, the U.S. Army Trial Defense Service; counsel is provided to soldiers in the many administrative and nonjudicial proceedings that can affect them; and judge advocates are now authorized, in some jurisdictions, to take an active role in assisting soldiers with personal legal problems under the Expanded Legal Assistance Program.

The Judge Advocate General School, located on the grounds of the University of Virginia at Charlottesville adjacent to the prestigious University of Virginia Law School, is where all new judge advocate officers receive initial orientation into the Army and where other members of the corps return for career and specialized courses. The Judge Advocate General School is the home of the Judge Advocate General Corps Regiment. Its students include other uniformed judge

The Army's legal arm traces its insignia to the year 1890. The crossed pen and sword symbolize the recording of testimony and the military character of the branch. The wreath is the traditional symbol for achievement.

advocates and civilian U.S. government attorneys from many agencies and some legal officers from the armed forces of foreign nations. The school's program of continuing legal education supplements and hones the general skills of the practicing attorney with graduate-level instruction in more than twenty fields of law necessary to fulfill the corps' complex mission. In addition to the basic and graduate courses, the school offers short courses to military judges, contracting officers, commanders, trial and defense counsel, international law specialists, and others concerned with Army administration.

The Judge Advocate General Corps is a special branch, and its services are required throughout every level of command. The opportunities for assignment anywhere the U.S. Army has troops are plentiful and diversified. Early in his or her service, a judge advocate is likely to be involved in legal assistance, claims, or courts-martial. They are basic missions of the corps, provide valuable trial experience, and are essential to the development of the well-rounded legal officer. Experienced judge advocates can look forward to challenging assignments at levels within the Army's structure where policy and decisions of importance are made. For more information on the JAG Corps, visit https://tjaglcs.army.mil.

MEDICAL CORPS

After Gen. George Washington had been appointed Commander in Chief of the Continental Army by the Continental Congress, he requested medical support for his troops, and on 27 July 1775, a Hospital Department was authorized, the forerunner of the original Army Medical Department.

From the end of the Revolutionary War until 1818, when the title of Surgeon General was authorized, the Chief of the Medical Department had little control over those physicians on duty with Army units, since they took their orders from the officers who commanded the units.

One of the Army's finest contributions to medical knowledge began in 1836, when Dr. Joseph Lovell started collecting medical books for the Surgeon General's Library. After the Civil War, Dr. John Shaw Billings was appointed librarian, and before he retired in 1895, the collection was expanded to several thousand volumes, and he was publishing the *Index Medicus* to make it possible for a physician to find a reference to everything ever published on any given medical subject. In 1956, the library was turned over to the Department of Health, Education, and Welfare (now Department of Health and Human Services) to become the National Library of Medicine. It is the most complete collection of medical literature in the world.

During the Civil War, Dr. William A. Hammond started a collection of morbid specimens for pathologic study. This was the beginning of the Army Medical Museum, which expanded to the Armed Forces Institute of Pathology, one of the foremost diagnostic, teaching, and research institutions in this country.

The busts of two Army physicians are included in New York University's Hall of Fame—Dr. Walter Reed, for conquering yellow fever, and Dr. William Crawford Gorgas, for using Reed's discovery to improve health conditions in Panama to make the building of that canal possible. As an added honor, the first Distinguished Service Medal ever issued went to Dr. Gorgas when he retired as Surgeon General of the Army in 1918.

The Medical Corps has contributed to civilian health and medicine in many other ways: Army doctors established the first American school of medicine; they published the first American textbooks on surgery, psychiatry, and bacteriology, and the first pharmacopoeia (encyclopedia of drugs and their uses); introduced the smallpox vaccination in this country; started the first systematic weather reporting; published the first summary on vital statistics; began the chlorination of water; and developed numerous vaccines against diseases of humans and animals.

Today the Army maintains some of the finest medical treatment facilities in existence. Medical centers, such as Walter Reed National Military Medical Center in Maryland, are accredited teaching institutions where internship and residency training are given, and medical and dental research are of world renown.

With centralized control vested in the Surgeon General, the Medical Department initiated many far-reaching projects that affected the medical profession in general and brought prestige and status to the Medical Corps. Military physicians held such titles as surgeon, assistant surgeon, or medical inspector and were not given military rank until 1847, when military rank was assigned to members of the Medical Corps then on active duty. An act of Congress in 1908 established the Medical Corps Reserve, the first such Army reserve group in this country, and the forerunner of the Officers Reserve Corps in 1916.

The all-physician Medical Corps is responsible for setting the physical standards for all individuals entering military service; maintaining their health while in service; and processing them for discharge or retirement. The clinical care of Army family members and retired personnel gives the Army doctor a well-rounded practice, but essentially military medicine is aimed at the care of troops.

First used as a cloth insignia in 1851, the caduceus in its present form was approved in 1902. Except for the Medical Corps, the caduceus worn by officers of the other Medical Department branches is superimposed with a letter or letters indicative of the specific corps. Rooted in mythology, the caduceus has historically been the emblem of physicians, symbolizing knowledge, wisdom, promptness, and skill.

Hence prevention of disease and injury is as important as rehabilitation. The control of the environment is becoming increasingly vital in the Army's relations with its civilian neighbors.

Since 1969, the Special Assistant to the Surgeon General for Medical Corps Affairs has served as the equivalent of the Chief of the Medical Corps. This position is currently authorized a brigadier general and entitled Chief, Medical Corps Affairs. This officer shares responsibility for the professional guidance, assignment, education, training, and career development of Medical Corps officers with the Commander, U.S. Army Total Army Personnel Command and the Commander, U.S. Army Medical Department Center and School. In addition, the Chief, Medical Corps Affairs is currently responsible for all professional health care operations for the U.S. Army Medical Command. He serves as the Deputy Commanding General for Health Care Operations at U.S. Army Medical Command, Fort Sam Houston, Texas. For the most current information on the Medical Corps, visit www.armymedicine.health.mil or see Stackpole's *Army Medical Officer's Guide*.

DENTAL CORPS

All specialties of dentistry are represented in the Army Dental Corps. The mission of the corps is one of providing all levels of dental care necessary to preserve the oral health of the Army in support of its fighting strength.

The Chief of the Dental Corps, who also is designated as the Assistant Surgeon General for Dental Services, holds the grade of major general. He or she serves as the principal adviser to the Surgeon General of the Army on all matters concerning the Dental Corps and dental services, establishes professional standards and policies for dental practice, and initiates and reviews recommendations relating to dental doctrine and organization.

Prior to 1911, dental support of the military was performed by civilians under contract. The Dental Corps was established on 3 March 1911 and consisted of sixty dental surgeons. In 1916, along with most of the Medical Department, the Dental Corps was reorganized, expanded, and became fully established as a needed and valuable contributor to the health and welfare of the entire Army. Since 1978, dental personnel have been organized into Dental Activities (DENTACs) and Area Dental Laboratories (ADLs), which are commanded by Dental Corps officers. On 1 November 1993, the U.S. Army Dental Command (Provisional) was established and later was activated as a major subordinate command of the U.S. Army Medical Command on 2 October 1994. As part of this ongoing reorganization of the U.S. Army Medical Department, Dental Clinic Commands, subordinate to DENTACs, were established at smaller posts and installations. The U.S. Army Dental Research Detachment, which has provided significant contributions to health care delivery, conducts research and assists in the development of materials and techniques related to combat dentistry.

Officers are appointed to the Dental Corps upon graduation from a school of dentistry accredited by the American Dental Association, and after being awarded a degree of either Doctor of Dental Surgery (DDS) or Doctor of Dental Medicine (DMD). During his or her Army career, a dental officer can expect a variety of assignments throughout the world. Early in their service, officers are normally assigned to dental clinics in direct patient care. Opportunities for advanced education and accredited residency training in a dental specialty are offered, on a competitive basis, usually after four to five years of active service. Other positions that may be available later are clinician, residency mentor or director, dental researcher, clinic chief, dental staff officer, dental clinic commander, or dental activity commander. For the most current information on the Dental Corps, visit www.armymedicine.health.mil.

VETERINARY CORPS

Veterinarians have been associated with the nation's military services since the mid-1800s, when veterinary surgeons were authorized for each cavalry regiment. The Veterinary Corps was made a part of the Medical Department on 3 June 1916, to centralize control of the veterinary personnel caring for the Army's animals and inspecting food supplies.

With the evolution of the mechanized cavalry and technological changes in food processing, the Veterinary Corps officer assumed new roles. By virtue of education and training, the veterinarian is eminently prepared to function not only in animal medicine but also in matters of public health and comparative medicine as an integral member of the military community health team. His or her professional services encompass food hygiene, veterinary public health and preventive medicine, and veterinary medical care of military animals. The veterinarians' services are vital to the management and care of the extensive laboratory animal resources and to military research and development. The corps' primary mission is to protect and preserve the health of people in the armed forces.

The Chief of the Veterinary Corps is a colonel who is also designated as the Director, U.S. Army Veterinary Services Activity. He or she participates in assignment and career planning of Veterinary Corps officers.

All members of the corps are veterinarians who have graduated from an accredited college of veterinary medicine after being awarded either a Doctor of Veterinary Medicine (DVM) or Veterinary Medical Doctor (VMD) degree. Many positions in the Army now require postgraduate specialized training, and the Veterinary Corps takes this into account in matching talents and interests in career planning. Veterinary officers are assigned wherever food hygiene and nutritional quality control, preventive medicine, animal medicine, or research is conducted. Veterinary Corps specialties include veterinary public health and preventive medicine, laboratory animal medicine, veterinary pathology, veterinary microbiology, and veterinary comparative medicine.

In 1980, the Air Force Veterinary Service was disestablished and the Army Veterinary Corps became the executive agent for Department of Defense Veterinary Services. In 1981, a program was established to train enlisted personnel to be veterinary food inspection technicians. Graduates are appointed as warrant officers. They assist the Army Veterinary Corps in the greatly expanded mission of providing veterinary services support throughout the Department of Defense. For the most current information on the Veterinary Corps, visit www.army medicine.health.mil.

MEDICAL SERVICE CORPS

There are two distinct purposes for the existence of the Medical Service Corps (MSC). One is to provide scientists and specialists in the specialties allied to medicine, and the other is to provide officers technically qualified to make the Medical Department self-sustaining in the areas of administration, supply, environmental sciences, mobilization preparedness, readiness training, and engineering activities.

The chief of the corps, a brigadier general, serves as an adviser and consultant to the Surgeon General and participates in the assignment of and career planing for MSC officers and Medical Department warrant officers. Within the allied sciences, pharmacists, optometrists, biochemists, physiologists, podiatrists, audiologists, and many other specialists are commissioned to support the full range of health care services available to all members of the Army, their dependents, and other beneficiaries.

As it is now constituted, the Medical Service Corps was established in 1947 to replace the Medical Administrative Corps, Sanitary Corps, and Pharmacy Corps. It is presently organized into four sections: Pharmacy, Supply, and Administration (PS&A); Medical Allied Sciences; Sanitary Engineering; and Optometry. The four sections are further divided into twenty-five distinct career fields or areas of specialization. The PS&A section includes positions related to personnel management, financial management, pharmacy operations, supply management, and patient administration. MSC officers are also assigned to units equipped for medical evacuation by helicopter. The specialties included in the Sanitary Engineering section are sanitary engineering, environmental science, medical entomology, and nuclear medical science. Except for the PS&A section, where changes do occur, officers commissioned in the Medical Service Corps usually remain within their specialties for their entire careers. The specialties that are included in the Medical Allied Sciences section are audiology, medical laboratory sciences (microbiology, biochemistry, immunology, parasitology, and related laboratory sciences), psychology, social work, optometry, and podiatry.

Along with those in other Medical Department corps, officers commissioned in the Medical Service Corps can expect to attend an orientation course at the U.S. Army Medical Department Center and School, located at Fort Sam Houston,

Texas. Upon completion of this course they are assigned to medical facilities or activities or other Army units requiring their particular skills.

At higher levels of operation, MSC officers are assigned to duties in research and development and as comptroller, plans and operations officer, personnel manager, materiel officer, deputy commander for administration in medical centers and community hospitals or medical units, and key staff adviser in major headquarters, Department of the Army, and Department of Defense. For the most current information on the Medical Service Corps, visit www.armymedicine.health.mil.

ARMY NURSE CORPS

The Army Nurse Corps, established in 1901 as a result of the devoted efforts of civilian contract nurses during the Spanish-American War, is the oldest military nursing corps and the first military organization to admit women. The mission of the Army Nurse Corps is to provide quality nursing service whether the need is for immediate deployment, sustained conflict, or any other health care mission. The Chief of the Army Nurse Corps, a brigadier general, serves as principal adviser and consultant to the Surgeon General on staff policies, activities, and other matters pertaining to nursing, nursing personnel, and the Army Nurse Corps. Army nurses contribute significantly to the Army Medical Department's team effort to provide high-quality, easily accessible, and cost-effective health care. They work in various clinical specialties—community health, psychiatric/mental health, operating room, anesthesia, obstetrics, and medical-surgical nursing, as well as functional areas—education, administration, and research. These officers serve in hospitals, in units in the United States, and overseas for humanitarian and nation-building operations. They also serve in staff positions in the Office of the Surgeon General and Department of the Army.

To provide well-educated nurses for the Army, the U.S. Army Cadet Command began commissioning nurses into the corps in 1982. This successful program relies on financial scholarships awarded to student cadets and the Nurse Summer Training Program (NSTP). The hands-on training allows nursing cadets to work in the hospital and in the field and provides cadets unique opportunities to develop and practice their leadership skills. To ensure a smooth transition for newly commissioned officers once they begin active duty, each Army hospital conducts a preceptor program that assigns the new nurse to a preceptor. This allows the new nurse officer to work one-on-one with an experienced nursing professional. This unparalleled clinical experience offers novice nurses unique opportunities to gain valuable professional skills under the direction of expert nursing practitioners. Army nurses further enhance their professional experience through a variety of continuing education, specialty courses conducted by the Army, and graduate and postgraduate study in civilian academic institutions at Army expense. For nurses who complete a master's degree in a clinical specialty, the corps offers advanced-practice nursing roles such as nurse anesthetists, nurse practitioners, clinical nurse specialists, and nurse midwives. Advanced-practice nurses have unique opportunities to integrate inpatient and outpatient care to follow the patient throughout an entire episode of care.

ARMY MEDICAL SPECIALIST CORPS

The Army Medical Specialist Corps, formed by enactment of Public Law 36 on 16 April 1947, is composed of four unique medical specialties: occupational therapists, physical therapists, dietitians, and physician's assistants. Although one of the younger corps of the Army Medical Department, the individual specialties of the Army Medical Specialist Corps have been contributing far longer; as early as the Spanish-American War, dietitians were serving as civilian practitioners. Occupational therapists, physical therapists, and dietitians played a large role in the rehabilitation of World War I and World War II casualties. Physician's assistants, formerly warrant officer specialists managed by the Medical Corps, were integrated into the Army Medical Specialist Corps in 1992 through a process of transition into commissioned officer grades.

The minimum educational qualification for accession is a bachelor's degree within the particular activity and the appropriate professional licensure, certification, or registration. However, the Army provides specialty postbaccalaureate training for those qualified accessions who must complete a dietetic internship or occupational therapy affiliation to become eligible for registration or certification. Physician assistants are accessed primarily through enlisted soldier participation in the baccalaureate-producing Military Physician Assistant Program at the Army Medical Department Center and School, Fort Sam Houston, Texas. Additionally, a Master's of Physical Therapy Degree program, affiliated with Baylor University in Waco, Texas, is conducted to provide basic, entry-level professional education for physical therapists. In addition to the provision of specialty training of qualified individuals or the direct accession of fully qualified practitioners, ROTC graduates, when professionally qualified, are also eligible for commission within the Army Medical Specialist Corps.

The Chief, Army Medical Specialist Corps, is a colonel who serves as an adviser and consultant to the Surgeon General on all matters pertaining to the corps. By statute, there are four assistant chiefs, one for each specialty area, who serve as consultants in their professional specialty areas and make recommendations on professional development within their specialty.

In October 1981, the Army Medical Specialist Corps implemented postgraduate training in clinical specialization paralleling current practice in civilian health care programs. These specialty training programs enhance the provision of patient care, support graduate medical education, and increase career opportunities and job satisfaction.

Army Medical Specialist Corps officers play a vital role not only in the maintenance of soldier readiness and direct support of warfighting operations but also in the comprehensive treatment and rehabilitation of patients and in the promotion of health and prevention of injury.

Since 1955, both men and women have been eligible for commission in the Army Medical Specialist Corps and are assigned to Army units in all echelons and theaters. They serve in all Army medical centers and Army community hospitals in the United States and in overseas commands. Army Medical Specialist Corps officers may also hold administrative and management positions at the Army Medical Department Center and School or may hold faculty positions at the school, teaching officer, enlisted and civilian personnel. Additionally, they may serve in positions at research and management information system agencies, in the Office of the Surgeon General, and in major command headquarters. For the most current information on the Army Nurse Corps, visit www.armymedicine.health.mil.

9

Army Posts and Stations

'Tis a good and safe rule to sojourn in many places, as if you meant to spend your life there; never omitting an opportunity to doing a kindness, or speaking a true word, or making a friend.

—John Ruskin

This chapter contains information of special interest to officers and their families about Army posts and stations in the United States. It has been developed as a dependable "first place to look" after receiving orders for change of station. The information was obtained through the generous cooperation of post commanders and their information and housing officers. It purposely has been made general in nature, designed to provide only a brief introduction to a particular post or station.

Shortly after receiving assignment orders, you can expect to be contacted by your sponsor on behalf of the commander of your new unit. Each incoming officer is assigned a sponsor of equal or greater rank (and usually of similar marital status) who is responsible for assisting with the transition to your new duty station. You should receive up-to-date, specific information about housing, schools, unit training and readiness requirements, and the quality of life on and off post, among other things.

Commanders want you to be fully and accurately informed about your new assignments and for your move to go well so that you arrive on time, ready for duty. If you do not hear from your sponsor by sixty days before your reporting date, call the Deputy Chief of Staff for Personnel (DCSPER), G-1, or S-1 of the unit to which you are being assigned and request to be contacted by a sponsor. He or she should be responsive to all your questions and reasonable needs to facilitate a smooth arrival at the new assignment.

An assignment to a new post or station should be approached positively. It offers a real opportunity not readily available to others in our society. At your new post you will see a new area of the country, perhaps different from any you have seen before, with the difference measured not only in terms of scenery, climate, or historic sites, but also in terms of people and their customs. You will also make new acquaintances, some of whom will become lifelong friends. The total opportunity of a new assignment is limited only by your ability to learn, to understand, to meet others and to be met by them, to participate in the activities of the station, and to pursue actively the off-duty recreational activities available. Your attitude is the major determinant in making the new assignment a rewarding and pleasant experience.

The typical, established Army post is a pleasant place to live. Where there are quarters for families, the activities are substantially the same as in civilian communities. The ties of neighborliness and friendship are especially noteworthy because all are engaged in the common mission of the nation's security. The vast majority of Army members are young and active, and there are extensive off-duty athletic, cultural, recreational, and social activities available for officers, noncommissioned officers, and soldiers and their spouses and children.

There is a wide variation in the living and recreational facilities of Army stations. Some posts have been in use for decades and have been progressively developed so that the resulting environment is highly pleasing with excellent athletic, cultural, social, and recreational facilities. Others, established during World War II or later, are less extensively developed. The variation in number and adequacy of family quarters is wide; inadequate housing is considered by some officers to be the most serious objection to an Army career. This chapter does not seek to minimize it. However, the Army's leaders have fought hard for action, and there has been considerable construction and improvement in the situation. The progress must continue until all of our permanent posts and stations have adequate housing that is up to the standards of professional people in other walks of life. The complete environment must be a good place for Army members and their spouses and children to live.

FACILITIES

The following facilities are common to Army posts:

Commissaries. Most installations are authorized to have a commissary where groceries and household supplies, similar to those sold in commercial supermarkets, may be purchased. Merchandise is sold at cost, and the savings are substantial. By shopping in the commissary, customers save an average of about 25–30 percent when compared with commercial supermarket prices. Those who take advantage of special promotional sales and use coupons save even more. An identification card is required for entrance, and personal checks and credit cards may be used for purchases. A 5 percent surcharge is added to the total bill at the checkout counter. Surcharge funds are used for building new commissaries, renovating old ones, and paying for such things as grocery carts, meat-slicing machines, and paper bags. These nonprofit facilities contribute greatly to the quality of life of active-duty and retired soldiers and their families and other personnel authorized to shop in them. The commissary benefit is the second most important form of nonpay compensation for soldiers and family members.

Post Exchanges. Excellent post exchanges are provided at Army stations. The Army and Air Force Exchange Service (AAFES) prescribes the items and services that may be sold, generally comparable to goods and services available in drug and department stores. Profits are used for athletic and recreational programs, including athletic fields and other facilities. Post exchanges add to the convenience of post life and provide economies for all active and retired personnel who use them. An ID card is required for entrance and purchase of merchandise. Credit cards, ATM cards, and checks are accepted forms of payment at most exchanges. Additionally, most exchanges will cash checks for you. For more information, visit www.aafes.com/exchange-stores/Movie-Guide/.

Theaters. Most stations in CONUS and overseas have excellent theaters. Through the cooperation of the motion picture industry and the good work of the AAFES, the best and most recent films are shown. Like at the post exchange, the profits are returned as dividends for athletic, recreational, and other programs for the general improvement of post life.

Athletic Facilities. Indoor and outdoor recreational and physical development facilities are provided at almost every station. At many of these stations, the facilities are outstanding. They include playing fields, some of which have bleachers and dressing rooms for teams; indoor and outdoor swimming pools; fitness centers with extensive free weights, a variety of strength development machines, and cardiovascular fitness apparatuses; bowling alleys; tennis courts, many of which are equipped for night usage; and field houses for basketball and other indoor sports. Many posts have excellent golf courses.

Religious Activities. Most Army stations are provided with religious facilities and programs. The services of chaplains of the Protestant, Catholic, and Jewish faiths are almost universally available. Religious programs are conducted in much the same manner as in civilian communities.

Libraries. Most permanent posts have good libraries, although some posts were forced to eliminate them or curtail their services during the 1990s. Major TRADOC installations still possess excellent technical libraries, useful for military research, and offer free Wi-Fi and Internet access.

Schools. All Army stations at which dependents are authorized are provided with school facilities for children of Army members. In general, children attend on-post schools from kindergarten through the sixth grade, and junior and senior high school students attend local schools near the post. There are exceptions, of course. Where any distance is involved, free bus transportation is provided. Parochial or other private schools are available near many posts and may be used if desired. Consult the local authorities.

Medical Facilities. Most large Army posts have a station hospital; smaller posts may be equipped only with a clinic or dispensary, relying on a nearby military hospital for more extensive medical services. In any case, medical care is available to Army personnel and generally is available to family members, including those residing in off-post housing. See chapter 19, Personal and Financial Planning, for a complete description of medical care available and the procedures for obtaining it.

Family Quarters. As mentioned earlier, the problem of providing adequate quarters for Army members and their families is serious. The actual housing situation varies not only from post to post but also at any particular post, depending on the numbers of arriving and departing personnel. A post with an adequate number of family quarters today could find itself woefully short tomorrow when a major new unit is assigned.

In general, you should assume that there will be a waiting period for on-post quarters at any new post. While you are waiting, you will have to rent, lease, or perhaps purchase housing in the nearby civilian community. The actual waiting period may vary from a few days to many months. In some cases, such as the Washington, D.C. area, there is virtually no post housing available. Take this in stride. It is a recognized disadvantage of Army life, but it should be viewed in total context along with the many advantages.

In all cases, be realistic and businesslike in your approach to providing housing for your family. It always is advisable to obtain the latest quarters information from the housing officer at your new post. If you do not receive this information within a few weeks after assignment orders, be sure to request it from your sponsor. Although advice from those who have previously been stationed at your new post is always welcome, remember that their information about quarters availability almost surely is out of date. Get the correct, up-to-date information about both on-post and off-post housing from the post housing officer and then make your personal plans accordingly.

One note of caution: When arriving at a new post, be sure to check with the housing office before making any commitments for lease or purchase of off-post quarters. Although policies vary, in most cases you must have the permission of the housing officer to reside off post.

Most stations have guest houses or other temporary accommodations, which you can use for short periods (five to ten days usually) while making arrangements for permanent quarters. These facilities generally are limited in number, and reservations well in advance are in order. Write to the post housing officer.

Activities for Spouses and Children. Most Army stations have organizations such as the officers' wives' club, Parent-Teacher Association, and youth activities of many kinds, including Boy Scouts, Girl Scouts, Teenage Clubs, and others. Active participation in these organizations is a good way for newcomers to become acquainted and to add their talents to the life of the post.

BASE REALIGNMENT AND CLOSURE (BRAC)

In the last few years, a number of Army posts and stations (and Air Force and Naval bases) have been identified for closure or mission realignment as a cost-saving measure. Selections have been made for the Secretary of Defense by the Commission on Base Realignments and Closures under the terms of 1988 legislation that mandated acceptance of all the commission's recommendations unless the entire package was rejected by both houses of Congress.

POST AND STATION DESCRIPTIONS

Billeting and Housing are two important points of contact for incoming officers and their families. Phone numbers for these offices are included below. Additionally, reservations for most guest house accommodations Army-wide can be made by calling (800) GO-ARMY1 (800-462-7691).

Most installations have websites, which contain more detail than can be included here. These websites are listed below or you can link to them from the Army's website: www.army.mil.

Designed to help reduce the stress of transfers, Relocation Assistance Centers have been established on most installations. Each has a library of welcome packages and videos from installations worldwide. A relocation assistance staff member can prepare a booklet for you or request a Welcome Aboard package or sponsor from your gaining command.

ARMED FORCES HOSTESS ASSOCIATION

The Armed Forces Hostess Association (Room 1E541, The Pentagon, Washington, D.C. 20310; Commercial: [703] 614-0350/0485) provides a unique service for all newcomers in the Washington metropolitan area. Local files include information on animal care, camps, entertainment, furniture repair, schools, vacation, touring, and other helpful items.

Note: In 2021 Congress established the Naming Commission to take stock of all military references to the Confederacy including those bases and posts named after Confederate Officers. The new name, if applicable, will appear in parentheses next to the current name. Renaming should be completed in 2024. Go to www.thenamingcommission.gov to learn more.

ABERDEEN PROVING GROUND, MARYLAND 21005
DSN: 298-5201; Commercial: (410) 278-5201

Aberdeen Proving Ground, established in 1917, consists of 72,500 acres on two post areas separated by the Bush River. The main post is located 20 miles northeast of Baltimore, on the Chesapeake Bay, adjacent to Aberdeen, Maryland. It is reached via I-95 or U.S. 40.

Major Activities. U.S. Army Communications-Electronics Command; U.S. Army Combat Capabilities Development Command; U.S. Army Test and Evaluation Command; U.S. Army Futures Command Network Cross-Functional Team; Aberdeen Test Center; 20th Chemical, Biological, Radiological, Nuclear and Explosives Command; U.S. Army Chemical Materials Agency; U.S. Army Medical Research Institute of Chemical Defense; Army Public Health Center; Army Research Laboratory; and program executive offices (PEOs), including Command, Control and Communications-Tactical; Intelligence, Electronic Warfare and Sensors; Assembled Chemical Weapons Alternatives; and Joint PEO Chemical, Biological, Radiological and Nuclear Defense. **Billeting.** Comm: (410) 278-5148

Housing Office. Comm: (410) 305-1076
Website. https://home.army.mil/apg

ANNISTON ARMY DEPOT, ANNISTON, ALABAMA 36201
DSN: 571-1110; Commercial: (256) 235-7501

Opened in 1941, Anniston Army Depot consists of 15,000 acres adjacent to Pelham Range, 10 miles west of Anniston, Alabama.

Major Activities. Repairs and retrofits combat tracked vehicles, artillery, and small arms; receives and stores general supplies, ammunition, missiles and small arms, and strategic materials.

Billeting. None.
Housing Office. None.
Website. www.anad.army.mil

FORT A. P. HILL (FORT WALKER), VIRGINIA 22427
DSN: 578-8324/8120; Commercial: (804) 633-8585

Established in 1942, Fort A. P. Hill is a part of the Military District of Washington, D.C., and consists of 76,000 acres. Located 20 miles southeast of Fredericksburg and about 40 miles north of Richmond, the post is easily accessible from Route 301 or I-95.

Major Activity. Fort A. P. Hill is used year-round for all-purpose military training with more than 230,000 active and reserve troops training there annually.

Billeting. Comm: (804) 633-8335; DSN: 578-8335
Housing Office. Comm: (804) 633-8445; DSN: 578-8445
Website. www.army.mil/aphill

FORT BELVOIR, VIRGINIA 22060
DSN: 685-5001; Commercial: (703) 805-5001

Fort Belvoir, established in 1912, occupies 8,656 acres located astride U.S. 1, 11 miles southwest of Alexandria, Virginia, and about 16 miles southwest of Washington, D.C. The fort has undergone a major expansion of tenant organizations following the latest BRAC.

Major Activities. National Geospatial-Intelligence Agency; Defense Intelligence Agency; Defense Acquisition University; U.S. Army Cyber Command; Office of the Chief Army Resrve; U.S. Army Intelligence and Security Command; Night Vision and Electronics Sensors Directorate; U.S. Army Legal Services Agency; Defense Logistics Agency; Defense Contract Audit Agency; Davison Army Airfield; DeWitt Army Community Hospital; 29th Infantry Division, Virginia Army National Guard; National Museum of the United States Army.

Billeting. Comm: (703) 704-8300
Housing Office. Comm: (703) 806-3019
Off-Post Housing. Comm: (703) 805-4590
Website. https://home.army.mil/belvoir

FORT BENNING (FORT MOORE), GEORGIA 31905
DSN: 835-2011; Commercial: (706) 545-2011

Fort Benning was established in 1918. The post occupies 182,000 acres, with the main gate located 9 miles south of Columbus, Georgia, off U.S. 27.

Major Activities. U.S. Army Maneuver Center of Excellence; 75th Ranger Regiment; 1st Security Force Assistance Brigade; 199th Infantry Brigade; 197th Infantry Brigade; 192nd Infantry Brigade; 198th Infantry Brigade; 194th Armored Brigade; 316th Cavalry Brigade; Airborne and Ranger Training Brigade; Western Hemisphere Institute for Security Cooperation; Maneuver Capabilities Development and Integration Directorate; Benning Martin Army Community Hospital; 1st Battalion, 28th Infantry Regiment; U.S. Army Marksmanship Unit; Soldier Lethality Cross-Functional Team; 98th Training Division (Reserve).

Billeting. Comm: (706) 689-0067
Housing Office. Comm: (706) 545-3921 DSN: 835-3921
Website. www.benning.army.mil

FORT BLISS, TEXAS 79916 AND 79918
DSN: 9782121; Commercial: (915) 568-2121

Fort Bliss was established in 1849. It occupies 1.1 million acres north of the city of El Paso, with the main gate located off I-10 or Highway 54.

Major Activities. 1st Armored Division, U.S. Northern Command's Joint Task Force North; U.S. Army Sergeants Major Academy; Joint Modernization Command; William Beaumont Army Medical Center; 32nd Army Air and Missile Defense Command; 11th Air Defense Artillery Brigade; 5th Armored Brigade; 402nd Field Artillery Brigade.

Billeting. Comm: (877) 711-8326 (915) 565-7777
Housing Office. Comm: (915) 568-2653
CHRRS. Comm: (915) 568-2898; DSN: 978-2898
Website. https://home.army.mil/bliss

FORT BRAGG (FORT LIBERTY), NORTH CAROLINA 28310
DSN:236-0011; Commercial (910) 396-0011

Fort Bragg was established in 1918 and occupies 148,609 acres. The main gate is located 10 miles northwest of Fayetteville, North Carolina, adjacent to Highway 24.

Major Activities. U.S. Army Forces Command; U.S. Army Reserve Command; XVIII Airborne Corps; U.S. Army Special Operations Command; Joint Special Operations Command; 82nd Airborne Division; 1st, 2nd and 3rd Brigade Combat Teams; 82nd Combat Aviation Brigade; 82nd Sustainment Brigade; 82nd Airborne Division Artillery; Security Force Assistance Command; 3rd Expeditionary Sustainment Command; 1st Special Forces Command; 3rd Special Forces Group (Airborne); U.S. Army John F. Kennedy Special Warfare Center and School; U.S. Army Civil Affairs and Psychological Operations Command; U.S. Army Aviation Command; 43rd Air Mobility Operations Group; 4th Training Brigade (ROTC); 20th Engineer Brigade; 108th Air Defense Artillery Brigade; 44th Medical Brigade; 16th Military Police Brigade; 525th Battlefield Surveillance Brigade; U.S. Army Parachute Team (Golden Knights); Womack Army Medical Center

Billeting. Comm: (910) 396-7700; DSN: 236-7700. Reservations: Comm: (910) 396-9510; DSN: 236-9510. Guest housing: Comm: (910) 396-7700; DSN: 236-7700

Housing Office. Comm: (910) 396-1022; DSN: 236-1022

CHRRS. Comm: (910) 396-2413; DSN: 236-2413

Website. www.bragg.army.mil

FORT BUCHANAN, PUERTO RICO 00934
DSN: 740-3400; Commercial: (787) 707-3400

Major Activities. Supports U.S. Army South.

Billeting. Guest house: Comm: (787)DSN: 740-3367/

Housing Office. Comm: (787) 707-3367/2317/3153/3256; DSN: 740-3367

CHRRS. (787) 707-5811/3601

Website. https://home.army.mil/buchanan/

FORT CAMPBELL, KENTUCKY 42223
DSN: 635-9467; Commercial: (270) 798-9467

Fort Campbell opened in 1942 and is 105,000 acres in size, split by the Kentucky-Tennessee state boundary. The main gate is located off Highway 41A, 116 miles south of Hopkinsville, Kentucky, and adjacent to Clarksville, Tennessee.

Major Activities. 101st Airborne Division (Air Assault); 5th Special Forces Group (Airborne); 160th Special Operations Aviation Regiment (Airborne); Air Assault School; 52nd Florence Blanchfield Army Community Hospital.

Billeting. Comm: (270) 798-5281/6709; DSN: 635-5281/6709. Turner Guest House: Comm: (270) 439-2229; VOQ: Comm: (270) 798-5618 or (615) 431-4496; DSN: 635-5618

Housing Office. On-post housing: Comm: (270) 798-3808; DSN: 635-3808

CHRRS. Comm: (270) 798-3808/2140; DSN: 635-3808/2140

Website. https://home.army.mil/campbell

CARLISLE BARRACKS, PENNSYLVANIA 17013
DSN: 242-3131; Commercial: (717) 245-3131

Carlisle Barracks was established in 1757. The post is 400 acres in size, adjacent to the eastern edge of the city of Carlisle, on U.S. Highway 11, 18 miles west of Harrisburg, Pennsylvania.

Major Activities. U.S. Army War College; Strategic Studies Institute; Center for Strategic Leadership; U.S. Army Heritage and Education Center; U.S. Army Peacekeeping Institute; Dunham Army Health Clinic.

Billeting. Comm: (717) 245-4245; DSN: 242-4245

Housing Office. (717)243-7177; DSN: 242-7177

Website. https://home.army.mil/carlisle

FORT CARSON, COLORADO 80913
DSN: 691-5811; Commercial: (719) 526-5811

Fort Carson, established in 1945, is 137,000 acres in size. The main gate is located 4 miles south of the city limits of Colorado Springs, off Colorado 115. Exit 135 off I-25 also provides easy access. The post maintains a 244,000-acre maneuver area, Pinon Canyon Maneuver Site, east of Trinidad, Colorado, about 155 miles south of Fort Carson. PCMS is used for brigade-sized exercises.

Major Activities. 4th Infantry Division; 10th Special Forces Group (Airborne); 4th Security Force Assistance Brigade; 627th Hospital Center; 4th Engineer Battalion; 759th Military Police Battalion; 71st Ordnance Group (Explosive Ordnance Disposal); Medical Department Activity-Fort Carson; World Class Athlete Program; 13th Air Support Operations Squadron; 1st Space Brigade.

Billeting. Comm: (719) 526-4832; DSN: 691-4832

Housing Office. On-post: Comm: (719) 526-2323; DSN: 691-2323.

Website. https://home.army.mil/carson

FORT DETRICK, MARYLAND 21702
DSN: 343-8000; Commercial: (301) 619-8000

Fort Detrick, established in 1943, occupies 1,200 acres just off U.S. 15 in Frederick, Maryland. It is approximately 50 miles from both Baltimore and Washington, D.C.

Major Activities. The post houses more than 50 tenant organizations representing five Cabinet-level agencies and all armed services; major areas are medical research, strategic communications (signal) and defense medical logistics.

Billeting and Housing. Comm: (301) 619-3224154: DSN: 343-3224

Website. https//home.army.mil/detrick

JOINT BASE MCGUIRE-DIX-LAKEHURST, FORT DIX, NEW JERSEY 08640

DSN: 650-1100; Commercial: (609) 754-1100

Fort Dix was established in 1917. It is a 31,000-acre post located off Exit 7, New Jersey Turnpike, near Wrightstown, New Jersey, on State Highway 68, 17 miles south of Trenton. It is co-located with McGuire Air Force Base.

Major Activities. DoD's only triservice base; Army Reserve's 99th Regional Support Command, 174th Infantry Brigade; various agencies involved with reserve component training.

Billeting. Comm: (609) 754-4667; DSN: 944-4667

Housing Office. Comm: (609) 754-3662; DSN: 944-3662

Website. https://www.jbmdl.jb.mil

FORT DRUM, NEW YORK 13602

DSN: 772-5461; Commercial: (315) 772-5461

Fort Drum was established in 1908. It consists of 107,000 acres and is located 9 miles west of Watertown, New York, off I-81 at Exit 48, approximately 70 miles north of Syracuse and 25 miles south of the Canadian border.

Major Activities. 10th Mountain Division (Light Infantry). The post continues its role as a major training center in the northeastern United States and as a cold-weather training site for active Army, Marine, and Canadian forces; supports reserve and National Guard forces; New York and New Jersey Army National Guard equipment concentration sites; Wheeler-Sack Army Airfield.

Billeting. Comm: (315) 773-7777; DSN: 341-9726/5427

Housing Office. Comm: (315) 836-4168; DSN: 341-6380

CHRRS. Comm: (315) 772-6556; DSN: 341-6556

Website. https://www.home.army.mil/drum

DUGWAY PROVING GROUND, UTAH 84022

DSN: 789-2929; Commercial (435) 831-2929

Dugway Proving Ground was established in 1942. It covers 1,300 square miles in west-central Utah, about 85 miles southwest of Salt Lake City. It may be reached via I-80 and an unnumbered county road, or via State Routes 36 and 199 from Tooele, Utah.

Major Activities. Dugway Proving Ground's mission is to test Army equipment to provide protection to U.S. military personnel in the field. Testing includes battlefield smoke and obscurant testing and production qualification testing of mortar and artillery munitions. It is the central point for DoD chemical and biological defensive testing.

Billeting. Comm: (435) 831-6500; DSN: 789-6500

Housing Office. Comm: (435) 831-2116; DSN: 789-2116

Website. https://www.army.mil/dugwaygarrison

JOINT BASE LANGLEY-EUSTIS, FORT EUSTIS, VIRGINIA 23604
DSN: 826-1212; Commercial: (757) 878-1212

Fort Eustis was established in 1918. This 8,200-acre post is located in southern Virginia between Williamsburg and Newport News. It may be reached via U.S. Highway 60 and is just off I-64.

Major Activities. U.S. Army Training and Doctrine Command; U.S. Army Training Support Center; Joint Task Force Civil Support; 7th Transportation Command (Expeditionary); 128th Aviation Brigade; 93rd Signal Brigade; 597th Transportation Brigade; U.S. Army Center for Initial Military Training; Army Training Support Center; Technology Development Directorate-Aviation Technology; Systems Integration and Demonstration; Army's Futures and Concepts Center; McDonald Army Community Hospital.

Billeting. Comm: (757) 764-4667, ; DSN: 826-4667

Housing Office. Comm: (757) 764-5040; DSN: 826-5040

Website. https://www.jble.af.mil

GILLEM ENCLAVE, GEORGIA 30050 – UNDER FORT GORDON/EISENHOWER

FORT GORDON (FORT EISENHOWER), GEORGIA 30905
DSN: 780-1110; Commercial: (706) 791-0110

Established in 1941, Fort Gordon occupies 56,000 acres about 9 miles southwest of Augusta, Georgia, between U.S. Highways 1 and 78.

Major Activities. U.S. Army Cyber Center of Excellence (includes cyber, signal, and network); Dwight D. Eisenhower Army Medical Center.

Billeting. Comm: (706) 791-2277; DSN: 780-2277

Housing Office. Comm: (706) 791-2473/4502; DSN: 780-2473/4502

Website. www.gordon.army.mil

FORT HAMILTON, NEW YORK 11252
DSN: 232-4780; Commercial: (718) 630-4780

Fort Hamilton, established in 1825, occupies 170 acres in the Bay Ridge section of Brooklyn, New York, near the Verrazano-Narrows Bridge.

Major Activities. New York City Recruiting Battalion; Corps of Engineers, North Atlantic Division; 179th Deployment Support Brigade; New York National Guard Task Force Empire Shield. Provides administrative and logistical support for Army and defense activities in the New York metropolitan area.

Billeting. Comm: (718) 439-2359; DSN: 232-2359

Housing Office. Comm: (718) 333-5815; DSN: 232-5815

Website. https://www.home.army.mil/hamilton

FORT HOOD (FORT CAVAZOS), TEXAS 76544
DSN: 737-1110Commercial: (254) 287-1110

Fort Hood was established in 1942. It occupies 217,337 acres about 60 miles north of Austin and 50 miles south of Waco, with the main gate adjacent to Killeen. U.S. Highway 190 provides four-lane controlled access to the post from I-35, the main north-south route through central Texas.

Major Activities. HQ, III Corps; 1st Cavalry Division, First Army Division West; 3rd Air Support Operations Group; 13th Sustainment Command (Expeditionary); 1st Medical Brigade; 3rd Cavalry Regiment; 3rd Security Force Assistance Brigade; 89th MP Brigade; 504th Military Intelligence Brigade; 36th Engineer Brigade; 407th Field Support Brigade; 48th Chemical Brigade; 69th Air Defense Artillery Brigade); 418th Contracting Support Brigadel Darnall Army Medical Center.

Billeting. Comm: (254) 532-8233; DSN: 737-2700/3815

Housing Office. Comm: (254) 220-4799; DSN: 737-4799

Website. https://www.home.army.mil/hood

FORT HUACHUCA, ARIZONA 85613
DSN: 821-7111; Commercial (520) 533-7111

Fort Huachuca (pronounced Wah-CHOO-kah) is in Cochise County, 70 miles southeast of Tucson, Arizona, on State Highway 90, 28 miles south of I-10, near the Benson exit. The main gate is adjacent to the city of Sierra Vista. The post was established in 1877 and comprises 73,000 acres.

Major Activities. U.S. Army Intelligence Center of Excellence; Network Enterprise Technology Command; U.S. Army Electronic Proving Ground; U.S. Army Information Systems Engineering Command; Joint Interoperability Test Command; 11th Signal Brigade.

Billeting. Comm: (520) 458-9066; DSN: 821-9066

Housing Office. Comm: (520) 515-9000; DSN: 821-9000

Website. https://www.home.army.mil/huachuca

HUNTER ARMY AIRFIELD, GEORGIA 31409
DSN: (912) 977-7947; Commercial: (912) 315-2588

Hunter Army Airfield is adjacent to Savannah, Georgia. It covers about 5,400 acres. It is a subinstallation of Fort Stewart, which is 40 miles away.

Major Activities. Supports the 3rd Infantry Division (Mechanized), Combat Support Aviation Brigade; 1st Battalion, 75th Ranger Regiment, 3rd Battalion, 160th Special Operations Aviation Regiment; 224th MI Battalion; USMC Reserve Center; 6th ROTC Brigade; USCG Air Station Savannah; 3rd Military Police Group; USMC Reserve Center; 117th Air Control Squadron; Georgia Army National Guard.

Billeting (Fort Stewart). Comm: (912) 369-6962; DSN: 475-6962
Housing Office. Comm: (912) 408-2480
Website. https://www.home.army stewart.mil/

FORT IRWIN, CALIFORNIA 92310
DSN: 470-3369; Commercial (760) 380-3369

Fort Irwin, established in 1940, is a 642,000-acre post located about 37 miles northeast of Barstow, California, halfway between Las Vegas and Los Angeles.

Major Activities. U.S. Army National Training Center; 11th Armored Cavalry Regiment; Weed Army Community Hospital.

Billeting. Landmark Inn Hotel: (760) 386-4040
Housing Office. Comm: (760) 386-4663; DSN: 470-4663.
Website. https://www.home.army.mil/irwin

FORT JACKSON, SOUTH CAROLINA 29207
DSN: 734-1110; Commercial (803) 751-1110

Fort Jackson was established in 1917. It is a 52,000-acre post located within the corporate limits of Columbia, South Carolina, between I-20 and U.S. 26.

Major Activities. U.S. Army Training Center; U.S. Army Soldier Support Institute (Adjutant Generals School, Finance School); Armed Forces Chaplaincy Center; Army Drill Sergeant School; Finance School; 165th, 171st, and 193rd Infantry Brigades; 81st Regional Readiness Command; Moncrief Army Community Hospital.

Billeting. Comm: (803) 782-9802
Housing Office. Comm: (803) 738-8275; DSN: 734-8275
Website. https://www.home.army.mil/jackson

JOINT FORCES STAFF COLLEGE, VIRGINIA 23511
DSN: 646-6124; Commercial: (757) 443-6124

The Joint Forces Staff College is located off Hampton Boulevard, 2 miles south of the main gate, Norfolk Naval Base, Virginia.

Major Activity. Under the direction of the Joint Chiefs of Staff and as part of the National Defense University, the college educates midcareer U.S. military officers and selected allied officers for joint and combined staff duty.

Billeting and Housing. Navy Gateway Inn and Suites (757) 394-9054
Website. https://www.jfsc.ndu.edu

FORT KNOX, KENTUCKY 40121
DSN: 464-1000; Commercial (502) 624-1000

Fort Knox was established in 1918. It is a 110,000-acre post located about 45 miles south of Louisville, Kentucky.

Major Activities. U.S. Human Resources Command; U.S. Army Cadet Command; U.S. Army Recruiting Command; V Corps Headquarters; U.S. Army Recruiting and Retention College; First U.S. Army Division East; 4th Cavalry Brigade; 1st Theater Sustainment Command; 84th Training Command; 100th Division; 83rd Army Reserve Readiness Training Center; U.S. Army Reserve Aviation Command; 19th Engineer Battalion; Ireland Army Hospital; George S. Patton Museum of Leadership. The U.S. Bullion Depository is located at Fort Knox.

Billeting. Comm: (502) 624-3491; DSN: 464-3491

Housing Office. Comm: (502) 378-3708; DSN: 464-3708

Website. https://home.army.mil/knox

FORT LEAVENWORTH, KANSAS 66027
DSN: 552-4021; Commercial: (913) 684-4021

Established in 1827, Fort Leavenworth occupies 5,634 acres just north of the city of Leavenworth, Kansas, about 35 miles northwest of Kansas City, Missouri.

Major Activities. U.S. Army Combined Arms Center; Army University; Mission Command Center of Excellence; Mission Command Training Program; U.S. Army Combined Arms Training Center; U.S. Army Command and General Staff College; Combined Arms Doctrine Directorate: Center for Army Lesson Learned; U.S. Disciplinary Barracks; Munson Army Community Hospital.

Billeting. Comm: (913) 364-1301; DSN: 552-1301

Housing Office. Comm: (913) 392-7681; DSN: 552-4921

Website. https://www.home.army.mil/leavenworth

FORT LEE (FORT GREGG-ADAMS), VIRGINIA 23801
DSN: 687-7451; Commercial: (804) 734-7451

Fort Lee, which opened in 1941, is a 5,575-acre post located 3 miles east of Petersburg, Virginia, off State Highway 36.

Major Activities. U.S. Army Combined Arms Support Command and Sustainment Center of Excellence; U.S. Army Logistics Management University; U.S. Army Quartermaster, Ordnance, and Transportation Schools; Army Logistics University and Soldier Support Institute; Defense Commissary Agency; Defense Contract Management Agency; Kenner Army Community Hospital.

Billeting. Comm: (804) 733-4100

Housing Office. Comm: (804) 566-3300/734-5091/5004; DSN: 539-5091

Website. https://www.home.army.mil/lee

FORT LEONARD WOOD, MISSOURI 65473
DSN: 581-0131; Commercial: (573) 596-0131

Fort Leonard Wood was established in 1940. This 63,000-acre post is located near Waynesville, Missouri, some 29 miles southwest of Rolla and 130 miles southwest of St. Louis off I-44.

Major Activities. U.S. Army Maneuver Support Center of Excellence; U.S. Army Engineer Center and School; U.S. Army Chemical, Biological, Radiological, and Nuclear School; U.S. Army Military Police School; 4th Maneuver Enhancement Brigade; 94th Engineer Battalion; 5th Engineer Battalion; 92nd MP Battalion; 193rd Brigade Support Battalion; General Leonard Wood Army Community Hospital.

Billeting. Comm: (573) 596-0999; DSN 581-0999

Housing Office. Comm: (573) 329-0122 DSN: 581-0965

Website. https://www.home.army.mil/wood

JOINT BASE LEWIS-MCCHORD, FORT LEWIS, WASHINGTON 98433
DSN: 357-1110; Commercial (253) 967-1110

Fort Lewis was established in 1917. The 86,000-acre post is located next to Puget Sound, 50 miles south of Seattle, about midway between Olympia and Tacoma, Washington, off I-5.

Major Activities. I Corps; HQ, 7th Infantry Division; 62nd Airlift Wing; 446th Airlift Wing; 593rd Expeditionary Sustainment Command; 1st Brigade, 2nd Infantry Division; 2nd Brigade, 2nd Infantry Division; 5th and 6th Military Police Groups (CID); 16th Combat Aviation Brigade; 22nd Signal Brigade; 201st Expeditionary Military Intelligence Brigade; 62nd Medical Brigade; 42nd Military Police Brigade; 555th Engineer Brigade; 1st Special Forces Group (Airborne); U.S. Army Cadet Command's 8th ROTC Brigade; 2nd Battalion (Ranger), 75th Infantry; 1st Multi-Domain Task Force; 5th Security Force Assistance Brigade; 66th Theater Aviation Command; 4th Battalion, 160th Special Operations Aviation Regiment (Airborne); 22nd Special Tactics Squadron; 404th Army Field Support Brigade; Madigan Army Medical Center. Yakima Training Center and Vancouver Barracks are subposts.

Billeting. Comm: (253) 967-2815/5051; DSN: 357-2815/5051

Housing Office. Comm: (253) 966-4663; DSN: 357-4082

Website. https://www.home.army.mil/lewis-mcchord

FORT MCCOY, WISCONSIN 54656-5000
DSN: 280-1110; Commercial: (608) 388-2222

Fort McCoy was established in 1909. It occupies 60,000 acres in west-central Wisconsin, midway between the cities of Sparta and Tomah, off State Highway 21.

Major Activities. Fort McCoy is a Warfighting Training Center and is the Army's largest mobilization site for reserve component units. The post's primary mission is to provide for the training and ensure the readiness of reserve and active component forces from all services. Major on-post tenant activities include the Army Reserve Readiness Training Center; Readiness Group-McCoy.

Billeting. Comm: (608) 388-2107; DSN: 280-2107

Housing Office. Comm: (608) 388-3704; DSN: 280-3704

Website. https://www.home.army.mil/mccoy

JOINT BASE MYER-HENDERSON HALL, FORT MYER, VIRGINIA 22211

DSN: 426-3283; Commercial (703) 696-3283

Composed of Fort Myer, Virginia, Fort McNair, Washington, D.C, and Henderson Hall, Virginia.

Major Activities. Fort Myer: 3rd Infantry Regiment (The Old Guard); the U.S. Army Band (Pershing's Own). Fort McNair: HQ, Military District of Washington (MDW); National Defense University, including the National War College and the Industrial College of the Armed Forces; Inter-American Defense College; Center of Military History.

Billeting Comm: (703) 696-3576/3577; DSN: 426-3576/3577

Housing Office Comm: (703) 696-3557/3558; DSN: 426-3557/3558

Website. https://home.army.mil/jbmhh

FORT MEADE, MARYLAND 20755

DSN: 622-2300; Commercial: (301) 677-2300

Fort Meade was established in 1917. Post size is 13,500 acres. Fort Meade is home to approximately 9,350 military personnel as well as 31,669 civilian employees. The installation lies 4 miles east of I-95 and one-half mile east of the Baltimore-Washington Parkway, between Maryland State Route 175 and 198. Fort Meade is situated within the communities of Odenton, Laurel, Severn, and Columbia.

Major Activities. National Security Agency; U.S. Cyber Command; Defense Information Systems Agency; Defense Media Activity; and 80 other installation partners.

Billeting. Comm: (301) 677-6529, (301) 674-7700, (301) 677-5660

Housing Office. Comm: (301) 677-9390

Website. https://home.army.mil/meade

PICATINNY ARSENAL, NEW JERSEY 07806
DSN: 880-4021; Commercial: (973) 724-6364

Picatinny Arsenal was established in 1880. It is sited on 6,500 acres at Dover, New Jersey, about 45 miles west of New York City.

Major Activity. Joint Center of Excellence for Guns and Ammunition; Combat Capabilities Development Command Armaments Center; Joint PEO Armaments and Ammunition; Project Manager Soldier Lethality.

Billeting. Comm: (973) 724-3506/8855; DSN: 880-3506/8855
Website. https://www.pica.army.mil

PINE BLUFF ARSENAL, ARKANSAS 71602
DSN: 966-3000; Commercial: (870) 540-3000

Established in 1941, Pine Bluff Arsenal is sited on 15,000 acres about 8 miles northwest of the city of Pine Bluff, Arkansas.

Major Activities. Design, manufacture, and demilitarization of smoke, riot control, and incendiary munitions, as well as chemical and biological defensive items.

Billeting. Comm: (870) 540-3008; DSN: 966-3008
Housing Office. Comm: (870) 540-3989; DSN: 966-3989
Website. www.pba.army.mil

JOINT READINESS TRAINING CENTER AND FORT POLK (FORT JOHNSON), LOUISIANA 71459
DSN: 863-1344; Commercial: (337) 531-1344

Fort Polk was established in 1941. The post is 198,000 acres in size, located in west-central Louisiana some 9 miles south of Leesville and about 50 miles west of Alexandria. It can be reached via U.S. 171, which connects with I-20 at Shreveport, about 20 miles to the north, and with I-10 at Lake Charles, about 70 miles to the south.

Major Activities. Joint Readiness Training Center; 3rd BCT; 10th Mountain Division; 46th Engineer Battalion; 519th Military Police Battalion; 1st Battalion, 5th Aviation Regiment; 32nd Hospital Support Center; Bayne-Jones Army Community Hospital.

Billeting. Guest house: Comm: (337) 531-9200; DSN: 863-9200
Housing Office. Comm: (337) 537-5000
Website. https://home.army.mil/polk

U.S. ARMY GARRISON, PRESIDIO OF MONTEREY, CALIFORNIA 93944

DSN: 768-6604; Commercial: (831) 242-6604

The Presidio of Monterey occupies 392 acres on a hill overlooking Monterey Bay and the city of Monterey. It was originally established by the Spanish in 1770; the United States first occupied it from 1846 until 1852. Reactivated in 1902, it became the home of the 11th Cavalry Brigade between the two world wars. Since 1946, it has been the home of the Army Language School, now known as the Defense Language Institute Foreign Language Center.

Major Activity. Defense Language Institute Foreign Language Center.

Billeting. Comm: (831) 242-5091; DSN: 768-5091

Housing Office. Comm: (831) 656-2321, (831) 644-0400, (800) 334-9168; DSN: 768-2321

Website. https://home/army/mil/monterey

REDSTONE ARSENAL, ALABAMA 35898

DSN: 746-2151; Commercial: (256) 876-2151

Redstone Arsenal was established in 1941. It is located on a 38,248-acre site adjacent to the southwest limits of the city of Huntsville, Alabama.

Major Activities. U.S. Army Material Command; HQ, U.S. Army Space and Missile Defense Command; U.S. Army Aviation and Missile Command; U.S. Army Contracting Command; U.S. Army Security Force Assistance Command; Missile Defense Agency; PEO Missiles and Space; PEO Aviation; U.S. Army Combat Capabilities Development Command Aviations and Missile Center; U.S. Army Rapid Capabilities and Critical Technologies Office; Future Vertical Lift Cross-Functional Team; Fox Army Hospital.

Billeting. Comm: (256) 876-5713, (256) 837-4130; DSN: 746-5713

Housing Office. Comm: (256) 842-2449; DSN: 746-2449

Website. https://home.army.mil/redstone

JOINT BASE ELMENDORF-FORT RICHARDSON, ALASKA 99505

DSN: (317) 552-1110; Commercial: (907) 552-1110

Fort Richardson was established in 1940. It occupies 62,500 acres adjacent to Anchorage, Alaska. (See also chapter 10, Foreign Service.)

Major Activity. 11th Air Force; Alaskan Command; 11th Airborne Division; 2nd BCT (Airborne), 11th Airborne Division; Alaska National Guard HQ.

Billeting and Housing Office. Comm: (907) 384-0133; DSN: 317-384-0133

Website. https://www.jber.jb.mil

FORT RILEY, KANSAS 66442
DSN: 856-3911; Commercial: (785) 239-3911

Fort Riley was established in 1853. This 100,000-acre post has its main gate located just 2 miles east of Junction City, Kansas, off State Highway 18. I-70 and U.S. 40 connect with Fort Riley.

Major Activities. 1st Infantry Division; 1st and 2nd Armored Brigade Combat Teams; 1st Infantry Division Artillery; 1st Combat Aviation Brigade; 1st Infantry Division Sustainment Brigade; Irwin Army Community Hospital.

Billeting. Comm: (785) 239-2830/8882; DSN: 856-2830/8882

Housing Office. Comm: (785) 239-3525, (800) 643-8991; DSN: 856-3525

Website. https://home.army.mil/riley

ROCK ISLAND ARSENAL, ILLINOIS 61299
DSN: 793-6001; Commercial: (309) 782-6001

Established in 1862, the arsenal occupies a 946-acre island in the Mississippi River between Moline, Illinois, and Davenport, Iowa.

Major Activities. HQ, Army Sustainment Command; HQ, First Army Command; HQ, Joint Munitions Command; Rock Island Arsenal Joint Manufacturing and Technology Center.

Billeting. Comm: (309) 782-0833; DSN: 793-0833

Housing Office. Comm: (309) 782-2376; DSN: 793-2376

Website. https://home.army.mil/ria

FORT RUCKER (FORT NOVOSEL), ALABAMA 36362
DSN: 558-1110; Commercial: (334) 255-1110

Fort Rucker was established in 1942. This 64,500-acre post is located 20 miles northwest of Dothan, Alabama.

Major Activities. U.S. Army Aviation Center of Excellence; Army Warrant Officer Career College; U.S. Army Combat Readiness Center; U.S. Army Aeromedical Center; U.S. Army Aeromedical Research Laboratory; U.S. Army School of Aviation Medicine.

Billeting. Comm: (334) 598-5216; DSN: 558-2626

Housing Office. Comm: (334) 255-9230/1205; DSN: 558-9230/1205

CHRRS. Comm: (334) 255-3705; DSN: 558-3705

Website. https://home.army.mil/rucker

FORT SAM HOUSTON, TEXAS 78234

DSN: 471-1211; Commercial: (210) 221-1211

Fort Sam Houston was founded in 1876. Its 2,900 acres lie within the city limits of San Antonio, Texas. However, another 27,000 acres are located at its subinstallation, Camp Bullis, 24 miles to the northwest.

Major Activities. HQ, U.S. Army North/Fifth Army; U.S. Army South; HQ, U.S. Army Medical Command; Command 5th Recruiting Brigade; 12th ROTC Brigade; U.S. Army Installation Command; U.S. Army Veterinary Command; U.S. Army Medical Center of Excellence; Brooke Army Medical Center.

Billeting. Comm: (210) 357-2705, ext. 2000; (800) 462-7691

Housing Office. Comm: (210) 221-0833/2341; DSN: 471-0833/2341

Website. https://www.jbsa.mil

SCHOFIELD BARRACKS, HAWAII 96857

DSN: (315) 456-1110; Commercial: (808) 449-7110

Schofield Barracks, established in 1908, is situated on 14,000 acres about 17 miles northwest of Honolulu, Hawaii. Subinstallations include Wheeler Army Airfield and Helemano Army Reservation. (See also chapter 10, Foreign Service.)

Major Activities. 25th Infantry Division (Light); U.S. Army Garrison Hawaii.

Billeting. The Inn at Schofield, (800) 490-9638; DSN: 815-455-5036

Housing Office. Comm: (808) 655-1060. Unaccompanied Personnel Housing Office: Comm: (808) 655-4249

Website. https://home.army.mil/25id

FORT SHAFTER, HAWAII 96858

DSN: (315) 456-7110; Commercial: (808) 449-7110

Fort Shafter, established in 1905, is located on 1,400 acres near Honolulu, Hawaii, off H-1 Freeway. (See also chapter 10, Foreign Service.)

Major Activities. HQ, U.S. Army Pacific (USARPAC); 8th Theater Sustainment Command; 311th Signal Command; 9th Mission Support Command; 94th Army Air and Missile Defense Command; 196th Infantry Brigade; Tripler Army Medical Center.

Housing Office. Comm: (808) 438-5063. Unaccompanied Personnel Housing Office: Comm: (808) 839-2336

CHRRS. Comm: (808) 474-1972/1973/1974/1975

Website.https://www.army.mil/usarpac(USARPAC Website)

FORT SILL, OKLAHOMA 73503
DSN: 639-4500; Commercial: (580) 442-4500

Fort Sill, established in 1869, is 94,220 acres in size. The main gate is about 3 miles north of Lawton, Oklahoma, off I-44.

Major Activities. Fires Center of Excellence; U.S. Army Air Defense Artillery School; U.S. Army Field Artillery School; 428th, 434th, and 75th Field Artillery Brigades; 30th and 31st Air Defense Artillery Brigades; 95th Training Division; Long Range Precision Fires Cross-Functional Team; Air and Missile Defense Cross-Functional Team; Henry Post Army Airfield; Reynolds Army Community Hospital.

Billeting. Comm: (580) 442-5000, (877) 902-3607; DSN: 639-5000. Guest house: Comm; (580) 442-3214; DSN: 639-3214

Housing Office. Comm: (580) 442-2813, (800) 695-1084; DSN: 639-2813

Website. https://sill-www.army.mil

SOLDIER SYSTEMS CENTER, NATICK, MASSACHUSETTS 01760
Public Affairs Office: Commercial: (508) 206-4300

Established in 1953, "Natick Labs" is located on 75 acres at Natick, 20 miles west of Boston.

Major Activities. Responsible for R&D, testing, and evaluation of textiles, technology, interactive textiles, nano technology, biotechnology, airdrop technology, food science, human physiology and warrior systems integration for all services.

Billeting and Housing. Comm: (508) 233-5409/5216; DSN: 256-5409/5216

CHRRS. Comm: (508) 233-5216; DSN: 256-5216

Website. https://www.army.mil/natick

FORT STEWART, GEORGIA 31314
DSN: 475-9879/9874; Commercial: (912) 435-9879/9874

Established in 1940, the Fort Stewart reservation consists of 279,000 acres near Hinesville, about 40 miles southwest of Savannah. State Route 119 and 144 pass through the reservation. Hunter Army Airfield is a subinstallation.

Major Activities. 3rd Infantry Division; Winn Army Community Hospital.

Billeting. Comm: (912) 368-4184; (912) 767-8384; DSN: 870-8384

Housing Office. Comm: (912) 767-2127; DSN: 870-2127

G.M.H. (912) 408-2460

Website. https://home.army.mil/stewart

TOBYHANNA ARMY DEPOT, PENNSYLVANIA 18446

DSN: 795-7000; Commercial: (570) 615-7000

Established in 1953, Tobyhanna is the DoD's premier facility for full life cycle support of all command, control, communications, computers, intelligence, surveillance and reconnaissance (C4ISR). It is located 20 miles southeast of Scranton.

Major Activities. Army Center of Industrial and Technical Excellence for C4ISR, and electronics, avionics and missile guidance and control systems; Air Force Technology Repair Center for rigid wall shelters and tactical missiles; DoD's Worldwide C4ISR readiness provider.

Billeting. (570) 615-6343

Housing Office. (570) 615-7888

Website. www.tobyhanna.army.mil

TOOELE ARMY DEPOT, UTAH 84074

DSN: 790-2211; Commercial: (435) 833-2211

Established in 1942, Tooele Depot is DoD's Western region conventional ammunition hub. It occupies 43,300 acres and is located 35 miles southwest of Salt Lake City.

Major Activities. Supports warfighter readiness through receipt, storage, issue, demilitarization and renovation of conventional ammunition.

Billeting. (435) 833-2056

Housing Office. On-post housing not available.

Website. www.tooele.army.mil

JOINT EXPEDITIONARY BASE LITTLE CREEK-FORT STORY, VIRGINIA 23459

DSN: 253-7385; Commercial: (757) 462-7385/7386

Fort Story was established in 1914. It occupies 1,451 acres at the mouth of Chesapeake Bay, adjacent to Virginia Beach.

Major Activities. Amphibious training site for active and reserve components of all services, and the Army's Logistics-over-the-Shore testing and training site.

Billeting. Comm: (757) 422-8818

Housing Office. Comm: (757) 422-7321; DSN: 438-7321

Website. www.eustis.army.mil/fort_story/

UNITED STATES MILITARY ACADEMY, WEST POINT, NEW YORK 10996

DSN: 312-688-2022; Commercial (845) 938-3808

West Point was established in 1802. The reservation includes 15,900 acres on the west bank of the Hudson River, about 50 miles north of New York City and 15 miles south of Newburgh, New York, on U.S. 9W.

Major Activities. The United States Military Academy.

Billeting. (845) 446-1028/1034; (845) 938-6816; DSN: 688-6816

Housing. Comm: (845) 938-4500/3942; DSN: 688-4500/3942

CHRRS. Comm: (845) 938-5948; DSN: 688-5948

Website. www.usma.edu

FORT WAINWRIGHT, ALASKA 99703

DSN: (317) 353-1110; Commercial: (907) 353-1110

Fort Wainwright is sited on 963,000 acres about 1 mile east of Fairbanks, Alaska. Originally established as an Army Air Field in 1940 and later designated Ladd Air Force Base, Fort Wainwright was so named in 1961 when the Army assumed command of the post. (See also chapter 10, Foreign Service.)

Major Activities. 1st BCT, 11th Airborne Division; 1st Battalion, 52nd Aviation Regiment; 1st Battalion (Attack), 25th Aviation Regiment; Medical Department Activity-Alaska; U.S. Army Garrison, Alaska.

Billeting. Comm: (907) 353-7291; DSN: 317-353-3800

Housing Office. Comm: (907) 353-1655/1666; DSN: 317-353-1655/1666

CHRRS. Comm: (907) 353-1660; DSN: 317-353-1660

Website. https://home.army.mil/alaska

WALTER REED NATIONAL MILITARY MEDICAL CENTER, BETHESDA, MD 20889

Commercial (301) 295-4000

Walter Reed Army Medical Center was reestablished in Bethesda in 2011. As the largest military medical center in the U.S., it provides services in more than one hundred clinics and specialties.

Website. www.wmmc.capmed.mil

WATERVLIET ARSENAL, NEW YORK 12189-4050

DSN: 374-5111; Commercial: (518) 266-5111

Watervliet Arsenal was established in 1813. It is located on 42 acres at Watervliet, about 6 miles north of Albany, New York. It is the oldest continuously operating manufacturing arsenal in the United States.

Major Activity. Watervliet Arsenal is the nation's only cannon manufacturing facility and is the home of the Army's Benet Laboratories.

Housing Office. Comm: (518) 266-5306; DSN: 924-5306

WHITE SANDS MISSILE RANGE, NEW MEXICO 88002
DSN: 638-3205; Commercial (928) 328-2151

Established in 1945, White Sands Missile Range occupies 1,874,666 acres, with the main facilities located about 26 miles east of Las Cruces on Highway U.S. 70.

Major Activities. Missile and weapons testing for the Army, Navy, and Air Force as well as NASA and commercial activities.

Billeting. Comm: (505) 678-4559; DSN: 258-4559

Housing Office. Comm: (505) 678-5110; DSN: 258-5110

YAKIMA TRAINING CENTER, WASHINGTON 98901
DSN: 638-3205; Commercial: (509) 577-3205

Yakima Training Center was established in 1941. It is a subinstallation of Joint Base Lewis-McChord, occupying 323,651 acres in eastern Washington, with the main gate about 7 miles north of the city of Yakima off I-82. It is the Army's primary training area in the Pacific Northwest.

Major Activity. Provides ranges and maneuver areas to all branches of the active and reserve components of the U.S. military and allied forces.

Billeting. Comm: (509) 577-3418; DSN: 638-3418

Housing Office (through Fort Lewis). Comm: (253) 967-4082; DSN: 357-4082

Website. https://home.army.mil/yakima

YUMA PROVING GROUND, ARIZONA 85365
DSN: 899-2151; Commercial: (928) 328-2151

Modern military equipment testing in Yuma can be traced back to 1943, when the U.S. Army Corps of Engineers opened the Yuma Test Branch. The Yuma Proving Ground occupies about 1,300 square miles, with the main facilities located about 25 miles northeast of Yuma, Arizona.

Major Activities. Yuma Proving Ground is part of the U.S. Army Test and Evaluation Command and is one of only two general purpose proving grounds within the command.

Billeting. Comm: (928) 328-2129; DSN: 899-2129

Housing Office. Comm: (928) 328-2127/3766; DSN: 899-2127/3766

Website. www.yuma.army.mil

HEADQUARTERS, U.S. ARMY FUTURES COMMAND, AUSTIN, TEXAS 78701

Commercial: (512) 726-4117

Established 2018. The Army's fourth major command engages with experts and innovators from academia, industry, and government to envision future battlefields, draft informative concepts, requirements and designs, accelerate transformational science and technology gains and converge advanced capabilities across the joint force, enabling overmatch against any adversary in any domain.

Major Activities. U.S. Army Combat Capabilities Development Command, Aberdeen Proving Ground, Maryland; Medical Research Development Command, Fort Detrick, Maryland; Futures and Concepts Center, Joint Base Langley-Eustis, Virginia; Artificial Intelligence Integration Center, Pittsburgh; Long-Range Precision Fires Cross-Functional Team (CFT) and Air and Missile Defense CFT, Fort Sill, Oklahoma; Future Vertical Lift CFT, Redstone Arsenal, Alabama; Soldier Lethality CFT, Fort Benning, Georgia; Next-Generation Combat Vehicle CFT, Warren, Michigan; Network CFT, Aberdeen Proving Ground, Maryland; Assured Positioning, Navigation and Timing/Space CFT, Redstone Arsenal; Synthetic Training Environment CFT, Orlando, Florida.

MAJOR RESERVE COMPONENT TRAINING SITES

In addition to the posts and stations listed, a number of semiactive or nonactive installations are capable of supporting maneuver training for units of brigade size or larger. Most are federally owned but operated by the state in which they are located; others are entirely state owned and operated.

Commercial telephone numbers are for operator assistance at sites listed; DSN numbers are for military points of contact.

Atterbury-Muscatatuck Training Center, Edinburgh, IN 46124. DSN: 569-2499; 812-526-1499.

Camp Blanding Joint Training Center, Starke, FL 32091. 904-682-3355.

Camp Bowie, Level 3 Training Center, Brownwood, TX 76801. 325-643-3055.

U.S. Army Garrison-Fort Buchanan, PR 00934. DSN: 740-4486; 787-707-4486.

Fort Chaffee Joint Maneuver Training Center, Fort Chaffee, AR 72905. DSN 312-962-2121; 479-484-2121.

U.S. Army Garrison, Fort Devens, MA 01434-4424. 978-615-6021. Joint Base McGuire-Dix-Lakehurst, NJ. See Joint Bases.

Camp Edwards, Joint Base Cape Cod, MA 02542. 339-202-9309.

Camp Grayling Joint Maneuver Training Center (Heavy), MI 49739. 989-344-6100.

Camp Gruber Training Center, Braggs, OK 74423. 918-549-6001.

Camp Guernsey Joint Training Center, Guernsey, WY 82214. DSN: 344-7810; 307-836-7834.

U.S. Army Garrison-Fort Hunter Liggett, CA 93928. 831-386-2530.

Fort Indiantown Gap-Army National Guard Training Center, Annville, PA 17003. DSN: 491-2000; 717-861-2000.

Camp James A. Garfield Joint Military Training Center, Ravenna, OH 44444. 614-336-6660.

Camp Joseph T. Robinson (Robinson Maneuver Training Center), North Little Rock, AR 72199. DSN: 318-962-5098; 501-212-5098.

Joint Forces Training Base, Los Alamitos, CA 90720. DSN: 972-2090; 562-795-2090.

Parks Reserve Forces Training Area, CA 94568. 925-875-4398.

Perry Joint Training Center, Port Clinton, OH 43452. 419-635-4021.

Fort Pickett-Army National Guard Maneuver Training Center, Blackstone, VA 23824. DSN: 441-8621; 434-292-8621.

Camp Rilea, Warrenton, OR 97146. DSN: 355-4052; 503-836-4052.

Camp Ripley-Minnesota National Guard Training Center, Little Falls, MN 56345. DSN: 871-2709; 320-632-7000.

Camp Roberts Maneuver Training Center, San Miguel, CA 93451. DSN: 949-8356; 805-238-8356.

Camp Santiago Joint Training Center, Salinas, PR 00751. 787-289-1400, ext. 7001, 7002, 7004.

Camp Shelby Joint Forces Training Center, MS 39407. DSN: 558-2000; 601-558-2000.

Camp Sherman Joint Training Center, Chillicothe, OH 45601.614-336-6460.

Camp Swift, Level 3 Training Center, Bastrop, TX 78602. 512-321-4122.

Camp W.G. Williams, Bluffdale, UT 84065. DSN: 766-5400; 801-878-5400.

TRAVEL AND RECREATION

Military posts or facilities, either of the Army or of its sister services, are located in each of the fifty states and around the world where there are concentrations of U.S. military personnel. As a general rule, each of these facilities has some provision to accommodate visitors, although the quality of the accommodations may vary from austere to modern. In addition, many of the military bases operate recreation areas, either on the station or close to nearby attractions. As with the guest quarters, the recreation facilities vary widely. Some are equipped only for daytime use; others have accommodations ranging from campsites to dormitory rooms to individual cottages.

All of these temporary quarters and recreation areas have one thing in common: they are available for the use of military personnel, active or retired, and their families. Proper identification is required. In some cases there are priorities for occupancy, and there generally are limits on how long the facilities may be used. However, by making proper inquiry and planning ahead, it is possible for a military family to travel around the country and to vacation in the mountains or at the seashore and at many attractive spots in between, all at prices substantially below the cost of commercial facilities. It is an important benefit that accrues to all military personnel.

A good source of information on Space-A military air travel, temporary military lodging, and "R&R" is Military Living Publications. This company is in its forty-eighth year of helping military families "travel on less per day . . . the military way!" It publishes a travel newsletter—the *R&R Space-A Report*—and travel books, maps, and atlases. Military Living's books include *Military Space-A Air Basic Training*; *Military Space-A Air Opportunities Around the World*; *Temporary Military Lodging Around the World*; *Military R.V. Camping and Outdoor Recreation Around the World, Including Golf Courses and Marinas*; *U.S. Forces Travel Guide to U.S. Military Installations*; and *U.S. Forces Travel Guide to Overseas U.S. Military Installations*.

The books are supported by a number of maps and atlases, including the *United States Military Road Atlas*; the *European U.S. Military Road Atlas*; *U.S. Military Installation Road Map*; and a series of seven state maps detailing U.S. military installations, including Alaska/Washington; California; Delaware, Maryland, Virginia, and D.C.; Florida and Puerto Rico; Hawaii and Guam; New Jersey, Pennsylvania, and New York; Georgia, Texas, and North and South Carolina.

Desert Shield commemorative maps are also available. The *R&R Space-A Report* contains current information on temporary military lodging, space-available travel on U.S. military aircraft, military recreation areas, and more. The *R&R Space-A Report* serves as a clearinghouse for information provided by military readers sharing their travel experiences around the world. It is published six times a year and is available by subscription. It is a valuable source of information for individuals interested in enjoying the better life at bargain rates.

Readers can save by buying these books, maps, and atlases at a military exchange. More information is available on Military Living's website: www.militaryliving.com. Military Living Publications' address and phone numbers are: P.O. Box 231, Ophelia, Virginia, 22530; voice: (703) 237-0203; fax: (703) 237-2233. Phone and fax orders are accepted with major credit cards or go to www.militaryliving.com.

10

Foreign Service

Where a man can live, there he can also live well.

—Marcus Aurelius

This chapter is provided as a source of "first information" for Army families under orders for overseas assignment. It includes information extracted from official publications, gathered from the experiences of Army members who have provided suggestions, and supplied by overseas commands. There are informative websites' official pamphlets on most overseas commands, which are supplied to officers after they receive orders for change of station. The overseas commands also mail informative pamphlets that they have prepared for incoming personnel. These pamphlets, which are also available online as PDFs, are especially important, since they include recent developments that may be of great interest.

AR 55-46, *Travel Overseas*, is of special importance and should be studied if you are on orders for an overseas assignment.

THE OPPORTUNITY AND THE RESPONSIBILITY

During the span of a normal Army career, you may expect several overseas assignments. During the Cold War, when approximately one-third of the Army was serving overseas, an officer could expect to spend about one-third of his or her service in overseas assignments. The majority of these assignments were for tours of three years in areas where the officer could be accompanied by family members, although some assignments were to short-tour areas and families had to be left behind. Army strength in overseas commands has been sharply reduced in recent decades, but overall Army strength also has been reduced. It is likely that a large portion of an officer's career will be spent in foreign service or on deployment.

Here is an excellent opportunity to observe and learn the history, culture, language, economics, religion, and ways of life of other cultures. Army people, including Army children, enjoy a unique opportunity to broaden their knowledge of the people and problems of these nations. You and your family will have many opportunities to enrich your interest in and your understanding of the flood of events that shape our world.

Army officers and their families who have served the conventional overseas assignments have become travel sophisticates who know their way around our nation and our world—what to do, how to do it, and also what not to do and why. Most have taken advantage of their travel opportunities. They have taken time to learn enough of the history and culture of the countries visited to be well informed. Some have learned the language of the host nation. Most have made lasting friendships, with memories to be cherished. These well-traveled, well-informed people are a national asset, and they have earned a wider recognition than has been extended. They have done well and deserve credit for it.

There is one phase of family life in the service that is best faced with candor. Soldiers must be sent into international trouble spots to discharge the nation's responsibilities. In addition, there are regions of the world where facilities for families are so scanty or so primitive that the presence of Army families is inadvisable. The Global War on Terror, coupled with multiple operational, peacekeeping and humanitarian missions of recent years are examples of the types of assignments that may be expected in the future. Assignment to these primitive or troubled areas is indeed a hardship, but the Army is careful to distribute such hardship tours to make family separations as few and as equitable as possible. Nevertheless, in a troubled world, with the vast responsibilities faced by our nation, the situation must be understood. Army families must look to the civilian leadership of our government to keep family separations at a minimum.

FOREIGN SERVICE TOURS

The policies regarding foreign service for officers are stated in AR 614-30, *Overseas Service*, and are summarized below.

The Broad Policy Selection. The paramount consideration in selecting an officer for service outside the continental United States (CONUS) is the existence of a valid requirement coupled with the officer's military qualifications to perform the duties required. To the maximum extent practicable, overseas tours are alternated between long- and short-tour areas, and attempts are made to achieve geographic and climatic variety. CONUS and Hawaii are the sustaining bases for all overseas assignments, and officers can normally expect a minimum of twelve months in CONUS or Hawaii after completion of an overseas tour. However, the requirements of the Army dictate the length of a tour in CONUS prior to return overseas. The policy is that each officer will receive a proportionate share of foreign service. The overseas commands have the option to fill vacancies either with qualified individuals already in their commands or through requisition upon the forces in the continental United States. As to an assignment in the continental United States, some are of fixed duration from which an officer may not be released prior to its completion, even for an overseas assignment. So it goes. In principle, the officer with the right grade and the right qualifications with the least credit for overseas service will be the next to fill a requisition from an overseas command.

Officer Volunteers. Officers may volunteer for overseas assignments ahead of their normal expectancy for such duty. They may state a preference for a specific overseas command or for several commands in order of preference. Officers who volunteer are considered to be available when HQDA approves their applications and they have been assigned in CONUS at least twelve months. However, officers serving in stabilized positions will not be voluntarily reassigned until they have completed their stabilized tours.

Temporary Deferment of Overseas Assignment. When the overseas movement of an officer would cause serious hardship, it is possible to secure temporary deferment for compassionate reasons. Officers who find it necessary to seek deferment are advised to study AR 614-30 before making a request.

Length of Foreign Service Tour. Normal tours of foreign service for all Army personnel are shown in AR 614-30. For purposes of departmental records, foreign service tours commence on the day an individual departs from an ocean or air terminal in the United States and terminate on the date of return to such a U.S. terminal. The day of return is counted as a day of foreign service.

Personnel who are accompanied or joined by their dependents serve the tour prescribed for those "With Dependents" or twelve months after arrival of dependents, whichever is longer. The tour prescribed for "All Others" is served by personnel who elect to serve overseas without their dependents, are serving in an area where movement of dependents is restricted, are not authorized movement of their dependents at government expense, or do not have dependents. There are special rules and provisions for Army members married to each other. In specified European countries, and in Japan and Canada, bachelor officers and enlisted personnel, male and female, serve the "With Dependents" tour.

Tours normally are uniform for personnel of all services at the same station. The service having primary interest (the most personnel) develops, in coordination with the other affected services, a mutually satisfactory tour length applicable to all military personnel in the locality.

There are provisions for reassignment between overseas commands without intervening tours in CONUS. The rules for computing tour lengths in this case are complex and can easily be misunderstood. The first delineation involves voluntary versus involuntary reassignment. Officers are advised to consult their personnel officers and AR 614-30 for details. An officer who is voluntarily reassigned will serve the complete prescribed tour in both areas. An officer who is involuntarily reassigned between overseas commands or between areas within an overseas command will serve a prorated tour in the new command in accordance with the formula in AR 614-30. The tour in the new area will be adjusted to give credit for that portion of the normal tour in the new area corresponding to the portion of the normal tour already served in the area from which reassigned. For example, an officer who completed one-third of another tour would be credited with having served four months of the normal twelve-month tour in Korea.

Geographic Areas and Tour Lengths. Consistent with Army requirements, overseas assignments are made to provide variety in the geographic areas to which assigned. Thus, an officer who has spent a tour in Europe should fully expect the next overseas tour to be in a different geographic area, such as Korea.

Similarly, the Army attempts to equalize the burden of family separations. In the above example, if the European tour had been a thirty-six-month "With Dependents" tour, the officer should not be surprised if the subsequent Korean tour were a twelve-month unaccompanied tour, as prescribed for "All Others."

AR 614-30 provides a complete list of overseas assignment areas and tour lengths. You should consult the regulation for specific details regarding an overseas assignment. However, in general, the "With Dependents" tour in European countries and other modern areas is thirty-six months, decreasing to twenty-four months in less desirable locations, whereas the corresponding "All Others" tour lengths range from twenty-four months to a minimum of twelve months.

PREPARATION FOR OVERSEAS TRAVEL

Action Before Departure. Official orders of the Department of the Army assigning you to an overseas command prescribe the timing, method of transportation, and other essential information about the journey. At the outset you should receive, or should have prepared, about fifty copies of your orders, needed on many occasions incident to the movement. For example, six copies are required with each shipment of personal property. Orders are also required for pay, travel allowances, and other matters en route and after arrival.

Upon receipt of orders, you should immediately report to the personnel officer, finance officer, quartermaster, transportation officer, and surgeon to obtain detailed instructions regarding the pending move. Be certain to understand these instructions and follow them carefully. You must set your official and personal house in order so that no unfinished business will arise at the last minute before departure or, worse, after departure. All personal bills or obligations must be paid, or definite arrangements made for future payment. See chapter 19, Personal and Financial Planning, on wills, insurance, and check-up suggestions for personal affairs.

Concurrent Travel of Families. Families may accompany you to an overseas station (concurrent travel), or they may be obliged to join you after a period of delay when it has been determined that there will be quarters available upon their arrival. The regulation of general information is AR 55-46, which should be consulted upon receipt of orders for overseas movement.

Shipment of Household Goods. The amount of household goods that may be shipped overseas on permanent change of station is stated in chapter 21, Travel. Area restrictions or other temporary reductions in allowances may occur. Consult the transportation officer. For duty at those overseas stations where furniture and equipment are supplied to a military family, the authorized allowances for shipment of personal belongings are sharply reduced.

With careful planning, and consideration of the advice you have received, you should be able to ship all the necessities plus a few nice-to-have items within the weight limit prescribed.

Shipment of Automobile. In most overseas assignments, the use of a privately owned automobile is essential. Consider at once the suitability of the car owned. Service for the car, including repair parts, is the thing to consider.

If shipment of a car is desired, consult the transportation officer at once. He or she will give you detailed information that may assist in having your car arrive at the overseas destination in much less time than would be the case if one essential step is missed along the line.

See to it that your orders permit travel to the port in your own car, if such travel is your desire. Be sure to deliver the car at the port no later than the time prescribed. Further, since cars may be loaded in order of arrival, it may be wise to get there a day early. The same advice applies to shipment of a trailer. Have ready several copies of your travel orders, a certificate of title to the car, or a statement that the vehicle is free from any legal encumbrance that would preclude its shipment.

Ask your transportation officer about buying marine insurance, which generally is regarded as desirable but is not required.

It is not unusual for a car to arrive overseas as much as a month behind the owner. Careful attention to the requirements that may permit it to be shipped on an earlier transport may save lots of walking.

Shipment of Pets. If you own pets, you must ascertain definite information about the regulations governing their shipment to the overseas destination, as well as the laws or regulations governing their reentry into the United States upon termination of tour. In some locations shipment of pets is prohibited, so be sure to check the rules for that location.

The shipment of pets must be planned with the same care as the travel of the rest of the family to avoid a family crisis. Pets must be vaccinated, placed in crates on some occasions, fitted with muzzles as required, fed, exercised, and, for some destinations, shipped separately.

Passports. You should ascertain at your home station whether passports will be required either for yourself or for your family. DoD 1000.21-R, *Passport and Passport Agent Services Regulation*, governs the procedures to obtain passports. See also par. 18, AR 55-46. Consult chapter 21, Travel. If passports are required, take action early to obtain them, since considerable time is often required.

Mode of Travel. The standard mode of travel for Army members and their dependents to, from, and between overseas commands is by air. Sea travel is authorized only as an exception when the soldier or the member's dependents, for medical reasons, cannot travel by air. Still in the military vocabulary, however, are many terms left over from the days of sea transport, such as hand baggage, hold baggage, and port of embarkation. These terms generally are applied to air transport just as they were in the past to sea transport.

Hand Baggage and Hold Baggage. Hand baggage is that baggage accompanying the passenger on board the aircraft, either in the flight cabin or checked for stowage in the cargo compartment. For travel outside the continental United States, hand baggage normally is limited to sixty-five pounds per person.

Hold baggage is that additional baggage authorized to be shipped with the traveler, as contrasted to household goods, which generally follow later. Hold baggage is limited to 600 pounds per person and generally is packed and shipped prior to the date of departure to be available at the destination at about the time the traveler arrives.

Judicious use of the weight allowances for the hand and hold baggage will enable you and your family to have on hand at the overseas destination those clothes and other essentials you will need during the month or more before arrival of your household goods.

Action at Port of Embarkation (POE). The orders for overseas movement will prescribe the point of departure and the date and hour that arrival at the POE is required.

Arrive on time. Report to the prescribed headquarters or to the office charged with processing transient personnel. Receive detailed and complete instructions and study them carefully. Quarters will be provided while awaiting departure, including quarters for families. It is considered more satisfactory to use furnished quarters than to choose to await departure at a hotel or the home of friends. Sometimes you must go to a local hotel because the terminal facilities are crowded; if you arrive early enough in the day, the chances for space assignment are increased. The more children and pets you have, the more time you must allow for the terminal personnel to help you. Although the government quarters furnished may not be luxurious, they will certainly be comfortable, clean, and adequate for the short stay required. Recreation facilities are provided; you and your family can be reached promptly for medical or other processing requirements; and costs are low. Terminal commanders take pride in these facilities, as well as the smoothness of their operation.

Travel Aboard Aircraft. The Air Mobility Command (AMC) provides overseas transport service for personnel, including family members, and freight, serving all military services. As noted earlier, except as may be authorized for medical reasons, all personnel movement is by aircraft.

Travel aboard airplanes of the AMC is substantially the same as aboard commercial airlines—in fact, AMC uses many aircraft chartered from commercial airlines. Flight attendants are provided. Special diets for babies and young children must be taken aboard by the sponsor.

Although Military Sealift Command (MSC) transport may be involved with troop transport to areas of conflict, it can be expected that air transport will be the standard mode. AMC has a proud tradition of service and will provide outstanding transportation to those who follow its advice and regulations. If you plan and think through your air travel and follow instructions, you and your family should have a delightful, but short, travel experience and arrive overseas in the best condition.

TIPS FOR FAMILY TRAVEL

High standards of personal conduct, including that of children, with due regard for the rights of others are a necessity. Thoughtfulness and consideration for other travelers will do much to make the trip a pleasant experience for all.

You will be informed by the terminal commander regarding when your presence is required for processing prior to boarding the aircraft. For long overseas flights, this normally is two hours before flight time, to enable complete processing of passengers and their baggage. Transportation from your temporary quarters to the terminal will be provided. It is necessary only that you be at the starting point on time, with hand baggage, official orders, passports (if required), immunization records, and any other pertinent papers, ready to board the aircraft.

After the short delay for check-in, which is necessary to ensure that all personnel are present and ready, you will board the aircraft. Families normally board first so that they can sit together, followed by any VIPs who may be on the flight, followed by other military personnel traveling alone.

It is possible for the number of children aboard an aircraft to exceed the number of other passengers. Try to have on hand items that will entertain and satisfy the needs of your children. A large, all-purpose bag or handbag is handy for carrying the personal items that will be needed during the journey. Do not overlook a few books, toys, or small electronic games that can help the children pass the hours.

As an aside to readers who may be traveling without dependents, you may notice at the air terminal a mother with several small children, proceeding overseas to join her husband. She will be encumbered with bags and, more often than not, the processing procedures will be strange to her. It is perfectly proper to offer to help her wrestle bags or children, or both, and to help make their journey a more pleasant one. Such a practice will pay rich dividends, not only in terms of the gratitude of the mother involved but also in terms of the day when your family may be traveling overseas to join you and a stranger offers a helping hand. Such acts of courtesy and thoughtfulness are in keeping with service customs and the officer's code.

Except when there are unusual circumstances of health, the care of your family at air terminals (POEs), on AMC aircraft, and at the port of debarkation overseas will present no great problems if you have thought out your plans well in advance and you follow instructions carefully. If you understand the requirements well, if facilities are adequate, and if you follow the standing operating procedures as to events, time, and place, everything will proceed smoothly and pleasantly. Personnel operating the POEs and aircraft have been confronted by nearly all possible problems and generally have a solution ready to apply promptly and with good grace. But if there are special circumstances, such as the need for medical supplies for a child or special foods or other unusual matters, you should make all necessary plans and arrangements in advance and in person.

Foresight at the home station before departure will pay rich dividends. What to take and what to leave behind in storage is an important matter. Considerations of the climate to be encountered will answer many questions. Within limitations of baggage and freight allowances, in case of doubt, it is wiser to take questionable articles of personal property than to leave them behind. Fragile or valuable nonessentials are best left behind in storage; even under the assumption that packing and crating will be done perfectly, which is an overly generous assumption, these shipments are subject to much handling, and there is considerable risk of damage and pilferage. However, do not relegate your best things to storage just because they are your best. Remember that you will probably be spending 10 percent or more of your service career at the overseas station, and you will want to have your best things with you.

SATISFACTION OF FAMILY NEEDS IN OVERSEAS COMMANDS

The Army is keenly concerned about physical facilities to serve the needs of its officers and enlisted personnel and their families overseas. This concern provides peace of mind to soldiers, which permits them to devote their full energies to their duties.

Grade school and high school facilities are provided in all established commands. Standards are closely supervised. Considering the intangible values to be derived from travel and life overseas for a few years, most parents believe that their children benefit from the experience. In the unusual case in which children's schools are not provided under military control, such as when on attaché duty, parents should investigate facilities used by American citizens residing in the area or ascertain the possibility for instruction by mail with home tutoring. Unless there are unusual circumstances indeed, there is no strong reason for Army families to be separated merely because of the school situation, because educational facilities for children have been provided in a manner that is acceptable to most parents.

Post exchanges in most overseas commands are not so sharply restricted as to what can be sold, as is the case in the United States, where civilian retail facilities are available. Although unusual needs must be anticipated and arrangements made for their satisfaction, the ordinary wants and necessities, with some items of luxury, can be obtained from the post exchange. Shopping on the internet is a way in which to obtain goods not readily available either in the foreign retail market or at the exchange. Before ordering, ensure that the vendor ships to APO addresses.

Officers' clubs and recreational facilities are provided at nearly all stations. Movies are provided by the Army and Air Force Exchange Service on a standard comparable to that in CONUS, and the pictures shown are the latest produced by the industry. Commissaries have the usual items that can be purchased at posts in the United States.

LIFE IN OVERSEAS COMMANDS

The following information about living conditions in overseas areas where significant numbers of U.S. Army personnel are stationed, accompanied or joined by their families, is presented as a helpful first reference for officers receiving foreign service orders. The discussions have been extracted from official publications, with up-to-date information provided by the headquarters of the overseas commands. In all cases, official information is sent as early as practicable to officers receiving assignment orders.

ALASKA

Headquarters, U.S. Army Alaska is at Joint Base Elmendorf-Richardson, adjacent to Anchorage. The major tactical unit in Alaska is the 11th Airborne Division.

An Army family bound for Alaska can look forward to a rich experience. Quarters, schools for children, and shopping facilities are available, and recreational, cultural, and sporting facilities and opportunities are abundant.

Excellent documents about service in Alaska are supplied to officers soon after they receive their assignment orders. The pamphlets are current, well illustrated, and informative.

Concurrent Travel. Officers should request concurrent travel to Alaska, for which see AR 55-46. If approved, apply to the nearest transportation officer for shipment of household goods. A sponsorship program assists incoming officers and their families, and direct communication with the sponsor is encouraged.

Automobile. Privately owned automobiles (one per servicemember), after authorization, are shipped from the port of Seattle at government expense.

Consult the transportation officer. It is recommended that cars be put in the best possible mechanical condition before shipment, as cost of repairs and maintenance is high, and that cars be winterized to -40 degrees for the Anchorage area and -75 degrees for the Fairbanks area. Recommended items include an engine block heater or a battery blanket.

Family Quarters. Government quarters are generally apartment style, frame two-story, or single or duplex style with a full basement. Basements are concrete, warm, and dry, with outlets for appliances. These quarters are available at Joint Base Elmendorf-Richardson.

All personnel electing to serve an accompanied tour are authorized a full household goods weight allowance. Single officers may request shipment of a full household goods weight allowance but should be aware that government storage for excess household goods is not available. Officers should plan to bring (or plan to purchase) all household goods (draperies, lamps, and so on) and furniture that will be needed for a three- or four-year stay. Quarters on post are equipped with a washer, dryer, refrigerator, and electric stove. Shipment of duplicate items is discouraged because of limited storage. Temporary issue of basic household goods can be arranged, if needed, and some furniture items may be available on a permanent basis. The ACS loan closets can supply temporary cooking utensils, silverware, and cribs.

The most common off-post housing is apartment style; two- and three-bedroom units are more readily available than larger units. Rental rates and utility costs are higher than in most CONUS locations. There is an additional quarters allowance for individuals authorized to occupy off-post housing.

Clothing Suggestions. Winter clothing can be purchased from local stores at slightly higher prices than normal, and from exchanges that supply practically any item normally available in CONUS. Mail-order service is satisfactory and utilized extensively. There are national chain-type department stores in Anchorage and Fairbanks. Quartermaster laundry and dry-cleaning facilities are available.

Schools for Children. On-post schools provide standard educational opportunities for children. Younger children, under age five, may attend preschool at Joint Base Elmendorf-Richardson. The minimum age for entrance to Alaska schools is five years, and this age must be reached by 15 August of the school year. A birth certificate is required for kindergarten, and evidence of promotion for all other grades. Immunization records for DPT, DT, or TD; polio; measles; and rubella (or a signed physician's waiver) are required for entry to any grade.

On-post school facilities are found at Joint Base Elmendorf-Richardson, kindergarten through sixth grade; seventh through twelfth grades attend schools in the local community, with bus transportation furnished by civilian authorities.

Recreation Facilities. Alaska has much the same opportunities for recreation as in CONUS. Clubs and fitness centers are found at each post. Craft, woodworking, and ceramics shops are provided. There are snowmobiles, flying, parachuting, and youth swim clubs, as well as drama and arts groups.

Individual sports include swimming, golf, tennis, bowling, skiing, skating, sled dog racing, and more.

Camping and hiking are popular activities, and you can hunt and fish to your heart's content at small expense.

HAWAII

The state of Hawaii, paradise of the Pacific, is "foreign service" only in the sense that it is outside the continental limits of the United States. Its climate resembles that of southern California and southern Florida. It is one of the few areas where local claims as to climate and scenery are justified by experience.

The Army's stations are Fort Shafter, Tripler Army Medical Center, and Schofield Barracks.

Major Activities. The U.S. Army Pacific (USARPAC) is headquartered at Fort Shafter and is one of the two major Army commands in the Asia-Pacific area. (The other is the U.S. Eighth Army located in Korea.) USARPAC is the Army component of the U.S. Pacific Command. USARPAC's area of operation covers 100 million square miles and includes 2.5 billion people in fifty different countries.

Major subordinate commands under USARPAC include U.S. Army Japan/9th Theater Army Area Command; 25th Infantry Division (Light); U.S. Army Alaska; and U.S. Army Chemical Activity, Pacific, on Johnston Atoll.

The 25th Infantry Division (Light), located at Schofield Barracks in Hawaii, is designed as a rapid deployment force of nearly 11,000 soldiers with the specific mission as the Pacific Command's ground combat reserve force.

The U.S. Army Chemical Activity on Johnston Atoll, 825 miles southwest of Hawaii, provides integrated command and management of all U.S. Army activities on the island and oversight for operations maintenance of the Johnston Atoll Chemical Agent Disposal System. This system disposes of obsolete and unserviceable chemical munitions under stringent security and safety standards.

Tripler Army Medical Center is a large general hospital facility serving all military forces and their dependents on the islands.

Family Quarters. Station and civilian facilities are at least equal to their mainland counterparts. Under normal conditions pertaining to size of the Army, the number of family-type housing units is short of the number needed for married families. Studio and one-bedroom apartments are in good supply. Civilian housing for families with more than four members is scarce. Rents are high. The U.S. Army Hawaii Housing Department at Fort Shafter will assist in securing temporary as well as permanent accommodations.

Temporary Lodging Allowance. Costs for temporary housing can be extremely high. There is a housing allowance, separate from the basic allowance for quarters, which can be obtained to defray excess housing costs pending assignment of government quarters or obtaining of permanent civilian housing. See AR 37-104 and the Joint Travel Regulations for the amounts, which are variable. At a maximum, this allowance can last no more than thirty days. *Caution:* Within one working day after arrival, application for government quarters should be made to qualify for payment of the temporary lodging allowance and establish priority for obtaining government quarters based on the date of departing CONUS.

Schools for Children. Adequate school facilities are available throughout the command. There are several outstanding private schools.

Automobile. A privately owned automobile is essential. Consult the transportation officer as to authorization for shipment.

Pets. Pets can be taken to Hawaii, but the rules are very complex. Know them thoroughly before you ship any animal. Animals are required to complete a 120-day confinement in the Hawaii State Animal Quarantine Section, at the expense of the owner. However, as a result of recent changes to Hawaii state law, if specific prearrival and postarrival requirements are met, animals may qualify for a five-day or less quarantine program, which has a provision for direct release at Honolulu International Airport after inspection. For specific instructions, contact your nearest Military Veterinary Service.

Recreation Facilities. Service in Hawaii provides for unusually varied and enjoyable recreation facilities. There are extensive facilities for team sports, golf, tennis, and water sports of all kinds, with outstanding programs for participation. Active organizations sponsor interesting activities for young people, and officers' clubs and open messes function in Hawaii as elsewhere.

The Armed Forces Recreation Center, Fort DeRussy, in the heart of Waikiki, is especially cherished as a service benefit. Individuals and families can secure accommodations for one day or several days and enjoy at small cost all the pleasures of the tourist paradise at Waikiki. Fort DeRussy has a fine beach for swimming, a picnic area, a good restaurant, and a club with entertainment. In addition, there is a post office, liquor store, post exchange with car rental, laundry, flower shop, and outdoor ice dispenser. A high-rise hotel, the Hale Koa, provides comfortable, modern temporary quarters. It can be contacted via a toll-free telephone number—(800) 367-6027.

In addition to the Hale Koa on Waikiki Beach, the Army has two other fine recreation sites on Oahu. The Pililaau Army Recreation Center on the west coast offers beautiful beaches, cabins, and dining facilities, in addition to the rental of almost anything you need to make your "getaway" more enjoyable. The Army's beach on the famed north shore at Mokuleia features a secluded swimming, sunbathing, and picnic area for those who don't enjoy the crowded beaches on the southern end of the island.

On the island of Hawaii is the Kilauea Military Camp, a recreation facility in Hawaii Volcanoes National Park. Military families go there at minimal expense for a stay of several days' duration. The camp is at an elevation of 4,000 feet, which provides a climate change—and a temporary need for warmer clothing than is worn on Oahu. Tours by bus are scheduled to points of beauty and interest. The accommodations are adequate and include a restaurant.

GERMANY

Service with the U.S. European Command (EUCOM), provides an opportunity for service of importance to the nation and an opportunity to travel and acquire an understanding of the people, the cultures, and the problems of the nations of the European Union.

Preparation for Travel.

Passports. Each member of your family will need a passport. For many assignments, passports are not needed by military personnel, but passports are required to take leave in certain overseas countries. Ask your installation passport agent for assistance. Failure to perform this task in a timely fashion could cause unnecessary delays in travel of family members.

Immunizations. No special immunizations are required for travel to the European theater, but you should check with your local immunization clinic. Animal immunizations before departure for Europe are required. Check with local veterinary clinics for information.

Hand Baggage and Hold Baggage. Plan these choices carefully. An earlier discussion in this chapter supplies details as to allowances. It is recommended that there be included in hold baggage a small supply of bed linens, bath linens, and cooking utensils for use prior to the arrival of household goods.

Travel Conditions en Route. Travel to Germany is via aircraft of the Air Mobility Command (AMC), in commercial aircraft chartered by AMC, or in commercial aircraft when AMC flights on AMC charter aircraft are unavailable. The accommodations are equivalent to commercial tourist class. Your designated sponsor will meet you at your destination.

Climate. The climate is mild. German summers are delightful, with warm days and cool nights. There are very few thunderstorms but considerable gentle rain. Winters are less severe than in our northern and central states; they compare with those of Maryland or Virginia. There is some snow, but except in the mountains, it does not remain long. In the mountainous areas, winters are much colder and the snow deeper, and winter sports are popular.

Housing Situation. Government family housing is no longer in short supply at most locations in Germany, and the situation is improving. More quarters are being added annually. Government housing generally is available within sixty days after arrival, depending on your grade, bedroom requirements, and location. Barring unusual circumstances, private rental housing generally can be obtained within one to eight weeks. Rental rates for private rental housing are generally high, and most landlords require a security deposit equal to a minimum of one month's rent. Soldiers who reside in private rental housing approved by the housing referral office (HRO) are paid an overseas housing allowance (OHA). The allowance paid is the difference between actual rental cost and the basic allowance for housing (BAH). Annual cost surveys are required to authorize the payment and to adjust the allowance.

Family Housing. Upon arrival in Germany, all personnel with present or future requirements for family housing must report to the local housing office. You will be advised of the local housing conditions and waiting periods for economy and government housing and assisted in applying for and locating housing. If your dependents did not accompany you, the housing office will assist you in sending for them after you have located housing. You will also be advised of your entitlements, including temporary lodging allowance (TLA). While at the housing office, you may place your name on waiting lists for temporary government quarters, permanent government quarters, and rentals on the economy. If you move into economy housing, you can still remain on the list for permanent government quarters and later move at government expense to the permanent quarters; however, EUCOM policy is that you must live in private rental housing for a minimum of one year before moving into government quarters.

Government quarters consist of both on-post housing and off-post government-leased housing. The on-post housing is primarily two-, three-, and four-bedroom units in multifamily apartment buildings. Laundry and drying rooms, complete with washers and dryers, are located in each building. The government-leased quarters are of varying sizes and styles. They may be small two- to six-family units, row houses, or apartments. There are even a few single houses. These leased quarters are all within commuting distance of your duty station and are assigned on the same basis as on-post quarters. Nearly all quarters, government and economy, are supplied with 220-volt, 50-cycle electrical current. Some renovated quarters have 110-volt current. Ask your sponsor if the housing in your community has been renovated.

Unaccompanied Officer Quarters. The same basic rules apply as stateside. Officers, commissioned and warrant, who are bona fide bachelors, and those not entitled to BAH at the with-dependent rate, are authorized to reside off post regardless of the availability of quarters, except for reasons of military necessity or when serving in a dependent-restricted area, unless adequate government-controlled quarters are not available. Bachelor quarters vary in size, and some include community kitchens. Availability of private rental housing and waiting periods for government housing are similar to those for married personnel.

Private Rental Housing. German landlords own private rental housing and lease to American soldiers and civilians. Army policy requires that all personnel, soldiers and civilians alike, register with the local HRO before entering into a lease agreement. Do not sign a lease until it has been checked by your HRO representative. This is for your benefit and protection, as well as being a requirement. In addition to approving private rentals, HROs maintain lists of available apartments and houses, assist in locating suitable housing, provide interpreter assistance, provide transportation to view apartments, and provide a multitude of other services as you search for and then occupy private rental housing. You also can go to a rental agent. Reimbursement for fees charged by rental agents may be authorized, but you must obtain approval from the local housing agent first. Although rents vary according to the size and location of the apartment or house, it is fair to say that rents are higher than for comparable housing in the United States. With deposits and initial start-up costs, it is not uncommon to face an initial cash outlay of $3,000 or more to get set up in private rental housing. Some of these costs may be reimbursable by MIHA (move-in housing allowance). Your housing office representative will help you determine eligibility.

Persons who have experienced living on the German economy can tell you that there are significant differences between renting housing in Germany and renting in the States. The first thing one notices about German housing is that when they say unfurnished, they mean it. You may find no light fixtures, no stove, no kitchen cabinets, and no refrigerator, and it is a rare German apartment that has closets. Appliances such as refrigerators, stoves, and hot water heaters, when provided, are much smaller than American appliances.

You may not be able to get American TV, which is broadcast by the Armed Forces Network-TV, unless you are close to an Army installation. Cable television and satellite reception are available in most areas. On the brighter side, watching German TV is an excellent way to begin to understand the language and learn the German customs.

Living on the economy in rental housing can be a wonderful experience, and if you enjoy your privacy when you get off work, you most likely will find it living in a German neighborhood.

Helpful Hints. Your HRO will assist you with these matters, but here are a few things you should be aware of:

- Payment of rent is in European Union currency (Eurodollars).
- Under German law, a verbal agreement is a valid contract.
- The German rule requires three months' written notice to terminate a lease. It is strongly recommended that you use the approved bilingual

rental contract available at the HRO. This allows a military tenant to terminate a rental contract with thirty days' written notice, or fifteen days if required.

- German law has provisions for the protection of tenants. If you receive a notice to vacate or a rent increase, consult your HRO.
- It may be advisable for you to obtain insurance in case you are sued for apartment damage or your pet bites someone. Check with your legal assistance office. For your protection and the protection of your own furniture, you will probably want to obtain insurance, which can be combined with rental insurance coverage for your household goods.
- Most Germans respect private and public property; they will expect you to do the same.
- Landlord house rules may limit your social activities, and late-night entertaining is usually not permitted. Loud playing of radios, TVs, or stereos or other excessive noise is prohibited by law, particularly from 1300 to 1500 hours and after 2200 hours.

Furniture and Household Equipment. The type of family housing, furniture, and household equipment support provided to you in Europe depends largely on your eligibility and the location of your duty station. If you are assigned to a duty station in Germany, the Netherlands, Belgium, or Italy, you usually are authorized to ship your full Joint Federal Travel Regulation (JFTR) weight allowance. (See chapter 21, Travel.) If your orders indicate that you are authorized to ship your full JFTR weight allowance, you can plan on receiving a refrigerator, range, washer, dryer, and dishwasher for your entire tour. You also can plan on using a loaner set of essential furniture items for up to ninety days inbound and sixty days outbound, pending the arrival or after the shipment of your own household goods. You also will receive curtains, wardrobes, kitchen cabinets, and light fixtures if they are not built in or otherwise provided by your landlord.

If you plan to live in private rental housing, you should bring anything you can use for storage space, such as small closets, kitchen cabinets, portable shelves, and hutches. Private rentals are usually not as well equipped for storage space as in CONUS.

Most Army Community Service Centers in Europe can lend you such items as cookware, kitchen utensils, flatware and chinaware, irons, ironing boards, high chairs, baby cribs, and transformers until your own household goods arrive.

One bit of advice: If you haven't been to Europe before, contact someone who has for advice and assistance in making your shipping arrangements. Be sure to contact your sponsor. Since government furniture stock varies from community to community, your sponsor can advise you about furniture availability at your new station. Your sponsor also should be able to assist with information regarding typical apartment size and configuration, floor plans of standard government quarters, and advice about household goods you should ship or leave behind.

Electrical Appliances. Electrical power in Europe is 220-volt, 50-cycle, instead of the 110-volt, 60-cycle power we have in the United States. Because of this difference, caution must be exercised in the use of U.S. electrical appliances. Most "smart" devices can automatically adapt (for example, smartphones all have the ability to detect the difference between voltage); however, double-check your power supply to ensure it handles both 110 and 220 volts. You will need a transformer to step down the 220-volt power to operate 110-volt appliances. Check with your sponsor. However, the transformer adjusts only the voltage, not the frequency. Thus, motor-driven appliances designed for use in the United States will operate at only five-sixths normal speed in Europe. Some appliances such as vacuum cleaners, electric razors, mixers, and blenders will operate satisfactorily at this reduced speed but not at peak efficiency. Others will not work at all. For example, an electric clock designed for use in the United States would lose 10 minutes every hour when operating on European electric current. Most appliances with heating elements, such as toasters and irons, operate satisfactorily with a transformer. However, transformers are inconvenient, particularly in the larger sizes. Therefore, it is easier to leave at home those appliances that require a large transformer, such as coffeemakers, toasters, hair dryers, or irons, and purchase 220-volt appliances in Europe. These items are sold at military exchanges, thrift shops, and European retailers.

Following are a few general guidelines for the use of lamps and small appliances.

Lamps. Floor, desk, and table lamps that use incandescent bulbs can be used by changing to a 220-volt bulb and adding an inexpensive adapter plug to match the electrical outlet. (Bulbs and adapters are available at the PX.)

Televisions. Cable TV service is available in Germany just as it is in CONUS. Most bases have the cable service provider located with the exchange facility. There, you can sign up for cable, and the provider will supply you with the necessary equipment for a nominal monthly fee.

DVDs. American movies on DVD are available through video clubs and Army stores. European DVDs are recorded for Zone 2 players and do not generally play on American (Zone 1) DVD players. Only specially designed DVRs or video recorders are equipped to record German satellite signals.

Personal Computers. These will normally operate on their own, but check your power supply to see if you require a transformer. You will need a standard plug adapter for your charger.

Radios and Stereos. Radios and stereos work satisfactorily with transformers and receive both AFN and European stations.

Microwave Ovens. As a general rule, microwave ovens for use with 60-cycle power cannot be used at 50 cycles and will not operate properly with a transformer. A 60-cycle microwave oven may be damaged when used on 50-cycle power. Special microwave ovens that can be used in Europe and converted for use in the United States are available at the exchange.

Clothing. You will need the same uniforms as would be required for duty in the United States. Uniforms can be purchased in the exchange.

Civilian clothing is authorized for wear by all military personnel in Germany when off duty. It can be purchased at post exchanges or at German stores. Many excellent tailors are to be found. English textiles can readily be obtained.

Dependents are advised to take with them a complete wardrobe. Evening clothing is worn on occasion by both men and women. Civilian clothing for men, women, and children is also available at post exchanges.

Cost of Living Allowance. A cost of living allowance (COLA) is paid to all soldiers. The COLA, which is determined by base pay, number of command-sponsored dependents, and duty location, is paid to give soldiers the same spending power as CONUS-based soldiers of comparable grade, years of service, and number of dependents. An overseas housing allowance (OHA) is paid to assist in renting or leasing private rental quarters. The amount is based on the actual rent paid, up to a ceiling determined by grade and duty location. Both allowances are reviewed twice monthly by the DoD Per Diem Committee. If the currency rate of exchange changes by 3 percent or more, the OHA and COLA are adjusted accordingly. The official exchange rate is fixed on a daily basis.

Army and Air Force Exchange Service (AAFES). AAFES, Europe, operates an extensive chain of retail stores plus automotive, food, and personal services stores. Concessionaire-operated specialty shops featuring fine European crystal and china, furniture, gift items, and a host of other merchandise augment the main retail stores. Personal services include laundry and dry cleaning; barber, beauty, optical, floral, and tailor shops; car and equipment rentals; and much more. Overall, authorized AAFES customers are offered a wider selection of merchandise and services than their stateside counterparts because fewer congressional constraints apply to overseas exchanges.

Automobile. An automobile for a family in Germany is a must. It can be shipped from the United States or bought secondhand from another American; you can purchase one of the many excellent English or European cars; or you can purchase certain American models through AAFES. A foreign car may be imported to the United States for personal use without paying an import tax. The individual, however, must pay transportation charges only if the purchase exceeds the one car entitlement or no car was shipped from CONUS.

AAFES, Europe, operates gasoline stations throughout Germany. Payment at AAFES stations can be made in U.S. dollars or with coupons purchased through AAFES. The coupons can also be redeemed at selected stations or autobahn locations throughout Germany.

The operator is required to have the official USAREUR operator's license and meet the requirements as to physical condition, knowledge, and driving ability. Vehicles must meet safety standards and be registered. No vehicle may be registered unless covered by recognized liability insurance; in this connection, one of the several authorized companies is our own (mutual) United Services Automobile Association. (See chapter 19, Personal and Financial Planning.) The

International Insurance Certificate ("green card") is required if you plan to travel in countries other than Germany. An international driver's license may also be required.

Germany provides good driving conditions on the autobahn, which is comparable to our best interstate highways, and on secondary roads, most of which are asphalt. Operators must learn the international road signs.

Elementary and Secondary Schools. The Department of Defense Educational Activity (DoDEA) provides education from kindergarten through grade 12 to eligible dependents of military and civilian personnel stationed overseas. Transportation to and from school is provided within established commuter zones. The educational program is accredited by the North Central Association of Colleges and Schools, the largest stateside accreditation agency.

College-Level Education. Opportunities abound for officers stationed in Germany to continue their personal and professional development. College-level courses are offered at Army Education Centers at most EUCOM installations. Courses are offered by many colleges and universities and include programs for certificates as well as associate, baccalaureate, and advanced degrees. There are opportunities to take vocational, academic, and technical courses. Credit hours acquired may be applied to degrees to be received overseas or used as transfer credit to colleges and universities in the United States.

Medical and Dental Facilities. Medical facilities comparable to those in the United States are available to military personnel and family members. It is advised that dental treatment be completed before travel to Germany. Dental care is provided, but it is on a space-available basis for family members. All family members of EUCOM military personnel, including those family members who are not command sponsored, are entitled to free medical care. Civilian medical care reimbursable under Tricare (see chapter 19) is available for family members under certain circumstances.

Recreation and Club Activities. A tour in Germany provides an exciting opportunity to encounter different people and cultures, visit some of the world's most historic and scenic places, attend colorful folk festivals, sample delicious food and drink, or ski and mountain climb in the Alps. In addition to these unique aspects of overseas life, military personnel and their families can participate in most of the same recreation and club activities that are available back home. There is a full range of on-base leisure activities, including team sports, theater, libraries, rod and gun clubs, multicrafts, and so on. The U.S. Forces also operates a vacation center in Garmisch, where the cost of staying in a world-famous resort is significantly reduced.

Family Support Program. In Germany there are a number of programs designed to support the special needs of Army family members. Army Community Service, the community's social service agency, helps individuals and families in everyday living situations, as well as during times of special need. Among its many programs are relocation assistance, consumer affairs, financial counseling, family member employment assistance, foster care, outreach programs, and spouse and child abuse prevention. Other family support programs include Child Development Services and Youth Services. Child Development Services are aimed at providing quality child development care at affordable, standardized prices in centers or in certified homes. The Youth Services Program offers recreation activities to meet the interests and needs of school-age American youth in Germany.

JAPAN

The U.S. Army's major command in Japan is the U.S. Army Japan (USARJ). Most Army installations and troop units on mainland Japan are in the area southwest of Tokyo, convenient to both Yokohama and Tokyo. The climate is very similar to that of North Carolina. Snowfalls are rare; June and September are rainy; and the summers are hot and humid. The climate in Okinawa is more tropical.

Concurrent Travel. Except for officers in grades colonel and above, who are automatically authorized concurrent travel, approval of dependent travel at government expense is based on availability of government housing or approved private rental housing.

Family Quarters. With the exception of 83rd Ordinance Battalion, near Hiroshima, all government quarters in Japan are on or near the installations. Single and unaccompanied officers are housed in BOQs. Officers are required to occupy government housing when it is available. Family quarters are attractive and comfortable, although slightly smaller than the average American home. There are large, well-equipped play areas for children; chapels; theaters; clubs; and post exchange, recreational, and hobby facilities. The Armed Forces Radio and Television Service provides radio and television broadcasts. On-post cable TV is available in most locations.

Mainland Japan is a weight-restricted area; you will be allowed to ship 25 percent of your weight allowance. (See chapter 21, Travel.) Standard furniture items, stoves and refrigerators, washers and dryers, and air conditioners are government furnished. Infant beds and high chairs are available for issue.

The electrical current is 50 cycles, 100 volts. Many electrical appliances can be adjusted or are set by the manufacturer for 50/60 cycles and will run fine. Some 60-cycle items that do not operate correctly are clocks, some CD players, and some appliances with heating elements.

If you own a freezer, there is no restriction against bringing it, but remember that your quarters may not be as large as your present home, and you may wish to use your allowable weight for other items. Microwave ovens that can be converted from 60 to 50 cycles are readily available through the post exchange. If your microwave is not convertible, you may want to consider storing it during your tour.

Schools for Children. Your children will attend DoDEA schools, kindergarten through grade 12, at Camp Zama. You can find updated information about DoDEA at www.dodea.edu/Pacific/Japan.

Child-Care Facilities. Almost every Army installation in Japan has, or has access to, a certified child-care facility. Many of the facilities are brand new or recently renovated. The Army has also recently started a program certifying in-home caregivers. See Child Development Services upon your arrival.

Automobile. Japan has an embargo against foreign vehicles. Non-Japanese cars built after March 1976 are not allowed to enter the country. Used Japanese cars are readily available and are normally inexpensive. Driving in Japan is on the left side of the road, and it is much easier to learn to drive on the left with a car designed for that purpose. All privately owned vehicles must be registered with the Japanese government and with the Provost Marshal.

A Japanese compulsory insurance law requires vehicle owners to purchase Japanese insurance in addition to liability coverage. The cost is not excessive.

You may obtain a U.S. Forces Operator's Permit after passing a written test on local traffic laws and regulations. Family members under age 18 may not drive off post.

Currency and Banking. The U.S. dollar is the authorized currency on all U.S. installations in Japan. The local currency (yen) can be purchased at each installation for transactions on the economy.

Military banking facilities furnish most expected services, with the exception of safe-deposit boxes, although you may wish to retain your stateside account. Credit unions also are available.

Adult Education. U.S. installations in Japan offer educational opportunities at all levels. Students may complete their GED or pursue an associate's, bachelor's, or advanced degree through American colleges that offer classes overseas. The Education Center will answer all your questions and assist you in setting up an individual program.

Medical and Dental Care. Military health care facilities are comparable to those found in the United States. In addition to military personnel and their family members, health care is available to Department of the Army civilians and their families on a pay basis. Some specialty care (e.g., orthodontics) may be limited.

Pets. Pets may be shipped by commercial carrier at the owner's expense. However, some AMC flights also have this service available through contract air service. Check with your local transportation office about this option. The pet may accompany the owner or be shipped at a later time. Although there is no quarantine period, specific immunization and health certificate requirements must be met. Coordination with your sponsor is important to ensure that you get proper information about the requirements. Also, pets accompanying their owners must in most instances be boarded, due to restrictions, while in temporary government housing. On-post kennels and veterinary services are available on mainland Japan.

Passport. A passport is required by each family member. Servicemembers enter and leave Japan on their ID card under the status of forces agreement.

Civilian Clothing. Military personnel are authorized to wear civilian clothing after duty hours and as regulations permit.

Recreation and Culture. USARJ offers a wealth of activities. Facilities include fitness centers, lighted athletic fields, tennis courts, handball and racquetball courts, swimming pools, bowling alleys, skeet ranges, golf courses, and driving ranges. There are many associations and clubs, arts and crafts facilities, organized youth activities, and scouting. Travel officers are available to arrange trips in Japan and the Far East. The Army operates a seaside beach recreation area on Okinawa.

Family Member Employment. There are opportunities throughout USARJ for family member employment. Recent regulatory changes ensure that family members are given priority consideration in filling many vacancies. Although certain occupational specialties have few openings, volunteer positions exist that allow skills to be maintained while overseas.

Sponsorship Program. USARJ has an active sponsorship program. Shortly after receiving your orders, you will be receiving a letter and welcome packet from your sponsor. Make the program work for you—write to your sponsor immediately to acknowledge the letter and to ask for any further information you may need.

KOREA

In most of metropolitan Seoul, standards of living approach those in the West. The climate is similar to that of central New England, although the summers are longer and hotter in the southern part of the Korean Peninsula.

Many staff and troop duties involve close coordination with members of the ROK military. In fact, many ROK Army enlisted personnel are assigned to duty with the Eighth Army under the Korean Augmentation to U.S. Army (or KATUSA) program.

Command Sponsorship. Installation housing is similar to stateside government housing when available, but accompanied housing is in short supply, and servicemembers can anticipate significant waiting lists for accompanied housing.

Bringing family members on unaccompanied tours is discouraged, since it may cause extreme financial hardship and stress on all concerned.

Non-command-sponsored personnel who bring their family members to Korea do so at their own expense and are not eligible for transportation reimbursement or additional housing allowances. Depending on location, non-command-sponsored family members may receive medical and dental care only on a space-available or "emergency care" basis. Although non-command-sponsored family members may now enter the exchange and commissary for general-purpose shopping, they may not purchase controlled items. Single items valued at $50 or more, alcohol, tobacco products, and certain other controlled items may be purchased only by a ration-card holder (i.e., the active-duty member of the non-command-sponsored family).

Housing. Government family quarters are limited and are available only in Daegu, Camp Humphreys, Osan, Pusan, and Chinhae to locally assigned command-sponsored families. Waiting lists are kept for each grade category and are sub-divided into bedroom categories. Travel of command-sponsored family members to Korea is generally deferred until adequate housing is available.

Although many command-sponsored families live comfortably in off-post Korean communities, Western-style economy housing is limited and expensive. In Seoul, rents of more than $3,600 per month and security deposits ranging from two months' to one year's rent are common, but command-sponsored families may receive increased housing allowances to help offset the additional cost. Also, the finance office will loan money for the security deposit, which must be repaid prior to departure. Contact the housing office at your gaining unit for more detailed information.

Education. Sixteen Army Education Centers and three U.S. Air Force Education Centers serve U.S. Forces Korea personnel and their family members. These centers offer a variety of educational programs, ranging from basic skills to graduate level. Undergraduate and graduate contract educational institutions currently offering programs in Korea include University of Maryland Global Campus, Troy University, and Embry Riddle Aeronautical University. Because of size, limited troop strength, the mission, and other factors, graduate programs are normally offered only at Humphreys, Daegu, and Osan Air Base.

Command-sponsored children in kindergarten through grade 12 are guaranteed space-required, tuition-free schooling provided by the DoDEA system. These schools are located at Osan, Camp Humphreys, and Daegu. DoDEA schools are equal to or superior to CONUS public schools. All teachers are fully certified, and DoDEA-Korea schools are accredited by the North Central Association of Colleges and Schools (NCA). You can find updated information about DoDEA at www.dodea.edu/Pacific/Korea.

In addition to the usual extracurricular activities for students, Junior Reserve Officers Training Corps (JROTC) programs are available.

School-age children who are not command sponsored are authorized attendance at DoDEA schools only on a space-available basis. Prudent parents should not gamble on space-available openings at any school in DoDEA-Korea. Parents intent on bringing non-command-sponsored children to Korea should be aware than tuition is quite expensive, ranging from $4,000 to $7,000 per child per year. The U.S. government does not endorse any of the private institutions.

Banking and Currency. Besides the dollar, the Korean won is used in many military facilities and for all transactions off post except for specially designated facilities. The American banking facility provides full banking services on many military installations, and mobile banking vans provide service in other locations. The United Services of America Federal Credit Union also has branches on many installations. Check-cashing services are available at these institutions and at the PX and military clubs throughout the Republic.

Recreational and Cultural Activities. Each installation provides a wide range of activities for soldiers and their families throughout the Republic of Korea. Installations conduct extensive programs at both remote and large sites, providing entertainment, physical fitness, athletic competition, tour and travel, library services, recreation centers, arts and crafts centers, and other quality-of-life programs.

The athletic competition for adults includes company-, post-, and service-level competition in a year-round sports program. Installations also conduct an extensive youth activities and sports program.

The Dragon Hill Lodge on Yongsan's South Post is a new, modern, luxury-class, nonappropriated fund (NAF)-owned hotel supporting R&R needs and personnel in- and out-processing. Hotel amenities include restaurants, lounges, an exchange, and banking, shopping, and banquet facilities.

Tour and travel centers throughout the peninsula offer low-cost package tours to both in-country and out-of-country locations. The Army club system offers social and leisure activities, good food at reasonable prices, and both local and stateside entertainment.

There are many opportunities for cultural exchange, understanding, and friendship. These include subsidized tours around Korea, cultural shows, outdoor activities, visits to the homes of Korean families, and a Reunion in Korea program that allows family members to visit servicemembers stationed in Korea.

Americans have access to up-to-the-minute news and entertainment through American Forces Korea Network (AFKN) radio and TV and the *Pacific Stars and Stripes* daily newspaper. *KORUS Magazine* covers newsworthy special events of interest to U.S. Forces throughout Korea. Most military units also publish a newspaper to keep their audiences informed of on-post activities and military news.

OKINAWA

The 10th Area Support Group is the senior Army command on Okinawa and is a major subordinate command of USARJ. The 10th ASG is located on Torii Station, on the western coast of the southern half of Okinawa, about 20 miles north of Naha, the capital city. Temperatures are moderate year-round, with summer highs in the 90s and rare lows occasionally falling to the mid-50s.

Family Quarters. For accompanied personnel, family housing on Okinawa is managed by the Air Force housing office, located at Kadena Air Base. Visit www.housing.af.mil/Okinawa for more information.

Schooling for Children. Most children of military members attend DoDEA schools. All are located within family housing areas. There are thirteen DoDEA schools on the island, including two high schools, three junior high schools, and eight lower-level schools. You can find updated information about DoDEA at www.dodea.edu/pacific/south.

Child Care. Several excellent child-development centers are located within family housing areas. Full-day, part-day, and drop-in programs are available for children from six weeks to eleven years old. A list of certified in-home caregivers is also available.

Pets. The only military veterinarian service available is located on Kadena Air Base.

Other Concerns. On the subjects of concurrent travel, weight allowances, private automobiles, currency and banking, adult education, medical and dental care, domestic help, passports, civilian clothing, recreation and culture, family member employment, and the sponsorship program, Okinawa follows the same guidelines as Japan.

ARMY FORWARD DEPLOYED LOCATIONS

There are many locations worldwide where units deploy on a rotational basis. Organizations assigned to these missions rotate in and out over a period of months (often six to nine). Soldiers assigned to these units receive imminent-danger pay, family separation allowance, and other financial incentives. Currently, Army units are assigned or rotate as part of operations at Guantanamo Bay, Cuba; Counter-Drug Operations in South America/Caribbean; the Sinai Peninsula; Qatar; Africa and the Horn of Africa; Kuwait; Kosovo; Korea; and Central and Eastern Europe.

PART II

Service as an Army Officer

11

Professional Development

> [T]here is no room for a lack of integrity or for those who place self before duty or self before comrades, or self before country. Careerism is the one great sin, and it has no place among you.
>
> —Gen. John W. Vessey Jr., Chairman, Joint Chiefs of Staff

The Army officer should expect to have a full spectrum of developmental opportunities for a successful career and be able to determine the value and satisfaction of his or her service to the Army. This chapter provides information about the guidelines and policies which can be used as a professional development guide for all officers. It prescribes professional development goals and tenets that allow all officers to make individual personal and professional decisions throughout their Army service. It also assists officers in setting practical goals for their service, evaluating achievement, and resetting those goals while in the Army; such vision provides a solid foundation to assess the ultimate value of selflessness and commitment to the country and the Army.

An *Army leader* is anyone who by virtue of assumed role or assigned responsibility inspires and influences people to accomplish organizational goals. They motivate people both inside and outside the chain of command to pursue actions, focus thinking, and shape decisions for the greater good of the organization. Everything begins with the Warrior Ethos, which compels soldiers to "fight through all conditions to victory." Therefore, leader development is critical to all commissioned, warrant, and noncommissioned officers. The differences in the professional development programs of each type are based on their duties and official responsibilities. Most commissioned officers' roles and responsibilities are mandated by law, namely, Title X of the United States Code. It outlines official responsibilities for which commissioned officers are held accountable. Generally, the commissioned officer is the strategic, operational, and tactical planner whose "big picture" perspective is important in the integration, synchronization, and prosecution of battle.

Warrant officers' roles and responsibilities have evolved from their inception on 7 July 1918 through the Total Warrant Officer Study (1985) to the current definition established by the Warrant Officer Personnel Management Study (2000).

The warrant officer is a self-aware and adaptive technical expert, combat leader, trainer, and advisor, who administers, manages, maintains, operates, and integrates Army systems and equipment across the full spectrum of Army operations.

Noncommissioned officers' roles and responsibilities are established by Army policy and guidelines in AR 135-205, Enlisted Personnel Management System. NCOs are the Army's first-line trainers, and their tactical and professional competence ensures the maintenance of performance standards, both for individuals and for units on the battlefield. NCO professional development is discussed in *The NCO Guide,* Stackpole Books.

The official document DA PAM 600-3, *Commissioned Officer Professional Development and Career Management*, pertains to commissioned officers and is accessible at https://armypubs.army.mil/.

OFFICER DEVELOPMENT SYSTEM

> I believe in what I'm doing. This outfit of mine, they look to me for—well, for help and advice, how to be better soldiers, better men in general. . . . I'm the one who has to lead him into that filthy, endless horror and try to bring him out of it again. I know I'm the one.
>
> —Sam Damon in Anton Myrer's *Once an Eagle* (1968)

The executive agents of officer development and management are the Department of the Army, Deputy Chief of Staff for Personnel, and Commander, U.S. Army Human Resources Command (HRC). In 1998, the Army instituted a new personnel management system. This effort (OPMS 3) has three major components. First, it created the Officer Development System (ODS) to integrate and synchronize officer management, leader and character development, and evaluation systems of the Army. Second, it adopted a holistic and strategic approach to human resource management pertaining to officer development and management for the twenty-first century. Third, it established a management system with four career fields: operations, operations support, institutional support, and information operations.

The overarching focus of the Army's professional development programs is *warfighting.* The professional development of an Army officer is the result of collective efforts of the institution, commanders, and the individual officer. Each is responsible for establishing clear, succinct guidelines that reinforce leader and character development, consistent with the institution's corporate values of loyalty, duty, respect, selfless service, honor, integrity, and personal courage (collectively referred to by the acronym LDRSHIP).

The approach to officer development is cooperative and holistic. The Army wants its policies regarding officer evaluation, utilization, and promotion to be consistent with its warfighting and leadership doctrine. Leaders in the field, especially commanders, are to provide accurate and candid assessments of their subordinate officers' performance and future value to the Army. The individual officer is to subscribe to and live by the Army values daily, while continually seeking to realize his or her maximum potential in the Army. The following paragraphs discuss the major activities that establish and influence officer development guidelines and policies.

Officer Life Cycle Development Model. There are four distinct stages related to the military grade and experiences of an officer. Each stage reflects the education and training, operational assignments, and self-development goals required by an officer's branch, functional area, and career field. These stages are flexible to accommodate an officer's own capabilities, demonstrated performance, and needs of the Army. A generic life cycle model is shown below.

Company Grade. This period of service consists of two phases—lieutenant and captain. During this period, the officer is required to understand and supervise the day-to-day operations of the Army. It is a period of intense learning provided through institutional training and education, operational assignment and utilization, and self-development and reflection. This period culminates in company command and/or an initial position as a staff officer.

The Basic Officer Leaders Course (BOLC) occurs in two phases. BOLC A is a pre-commissionin training conducted by the traditional precommissioning sources. Once a cadet graduates, he/she is commissioned as a second lieutenant. Officers then attend the Basic Officer Leader Course Phase B at the appropriate branch school, which provides instruction related to the overall mission and function of the officer's branch. Officers may request additional skill training such as Ranger or Airborne Schools and may be selected for this type of training based on the needs of the Army. During this phase, officers should serve in their initial assignments and apply branch course training; seek tactical and technical expertise from experienced subordinates, peers, and superiors; and take the opportunity to improve their basic leadership skills of planning, organizing, communicating, and supervising. The initial assignment is normally with troops to provide the officer the opportunity to understand and apply the intricate relationships between leader and led, officer and NCO, and among combat, combat support, and combat service support branches of the Army. Officers are expected to be the example of soldiering and leadership to their enlisted subordinates. Officers should set goals to understand and apply the dynamics of team building from team to company levels, as well as use of prescribed troop leading procedures. They should also subscribe to and strive to exhibit the Army LDRSHIP values and prepare themselves tactically and technically for company command.

Captain Phase. This phase normally extends from about four to eleven years of commissioned service. It includes attendance at the branch captain's career course, company command and/or service as a staff officer at battalion or higher staff levels, and designation of a functional area. Some officers may request attendance at advanced civil schooling to pursue master's degrees or training with industry (TWI) and may be selected for this based on the requirements of their functional areas or the needs of the Army. Depending on other dynamics within the Army, such as expansion or downsizing, officers considering this option should carefully weigh the time they will spend away from more "main-stream" assignments against the advantages of graduate schooling. For some branches and specialties, advanced civilian academic pursuits are a natural—even mandatory—part of professional development. For others, it should be seen as part of the vision the officer is building for the type and scope of contributions he or she wants to make during service as an officer. Promotion to captain takes about four years.

During this phase, officers normally serve in their second or third assignment to apply their tactical, technical, and leadership skills in commanding at the company level or serving in subsequently higher staff levels. They should seek tactical and technical experience as a battalion, brigade, division, or general staff officer and continue to take advantage of every opportunity to refine their leadership skills, especially in communicating and supervising. During this phase, officers are expected to set the example of soldiering and leadership to junior officers, as well as enlisted soldiers.

The officer should set goals to successfully command at the company level, tactically and technically prepare for service as the operations/executive officer at the battalion level or as a staff/action officer at levels of battalion or higher, and continue to be an exemplar of Army values.

Field Grade. This period of service consists of three phases—major, lieutenant colonel, and colonel. During this period, the officer is required to develop and supervise plans for the future of the Army, besides understanding its day-to-day operations. It is a period of varying assignments, including duty with troops and assignments at operational, Army, or joint duty staffs. During this phase, officers have an opportunity to select a career field in operations, operations support, information operations, or institutional support. This period includes battalion and brigade command or field-grade utilization in a designated career field.

During this phase, officers normally serve as battalion operations/executive officers in preparation for battalion command within the operations career field or pursue technical and academic education in preparation for assignments in the other three career fields (operations support, information management, and institutional support).

It includes attendance at Intermediate Level Education (ILE) Courses, selection of a career field, service as an operations/executive officer at battalion level, aviation troop command (if applicable), and assignment as a staff/action officer at levels of brigade or higher. Some officers may request advanced civil schooling to pursue a Ph.D. or executive-level training and may be selected based on their career field designation (CFD) and the needs of the Army.

During this phase, officers may be assigned as a Battalion Commander in charge of hundreds of soldiers or a general staff officer in a division or corps. Many will serve in high visibility positions in their branch, FA or JIIM and a possible assignment to a cross-branch/FA development positions. Outstanding performance will merit more and more challenging positions. Some officers are

Years of Service

0 5 10 15 20 25 30

LT	CPT	MAJ	LTC	COL
BOLC	Captains Career Course	Intermediate Level Education	Senior Service College	
	☆ *Functional Area Decision*	☆ *Career Field Decision*		
Initial Branch Duty with Troops	Branch Assignment	Branch Assignment	Branch Assignment	Branch Assignment
Additional Skills Training	Branch Qualification	Branch Qualification	Branch/ Functional Area/ Generalist Assignment	Branch/ Functional Area/ Generalist Assignment
	Company/ Battery/Troop Command	Branch/ Functional Area/ Generalist Assignment	Functional Area Qualification	Brigade Command
	Functional Area Assignment	Functional Area Qualification	Battalion Command	Joint Duty
	ASC/TWI Assignment	ACS/TWI Assignment	Joint Duty	
		Joint Duty		

Commissioned Officer Career Development Model

selected for the Army War College or an equivalent, where they become "experts" at their profession.

Officers should set goals to prepare for battalion command in the operations career field or technically and academically prepare for an initial utilization in the other three career fields. Also, the officer should set the goal of preparing and mentoring company-grade officers for service as field-grade officers.

Lieutenant Colonel Phase. During this phase, officers serve in assignments that may include command at the battalion level. All should seek strategic and operational experience in each CFD utilization, as well as continue to seek every opportunity to provide mentorship of the officer corps.

Officers should set goals to successfully command at the battalion level and/or strategically and operationally prepare for command at the brigade level or service as senior advisers at operational, Army, or joint command. All should be stalwart examples and promoters of Army values.

Colonel Phase. This phase normally extends from the twenty-second to the thirtieth year of commissioned service. It includes multiple CFD assignments in the senior field-grade rank of the Army.

During this phase, officers serve as chief advisers, planners, and analysts for the senior leadership of the Army. They continue to seek strategic and operational experience in each CFD utilization and prepare and mentor more junior field-grade officers for service as senior field-grade officers.

Officers at this level should set goals to selflessly provide honest and forthright advice and analysis to the senior leaders of the Army to help ensure a vibrant, relevant, and principled organization in the future.

Planning the Future. Throughout an officer's service, he or she will have opportunities to influence how he or she serves the Army. It begins before the officer's oath of commission and may end in final assignments in one of the four career fields of the ODS. Selection of branch, functional area, and career field requires an assessment and evaluation by each individual officer of personal strengths, military and civilian schooling desires, and operational assignments and experience. The ultimate outcomes are influenced by the needs of the Army, officer qualification, and officer preference, in that order.

Branch Selection. An officer's Army service begins with branch selection during precommissioning. Upon commissioning, officers are designated in a basic branch from entry on active duty, training, and utilization. Some officers in combat support and combat service support are branch-detailed. These officers attend their detailed branch basic courses and are utilized and trained in their detailed branches. Upon completion of their detail, normally two or four years, these officers attend either a four-week transition training course (two-year detail) or transition training in conjunction with their enrollment into the Captain's Career Course. All officers are required to devote the first eight years to branch qualifying assignments. The branches that can be selected are listed below with their branch code. Note that some branches have restrictions.

BRANCHES AND FUNCTIONAL AREAS

Operations Division

Infantry[1] (11)	Engineer (21)	Field Artillery (13)
Aviation[1] (15)	Military Police (31)	Air Defense Artillery (14)
Armor[1] (19)	Chemical (74)	

Branch Transfer. Some officers may voluntarily request to transfer to another branch. A branch transfer permanently changes an officer's branch, component, or department. All transfers are processed in accordance with AR 614-100. Officers who are serving between their fourth and seventh years of commissioned service may volunteer for Special Forces branch and may be transferred only after completion of specified Special Forces training.

Functional Area Designation. Between the fifth and sixth years of commissioned service, an officer has the opportunity to request a functional area. This process begins with preference statements mailed to each officer in a cohort year group. Functional areas are designated by an Army centralized board that matches the officer's preference, military and academic education, and operational experience to the needs of the Army. Some officers may be designated earlier than their cohort group if they possess special skills or have been awarded graduate degrees supportive of critical functional areas.

Career Field Designation. At the tenth year of commissioned service, in conjunction with the majors' promotion board, officers have the opportunity to submit their preferences concerning branch or functional area priorities. Immediately following selection for promotion to major, officers are channeled into a career field for future branch or functional area utilization. The CFD process is implemented by an Army centralized board that considers the officer's preferences; education, training, and experience; demonstrated performance; and the needs of the Army. Each career field has its own unique characteristics and development track that reflect the future field-grade officer requirements for the Army. The functional areas aligned by career field that can be selected are listed below with their functional area codes.

BRANCHES AND FUNCTIONAL AREAS

Maneuver, Fires and Effects (MFE)

ArmySpecial Operations Forces (ARSOF)	Force Sustainment	Health Services	Operations Support	Functional Areas
Special Forces (18)	Adjutant General (42)	Army Medical Corps	Signal Corps (25)	FA 26-Network/Information Engineer
Civil Affairs (38)	Finance (36)	(60/61/62)	Military Intelligence (35)	FA 30-Information Operations
Psychological	Transportation (88)	Army Dental (63)	Cyber (17)	FA 34-Strategic Intel
Operations (37)	Ordnance (89/91)	Army Veterinary (64)	Information Operations	FA 40-Space Operations
	Quartermaster (92)	Army Medical		FA 46-Public Affairs
	Logistics (90)	Specialist (65)		FA 47-Academy Professor
	Acquisition	Army Nurse (66)		FA 48-Foreign Area Officer
		Medical Service (67)		FA 49-Operations Research/Systems Analysis
				FA 50-Force Management
				FA 52-Nuclear and Counter Proliferation
				FA 57-Simulation Operations
				FA 58-Enterprise Marketing and
				Behavioral Economics
				FA 59-Strategic Plans & Policy

OFFICER LEADER DEVELOPMENT AND CAREER MANAGEMENT

Leadership is intangible; no weapon ever designed will replace it.

—General of the Army Omar N. Bradley

Officer Leader Development is a process which enhances leader capabilities so leaders can assume positions of greater responsibility. There are three domains—institutional training, operational assignments, and self-development—which define and engage a continuous cycle of education, training, selection, experience, assessment, feeding, reinforcement, and evaluation. The purpose of the Leader Development process is to develop self-aware and adaptive leaders of character and competence who act to achieve results and who understand and are able to exploit the full potential of current and future Army doctrine. The following sections discuss how the Army's formal schooling (institutional training), assignment and utilization (operational assignments), and efforts of the individual officer fit into this integrated, progressive, and sequential process.

INSTITUTIONAL TRAINING

Institutional Training provides the solid foundation upon which all future development rests. The schools and training centers within this domain are where officers learn the knowledge, skills, and attributes essential to high-quality leadership while training to perform critical tasks. The small group instructional training format affords all officers the opportunity to share their experiences and participate in the learning process in small groups. Education and training are focused on maintaining an officer corps grounded in solid warfighting and leadership doctrine. It is comprehensive and spans the entire officer life cycle.

Precommissioning. The Army has three primary ways of educating and training, providing experience, and assessing and evaluating the individual officer. They are the U.S. Military Academy (USMA), West Point, New York; the Reserve Officers Training Corps (ROTC) under Cadet Command, headquartered at Fort Knox, Kentucky; and the Officer Candidate School (OCS) at Fort Benning, Georgia. Each experience includes evaluation of suitability for service as an officer and lays the foundations of core competencies to begin service in the Army.

Company Grade. All officers must focus their company-grade years on branch qualification, regardless of the functional area and career field they will later enter. The value an officer brings to a specialized functional area is dependent on experience gained by leading soldiers and mastering basic branch skills. Most branches prefer that an officer command at the company, battery, or troop level for at least eighteen months following the Captains Career Course.

Command is the essence of leadership development at this stage of an officer's career. The number of company commands within a specific branch may not afford all officers the opportunity to command at the captain level, or at least not for the desired length of command tours. Therefore, key staff positions have been identified that lead to branch qualification. Company command opportunities are in traditional tables of organization and equipment (TOE), or "line" units, or tables of distribution and allowances (TDA) units in training, garrison, and headquarters organizations. Generally, by the eighth year of service, an officer should have achieved branch qualification.

Basic Officer Leaders Course. BOLC's goal is to develop competent and confident leaders imbued with a Warrior Ethos who, regardless of branch, are grounded in fieldcraft and are skilled in leading soldiers, training subordinates and employing and maintaining equipment. BOLC consists of two phases:

- BOLC-A (Precommissioning) includes education and training at existing precommissioning sources, USMA, ROTC, and OCS. The curricula for these sources will be synchronized to be consistent with the other phases of BOLC, but otherwise will not change.
- BOLC-B (Branch Technical and Tactical Training) will be conducted by branch schools to ensure all officers receive a basic foundation of technical branch skills and tactical skills in order to gain the knowledge to succeed in their first units of assignment.

Captains Career Course (CCC). CCC provides captains an opportunity to learn the leader, tactical, and technical tasks, and the supporting skills and knowledge needed to perform as company commanders and battalion level staff officers. It includes a distance learning phase that includes topics such as the following: effective communication, drug and alcohol, task management and environment laws.

During the resident phase of CCC instruction, captains learn, among other topics, how to

- Establish and maintain a disciplined command climate.
- Execute the unit's assigned missions.
- Command, control, lead, supervise, discipline, train, and develop subordinates and care for their families.
- Develop the unit Mission Essential Task Lists (METL) and training plan.
- Plan, supervise, train, and evaluate unit leader management of personnel, administration, supply, maintenance, safety, and security actions.
- Plan and supervise the safe use, maintenance, security, and accountability of unit equipment and materiel.
- Administer the UCMJ at the company level.

Additionally, captains learn to function as staff officers by analyzing and solving military problems, communicating, and interacting as members of a staff for matters related to training, mobilization, unit deployment, and combat operations. Advanced instruction in technical branch skills is interspersed throughout the course.

Field Grade. Officers have many decisions to make that will determine their assignments and schooling for the remainder of their Army service. The major focus is continued professional development that supports implementation and support of operational warfighting, regardless of branch or functional area.

Intermediate Level Education (ILE). The Army's ILE program provides all mid-grade officers a higher level of professional military education and additional leader development training. It also prepares leaders to operate in multidomain, joint, multinational, and interagency environments, and for duty as field-grade commanders and staff officers. ILE consists of two phases which vary depending on when an officer attends schooling and what career field the officer is serving in at the time of schooling.

Part I for all officers who do not attend Sister Service, Foreign, WHINSEC, NPS, or NIU will attend a common core course that trains field-grade officers with a common warrior ethos and warfighting focus needed in senior leadership positions. Officers selected to attend the ten-month resident CGSOC course at Fort Leavenworth will complete both the Part 1 (Common Core) and Part 2 (AOC) back to back while in residence. All other officers will complete the CGSOC Common core either at one of the three satellite locations (Fort Belvoir, Fort Leavenworth, and Redstone Arsenal) or through Distance Learning (DL).

Part 2 consists of branch or functional area specific training for each officer. For basic branches and FA 57 officers, this training is the Advanced Operations Course (AOC) taught in conjunction with the common core at Ft. Leavenworth. The purpose of AOC is to develop operational career field officers with a warfighting focus for battalion and brigade command, capable of conducting multidomain operations in JIIM environments with the requisite competencies to serve successfully as division through echelons above corps staff officers. Officers who do not attend the resident course at Fort Leavenworth must complete AOC through either distance learning or the satellite location at Fort Leavenworth. Basic branch officers selected for a blended (satellite) ILE may elect to attend Part 1 or Part 2 of ILE in residence and complete the other part via DL. They may not attend both the Common Core and AOC satellite locations. For officers serving in functional areas (except FA57), Part 2 (typically called ILE qualification) training will vary depending on what functional area the officer is serving in. It may consist of graduate-level education or other functional area specific courses (Qualification Courses) as determined by the Functional Area Proponent. Officers must complete both Part 1 and Part 2 to be an ILE graduate and MEL 4 complete.

Under Optimized Universal ILE, all officers in cohort year group 2004 and later will be board selected for both residential and satellite ILE attendance (officers in Cohort Year Group 2003 and earlier will complete ILE via DL after January 2018). If an officer in Cohort Year Group 2004 and later is not selected to attend either resident or satellite ILE, then they must complete ILE via distance learning.

In order to be considered by the ILE board, an officer must first be selected for promotion to major. This includes officers selected Primary Zone (PZ), Early Consideration, and Above the Zone (AZ). An officer who is non-select to the rank of major will not be considered by the ILE board. This preserves the officer's one look in the event that the officer is selected AZ to major the following year. Officers must be in the grade of CPT/P or MAJ with at least eight years of

Active Federal Commissioned Service (AFCS) as of the start date of the course in order to attend. Additionally, all officers must have graduated from (or have credit for completing) the Captains Career Course and have met their professional development requirements in their basic branch.

The majority of officers eligible for resident ILE will attend Army ILE schools; however, some officers attend other service, joint, or foreign schools that have been designated as approved equivalents. Attendance at any ILE course incurs a two-year active duty service obligation.

ILE graduates have the opportunity to apply for and attend additional Army, Navy, and Air Force schools, such as Marine Corps' School of Advanced Warfighting or the Army's School of Advanced Military Studies. Information on these is found at the schools' websites or with your assignment officer.

Senior Service College (SSC). The SSC is the final major military educational program available to prepare officers for the positions of greatest responsibility in the Department of Defense. These include attendance at the Army War College (AWC), the Eisenhower School, the National War College (NWC), other service colleges, and resident fellowships at governmental agencies and academic institutions. Officers are assigned to duty with the Army Staff (ARSTAF), Joint Chiefs of Staff (JCS), Secretary of Defense (SECDEF), major Army command (MACOM), and to the staffs of Combatant Commander as branch, functional area, branch/functional area generalist, or joint service officers.

Other Schooling. The Army affords officers opportunities to seek other technical or advanced schooling based on the officer's preference and performance and on the needs of the Army. In cases in which language fluency is essential, officers may be offered an opportunity for foreign language training; most officers attend the Defense Language Institute (DLI) at the Presidio of Monterey, California. The length of training is dependent on the language requirement.

Advanced Civil Schooling. An officer may request to attend a civilian academic institution to obtain master's and doctoral degrees in designated disciplines. Selection is dependent on the officer's branch or functional area skill, and academic proficiency is measured by undergraduate performance and scores from the Graduate Record Examination (GRE) or Graduate Management Admission Test (GMAT). As always, the needs of the Army are paramount. Guidelines for this program are available in AR 621-1.

Training with Industry (TWI). Some officers may be selected to participate in TWI, where officers are assigned to a civilian industry to observe and learn the technical and managerial aspects of that field. Selection for this program is based on an officer's branch qualification, demonstrated performance, and the needs of the Army. Guidelines for this program are outlined in AR 621-1.

SELF-DEVELOPMENT

Self-development is a planned, competency-based, progressive, and sequential process that officers use to improve performance and achieve developmental goals. It is a continuous process that occurs during institutional training and operational assignments. Self-development is a joint effort between leader and led, superior and subordinate, senior and junior officers.

There are myriad means of self-development, from taking Army correspondence courses to visiting museums and studying battlefields (and reflecting upon what you see and feel when you go) to reading books and viewing films. Hobbies and other distractions are important, as they provide much-needed mental and emotional excursions from the professional "grind," but here's something to think about: If *everything* about the Army seems to be a grind for you, that is telling you something worthy of deep consideration and contemplation.

This edition of the *Army Officer's Guide* includes not only a professional reading list but also a list of movies available. Read and watch them when you can; immerse yourself in the ethos of the soldier; reflect on its meaning. Discuss your thoughts with your family, friends, and peers. Even better, watch some of these films with your subordinates or fellow officers and listen to and observe their reactions. These are opportunities you are unlikely to get at work, especially in the frenetically paced schedule of Army units, which are inevitably trying to do much more with less.

Evaluation. An officer's evaluation is normally from his or her supervisors in the immediate chain of command. Evaluations may be formal, such as the Officer Evaluation Report (OER), or informal, such as feedback and advice from superiors, peers, or subordinates. Officers should accept all feedback, carefully and candidly assess what is valid, and develop a plan of action to continue their strengths and correct their weaknesses.

Self-Assessment. Any self-development plan begins with a competency-based assessment of an officer's leadership skills, attributes, and behaviors. An officer's self-reflection on schooling, experience, and individual performance becomes critical in identifying strengths and weaknesses within the leadership construct of leader, led, situation, and communications. Comprehensive, honest assessments build the foundation for a practical, realistic set of goals that maximize an officer's strengths and minimize or correct an officer's weaknesses.

Plan of Action. An officer should periodically establish a set of developmental goals that include continuing civilian education; enrolling in and attending institutional training; participating in a self-study and reading program; increasing military knowledge of history, doctrine, and tactics; and seeking mentorship from experienced superiors or peers.

Human Resources Command. Officers should periodically visit the Officer Personnel Management Directorate (OPMD). The primary purpose of such visits is to review one's individual official records (listed below) to ensure their accuracy and completeness. The officer should also take the time to discuss his or her preferences for future assignments and utilization with the assignment officer and gain an assessment of his or her standing within the year group cohort.

Officer Record Brief (ORB). This is a word picture of an officer's service to date that contains his or her civilian and military schooling and dates, assignments, and duty positions held. It also includes promotion dates, source of commission, physical profile, decorations and badge data, and other objective items about an officer's service. If it is not accurate, board members will have no way of knowing what you have really accomplished, so keep it up-to-date and accurate.

Official Photo. This is a presentation that shows whether an officer can wear the Class A uniform, can follow regulations, and appears healthy and fit. Centralized Army boards *do not* use the photo.

Official Military Personnel File (OMPF). Each officer has an OMPF that contains copies of all OERs, promotion orders, and decorations and citations. Submission and processing to the OMPF is in accordance with AR 600-201. Officers are responsible for periodic review of the OMPF to ensure its completeness and accuracy. Boards mainly use OERs to determine officer quality, and board procedures are such that each OER generally gets a very quick appraisal. Incorrect names or other data will probably not be noticed, but negative information will be. Although it may sound redundant to the uninitiated, make sure that the reports in your file are all yours and that they have been correctly marked and accurately reproduced on your microfiche.

HRC Online. The U.S. Army Personnel Command provides an online Internet service to its officers. Visit the home page at www.hrc.army.mil/.

OFFICER EVALUATION SYSTEM (OES)

As part of the OPMS XXI Study, the Officer Evaluation System (OES) was established to be integrated into ODS. OES identifies those officers most qualified for advancement and assignment to positions of increased responsibility. Under this system, officers are evaluated on their performance and potential through duty evaluations, school evaluations, and HQDA evaluations (both central selection boards and PERSCOM officer management assessments).

The assessment of an officer's potential is based on officer performance documented in the OER. It reflects an officer's background in terms of experience and expertise and includes such items as branch and specialty qualification, performance in demanding positions, civil and military schooling, and physical fitness. It also contains an assessment of an officer's execution of tasks in support of the organization or Army missions. The primary focus is the accomplishment of tasks, but also important is how an officer adheres to the core values of the Army.

In January 2014, the Army implemented a new OER system designed to strengthen rater accountability and reflect current Army leadership doctrine. Three OER grade plates were established: one for company grade officers (DA 67-10-1), one for field-grade officers (DA 67-10-2), and one for colonels and general officers (DA 67-10-3). Brigadier generals and chief warrant officers 5 will not receive box checks; major generals and above will not receive evaluations.

Each set of grade plates identifies performance attributes that reflect the potential appropriate for the individual officer's grade. Evaluations must be entered and submitted through a Web-based system called Evaluation Entry System (EES), https://evaluations.hrc.army.mil. The Army-issued Common Access Card is needed to access the EES. The system includes an enhanced wizard for

HQDA#:

Attachments Menu

OFFICER EVALUATION REPORT SUPPORT FORM

For use of this form, see AR 623-3 ; the proponent agency is DCS, G-1.

SEE PRIVACY ACT STATEMENT IN AR 623-3

PART I - ADMINISTRATIVE (Rated Officer)

a. NAME (*Last, First, Middle Initial*)	b. SSN (or DOD ID No.)	c. GRADE/ RANK	d. DATE OF RANK (*YYYYMMDD*)	e. BRANCH	f. COMPONENT (*STATUS CODE*)

g. UNIT, ORG., STATION, ZIP CODE OR APO, MAJOR COMMAND	h. UIC CODE	i. THRU DATE OF LAST COMPLETED EVALUATION

j. RATED OFFICER'S EMAIL ADDRESS (*.gov or .mil*)

PART II - AUTHENTICATION

a1. NAME OF RATER (*Last, First, Middle Initial*)	a2. SSN (or DOD ID No.)	a3. RANK	a4. POSITION	a5. EMAIL ADDRESS (*.gov or .mil*)
b1. NAME OF INTERMEDIATE RATER (*Last, First, Middle Initial*)	b2. SSN (or DOD ID No.)	b3. RANK	b4. POSITION	b5. EMAIL ADDRESS (*.gov or .mil*)
c1. NAME OF SENIOR RATER (*Last, First, Middle Initial*)	c2. SSN (or DOD ID No.)	c3. RANK	c4. POSITION	c5. EMAIL ADDRESS (*.gov or .mil*)
c6. SENIOR RATER'S ORGANIZATION	c7. BRANCH	c8. COMPONENT	c9. SENIOR RATER PHONE NUMBER	
d1. INDIVIDUAL TO PERFORM SUPPLEMENTARY REVIEW (*Last, First, Middle Initial*) - (*IF REQUIRED*)		d2. RANK	d3. POSITION	d4. EMAIL ADDRESS (*.gov or .mil*)

PART III - VERIFICATION OF FACE - TO - FACE DISCUSSION

MANDATORY RATER/RATED OFFICER INITIAL FACE-TO-FACE COUNSELING ON DUTIES, RESPONSIBILITIES AND PERFORMANCE OBJECTIVES FOR THE CURRENT RATING PERIOD TOOK PLACE ON (DATE) ____ RATED OFFICER INITIALS ____ RATER INITIALS ____ SENIOR RATER INITIALS ____

RATED OFFICER ACCESS TO SUPPORT FORMS PRIOR TO INITIAL COUNSELING: RATER (Date ____) SENIOR RATER (Date ____)

PERIODIC RATER / RATED OFFICER FOLLOW-UP FACE-TO-FACE COUNSELINGS:

DATE ____ RATED OFFICER INITIALS ____ RATER INITIALS ____ SENIOR RATER INITIALS ____

DATE ____ RATED OFFICER INITIALS ____ RATER INITIALS ____ SENIOR RATER INITIALS ____

DATE ____ RATED OFFICER INITIALS ____ RATER INITIALS ____ SENIOR RATER INITIALS ____

PART IV - RATED OFFICER - DUTIES AND RESPONSIBILITIES

a. PRINCIPAL DUTY TITLE:

b. POSITION AOC/BRANCH:

c. STATE YOUR SIGNIFICANT DUTIES AND RESPONSIBILITIES:

PART V - PERFORMANCE OBJECTIVES AND ACCOMPLISHMENTS

a. INDICATE YOUR MAJOR PERFORMANCE OBJECTIVES:

b. LIST SIGNIFICANT CONTRIBUTIONS AND ACCOMPLISHMENTS:

DA FORM 67-10-1A, MAR 2019

Page 1 of 5

APD LC v1.00ES

Officer Evaluation Report Support Form

HQDA#:

PART V - PERFORMANCE OBJECTIVES AND ACCOMPLISHMENTS CONTINUED Describe adherence to leadership attributes and demonstration of competencies

A. CHARACTER: (Army Values, Empathy, Warrior Ethos/Service Ethos, Discipline - see ADRP 6-22)

INDICATE YOUR MAJOR PERFORMANCE OBJECTIVES:

LIST SIGNIFICANT CONTRIBUTIONS AND ACCOMPLISHMENTS:

B. PRESENCE: (Military and professional bearing, Fitness, Confidence, Resilience - see ADRP 6-22); (Safety/ Individual and unit deployment readiness/Support of behavioral health goals, AR 623-3 and Mission Command Principles, see ADP 6-0, addressed under fitness and resilience)

APFT GOALS: PU SU RUN HEIGHT/WEIGHT (ONLY AS NEEDED)

INDICATE YOUR MAJOR PERFORMANCE OBJECTIVES:

LIST SIGNIFICANT CONTRIBUTIONS AND ACCOMPLISHMENTS:

C. INTELLECT: (Mental agility, Sound judgment, Innovation, Interpersonal tact, expertise - see ADRP 6-22 and ADRP 6-0)

INDICATE YOUR MAJOR PERFORMANCE OBJECTIVES:

LIST SIGNIFICANT CONTRIBUTIONS AND ACCOMPLISHMENTS:

D. LEADS: (Leads others, builds trust, extends influence beyond the chain of command, Leads by example, Communicates-see ADRP 6-22 and ADRP 6-0)

INDICATE YOUR MAJOR PERFORMANCE OBJECTIVES:

LIST SIGNIFICANT CONTRIBUTIONS AND ACCOMPLISHMENTS:

E. DEVELOPS: (Creates a positive environment/Fosters esprit de corps, prepares self, Develops others, Stewards the profession - see ADRP 6-22)

INDICATE YOUR MAJOR PERFORMANCE OBJECTIVES:

LIST SIGNIFICANT CONTRIBUTIONS AND ACCOMPLISHMENTS:

F. ACHIEVES: (Gets Results - see ADRP 6-22 and ADRP 6-0)

INDICATE YOUR MAJOR PERFORMANCE OBJECTIVES:

LIST SIGNIFICANT CONTRIBUTIONS AND ACCOMPLISHMENTS:

DA FORM 67-10-1A, MAR 2019

Page 2 of 5

APD LC v1.00ES

Officer Evaluation Report Support Form *(continued)*

HQDA#:

PART VI - RATER SELF DEVELOPMENT GOALS

PART VII - SENIOR RATER COMMENTS

RATED SOLDIER -SIGNATURE AND DATE:

Continuation Section

DA FORM 67-10-1A, MAR 2019

APD LC v1.00ES

Officer Evaluation Report Support Form *(continued)*

PARTS I-IV INSTRUCTIONS. AR 623-3 outlines the administrative requirements necessary to complete these portions of the support form.
Some key **requirements: The rater will --**
a. Provide a copy of his or her support form (or equivalent), along with the senior rater's support form (or equivalent), to the rated Soldier at the beginning of the rating period.
b. Discuss the scope of the rated Soldier's duty description with him or her within 30 days after the beginning of the rating period. This counseling will include, as a minimum, the rated Soldier's duty description and the performance objectives to attain. The discussion will also include the relationship of the duty description and objectives with the organization's mission, problems, priorities, and similar matters.
c. Counsel the rated Soldier.
(1) If the rated Soldier is recently assigned to the organization, the rater may use the counseling to outline a duty description and performance objectives. This discussion gives the rated Soldier a guide for performance while learning new duties and responsibilities in the unit of assignment.
(2) If the rater is recently assigned, this first counseling may be used to ask the rated Soldier for an opinion of the duty description and objectives. By doing this, the rater is given a quick assessment of the rated Soldier and the work situation. It will also help the rater develop the best duty description and performance objectives for the rated Soldier.
d. Raters of CPTs, LTs, CW2s, and WO1s will also conduct quarterly follow-up counseling sessions to discuss performance, update and/or revise developmental tasks, as required, and assess developmental progress. Summary or key comments will be recorded for inclusion when preparing final OERs.
Senior raters and reviewing officials will --
(1) Ensure support forms (or equivalent) are provided to all rated Soldiers they senior rate at the beginning of and throughout the respective rating periods.
(2) Use all reasonable means to become familiar with a rated Soldier's performance. When practical, use personal contact, records and reports, and the information provided on the rated Soldier's support form.
The rated officer plays a significant role in counseling sessions and the evaluation process throughout the rating period. In the event of geographical separation, correspondence and telephone conversations will be used as alternatives to face-to face counseling followed by face-to-face discussions between the rated Soldier and the rater at the earliest opportunity.

PART V INSTRUCTIONS: ICW ADRP 6-22 and ADP 6-0 rated officer performance objectives will align with the attributes and competencies required for all officers. The overall definition of each attribute and competency is addressed in the base support form. Key points:

A. CHARACTER: Army Values, Empathy, Warriors Ethos/Service Ethos, and Discipline.
Army Values: Values are principles, standards, or qualities considered essential for successful leaders. Values are fundamental to help people discern right from wrong in any situation. The Army has seven values to develop in all Army individuals: loyalty, duty, respect, selfless service, honor, integrity, and personal courage. **Empathy:** The propensity to experience something from another person's point of view. The ability to identify with and enter into another person's feelings and emotions. The desire to care for and take care of Soldiers and others. **Warrior Ethos/Service Ethos:** The internal shared attitudes and beliefs that embody the spirit of the Army profession for Soldiers and Army Civilians alike. **Discipline:** Control of one's own behavior according to Army Values; mindset to obey and enforce good orderly practices in administrative, organizational, training, and operational duties. Personal beliefs related to upbringing, culture, religious backgrounds, and traditions are also central to character.

B. PRESENCE: Military and professional bearing, Fitness, Confidence, Resilience Military and professional bearing:
Possessing a commanding presence. Projecting a professional image of authority. **Fitness:** Having sound health, strength, and endurance that support one's emotional health and conceptual abilities under prolonged stress. **Confidence:** Projecting self-confidence and certainty in the unit's ability to succeed in its missions. Demonstrates composure and outward calm through control over one's emotions. **Resilience:** Showing a tendency to recover quickly from setbacks, shock, injuries, adversity, and stress while maintaining a mission and organizational focus.

C. INTELLECT: (Mental agility, Sound judgment, Innovation, Interpersonal tact, expertise) Mental agility: Flexibility of mind; the ability to break habitual thought patterns. Anticipating or adapting to uncertain or changing situations; to think through outcomes when current decisions or actions are not producing desired effects. The ability to apply multiple perspectives and approaches. **Sound judgment:** The capacity to assess situations shrewdly and draw sound conclusions. The tendency to form sound opinions, make sensible decisions and reliable guesses. The ability to assess strengths and weaknesses of subordinates, peers, and enemy to create appropriate solutions and action. **Innovation:** The ability to introduce new ideas based on opportunity or challenging circumstances. Creativity in producing ideas and objects that are both novel and appropriate. **Interpersonal tact:** The capacity to understand interactions with others. Being aware of how others see you and sensing how to interact with them effectively. Conscious of character, reactions and motives of self and others and how they affect interactions. Recognizing diversity and displaying self-control, balance, and stability. **Expertise:** Possessing facts, beliefs, logical assumptions and understanding in relevant areas.

D. LEADS: (Leads others, builds trust, extends influence beyond the chain of command, Leads by example, Communicates)
Leads others: 1. **Uses appropriate methods of influence to energize others.** Uses methods ranging from compliance to commitment (pressure, legitimate requests, exchange, personal appeals, collaboration, rational persuasion, apprising, inspiration, participation, and relationship building). 2. **Provides purpose, motivation and inspiration.** Inspires, encourages, and guides others toward mission accomplishment. Emphasizes the importance of organizational goals. Determines the course of action necessary to reach objectives and fulfill mission requirements. Communicates instructions, orders, and directives to subordinates. Ensures subordinates understand and accept direction. Empowers and delegates authority to subordinates. Focuses on the most important aspects of a situation. 3. **Enforces standards** - Reinforces the importance and role of standards. Performs individual and collective tasks to standard. Recognizes and takes responsibility for poor performance and addresses it appropriately. 4. **Balances mission and welfare of followers.** Assesses and routinely monitors effects of mission fulfillment on mental, physical, and emotional attributes of subordinates. Monitors morale, physical condition, and safety of subordinates. Provides appropriate relief when conditions jeopardize success of the mission or present overwhelming risk to personnel.
Builds Trust: 1. Sets personal example for trust. Is firm, fair, and respectful to gain trust. Assesses degree of own trustworthiness. **2. Takes direct actions to build trust.** Fosters positive relationship with others. Identifies areas of commonality (understanding, goals, and experiences). Engages other members in activities and objectives. Corrects team members who undermine trust with their attitudes or actions. **3. Sustains a climate of trust.** Assesses factors or conditions that promote or hinder trust. Keeps people informed of goals, actions, and results. Follows through on actions related to expectations of others.
Extends influence beyond the chain of command. 1. Understands sphere, means and limits of influence. Assesses situations, missions, and assignments to determine the parties involved in decision making, decision support, and possible interference or resistance. **2. Negotiates, builds consensus and resolves conflict.** Builds effective working relationships. Uses two-way, meaningful communication. Identifies individual and group interests. Identifies roles and resources. Generates and facilitates generation of possible solutions. Applies fair standards to assess options. Creates good choices between firm, clear commitment and alternatives to a negotiated agreement.
Leads by example. 1. Displays character: Sets the example by displaying high standards of duty performance, personal appearance, military and professional bearing, physical fitness and ethics. Fosters an ethical climate; shows good moral judgment and behavior. Completes individual and unit tasks to standard, on time, and within the commander's intent. Demonstrates determination, persistence, and patience. Uses sound judgment and logical reasoning. **2. Exemplifies the Warrior Ethos.** Removes or fights through obstacles, difficulties, and hardships to accomplish the mission. Demonstrates the will to succeed. Demonstrates physical and emotional courage. Shares hardships with subordinates. **3. Leads with confidence in adverse situations:** Provides leader presence at the right time and place. Displays self-control, composure, and a positive attitude. Is resilient. Remains decisive after discovering a mistake. Acts in the absence of guidance. Does not show discouragement when facing setbacks. Remains positive when the situation becomes confusing or changes. Encourages subordinates when they show signs of weakness. **4. Demonstrates technical and tactical competence** Meets mission standards, protects resources, and accomplishes the mission with available resources using technical and tactical skills. Displays appropriate knowledge of equipment, procedures and methods; recognizes and generates innovative solutions. Uses knowledgeable sources and subject matter experts. **5. Understands the importance of conceptual skills and models them to others.** Displays comfort working in open systems. Makes logical assumptions in the absence of facts. Identifies critical issues to use as a guide in making decisions and taking advantage of opportunities. Relates and compares information from different sources to identify possible cause-and-effect relationships. **6. Seeks diverse ideas and points of view.** Encourages honest communication among staff and decision makers. Explores alternative explanations and approaches for accomplishing tasks. Reinforces new ideas; demonstrates willingness to consider alternative perspectives to resolve difficult problems. Discourages individuals from seeking favor through tacit agreement.
Communicates: 1. Listens actively: Listens and watches attentively. Makes appropriate notes. Tunes in to content, emotion, and urgency. Uses verbal and nonverbal means to reinforce with the speaker that you are paying attention. Reflects on new information before expressing views.

DA FORM 67-10-1A, MAR 2019 Page 4 of 5
APD LC v1.00ES

Officer Evaluation Report Support Form *(continued)*

PART V INSTRUCTIONS CONTINUED: ICW ADRP 6-22 and ADP 6-0 rated officer performance objectives will align with the attributes and competencies required for all officers. The overall definition of each attribute and competency is addressed in the base support form. Key points:
Communicates (continued) - 2. **Creates shared understanding:** Shares necessary information with others and subordinates. Protects confidential information. Coordinates plans with higher, lower and adjacent organizations. Keeps higher and lower headquarters, superiors and subordinates informed. Expresses thoughts and ideas clearly to individuals and groups. Recognizes potential miscommunication. Uses appropriate means for communicating a message. 3. **Employs engaging communication techniques:** States goals to energize others to adopt and act on them. Uses logic and relevant facts in dialogue; expresses well-organized ideas. Speaks enthusiastically and maintains listeners' interest and involvement. Makes appropriate eye contact when speaking. Uses appropriate gestures. Uses visual aids as needed. Determines, recognizes, and resolves misunderstandings. 4. **Is sensitive to cultural factors in communication:** Maintains awareness of communication customs, expressions, actions, or behaviors. Demonstrates respect for others.
E. DEVELOPS: (Create a positive environment/Fosters esprit de corps, prepares self, Develops others, Stewards the profession)
Creates a positive environment/Fosters esprit de corps: 1. Fosters teamwork, cohesion, cooperation and loyalty (esprit de corps) Encourages people to work together effectively. Promotes teamwork and team achievement to build trust. Draws attention to the consequences of poor coordination. Integrates new members into the unit quickly. **2. Encourages fairness and inclusiveness.** Provides accurate evaluations and assessments. Supports equal opportunity. Prevents all forms of harassment. Encourages learning about and leveraging diversity. **3. Encourages open and candid communications.** Shows others how to accomplish tasks while respectful and focused. Displays a positive attitude to encourage others and improve morale. Reinforces the expression of contrary and minority viewpoints. Displays appropriate reactions to new or conflicting information or opinions. Guards against groupthink. **4. Creates a learning Environment:** Uses effective assessment and training methods. Encourages leaders and their subordinates to reach their full potential. Motivates others to develop themselves. Expresses the value of interacting with others and seeking counsel. Stimulates innovative and critical thinking in others. Seeks new approaches to problems. Communicates the difference between professional standards and a zero-defects mentality. Emphasizes learning from one's mistakes. **5. Encourages subordinates to exercise initiative, accept responsibility and take ownership:** Involves others in decisions and informs them of consequences. Allocates responsibility for performance. Guides subordinate leaders in thinking through problems for themselves. Allocates decision-making to the lowest appropriate level. Acts to expand and enhance subordinate's competence and self-confidence. Rewards initiative.
6. Demonstrates care for follower well-being: Encourages subordinates and peers to express candid opinions. Addresses subordinates' and families' needs (health, welfare, and development). Stands up for subordinates. Routinely monitors morale and encourages honest feedback.
7. **Anticipates people's on-the-job needs:** Recognizes and monitors subordinate's needs and reactions. Shows concern for how tasks and missions affect subordinate morale. 8. **Sets and maintains high expectations for individuals and teams:** Clearly articulates expectations. Creates a climate that expects good performance, recognizes superior performance, and does not accept poor performance. Challenges others to match the leader's example. **Prepares self. 1. Maintains mental and physical health and wellbeing:** Recognizes imbalance or inappropriateness of one's own actions. Does not allow emotion to unduly influence decision-making. Applies logic and reason to make decisions or when interacting with emotionally charged individuals. Recognizes the sources of stress and maintains appropriate levels of challenge to motivate self. Manages regular exercise, leisure activities, and time away. Stays focused on life priorities and values. **2. Expands knowledge of technical, technological and tactical areas:** Seeks knowledge of systems, equipment, capabilities, and situations, particularly information technology systems. Keeps informed about developments and policy changes inside and outside the organization. **3. Expands conceptual and interpersonal capabilities:** Understands the contribution of concentration, critical thinking, imagination, and problem solving in different task conditions. Learns new approaches to problem solving. Applies lessons learned. Filters unnecessary information efficiently. Reserves time for self-development, reflection, and personal growth. Considers possible motives behind conflicting information.
4. Analyzes and organizes information to create knowledge: Reflects on prior learning; organizes insights for future application. Considers source, quality or relevance, and criticality of information to improve understanding. Identifies reliable resources for acquiring knowledge. Sets up systems of procedures to store knowledge for reuse.
5. Maintains relevant cultural awareness: Learns about issues of language, values, customary behavior, ideas, beliefs, and patterns of thinking that influence others. Learns about results of previous encounters when culture plays a role in mission success. **6. Maintains relevant geopolitical awareness:** Learns about relevant societies experiencing unrest. Recognizes Army influences on unified action partners and enemies. Understands the factors influencing conflict and peacekeeping, peace enforcing and peacemaking missions. **7. Maintains self-awareness: employs self understanding and recognizes impact on others:** Evaluates one's strengths and weaknesses. Learns from mistakes to make corrections; learns from experience. Seeks feedback; determines areas in need of development. Determines personal goals and makes progress toward them. Develops capabilities where possible but accepts personal limitations. Seeks opportunities to use capabilities appropriately. Understands self-motivation under various task conditions. **Develops others. 1. Assesses developmental needs of others:** Determines strengths and weaknesses of subordinates under different conditions. Evaluates subordinates in a fair and consistent manner. Assesses tasks and subordinate motivation to consider methods of improving work assignments, when job enrichment would be useful, methods of cross-training on tasks and methods of accomplishing missions. Designs ways to challenge subordinates to improve weaknesses and sustain strengths. Encourages subordinates to improve processes. **2. Counsels, coaches and mentors:** Improves subordinate's understanding and proficiency. Uses experience and knowledge to improve future performance. Counsels, coaches and mentors subordinates, subordinate leaders, and others.
3. Facilitates ongoing development: Maintains awareness of existing individual and organizational development programs and removes barriers to development. Supports opportunities for self-development. Arranges training opportunities to help subordinates improve self-awareness, confidence, and competence. Encourages subordinates to pursue institutional learning opportunities. Provide subordinates information about institutional training and career progression. Maintains resources related to development. **4. Builds team or group skills and processes:** Presents challenging assignments for team or group interaction. Provides resources and support for realistic, mission-oriented training. Sustains and improves the relationships among team or group members. Provides feedback on team processes.
Stewards of the profession. 1. Supports professional and personal growth: Supports developmental opportunities for subordinates such as PME attendance, key developmental assignments in other organizations, and broadening assignments. **2. Improves the organization:** Makes decisions and takes action to improve the organization beyond their tenure.
F. ACHIEVES: (Gets Results). 1. Prioritizes, organizes and coordinates taskings for teams or other organizations structures/groups: Ensures the course of action achieves the desired outcome through planning. Organizes groups and teams to accomplish work. Ensures all tasks can be executed in the time available and that tasks depending on other tasks are executed in the correct sequence. Limits over specification and micromanagement.
2. Identifies and accounts for capabilities and commitment to task: Considers duty positions, capabilities, and developmental needs when assigning tasks. Conducts initial assessments to assume a new task or a new position. **3. Designates, clarifies, and deconflicts roles:** Establishes and employs procedures for monitoring, coordinating, and regulating subordinate's actions and activities. Mediates peer conflicts and disagreements.
4. Identifies, contends for, allocates and manages resources: Tracks people and equipment. Allocates adequate time for task completion. Allocates time to prepare and conduct rehearsals. Continually seeks improvement in operating efficiency, resource conservation, and fiscal responsibility. Attracts, recognizes, and retains talent.
5. Removes work barriers: Protects organization from unnecessary taskings and distractions. Recognizes and resolves scheduling conflicts. Overcomes obstacles preventing accomplishment of the mission.
6. Recognizes and rewards good performance: Recognizes individual and team accomplishments; rewards appropriately. Credits subordinates for good performance; builds on successes. Explores reward systems and individual reward motivations.
7. Seeks, recognizes and takes advantage of opportunities to improve performance: Asks incisive questions. Anticipates needs for actions; envisions ways to improve. Acts to improve the organization's collective performance. Recommends best methods to accomplish tasks; uses information and technology to improve individual and group effectiveness. Encourages staff to use creativity to solve problems. **8. Makes feedback part of work processes:** Gives and seeks accurate and timely feedback. Uses feedback to modify duties, tasks, procedures, requirements, and goals. Uses assessment techniques and evaluation tools (such as AARs) to identify lessons learned and facilitate consistent improvement. Determines the appropriate setting and timing for feedback.
9. Executes plans to accomplish the mission: Schedules activities to meet commitments in critical performance areas. Notifies peers and subordinates in advance of required support. Keeps track of task assignments and suspense's; attends to details. Adjusts assignments, if necessary.
10. Identifies and adjusts to external influences on the mission and organization: Gathers and analyzes relevant information about changing conditions. Determines causes, effects, and contributing factors of problems. Considers contingencies and their consequences. Makes necessary, on-the-spot adjustments.

PART VI RATER SELF DEVELOPMENT GOALS INSTRUCTIONS: These goals are beyond the current career progression that the officer is assigned. This area should focus on those branched detailed, functional designated, have completed branch qualification and anticipate broadening and advance civil schooling opportunities. The officer will also state self development objectives based on MSAF assessments (optional) or other identified areas required for development.

Officer Evaluation Report Support Form ***(continued)***

HQDA#: | Attachments Menu

COMPANY GRADE PLATE (O1 - O3; WO1 - CW2) OFFICER EVALUATION REPORT
For use of this form, see AR 623-3; the proponent agency is DCS, G-1.

See Privacy Act Statement in AR 623-3.

PART I - ADMINISTRATIVE *(Rated Officer)*

a. NAME *(Last, First, Middle Initial)* | b. SSN (or DOD ID No.) | c. RANK | d. DATE OF RANK *(YYYYMMDD)* | e. BRANCH | f. COMPONENT *(Status Code)*

g. UNIT, ORG., STATION, ZIP CODE OR APO, MAJOR COMMAND | h. UIC | i. REASON FOR SUBMISSION

j. PERIOD COVERED FROM *(YYYYMMDD)* THRU *(YYYYMMDD)* | k. RATED MONTHS | l. NON RATED CODES | m. NO. OF ENCLOSURES | n. RATED OFFICER'S EMAIL ADDRESS *(.gov or .mil)*

PART II - AUTHENTICATION *(Rated officer's signature verifies officer has seen completed OER Parts I-VI and the administrative data is correct)*

a1. NAME OF RATER *(Last, First, Middle Initial)* | a2. SSN (or DOD ID No.) | a3. RANK | a4. POSITION

a5. EMAIL ADDRESS *(.gov or .mil)* | a6. RATER SIGNATURE | a7. DATE *(YYYYMMDD)*

b1. NAME OF INTERMEDIATE RATER *(Last, First, Middle Initial)* | b2. SSN (or DOD ID No.) | b3. RANK | b4. POSITION

b5. EMAIL ADDRESS *(.gov or .mil)* | b6. INTERMEDIATE RATER SIGNATURE | b7. DATE *(YYYYMMDD)*

c1. NAME OF SENIOR RATER *(Last, First, Middle Initial)* | c2. SSN (or DOD ID No.) | c3. RANK | c4. POSITION

c5. SENIOR RATER'S ORGANIZATION | c6. BRANCH | c7. COMPONENT | c9. EMAIL ADDRESS *(.gov or .mil)*

c8. SENIOR RATER PHONE NUMBER | c10. SENIOR RATER SIGNATURE | c11. DATE *(YYYYMMDD)*

d. This is a referred report, do you wish to make comments?
☐ Referred ☐ Yes, comments are attached ☐ No

e1. RATED OFFICER SIGNATURE | e2. DATE *(YYYYMMDD)*

f1. Supplementary Review Required? ☐ Yes ☐ No

f2. NAME OF REVIEWER *(Last, First, Middle Initial)*

f3. RANK | f4. POSITION | f5. Comments Enclosed

f6. SUPPLEMENTARY REVIEWER SIGNATURE | f7. DATE *(YYYYMMDD)*

PART III - DUTY DESCRIPTION

a. PRINCIPAL DUTY TITLE | b. POSITION AOC/BRANCH

c. SIGNIFICANT DUTIES AND RESPONSIBILITIES

PART IV - PERFORMANCE EVALUATION - PROFESSIONALISM, COMPETENCIES, AND ATTRIBUTES *(Rater)*

a. APFT Pass/Fail/Profile: ____ Date: ____ Height: ____ Weight: ____ Within Standard? ____

Comments required for "Failed" APFT, or "Profile" when it precludes performance of duty, and "No" for Army Weight Standards? Reset Item a. APFT/Pass/Fail/Profile

b. This Officer's overall Performance is Rated as: *(Select one box representing Rated Officer's overall performance compared to others of the same grade whom you have rated in your career. Managed at less than 50% in EXCELS.)*

I currently rate ____ **Army Officers** in this grade.

A completed DA Form 67-10-1A was received with this report and considered in my evaluation and review: ☐ Yes ☐ No *(explain in comments below)*

EXCELS (49%) ☐ PROFICIENT ☐ CAPABLE ☐ UNSATISFACTORY ☐

Comments:

DA FORM 67-10-1, MAR 2019

Page 1 of 2
APD LC v1.00ES

Officer Evaluation Report (OER)

HQDA#:

NAME:	SSN (or DOD ID No.)	PERIOD COVERED: FROM *(YYYYMMDD)*	THRU *(YYYYMMDD)*

c. 1) **Character**: *(Adherence to Army Values, Empathy, and Warrior Ethos/ Service Ethos and Discipline. Fully supports SHARP, EO, and EEO.)*	
c. 2) **Presence**: *(Military and Professional Bearing, Fitness, Confident, Resilient)*	
c. 3) **Intellect**: *(Mental Agility, Sound Judgment, Innovation, Interpersonal Tact, Expertise)*	
c. 4) **Leads**: *(Leads Others, Builds Trust, Extends Influence beyond the Chain of Command, Leads by Example, Communicates)*	
c. 5) **Develops**: *(Creates a positive command/ workplace environment/Fosters Esprit de Corps, Prepares Self, Develops Others, Stewards the Profession)*	
c. 6) **Achieves**: *(Gets Results)*	

PART V - INTERMEDIATE RATER

PART VI - SENIOR RATER

a. POTENTIAL COMPARED WITH OFFICERS SENIOR RATED IN SAME GRADE (OVERPRINTED BY DA)

☐ **MOST QUALIFIED** *(limited to 49%)*

☐ **HIGHLY QUALIFIED**

☐ **QUALIFIED**

☐ **NOT QUALIFIED**

b. I currently senior rate ______ **Army Officers** in this grade.

c. COMMENTS ON POTENTIAL:

d. List 3 future **SUCCESSIVE** assignments for which this Officer is best suited:

DA FORM 67-10-1, MAR 2019

Page 2 of 2
APD LC v1.00ES

Officer Evaluation Report (OER) ***(continued)***

HQDA#:

Attachments Menu

FIELD GRADE PLATE (O4 - O5; CW3 - CW5) OFFICER EVALUATION REPORT
For use of this form, see AR 623-3; the proponent agency is DCS, G-1.

See Privacy Act Statement in AR 623-3.

PART I - ADMINISTRATIVE *(Rated Officer)*

a. NAME *(Last, First, Middle Initial)* | b. SSN (or DOD ID No.) | c. RANK | d. DATE OF RANK *(YYYYMMDD)* | e. BRANCH | f. COMPONENT *(Status Code)*

g. UNIT, ORG., STATION, ZIP CODE OR APO, MAJOR COMMAND | h. UIC | i. REASON FOR SUBMISSION

j. PERIOD COVERED: FROM *(YYYYMMDD)* THRU *(YYYYMMDD)* | k. RATED MONTHS | l. NON RATED CODES | m. NO. OF ENCLOSURES | n. RATED OFFICER'S EMAIL ADDRESS *(.gov or .mil)*

PART II - AUTHENTICATION *(Rated officer's signature verifies officer has seen completed OER Parts I-VI and the administrative data is correct)*

a1. NAME OF RATER *(Last, First, Middle Initial)* | a2. SSN (or DOD ID No.) | a3. RANK | a4. POSITION

a5. EMAIL ADDRESS *(.gov or .mil)* | a6. RATER SIGNATURE | a7. DATE *(YYYYMMDD)*

b1. NAME OF INTERMEDIATE RATER *(Last, First, Middle Initial)* | b2. SSN (or DOD ID No.) | b3. RANK | b4. POSITION

b5. EMAIL ADDRESS *(.gov or .mil)* | b6. INTERMEDIATE RATER SIGNATURE | b7. DATE *(YYYYMMDD)*

c1. NAME OF SENIOR RATER *(Last, First, Middle Initial)* | c2. SSN (or DOD ID No.) | c3. RANK | c4. POSITION

c5. SENIOR RATER'S ORGANIZATION | c6. BRANCH | c7. COMPONENT | c9. EMAIL ADDRESS *(.gov or .mil)*

c8. SENIOR RATER PHONE NUMBER | c10. SENIOR RATER SIGNATURE | c11. DATE *(YYYYMMDD)*

d. This is a referred report, do you wish to make comments?
☐ Referred ☐ Yes, comments are attached ☐ No

e1. RATED OFFICER SIGNATURE | e2. DATE *(YYYYMMDD)*

f1. Supplementary Review Required? ☐ Yes ☐ No

f2. NAME OF REVIEWER *(Last, First, Middle Initial)*

f3. RANK | f4. POSITION | f5. Comments Enclosed

f6. SUPPLEMENTARY REVIEWER SIGNATURE | f7. DATE *(YYYYMMDD)*

PART III - DUTY DESCRIPTION

a. PRINCIPAL DUTY TITLE | b. POSITION AOC/BRANCH

c. SIGNIFICANT DUTIES AND RESPONSIBILITIES

PART IV - PERFORMANCE EVALUATION - PROFESSIONALISM, COMPETENCIES, AND ATTRIBUTES *(Rater)*

a. APFT Pass/Fail/Profile: Date: Height: Weight: Within Standard?

Comments required for "Failed" APFT, or "Profile" when it precludes performance of duty, and "No" for Army Weight Standards? Reset Item a. APFT/Pass/Fail/Profile

b. THIS OFFICER POSSESSES SKILLS AND QUALITIES FOR THE FOLLOWING BROADENING ASSIGNMENTS

c. THIS OFFICER POSSESSES SKILLS AND QUALITIES FOR THE FOLLOWING OPERATIONAL ASSIGNMENTS

d1. Character:
(Adherence to Army Values, Empathy, and Warrior Ethos/Service Ethos and Discipline. Fully supports SHARP, EO, and EEO.)

DA FORM 67-10-2, MAR 2019

Page 1 of 2
APD LC v1.00ES

Officer Evaluation Report (OER) ***(continued)***

HQDA#:

NAME	SSN (or DOD ID No.)	PERIOD COVERED: FROM (YYYYMMDD)	THRU (YYYYMMDD)

d2. Provide narrative comments which demonstrate performance regarding field grade competencies and attributes in the Rated Officer's current duty position. *(i.e. demonstrates excellent presence, confidence and resilience in expected duties and unexpected situation, adjusts to external influence on the mission or taskings and organization, prioritizes limited resources to accomplish mission, proactive in developing others through individual coaching counseling and mentoring, active learner to master organizational level knowledge, critical thinking and visioning skills, anticipates and provides for subordinates on–the-job needs for training and development, effective communicator across echelons and outside the Army chain of command, effective at engaging others, presenting information and recommendations and persuasion, highly proficient at critical thinking, judgment and innovation, proficient in utilizing Army design method and other to solve complex problems, uses all influence techniques to empower others; proactive in gaining trust in negotiations, remains respectful, firm and fair. Fully supports SHARP and creates a positive command/workplace environment.)*

COMMENTS:

e. This Officer's overall Performance is Rated as: *(Select one box representing Rated Officer's overall performance compared to others of the same grade whom you have rated in your career. Managed at less than 50% in EXCELS.)*

I currently rate ____ **Army Officers** in this grade.

A completed DA Form 67-10-1A was received with this report and considered in my evaluation and review: ☐ Yes ☐ No *(explain in comments below)*

EXCELS (49%) ☐ **PROFICIENT** ☐ **CAPABLE** ☐ **UNSATISFACTORY** ☐

Comments:

PART V - INTERMEDIATE RATER

PART VI - SENIOR RATER

a. **POTENTIAL COMPARED WITH OFFICERS SENIOR RATED IN SAME GRADE (OVERPRINTED BY DA)**

☐ **MOST QUALIFIED** *(limited to 49%)*

☐ **HIGHLY QUALIFIED**

☐ **QUALIFIED**

☐ **NOT QUALIFIED**

b. I currently senior rate ____ **Army Officers** in this grade.

c. COMMENTS ON POTENTIAL:

d. List 3 future **SUCCESSIVE** assignments for which this Officer is best suited:

DA FORM 67-10-2, MAR 2019

Page 2 of 2
APD LC v1.00ES

Officer Evaluation Report (OER) ***(continued)***

HQDA#: Attachments Menu

STRATEGIC GRADE PLATE (O6) OFFICER EVALUATION REPORT
For use of this form, see AR 623-3; the proponent agency is DCS, G-1.

See Privacy Act Statement in AR 623-3.

PART I - ADMINISTRATIVE *(Rated Officer)*

a. NAME *(Last, First, Middle Initial)* | b. SSN (or DOD ID No.) | c. RANK | d. DATE OF RANK *(YYYYMMDD)* | e. BRANCH | f. COMPONENT *(Status Code)*

g. UNIT, ORG., STATION, ZIP CODE OR APO, MAJOR COMMAND | h. UIC | i. REASON FOR SUBMISSION

j. PERIOD COVERED FROM *(YYYYMMDD)* THRU *(YYYYMMDD)* | k. RATED MONTHS | l. NON RATED CODES | m. NO. OF ENCLOSURES | n. RATED OFFICER'S EMAIL ADDRESS *(.gov or .mil)*

PART II - AUTHENTICATION *(Rated officer's signature verifies officer has seen completed OER Parts I-VI and the administrative data is correct)*

a1. NAME OF RATER *(Last, First, Middle Initial)* | a2. SSN (or DOD ID No.) | a3. RANK | a4. POSITION

a5. EMAIL ADDRESS *(.gov or .mil)* | a6. RATER SIGNATURE | a7. DATE *(YYYYMMDD)*

b1. NAME OF INTERMEDIATE RATER *(Last, First, Middle Initial)* | b2. SSN (or DOD ID No.) | b3. RANK | b4. POSITION

b5. EMAIL ADDRESS *(.gov or .mil)* | b6. INTERMEDIATE RATER SIGNATURE | b7. DATE *(YYYYMMDD)*

c1. NAME OF SENIOR RATER *(Last, First, Middle Initial)* | c2. SSN (or DOD ID No.) | c3. RANK | c4. POSITION

c5. SENIOR RATER'S ORGANIZATION | c6. BRANCH | c7. COMPONENT | c9. EMAIL ADDRESS *(.gov or .mil)*

c8. SENIOR RATER PHONE NUMBER | c10. SENIOR RATER SIGNATURE | c11. DATE *(YYYYMMDD)*

d. This is a referred report, do you wish to make comments? ☐ Referred ☐ Yes, comments are attached ☐ No | e1. RATED OFFICER SIGNATURE | e2. DATE *(YYYYMMDD)*

f1. Supplementary Review Required? ☐ Yes ☐ No | f2. NAME OF REVIEWER *(Last, First, Middle Initial)*

f3. RANK | f4. POSITION | f5. Comments Enclosed

f6. SUPPLEMENTARY REVIEWER SIGNATURE | f7. DATE *(YYYYMMDD)*

PART III - DUTY DESCRIPTION

a. PRINCIPAL DUTY TITLE | b. POSITION AOC/BRANCH

c. SIGNIFICANT DUTIES AND RESPONSIBILITIES

PART IV - PERFORMANCE EVALUATION - PROFESSIONALISM, COMPETENCIES, AND ATTRIBUTES *(Rater)*

a. APFT Pass/Fail/Profile: Date: Height: Weight: Within Standard?

Comments required for "Failed" APFT, or "Profile" when it precludes performance of duty, and "No" for Army Weight Standards? Reset Item a. APFT/Pass/Fail/Profile

b. THIS OFFICER POSSESSES SKILLS AND QUALITIES FOR THE FOLLOWING STRATEGIC ASSIGNMENTS

c1. Character:
(Adherence to Army Values, Empathy, and Warrior Ethos/Service Ethos and Discipline. Fully supports SHARP, EO, and EEO.)

DA FORM 67-10-3, MAR 2019

Page 1 of 2
APD LC v1.00ES

Officer Evaluation Report (OER) *(continued)*

HQDA#:

NAME	SSN (or DOD ID No.)	PERIOD COVERED: FROM *(YYYYMMDD)*	THRU *(YYYYMMDD)*

c2. **Provide narrative comments which demonstrate performance and potential regarding strategic competencies in the Rated Officer's current duty position.** *(i.e. providing vision, motivation, and inspiration, negotiating within and beyond national boundaries, building strategic consensus, leading and inspiring change, dealing with uncertainty and ambiguity, creates a positive environment to prepare for the future, expanding knowledge in cultural and geopolitical areas, self-awareness and recognition of impact on others, building team skills and processes, allocating the right resources, capitalizing on unified action partner assets, capitalizing on technology, accomplishes missions consistently and ethically. Fully supports SHARP and creates a positive command/workplace environment.)*

A completed DA Form 67-10-1A was received with this report and considered in my evaluation and review ☐ YES ☐ NO *(explain)*

COMMENTS ON PERFORMANCE:

COMMENTS ON POTENTIAL:

PART V - INTERMEDIATE RATER

PART VI - SENIOR RATER

a. POTENTIAL COMPARED WITH OFFICERS SENIOR RATED IN SAME GRADE (OVERPRINTED BY DA)

☐ **MULTI-STAR POTENTIAL** *(limited to 24%)*

☐ **PROMOTE TO BG** *(25% to 49%)*

☐ **RETAIN AS COLONEL**

☐ **UNSATISFACTORY**

Note: Combined cumulative percentages of both "MULTI-STAR POTENTIAL" and "PROMOTE TO BG" must be less than 50%.

b. I currently senior rate ____ **Army Officers** in this grade.

c. COMMENTS ON POTENTIAL:

d. List 3 future **SUCCESSIVE** assignments for which this Officer is best suited:

DA FORM 67-10-3, MAR 2019

Page 2 of 2
APD LC v1.00ES

Officer Evaluation Report (OER) *(continued)*

HQDA#.

Attachments Menu

STRATEGIC GRADE PLATE GENERAL OFFICER EVALUATION REPORT
For use of this form, see AR 623-3; the proponent agency is DCS, G-1.

See Privacy Act Statement in AR 623-3.

PART I - ADMINISTRATIVE *(Rated Officer)*

a. NAME *(Last, First, Middle Initial)* | b. SSN (or DOD ID No.) | c. RANK | d. DATE OF RANK *(YYYYMMDD)* | e. BRANCH | f. COMPONENT *(Status Code)*

g. UNIT, ORG., STATION, ZIP CODE OR APO, MAJOR COMMAND | h. UIC | i. REASON FOR SUBMISSION

j. PERIOD COVERED: FROM *(YYYYMMDD)* THRU *(YYYYMMDD)* | k. RATED MONTHS | l. NON RATED CODES | m. NO. OF ENCLOSURES | n. RATED OFFICER'S EMAIL ADDRESS *(.gov or .mil)*

PART II - AUTHENTICATION *(Rated officer's signature verifies officer has seen completed OER Parts I-VI and the administrative data is correct)*

a1. NAME OF RATER *(Last, First, Middle Initial)* | a2. SSN (or DOD ID No.) | a3. RANK | a4. POSITION

a5. EMAIL ADDRESS *(.gov or .mil)* | a6. RATER SIGNATURE | a7. DATE *(YYYYMMDD)*

b1. NAME OF SENIOR RATER *(Last, First, Middle Initial)* | b2. SSN (or DOD ID No.) | b3. RANK | b4. POSITION

b5. SENIOR RATER'S ORGANIZATION | b6. BRANCH | b7. COMPONENT | b9. EMAIL ADDRESS *(.gov or .mil)*

b8. SENIOR RATER PHONE NUMBER | b10. SENIOR RATER SIGNATURE | b11. DATE *(YYYYMMDD)*

c. This is a referred report, do you wish to make comments?
☐ Referred ☐ Yes, comments are attached ☐ No

d1. RATED OFFICER SIGNATURE | d2. DATE *(YYYYMMDD)*

PART III - DUTY DESCRIPTION

a. PRINCIPAL DUTY TITLE | b. POSITION AOC/BRANCH

c. SIGNIFICANT DUTIES AND RESPONSIBILITIES

PART IV - PERFORMANCE EVALUATION - PROFESSIONALISM, COMPETENCIES, AND ATTRIBUTES *(Rater)*

a. APFT Pass/Fail/Profile: ______ Date: ______ Height: ______ Weight: ______ Within Standard? ______
Comments required for "Failed" APFT, or "Profile" when it precludes performance of duty, and "No" for Army Weight Standards? Reset Item a. APFT/Pass/Fail/Profile

b. COMMENTS ON CHARACTER & POTENTIAL:

PART V - SENIOR RATER EVALUATION

COMMENTS ON CHARACTER & POTENTIAL:

DA FORM 67-10-4, NOV 2015 APD LC v1.00ES

Officer Evaluation Report (OER) ***(continued)***

guidance in preparing an evaluation, a multi-pane dashboard that allows users to view data input and evaluation forms simultaneously, a built-in tool to view and manage rater and senior rater profiles, and a quick reference to AR 623-3 and DA PAM 623-3. A helpful instructional video created by the Army Human Resources Command can be found at www.youtube.com/watch?v=9btxdhwSXaU.

OER Support Form (DA Form 67-10-1A). OER Support Forms remain a primary tool, ensuring clear expectations and requirements between an officer and his or her rater. Procedures require senior raters to pass their own support forms two levels down the rating chain; that senior raters verify that initial counseling occurred between the rated officer and the rater by initialing the OER Support Form; and that all rating chains keep a record of all follow-up counseling on the front of the OER Support Form. Initial counseling is required to be conducted within the first 30 days of the rating period. Follow-up counseling is mandatory for lieutenants and warrant officers (WO1).

OER Grade Plate Template (DA Form 67-10-1, 2, 3). There are six parts to the OER. Part I details administrative data about the officer. Part II is the authentication, including e-signatures and email addresses of the rater, intermediate rater, and senior rater, and the signature for the rated officer. Part III details the duty description of the current job. Part IV is filled out by the rater and covers the professionalism, competencies, and attributes of the rated officer, along with narrative remarks. Part V is the intermediate rater narrative remarks section. Part VI is the senior rater narrative remarks section, which addresses both performance and future potential, including a block check comparison against the senior rater's population. The remaining section of the form discusses potential successive assignments recommended by the senior rater.

Managed Profile of the Senior Rater. The managed profile technique is an evolution of the OER system. In this field, the senior rater checks one of four boxes for company and field-grade officers, indicating they are rated as "Most Qualified," "Highly Qualified," "Qualified," or "Not Qualified," according to the senior rater's view of the officer's potential. This technique discourages senior raters from reducing the meaning of OERs by inflating the system. Senior raters never rate more than 50 percent of officers of a given rank as "Most Qualified." This has some serious implications for profiles with few ratings. HRC provides chain-teaching briefings to all officers to keep them informed of all critical, relevant changes.

OF SPECIAL INTEREST FOR RESERVE COMPONENT OFFICERS

Reserve Officer Personnel Management Act (ROPMA). ROPMA assists the Army National Guard and U.S. Army Reserve (USAR) in managing officers, other than warrant officers and commissioned warrant officers, who are not on the active-duty list.

Although the Reserve Components' promotion process (including the mandatory promotion boards and the Army National Guard unit vacancy and USAR position vacancy boards) is a key component of the act. ROPMA also ties accessions and separations together with promotions into a complete officer management system.

The mandatory promotion board "best qualified" selection criteria require establishing numeric selection objectives by grade, captain through colonel, for each competitive category. The Secretary of the Army determines the numeric objectives for mandatory promotion boards based on the needs of each of the Army Reserve Components—that is, the selection objectives for the Army National Guard and USAR will be established separately, based on each of those components needs.

ROPMA provides a degree of flexibility to officers recommended by mandatory promotion boards by permitting them to request voluntary delay of promotion. Any National Guard or USAR officer may request voluntary delay of a promotion until he or she finds and/or is assigned or attached to a position in the higher grade. An AGR officer for whom no position at the higher grade is available is automatically retained on a promotion list pending assignment or attachment to a position of the higher grade. A National Guard traditional drilling guardsman or USAR TPU officer who did not receive an assignment or attachment to a unit position in the grade to which recommended at the end of the approved delay period either declines the promotion (and is considered to have failed selection) or transfers to the individual ready reserve (IRR) to accept promotion, if otherwise qualified.

An officer serving in the grade of first lieutenant through major who receives two failures of selection for promotion by a mandatory board will be separated IAW applicable regulations, unless retained under another provision of law.

Of special interest is a requirement for officers who retire under the retired pay for non-regular service provisions of Chapter 1223 of Title 10, United States Code (USC) (retirement at age sixty with a twenty-year letter). Section 1370(d) of Title 10, USC, requires lieutenant colonels and above to serve three years' time in grade (TIG) to retire at that grade, except if the officer reaches mandatory removal date (MRD) for age or years of service after completing a minimum of six months TIG. This provision applies only to officers who are recommended for promotion to lieutenant colonel and above on or after 1 October 1996.

Reserve Component officers ordered to active duty, on or after 1 October 1996, during time of war or national emergency may remain on the reserve active-status list (RASL), as mandated by ROPMA, for up to twenty-four months. During this period, mobilized officers continue to be considered by mandatory promotion boards along with other eligible reserve officers not on the active-duty list. Additionally, mobilized officers no longer lose rank or date of rank when mobilized or demobilized.

Specific Changes Under ROPMA

TIG Requirements. Promotions are based on TIG requirements. These could change in coming years, based on the needs of the Army.

Mandatory promotion consideration is scheduled to meet the statutory maximum TIG for each grade. In the promotion zone, consideration for mandatory promotion occurs during the year prior to the year the officer attains maximum TIG in the current grade. The TIG requirement for promotion to first lieutenant is not the statutory minimum of eighteen months. It was reduced to two years from the pre–ROPMA standard of three years.

PROMOTION REQUIREMENTS

Grade	Position Vacancy Board: Minimum Years in Lower Grade	Mandatory Board: Maximum Years in Lower Grade
2LT to 1LT	2	2
1LT to CPT	2	5
CPT to MAJ	4	7
MAJ to LTC	4	7
LTC to COL	3	Announced Annually

Reserve Active-Status List (RASL). ROPMA requires the Secretary of the Army to maintain a single list of all reserve component officers who are serving in active status, not on the active-duty list. The list is maintained in order of grade and order of seniority in grade. The RASL is a management database for the reserve components' promotion system and is used to establish promotion zones and ensure that officers are properly considered for promotion and that officers considered by mandatory or position vacancy promotion boards are managed according to the recommendations of their respective boards.

Promotion Selection Quotas. ROPMA requires selection quotas to be determined annually for all mandatory promotion boards considering officers in the rank of first lieutenant and above. These quotas, to be determined by grade and competitive category, will also be forecast to ensure relatively similar selection objectives over a rolling five-year period.

"Best Qualified" Selection Criteria for Mandatory Promotion Boards. Under ROPMA, all mandatory promotion boards will use "best qualified" selection criteria.

Senate Confirmation for Colonel and Above. Senate confirmation is no longer required for promotion to lieutenant colonel. Officers recommended for promotion to colonel and general officer grades still require Senate confirmation before they are promoted.

Delay of Promotion. Officers may request voluntary delays of their promotions. Statute permits voluntary delays of up to three years. However, policy yet to be established will limit the length and manner in which requests for voluntary delays of promotion are approved. All such requests must be approved before becoming effective. Each approved voluntary delay of promotion reduces future promotion opportunities available for the next mandatory board of the same grade and competitive category by the number of approved promotion delays in effect.

An AGR officer who is recommended for promotion by a mandatory promotion board is automatically provided an indefinite delay of promotion if no position in the higher grade is available for the officer's assignment or attachment. This delay remains in force until (1) a higher-grade position for which the officer qualifies is available; (2) the officer transfers into the IRR and accepts promotion, if otherwise qualified; (3) the officer separates from the AGR program; or (4) the officer is either voluntarily or involuntarily removed from the promotion list.

Selective Continuation Board for Twice Failing Selection for Promotion. Based on the needs of the Army, as determined by the Secretary of the Army, selective continuation boards may be conducted to continue officers who have twice failed to be selected for promotion to major or lieutenant colonel beyond the date on which they would otherwise be separated.

Constructive Credit. ROPMA changes the methods by which constructive credit is awarded and by which total years of commissioned service for reserve officers are calculated. Prior to ROPMA, constructive service credit awarded at the time of initial appointment was included in an officer's total commissioned service. Subsequent to 1 October 1996, a reserve officer's years of commissioned service equates to time *actually served* as a commissioned officer in any armed service. Constructive service credited to a reserve officer upon initial appointment will no longer be considered in determining an officer's maximum years of commissioned service.

Types of Promotion Selection. The reserve components use two distinct promotion systems for officers. The first system is the mandatory promotion board, which considers all reserve component officers by grade and competitive category. The second system is the position vacancy promotion board. The National Guard and USAR have separate position vacancy promotion board systems. Officers in the National Guard are considered for position vacancies under the federal recognition process. Officers of the USAR are considered under the position vacancy board process.

Mandatory promotion boards consider both National Guard and USAR officers of the same grade and competitive category. Officers under consideration may be in the promotion zone, above the promotion zone, or, if the Secretary of the Army determines such a need, below the promotion zone. In-the-promotion-zone officers are those under consideration by a mandatory board during the year before they attain their maximum years of service in their current grade. Above-the-promotion-zone officers are those who were previously considered but not recommended by a mandatory board while in the promotion zone and are senior to the senior officer in the promotion zone. Below-the-promotion-zone officers are eligible for mandatory board consideration but are junior to the junior officer in the promotion zone.

Promotion recommendations under the position vacancy promotion system offer an immediate assignment to a position at the higher grade, as well as the potential for an accelerated promotion. To be eligible for a position vacancy promotion, the position must require a grade higher than the one an officer currently holds. In addition to meeting the grade requirement, an officer interested in a position vacancy system promotion must have attained the minimum civilian and military education called for at the higher grade.

Eligibility for both National Guard and USAR position vacancy boards is affected by the legislative change that prohibits consideration of an officer who was considered but not recommended (failed to be selected) by a mandatory promotion board, unless the Secretary of the Army determines that officer to be the only officer qualified to fill the vacancy. A second change for the USAR position vacancy boards is that any qualified, eligible officer on the RASL may compete for a position vacancy promotion.

Determination of Best Qualified Promotion Selection Quotas. A key feature of ROPMA is the requirement for the reserve components to annually establish mandatory promotion quotas for all grades within each competitive category. An additional requirement is that the mandatory promotion board requirements must be forecast not only for the current promotion year but also for the following four years. These requirements are intended to ensure that the manpower needs of the reserve components are filled, while at the same time providing relatively similar promotion opportunities to officers being considered by any mandatory promotion board within the five-year window. A number of critical variables will influence the calculations used in determining mandatory promotion quotas for each grade and competitive category. Force structure gains and losses, accessions and separations, historical selection rates for position vacancy promotion boards, and the number of delayed promotions must all be taken into consideration when mandatory promotion board selection requirements are determined.

MENTORSHIP

Leaders play a crucial and integral role in the development of their subordinates. Officers learn from their leaders' personal examples; thus, an officer bears immense responsibility to set junior officers on the right path, to teach junior officers the right ways, and to ensure that junior officers do the right things. This requires all leaders to possess the professional competence to appropriately train and educate their subordinates, the integrity to fairly evaluate them, and the personal courage to honestly discuss their performance and future in the Army. Although the following paragraphs discuss the role and responsibilities of the commander, the biggest leadership challenge for all officers is to develop junior officers when not in positions of formal authority. As the Officer's Code of Honor states, "An officer is responsible for the actions of his fellow officers. The dishonorable act of one diminishes the corps." However, the future viability of the Army and the officer corps rests not in the hands of commanders but in the hands of all fellow officers.

Commanders' Responsibilities. An officer's commander is critical in the professional development of the officer corps. The most important factor is command climate. The commander sets the climate that influences how his or her subordinates view the unit, the Army, and their future places in them. Good command climates inspire and reward outstanding performance and highly effective professional development. Commanders should succinctly establish the "non-negotiables" of their commands, and which mistakes they are willing to underwrite. They should then be consistent in the application of those policies. To develop our future warfighting leaders, commanders set the organizational climate that empowers junior leaders to seek challenging assignments; broaden their knowledge of history, doctrine, and tactics; and take the risks to holistically improve themselves.

Professional Development Counseling. Counseling is critical to the professional development of an officer. Counseling sessions provide fora for junior officers to interact with their commander or superior officers. These sessions should be an open exchange of ideas with some confidentiality and nonattribution, so that both officers can forthrightly discuss strengths, weaknesses, and plans to improve performance. The Junior Officer Development Support Form (JODSF) provides a tool to facilitate this type of counseling session. All officers should take the initiative to advise and help all other officers to continue a principled, duty-bound, and selfless officer corps in the future.

Mentorship Misunderstood. For a variety of reasons, to some officers, "mentorship" has come to be interpreted as cronyism—the practice by which a few officers artificially affect the attitudes, preferences, and (often sleazy) methods of their superiors in attempts to gain personal favors and special advancement. This is absolutely wrong. There is a large, thick line between the developmental process by which officers counsel, coach, and pass on the wisdom of many years' experience to their subordinates, and the practice of coattail riding. Like any organization that has experienced significant downsizing, the Army officer corps has, in recent years, experienced the effects of "careerism," or the unhealthy obsession with personal gain and ambition. There are some misguided officers who believe that the keys to success are not so much professional competence, inspirational and firm leadership, and selfless commitment to the Army and the Republic but rather personal maneuvering and intrigue to gain favor with particular, apparently upward-bound superiors. To these officers, this is what "mentorship" is about.

Such attitudes and actions are more appropriately relegated to organized crime syndicates and have no place in the U.S. Army. Naturally, any leader is most comfortable with subordinates who understand his or her philosophy, leadership style, and professional vision, but this in no way excuses cronyism.

True mentorship does not encourage aping the acts of the mentor but rather entails thoughtful discussion of how best to serve the unit and the Army. Mentors who have the best interests of the Army at heart in their communication of wisdom and demonstration of example will never go wrong. Mentored subordinates exposed to this spirit will profit enormously and become officers and stewards of Army traditions and values.

THE FREEDOM TO SUCCEED

Competition is integral to the Army in all it does, from warfighting to officer development. For the most part, recognition and advancement are based on personal merit; like American society at large, the Army must remain a meritocracy to derive the best from each for the good of all.

Gone are the days when officers could reasonably assume the attainment of certain ranks or assignments and measure their success thereby. In the uncertain, kaleidoscopically changing environment, there arc no safe ports, no guarantees of rank or privilege. In reality, there may never have been, but many certainly perceived that there were. As a result, many officers who entered the Army with preconceived notions of the metrics of "success" have been disappointed and disheartened by the unexpected developments of the post–War on Terror era.

Every officer should expect from his or her superiors the guidance that shapes what is acceptable and what is not. As officers develop and advance in rank and position, they must assume more responsibility for the future viability of the officer corps by mentoring the officers around them.

However, just like in any profession, individual preparation is key. Every officer is blessed with individual skills and talents, and each begins service in the Army at his or her own level. How far an officer can go and what he or she can achieve while serving in the Army is overwhelmingly dependent on his or her commitment and pursuit of excellence in a chosen field. Whether they become charismatic leaders, decisive tacticians, meticulous planners, or consummate analysts, all officers should challenge themselves to reach for the seemingly unreachable. The extent to which you do, and the extent to which you contribute to the best of your abilities, are the true measures of success as an officer. Our nation, our soldiers, and the legacy of those who have gone before require nothing less.

12

New Duty Assignments

WHAT YOU MAY EXPECT UPON ARRIVAL

Every new assignment brings professional and personal challenges and opportunities for greater service and growth. You may expect to be received at your new station with matter-of-fact courtesy and efficiency. Commanders generally regard the welcome of officers and their families into a unit or station complement as an important opportunity to strengthen organizational readiness and cohesion and to familiarize newcomers with the command climate and standards of the gaining organization. The reception of incoming personnel is carefully planned and should be conducted to reflect the highest levels of professionalism, care, and concern for the incoming officer and his/her family.

If your first station is on a temporary duty (TDY) assignment to attend a service school, you may expect to be one of many officers arriving on the same day at about the same time. Typically, signs will direct you to where you should report, but in case of doubt, don't hesitate to ask for directions. A reception center will process all incoming student officers as rapidly and efficiently as possible.

The needs of newcomers are usually anticipated, and information provided promptly. Enter with confidence, and you will be made welcome. If you should encounter a member of the reception detail who is thoughtless and seems uninstructed, do not let the episode unduly influence your conclusions about the post or your new unit. Read your instructions carefully and complete them promptly. If something is unclear, just inquire. People expect to help you, and they will help you as perhaps they were the new ones a few weeks earlier. Later, when you are helping to receive newcomers, do your part to be helpful and understanding.

ARRIVAL AT A PERMANENT DUTY STATION

Success upon arrival at any new permanent duty station depends on many factors, chief among them assistance by your sponsor; your preparation for arrival; and your conduct in the first days and weeks of your new assignment.

Under the provisions of AR 600-8-8, the *Total Army Sponsorship Program* (28 June 2019), commanders appoint a sponsor for each incoming officer and family. As soon as possible, the sponsor provides the incoming person with information about the post, unit, mission, standards, command philosophy, housing, local area, and, if appropriate, schools and jobs for family members. Upon arrival, they greet incoming personnel and assist with in-processing and procuring appropriate quarters. As experienced members of the unit, sponsors can be resources for information about how to be best prepared for successfully and swiftly assuming your new duties. Do not hesitate to ask them about the unit's mission, its METL (Mission Essential Task List), SOPs, short- and long-range training schedule, and physical training standards. Then, prepare yourself professionally and physically by learning all you can about the unit's operations and by training for the demands of the new job.

How to Report for Duty. Orders assigning an officer to a station for duty include a date of reporting. To comply with the order, you must reach the station and report to the proper official prior to midnight of the date prescribed. Try to arrive during the hours of the ordinary business day, if at all practicable, preferably before noon and as early as 0900 if you can. This will allow you a full day to complete many important official and personal arrangements. Try to report no later than 1600. If the date of reporting lies within your discretion, you should avoid arrival at a new station on Saturday, Sunday, or an official holiday.

Consider arriving in the area on the afternoon before you are due to report and staying at a convenient motel or on-post guest house. If you know your impending unit of assignment, get the right patch sewn on and acquire the appropriate distinctive unit insignia. While reporting in civilian clothing is not unheard of today, you should report in the duty uniform of your assigned unit. If you are unsure about what you should wear, email the unit S-1 or ask your sponsor.

Carry copies of your orders with you. Several copies will be needed for your travel pay voucher, and other copies will be needed for administrative purposes at the new station. Generally, you should arrive with no fewer than twenty-five copies. Don't forget to place a copy in your own personal records file.

Proceed to the post or station reception center to report for duty. In those cases in which a large number of officers report within a short period, as at a training center, port of embarkation, or other large troop concentration, the reception of arriving officers may be handled by a receiving committee. In such a case, the formality of reporting consists only of presenting yourself at the proper office (which is usually indicated by signs), presenting your orders, signing the register, and receiving instructions. A member of this committee usually handles quartering and messing arrangements and provides, often in written instructions, the information the newcomer requires. At a later time a meeting may be held at which the commanding officer or a representative addresses the group for purposes of organization or orientation. When this procedure is followed, be certain that you have received all the instructions that should be in your possession and that you understand them thoroughly.

At other times, the arrival of a single officer or a small group of officers is but an incident in the day's work, and no such elaborate arrangements will be made. Arrange to meet your sponsor at the post headquarters, and he or she will escort you to the adjutant's office. Remove your cap, knock, enter, and, if the adjutant is senior in grade to yourself, salute and report: "Sir, Lieutenant ——— reports for duty"; at the same time extend a copy of your orders. If the adjutant is junior in rank to you, as an Army custom it is proper to state: "I am Major ——— reporting for duty." The adjutant will welcome you to the garrison, give you the information you need, and arrange a time for you to call upon the commander. The adjutant will apprise you of your quarters assignment or whom to see to obtain this information. Most likely, he or she will arrange for the delivery of your baggage to your quarters. When you leave the adjutant, you should have no doubt in your mind about what you are to do initially and when and where you are to do it. Your first task, more than likely, will be to establish yourself in your quarters and prepare yourself for the duties to come.

Getting Established in Quarters. After reporting for duty, immediately get yourself established. If the assignment is for TDY, this may mean only determining the location of the bachelor officer quarters (BOQ) and moving into your assigned room. If you arrive on permanent change of station (PCS) orders with family members, getting established will involve the assignment and occupancy of government quarters or the rental or purchase of other housing.

Most Army posts operate guest houses, which are available for short periods to newly assigned personnel until they can find permanent housing. Rates are reasonable. Space is usually reserved, when available, by applying in advance of need to the housing officer. Your sponsor should assist you in these arrangements well in advance of your arrival.

Subject to some limitations, such as if the unit is going to the field or being deployed, the officer is normally given a reasonable period of time to arrange personal affairs. Your unit sponsor should be able to provide most of the information you will need. In the case of reporting to a major post for a school or for further assignment to a unit, you will want to arrive a few days earlier than required. A major post will have personnel, such as the housing officer, to assist you, and an Army Community Service Center, at which you can get a wealth of current information, pick up any essential household goods for temporary use, and receive many other kinds of assistance.

Collection of Travel and Transportation Allowances. Soon after arriving at the new station, submit vouchers to collect cash allowances, such as those for your personal travel, family travel, dislocation, per diem, or any other allowance, such as uniform, to which you may be entitled. Ask the personnel officer for the procedure and request assistance as needed. Copies of travel orders will be required.

Garrison Regulations. Carefully study the local garrison regulations. They usually contain useful information on local conditions, facilities, and conveniences, as well as requirements that will assist you in making adjustments to the new environment. Be sure to comply fully with the post's regulations as they vary from post to post, depending on the legal requirements of the state in which the post is located.

The Post Exchange. Soon after arrival, pay a visit to the post exchange, the community store operated under the supervision of the Army and Air Force Exchange Service (AAFES). At most stations, the exchange consists of a shopping center combining the features of a department store and a drugstore, plus a number of specialty shops (e.g., barber, florist, shoe repair, tailor, dry cleaner) and a food court or national franchise fast-food vendor. It supplies other services for the benefit of the officers and enlisted personnel of the garrison. Some of the profits from the operation of the exchange revert to the organization to spend on the troops through the central welfare fund.

The Commissary. Nearly every Army post has a commissary, operated by the Defense Commissary Agency, that is similar to a commercial supermarket. If your first assignment is on TDY, you may not feel the need to use the commissary. If your spouse and family are with you, however, you certainly will want to do the bulk of your grocery shopping at the commissary. Food and other household items may be purchased at the commissary at savings over many commercial, off-post stores.

Post Transportation Office. The post Transportation Office will assist you with any issues associated with the movement of your personal and household property and any claims occasioned by loss or damage.

Learn Your Way Around. Study a map of the post and locate the important buildings, roads, training areas, and recreational facilities. If your new assignment is overseas, ask your sponsor about local cultural customs and courtesies, laws and taboos, applicable Status of Forces Agreements (SOFAs), and off-post dress regulations. It is also a good idea to speak to the intelligence or security officer about the threat conditions at any off-post location to which you are considering traveling in your early days of your new assignment.

Reporting Time for Starting to Work. Instructions from the adjutant or your immediate commander will indicate how long you have to become established before reporting to work.

THE EARLY DUTY DAYS

Arrival at a first station and assignment to duty with a unit provide an exceptional, interesting challenge. Start with this initial conviction of personal confidence: The Army selects its officers carefully and trains them so that they will succeed. The education and the training programs at the USMA, at ROTC units at colleges and universities, at Officer Candidate Schools, and at the branch service schools are all carefully conducted. Although all newly appointed officers will have more to learn throughout their service, each officer should start with confidence in his or her preparation. The Army functions as a team, and it is in the team's interest that you are successful as an officer.

Commanders of companies and battalions are fully aware of the problems encountered by newly appointed, incoming officers and understand the additional instruction each will require. The success in command responsibility your new commander or staff supervisor seeks depends in part on your own success. Everyone gains when the incoming officer adjusts quickly and is identified by new associates as a trustworthy, capable addition to the command. Inversely, all stand to lose if the newcomer doesn't measure up or fails. Since no officer could possibly know everything upon arrival, a newcomer will certainly receive detailed information, probably some material for study, and time to complete his or her preparation. The procedure is based on infinite experience, free from mystery. Always, if you don't know, ask, and don't hesitate to ask officers of your own rank, your sponsor, or your subordinates.

MAKING THE BEST POSSIBLE FIRST IMPRESSION

Whenever you assume a new duty assignment, it follows that you will work under new superior officers and have new associates and new subordinates. Just as they will extend a true welcome and take for granted that you are competent, you must be sure to put your best foot forward at the outset to obtain the initial goodwill and confidence of the official group that you join.

Let it be known at once that the assignment is welcomed; find things to comment upon favorably and avoid criticism. Become acquainted promptly with associates, especially junior leaders. Learn your responsibilities and discharge them fully from the start, seeking information and assistance as circumstances require. Solicit the support of your new organization. Try to stimulate a feeling of confidence and enthusiasm in the minds of your associates.

Get Acquainted. Immediately start getting acquainted with your people, whether officers, enlisted soldiers, or civilians. Start with your principal assistants, be they platoon sergeant, first sergeant and executive officer, or the staff noncommissioned officer in charge (NCOIC). Learn their names and use them. Show a sincere interest in them, and learn their backgrounds, special training, service experience, and aspirations. Invite their cooperation and suggestions, which you will need. Neither belittle yourself nor boast in doing these things, but explain your background so that they can judge wisely how best to advise you. Similarly, meet and talk with other officers and NCOs, and quickly learn and use their names and ranks; learn their service training and backgrounds. As time is available, talk with every enlisted soldier in the same manner. These interviews should be informal, objective, and impersonal, but keep in mind that you are inquiring, searching, learning, and starting to build a new team. Knowing and looking after your soldiers is a basic principle of leadership.

When you start your new responsibilities, be sure that you understand the mission of today and tomorrow, next week, even next quarter. Your commander or immediate supervisor will inform you. Go over the mission with your principle subordinates and peers. After you have been around for a while, formally counsel each of your subordinate NCOs, and let them know exactly what you expect from them, in reasonable detail. This, coupled with your example and uncompromisingly fair dealing, will set the conditions for success and a solid professional relationship.

Peer Relationships. Meet and become acquainted with your fellow officers in your quarters or theirs, at the officers' club, on duty, at battalion meetings, or in the mess hall. You will certainly find some good-natured jesting but no period of hazing or probation. Your fellow officers will take for granted that you know your job and that you are determined to do it well. Good officers hope that you will succeed, for your success is to the unit's advantage, and thus helps everyone. Most likely, some of these officers will soon be counted among your closest friends.

Social Introduction. Most organizations hold periodic "Hail and Farewells," social events at which newcomers and their spouses, as appropriate, are introduced to the officers and spouses of the unit or staff. Generally, the commander or staff section chief will introduce you with brief remarks about your background, and will express the unit's sincere welcome. Often, no remarks by the new officer are necessary, but if you are asked to make some, be simple and brief and express sincere enthusiasm for the new assignment. Lengthy speeches are neither expected nor appropriate.

Some Useful Don'ts. While striving to make a favorable first impression, use equal care to avoid those things that may make a bad one. Don't talk too much, especially of yourself. Your standing among new associates will depend wholly on what you accomplish in the future, not your prideful past. While you are sizing up the new command and its members, your new associates are also sizing up their new officer. Given a chance, some will find occasion for critical comment.

Go slowly, too, with changes that affect the lives and likes of your personnel. They probably prefer things the way you find them. Later, with complete information at hand, you can make improvements that will be satisfying and enhance unit morale and readiness. But bide your time until you reach reasoned conclusions, fortified by discussions and recommendations of subordinates. The rate at which you affect change will inevitably be based on your commander's wishes, the unit mission and your subordinates' readiness for its execution, and your personality. Make improvements, make changes for the better, mold the organization to your special judgments—but exercise prudence, and don't try to do it all at the start. Change for its own sake is *never* appropriate because it is almost always disruptive and wasteful, and generally imparts suspicion of grandstanding among the unit's soldiers.

Never make a statement that could possibly be considered a poor reflection on your predecessor. As he or she has departed and cannot defend himself or herself, such acts are cowardly. Your predecessor probably has the respect and even the admiration of all or most of the troops, as most good officers do. For you as the new leader to make harsh or belittling remarks about your predecessor is to injure greatly and perhaps permanently your own chances of gaining the trust and respect of those same soldiers. Also, be careful about forming opinions regarding the former commander based solely on reports from subordinates. Draw your own conclusions from the facts at hand and keep your own counsel. Finally, beware of brash statements of the superiority of your former unit; such statements as "this isn't how we did it in the 53rd at Fort Swampy" are guaranteed to rouse the ire of your new charges, who almost assuredly think that their unit is far superior to your last one.

The leader holds a lonely position. You must put your best foot forward and keep it there. You must set the example. Thoughtless statements, careless attention to appearance, and other little violations, which might be forgotten when done by others, weigh heavily when done by a leader. The higher the command, the more they weigh. Similarly, the earlier in your assignment that you make them—before your soldiers have gotten to know you, respect you, and trust your judgment—the more difficult it will be to overcome the negative impact that such words and deeds can have.

First Actions. Whatever the nature of the new assignment, certain first actions are necessary to gain knowledge of the missions, responsibilities, and personnel. These first actions are keys to getting off to a good start.

What Is the Mission? Seek information from your predecessor, whenever possible, and always from your sponsor. Go to your new boss and listen carefully. Study the policy file and your predecessor's continuity file, if available. Discuss the tasks at hand with other officers, noncommissioned officers, or civilians concerned with the work of the organization.

What Are the Current Training or Work Projects? Ascertain at once what is being done to execute the missions: what are the principal immediate projects, who is working on them, what are their status and expected completion dates? If you did not get one from your sponsor prior to your arrival, procure a copy of the unit Mission Essential Task List (METL) and training schedule.

What Reports Are Required? Obtain a list of periodic reports, when they are due, and who is to prepare them.

Find Out Who Does What. Have each key individual write out his or her specific duties and responsibilities. Not only will you gain an appreciation and understanding of each subordinate's routine tasks and functions, but each individual will benefit by revisiting his or her duties and responsibilities.

Check on Handling of Classified Material and Rules of Physical Security. Find out quickly what classified material is held by the unit and learn the local regulations for the physical security of weapons, communications equipment (especially cryptological gear), Chemical, Biological, Radiological, Nuclear, and Explosives (CBRNE) defense items, and so forth. If carelessness is indicated, correct this at once. Many officers have been reprimanded (or worse) for inattention to the security of classified documents and for overlooking the procedures for their handling. It is especially important that you take nothing for granted regarding classified information, weapons, communications, and physical security. Always treat these as a matter of grave national security.

ASSUMPTION OF A STAFF POSITION

Staff officers are selected for assignment by the commander they serve. Although it is traditional, and, in many ways, healthy and normal, for junior officers to shun staff jobs, the junior officer who avoids staff duty in today's Army is practically an extinct species. Staff experience as a junior officer is not only virtually inevitable but an important step in the professional development process, laying the groundwork for greater understanding of how the Army works and for competence in future command and staff assignments.

Upon reporting for duty on a staff assignment, you should report to the chief of your section, branch, or division, and then as directed to the executive officer or chief of staff. All well-administered headquarters have staff manuals or standing operating procedures (SOPs), which set forth the organization of each staff section and the specific responsibilities of its members. Here you should find the specific duties of your own particular segment of the staff. As the purpose of this manual is to provide for teamwork and you have become a member of the staff team, your first effort must be to become thoroughly familiar with the manual. (See chapter 14, Staff Assignments and Procedures, for helpful information about responsibilities.)

Next, become acquainted with other members of the staff, particularly those individuals of other staff sections with whom you will work. The chief or executive of the branch or division will provide this information. At the same time, seek the acquaintance of the unit commanders and unit officers served by the staff. For example, as a new battalion adjutant, you should seek out and introduce yourself to the company commanders and junior officers of the battalion. Each staff officer must work continually with officers having similar spheres of responsibility in headquarters both junior and senior to his or her own. For example, battalion intelligence officers regularly work with the intelligence officers of other battalions and with one or several individuals of the brigade or division intelligence staff. He or she must establish and maintain a close working liaison with intelligence officers of other units of the division. Meet the commanders and staff officers with whom you work as soon as practicable after assuming a staff position.

Each staff has its own operating procedures and ways of getting things done, depending on the desires of the commander, chief of staff, or executive officer. Whether special office forms and routines, prescribed ways of writing reports or staff studies, or prescribed reports to be received and studied, learn any unique procedures and follow them.

Like American society at large, today's officer corps is increasingly "visually oriented." While an officer should *never* attempt to substitute appearance for substance in performing his or her duties, staff officers should be familiar with several computer programs. At a minimum, they should be competent with recent versions of Microsoft Word, Microsoft Outlook, and Microsoft PowerPoint. If you have not already learned how to use this software, be sure that you arrive at your first staff assignment with a working knowledge of at least these three staples of the Army staff world.

Since staff assignments provide excellent opportunities to learn the intimate details of the military profession, embrace the "hard jobs" as they are the training grounds—and sometimes the proving grounds—for command.

ASSUMPTION OF COMMAND

Command is the highest responsibility that can be given an officer. As an Army officer, you must anticipate that you may be assigned without warning not only to assume command of a unit consistent with your rank (i.e., a staff captain ordered to take command of a company) but also of larger ones (i.e., a company commander assuming command of a battalion). Familiarity with problems of command of the next higher unit should be part of individual development and planning. Be ready for such opportunities. In combat they are normal. Success in battle may depend on the speed and efficiency with which lost leaders are replaced.

Assumption of Command in Combat. In combat, fallen or otherwise incapacitated leaders must be replaced at once. In the absence of prior instructions to the contrary, seniority of the officers assigned to the unit determines the line of succession. Circumstances will govern without time for inquiry or the niceties of position. If you are the officer who discovers the situation, promptly assume command and adjust as necessary as time permits. Immediately notify your commander and leaders of subordinate and adjacent units of your assumption of command. Aggressiveness, promptness, and the need for control require the most available officer to take over the tasks of leadership. You must immediately learn your predecessor's mission and, if available, his or her plan to meet it. Next, you must learn the position of your new units. Thereafter you must meet whatever situations may arise with positive action.

Prior planning, professional development, and advanced experience are critical to successfully assuming command in battle. When assigned to a combat unit, develop a knowledge of the officers and noncommissioned officers of adjacent units. This information may prove useful if you suddenly become responsible for their command.

It should be clear that learning your own job is not enough. You must learn the job of your superiors—two levels up.

Assumption of Command Under Conditions Other Than Combat. Whenever the commander is detached from command of an organization for any reason, the senior officer present for duty automatically assumes command, pending different orders from higher authority. If you are that officer, you stand in regard to your duties in the same situation as your predecessor. Within the bounds of reason and common sense, you should avoid actions that amount to permanent commitments. You are the acting commander, not the next commander. Command prudently and in the spirit of the permanent commander.

Transfer of Command Responsibilities. An officer assuming command of an organization is required to take over the mission and all administrative, fiscal, logistical, training, and any other responsibilities of his or her predecessor. The changeover must be made so that the old commander is fully relieved and the successor fully informed, a process that may take three to five days (or nights) to complete. In peacetime, a complete inventory of property must be accomplished before assumption of command. This can take up to thirty days for large or complex companies. The organization must continue operating or training during the transfer period.

Transfer of the responsibilities of company command includes the following steps: the mission of the command; organizational records, including classified documents; property supplied the unit, as shown by the company property book; property in the hands of organization members; official funds or allowances; the unit fund, and property purchased from the fund; and a clear understanding of training or work in progress.

Suggestions to an Outgoing Commander. In advance of the transfer, the outgoing commander should make certain that subordinates have all their records posted correctly, that the number of actions pending is reduced to a minimum, and that property to be checked and funds to be transferred are all in correct order. Smooth transition of command requires the elimination of dangling, unfinished actions.

Company commanders face the special challenge of property transfer. If you have been properly performing your periodic inventories and your supply NCO has been doing his or her job, this procedure will not be overly stressful. Seek to resolve discrepancies by all legitimate means. *Never* conceal or mislead, however—it is tantamount to lying and may be effectively stealing as well.

The same principles apply to unit funds. Pay the outstanding bills, and determine any further obligations. Check the property that has been purchased from the fund. If there are shortages or broken or unserviceable items, take appropriate steps to clear the records or repair the articles. Be sure that all is in balance and all is clear.

As outgoing commander, you will wish to make a clean, complete severance. Never try to cover up or deceive the incoming commander, as it will all be disclosed later. One inescapable test of your competence as a commander—important because it will be the impression remembered—is the condition of your unit, disclosed upon your transfer from command.

Suggestions to the Incoming Commander. The incoming commander should simplify the transfer by finding out at the start exactly what is to be done. The old and new commanders must make a joint inventory and check all the matters involved. The transfer is often supervised by an officer designated by the appropriate higher commander. However, as the responsibility passes from the old commander to the new, understand the requirements of the transfer and pursue them thoroughly.

The Unit Mission. Discuss the status and mission of the organization with the commander being relieved and with the next higher commander. Discern the state of training, work or training in progress, and current problems. As soon as you assume command, you will be expected to execute the new responsibility effectively.

Administrative Records. As the incoming commander, you must learn about the company files, personnel records, and organization records, where they are kept, by whom, and the extent of your responsibilities. Inventory any classified documents and ensure that they are being correctly safeguarded.

Organization Property. All property in the organization (platoon, company, or battalion) is the responsibility of the commander, even though the Army system calls for company commanders to be primarily responsible, with hand receipts to individual officers and noncommissioned officers. They are responsible, too, but the commander is in no way relieved. You must know your soldiers' equipment and ensure its readiness. As soon as possible, check all major items of equipment on a serial number basis. Ask the battalion supply NCO to assist you on nomenclature. At the first inspection, have all the troops display their individual clothing and equipment, with supply personnel in assistance; again, the check must be reconciled against records for completeness and readiness. The basic principle is simple: All are responsible for seeing that the property is available, always serviceable, and ready for use.

A question may arise over the disposition of overages. The supply system cannot function properly unless units requisition items they need and are able to eliminate shortages of their authorized equipment. "Pack-rat" policies by a few units can be ruinous. These secret surpluses may prevent another unit from obtaining supplies or equipment desperately needed. Excess equipment should be turned in to the battalion supply system as soon as it is found.

In general, if an article charged to the unit is there, see it, examine it, and, if reasonably serviceable, check it off. Without being picayune, perform an accurate, swift, commonsense job with which you can live after your predecessor has departed. Never assume command until all supply and property deficiencies have been identified, proper accountability has been established, or corrective action has been initiated by the outgoing commander.

Know Your Soldiers. As the commander of a small unit, you must become intimately acquainted with your troops. In a small unit, such as a platoon or company, a wise commander will meet each soldier one-on-one sometime during the first few months. Review each individual's record during the interview, and go over the salient service, skills, and achievements of each one. Meeting individually with each soldier may be difficult to accomplish in larger companies or in companies that are spread over multiple locations, but it is very beneficial.

Prepare a card or a page of a notebook on which basic personal data for each individual can be prerecorded and on which you can make notes. Question each soldier about his or her duty assignment, aspirations, specialization, promotion concerns, and schools. Try to obtain sincere recommendations about improving performance, enhancing morale, and other unit matters, but don't allow a "complaint session." The spirit of the interview is to get to know the soldiers, letting them meet you, and showing them that you care about them individually and personally.

Brigade and division commanders meet members of their large commands at group meetings, in a theater or outdoor assembly, and hold as many individual meetings with senior officers and noncommissioned officers as is feasible and as time permits. The more subordinates who are interviewed personally or have personal contact with their senior commanders, the more confidence and cohesion the unit will develop. Such contact also contributes to the development of genuine loyalty (above and below), as well as the smooth operation that can only stem from familiarity with the commander's intent.

Most importantly, establish personal, face-to-face contact with members of the organization at the start. Skillfully done, this interaction will work wonders in securing a favorable reaction from new subordinates and give you personal knowledge on which to build.

Do Not Hasten to Change Things. Command of a unit is ultimately a personal responsibility. The techniques used successfully by one commander may be entirely unsuited to another. With the passage of time, adjustment of these matters is advisable. Such changes, however, should be made only after mature consideration. When appropriate garner input from your superiors, peers and/or subordinates. Work to get your subordinates to accept ownership of any changes, and direct the praise and credit for any successful changes upon your soldiers. This will alleviate any potential resentment and resistance, and can improve morale, motivation and a sense of team. Infrequently, a new commander unwisely sets forth the opinion that the organization was at a very low ebb of efficiency when he or she assumed command. Such statements are usually accompanied by the further observation that since his or her own unusual powers were applied, the highest standards have been achieved. Such an approach to the assumption of command is a cowardly assault on the previous commander and an insult to the soldiers of the command who, after all, were responsible for the unit's performance and character under the previous commander. Inevitably, those members of the unit who may have worked for its progress will feel resentment. Instead, study the methods in use, and when an opportunity for improvement is seen, do not hesitate to adopt it. Face each new assignment as a challenge to do better work and improve results.

See chapter 13 for a more detailed explanation of the responsibilities, principles, and techniques of command.

REASSIGNMENT

The nature of the Army mission and Army service make frequent reassignments inevitable. Over the course of twenty years of service, officers will attend several, perhaps even many, lengthy service schools. Even with the increasingly common force projection (versus forward deployment) posture of the Army of the early twenty-first century, overseas assignments are common and to be expected. Some assignments have a standard or maximum duration, while others are of an indeterminate length; many can even be extended or shortened unexpectedly in accordance with changes to tour policy during the tours themselves. For these reasons, among others, you should be prepared for unexpected change-of-station and duty assignments.

These periodic changes provide advantages. Your experience is broadened. Your acquaintances are multiplied. You have the opportunity to see and enjoy different sections of our country, as well as nations on other continents. Each change extends an opportunity to make a new start and earn a better record.

Action upon Receipt of Orders for Change of Station. When facing reassignment, take prompt steps for relief from current responsibilities and arrange personal affairs prior to departure. A clean break must be made from all responsibilities.

Make a Clean Break. Few shortcomings are more injurious to an officer's standing than leaving a residue of unfinished official business to annoy and confuse the officer's successor and his or her commander. Clearance must be obtained for all property and fund responsibilities. All unfinished transactions, such as disposition of unserviceable or lost property and vouchers for fund expenditures, must be set forth clearly and fully, for responsibility will not be finally terminated until these matters are completed. Individual and organizational records that require action by the responsible officer must be brought up to date before departure or be completed by mail after departure. A clear picture of the mission and responsibilities must be passed on to your successor so that he or she can start with essential information. Provide your successor with necessary information about your subordinates, but avoid unduly influencing his or her opinion.

Clearing Post. Personal bills and obligations must be satisfied and clearance from the post obtained. Most importantly, obligations to civilian merchants should be paid before departure or else, prior to that time, definite arrangements acceptable to the merchant are made for deferring payment. Plan for and take sufficient time to clear post. Although many posts now have "one-stop" clearing centers, some do not, and even those that do may require actions that cannot be resolved in a day or two. For example, the central issue facilities (CIFs) on some posts require advance reservations or have very limited hours for turning in field gear. Find out what is required for clearance (including the maintenance standards for the turn-in of issue items) as soon as you receive orders, and take action to clear post in an organized, efficient fashion.

Counseling and Rating Subordinates. Provide a final counseling to each of your subordinates, whether you are required to complete an official efficiency report on them or not. Put some thought into what you say and write; do not leave this requirement until the very end and produce "cookie-cutter" evaluations or feedback. You don't want your record to include meaningless or trite official evaluations, and you should *never* inflict this discourtesy on others, or deprive the Army of your honest, carefully considered evaluations. After you have presented each with their final report and counseled them appropriately, one technique for optimizing learning and professional growth is to invite their candid, straightforward recommendations of what you should sustain and what you should improve about your own duty performance. Take these comments to heart, and accept them in the same spirit of professionalism in which you provided your feedback to them.

Express Gratitude. While initiating formal military awards or decorations will probably not be appropriate for most of your subordinates at the end of your tour (unless they, too, happen to be coincidentally eligible for meritorious service awards), writing letters of appreciation—official or personal—is always proper for those subordinates whose service and professional support you especially valued. While public and private verbal expressions of your gratitude to subordinates are definitely in order, a written declaration of your appreciation of their services is almost always highly prized. A letter of appreciation requires much more time and thoughtfulness than spoken words, and the recipient of the kind words will remember your expressed professional sincerity the next time he or she is called upon to perform difficult or time-consuming duties for your successors.

A Gracious Farewell. Generally, the unit will provide at least one social opportunity to bid farewell to its outgoing commander with a "Hail and Farewell" social gathering. When given the opportunity to speak, be positive and professional in your remarks, recounting briefly what you have learned and singling out those to whom you are grateful for support and advice during your tour. Avoid "sour grapes," and instead, express your special gratitude for the support and loyalty of your soldiers, without which you would have failed in the performance of your duties. You should also express these sentiments to the soldiers themselves, whether at your final formation in command or on your last day of duty on the staff. Stop by at the various offices where you have had frequent official transactions and deliver a pleasant farewell to all.

Your Professional Development. Just as you will evaluate and counsel all of your subordinates prior to departure, your superiors should do the same for you. Assist them by completing and submitting your OER Support Form promptly, giving your raters plenty of time to reflect on your performance and consider your contributions. At the time this text is being written, official Army policy dictates that all efficiency reports be completed and provided to the officer prior to his or her departure; whether this policy is continued or not, doing so makes common sense and is professionally courteous. Counseling and other feedback are most effective if rendered immediately following your period of service in any given duty position. Take this written and verbal feedback quite seriously; even if your rating seems "perfect," seek your superiors' advice about how to improve your duty performance in the future.

No Excuse for Late Ratings. Just as you must conscientiously and carefully complete all required ratings of your subordinates before you leave—and provide concomitant verbal counseling—you should be provided likewise with a completed copy of your efficiency report and be counseled prior to departure. If you are assured that "your OER will follow," respectfully request that you be counseled and apprised of the specifics of your rating before you leave. There are several sound reasons for this practice. The longer the period between duty performance and feedback, the less effective the feedback, and your professional development will be needlessly stunted. Also, if clarification of any points about your duty performance arise, it is much easier to resolve the issues while you are physically present and close to your tour of duty. Finally, although extremely rare, some unscrupulous or morally weak officers may attempt to inflict an unfair or purposely inaccurate report on a subordinate. Since time and distance decrease the rated officer's opportunity to dispute the rating, and since such faint-hearted raters can rarely withstand the "pressure" of expressing their unprofessional sentiments to the rated officer's face, such raters will sometimes purposely avoid completing reports until the rated officer has departed, or avoid counseling their subordinates altogether. To the greatest extent you can, in consonance with military courtesy and professional decorum, use the chain of command to insist that you be formally rated and counseled prior to departing for your next duty station. Any more than you would leave signed blank checks, *never* leave signed, blank copies of DA Form 67-9 behind for others to fill in.

13

Responsibilities of Command

> Leadership in a democratic army means firmness, not harshness; understanding, not weakness; justice, not license; humaneness, not intolerance; generosity, not selfishness; pride, not egotism.
>
> —General of the Army Omar N. Bradley

Command is the highest responsibility that can be entrusted to an officer. Commanders (and platoon leaders) are responsible for everything their unit does or fails to do. The relationship between commanders and their soldiers is a unique one, with many mutual and complementary responsibilities. Although the requirements are many and serious, the experience is the most rewarding one of most officers' service.*

Duty with a troop unit involves a wide variety of responsibilities. There is the mission, or a series of missions; the training of members of the command or the training of incoming replacements; supply, care, and maintenance of equipment; provision for food service and its continual supervision; unit transport; unit administration; the overall requirement for good management; and finally, the infinite variety of human problems when one is responsible for men and women. The total requirement is leadership combined with professional competence.

All officers assigned to troop units need an understanding of Army command policy, and they need to know and apply sound methods of management. The service schools provide excellent instruction in the subjects an officer requires with a unit of his or her arm or service. But there is a distinct difference between "knowing about," in the academic sense, and "knowing how," which is gained by on-the-job experience. The capability of knowing how is the goal to seek.

The subjects discussed in this chapter are common to Army units of each arm and service. They are included for the officer made responsible for one or

*For the purposes of much of this chapter—except for those portions dealing specifically with the legal and statutory authorities of commanders—the principles and techniques discussed apply to leading platoons as well as commanding larger units, from detachment up.

more of these missions, often effective at once, who has the need for immediate reference information, suggestion, and guidance. What is stated here represents the experience of many officers and is provided as sound counsel. The brief discussions are meant to help you make a confident start. You will need to study the official manuals, of course, just as you must learn the standing orders in effect for your unit. These discussions are important for reference by officers on troop duty, and the sooner experience is gained in these essential duties the better.

ARMY COMMAND POLICY (EXTRACTS AND DEFINITIONS)

Army Command Policy and Procedure. AR 600-20, *Army Command Policy*, is of such importance that officers are urged to add a copy to their personal libraries for reference in routine as well as emergency situations.

Right to Command. Command is exercised by virtue of office and the special assignment of members of the armed forces holding military rank who are eligible by law to exercise command.

Assignment and Command. Members of the Army are assigned to stations or commands where their services are required, and are there assigned to appropriate duties by the commanding officer.

Warrant Officers. Warrant officers may be assigned duties as station, unit, or detachment commander, and, when so assigned, they are vested with all powers normally exercised by a commissioned officer except as indicated in AR 611-112. Warrant officers are now commissioned upon promotion to chief warrant officer (CW2, CW3, CW4, or CW5). As such, they are vested with the same powers as other commissioned officers.

Military Rank. Military rank is the relative position or degree of precedence bestowed on military persons. It marks their station and confers eligibility to exercise command or authority in the military service within the limits prescribed by law. Conferring honorary titles of military rank upon civilians is prohibited.

Chain of Command. The chain of command is the most fundamental and important organizational technique used by the Army. It is the succession of commanders, superior to subordinate, through which command is exercised. It extends from the President, as commander in chief, down through the various grades of rank to the enlisted soldiers leading the smallest Army elements and to their subordinates. Staff officers and administrative noncommissioned officers are not in the chain of command. A simple and direct command channel facilitates transmittal of orders from the highest to the lowest levels in a minimum of time and with the least chance of confusion. The command channel extends upward in the same manner for matters requiring official communication from subordinate to superior.

Each individual in the chain of command is delegated sufficient authority to accomplish assigned tasks and responsibilities.

Every commander has two basic responsibilities in the following priority: *accomplishment of mission* and *the care of personnel and property.*

A superior in the chain of command holds a subordinate commander responsible for everything his or her command does or fails to do. Thus, in relation to his or her superior, a commander cannot delegate any *responsibilities.* However, in relation to subordinates, an officer does subdivide assigned responsibility and authority and assigns portions to various commanders and staff members. In this way, an appropriate degree of responsibility and authority becomes inherent in each command echelon. The necessity for a commander or staff officer to observe proper channels in issuing instructions or orders to subordinates must be recognized. Constant and continuous use of the chain of command is vital to the combat effectiveness of any Army unit.

Temporary Command. In the event of the death, disability, or temporary absence of the commander of any element of the Army, the next senior regularly assigned commissioned officer, warrant officer, cadet, noncommissioned officer, specialist, or private present for duty and not ineligible for command will assume command until relieved by proper authority. A member in temporary command will not, except in urgent cases, alter or annul the standing orders of the permanent commander without authority of the next higher commander.

Emergency Command. In the event of emergency, the senior commissioned officer, warrant officer, cadet, noncommissioned officer, specialist, or private among troops at the scene of the emergency will exercise control or command of the military personnel present. These provisions are also applicable to troops separated from their parent units under battlefield conditions or in prisoner of war status. *(Caution by the* Army Officer's Guide: *This is a matter that officers should understand and be prompt to apply when emergencies occur. A natural catastrophe, such as a fire or tornado, a railroad wreck, a riot, or other unexpected, potentially dangerous situation requires the senior present to take charge of all troops present and to take prompt action as the situation demands.)*

COMBAT COMMAND AND LEADERSHIP

The ultimate duty of every officer is to lead American soldiers in combat. The challenges are similar to those encountered under other conditions, only of far greater immediate import and effect. While it is undeniably true that even seemingly minor decisions in peacetime—such as overlooking fighting positions with flimsy overhead cover or tolerating soldiers lolling about a rear area without helmets on—can have a mortal impact months or even years down the line, officers' actions in combat have a much more instantaneous and direct impact. In the "come as you are" situations that have characterized combat since 1974 and are likely to continue for the foreseeable future, there is little room for dramatic changes in leadership style between peace and combat. Time is just too short. Therefore, in peacetime, the wise officer leads in a style and with standards that can be naturally and logically transferred to combat conditions.

Leading by Example and Sharing Hardship. These two sides of the same coin are probably the most important actions for officers in combat, although they are by no means the only ones required for successful combat leadership. The stresses and strains of combat tend to strip away human veneers and reduce all to more basic behaviors and attitudes. The soldiers of a democracy do not give their trust and confidence to leaders who appear to be remote and unattached. Similarly, they are wary of those who appear to be unfamiliar with or insensitive to their needs. Thus, living by the same standards you expect of your soldiers, and living in the same conditions, will inspire their confidence in you and elicit trust in your judgment.

Listening and Explaining. These are always advisable traits, but they are indispensable for success in combat. Solicit and listen to advice, ideas, and feedback from your subordinates; their combined years of experience are far greater than yours, and their insights are likely to provide fresh ideas that will help bring about success on the battlefield. If, by your best judgment, you decide to reject them—the Army is not a democracy—take the time to explain why, whenever possible. Good soldiers do not expect you to do what they suggest every time, but they do appreciate knowing why you have decided to do otherwise. The same goes for orders you must pass on from higher up; make sure you understand them and the commander's intent when you receive them, and explain them to your soldiers to the best of your ability. This will also help build confidence in you and your judgment, so that when you are in combat and there is no time for lengthy explanations, your soldiers will trust your decision making.

Two Kinds of Courage. In combat, you must demonstrate not only the *physical courage* to lead, to be out front where the heat is most intense, but also the *moral courage* to do the right thing. The first type of courage is not always possible to demonstrate in peacetime—many assignments simply do not provide opportunities to lead at the risk of one's life. Yet this is precisely what you must do in combat if you expect your soldiers to do the same. All thoughts of patriotism, all intellectual considerations of value, disappear in combat, and if leaders are groveling in the dirt or hiding out in a bunker, they can only expect their subordinates to be doing largely the same thing. Demonstrate the courage to get up, get involved, and take the risks you expect your soldiers to, and they will join you and perform feats of courage that go beyond anything you expected or demanded.

The other kind of courage—moral courage—is something you must demonstrate every single day, in peace or in war. It is not something that can be gathered on a C-130 en route to a drop zone or aboard a chartered 747 inbound to an intermediate staging base. Moral courage must be built upon a sound and thorough understanding of what is right, which is based on the best enduring traditions of the Army and the Republic. Balanced with tact and a keen sense of proportion, moral courage must be exercised constantly, in everything you do. Candor with your soldiers, forthrightness with your peers and superiors, and discipline in your actions are essential all the time but are especially essential in combat. Without moral courage, you will not be trusted, you will not be respected, and, ultimately, you will not be effective in your combat mission.

LEADERSHIP AND DISCIPLINE

Captain Greenwalt never said, "Get going!" He always said, "Let's go!"

—Veteran of 276th Infantry Regiment, referring to his company commander in their first combat, January 1945

What Is Leadership? Consider these definitions: "A leader is a person fitted by force of ideas, character, or genius, or by strength of will or administrative ability to arouse, incite, and direct men in conduct and achievement." "Leadership is the art of imposing one's will upon others in such a manner as to command their respect, their confidence, and their whole-hearted cooperation."

Many thoughtful observers who have studied military leaders in the routine conduct of their duties have reached a similar conclusion: "Officers who are proficient in leadership have a high standard of discipline in their units, and in their day-to-day experience they encounter few disciplinary problems; also, their need to invoke or to use their power to punish under Article 15, or other powers stated in the *Manual for Courts-Martial*, is used with less frequency than in units with less-capable leaders." Continuing this same line of thought, under similar conditions in combat, the quality of leadership and the standard of discipline in a unit have a direct and predictable bearing on the accomplishment of mission and the number of casualties. When there is good leadership and good discipline, achievement of mission with minimum casualties is a standard expectancy.

The main goal of leadership and discipline is to produce *cohesion* in units. This is the quality that glues soldiers together, especially under stress. Cohesion is a product of soldiers' confidence in the commander and subordinate leaders, as well as shared experience over time. Leaders inspire the confidence of their subordinates by displaying technical and tactical competence and coolness under stress and by looking after their soldiers. They also develop cohesion by training their units to standard under demanding, challenging circumstances and treating everyone fairly. In the Army today, there is great emphasis on high-quality training, and rightfully so; junior officers will learn about it from their precommissioning experience all the way through their basic and career courses.

Fairness, however, is sometimes less emphasized but no less important. Fairness is a cornerstone of Army meritocracy. Soldiers expect it, possibly more than in any other sector of American society. Fairness does not mean identical treatment for all (see the section on Rights, Privileges, Restrictions, and Benefits in chapter 3, page 84), but it does mean reasonably equitable treatment in accordance with Army regulations, policies, and, most of all, what is best for the soldier and the Army. The establishment of a command climate characterized by fairness is one of the most difficult challenges of command, and also one of the most important. Explain your decisions—about promotions, punishments, unit policies, and so forth—to the appropriate personnel (your top NCO, XO, subordinate commanders, and staffs) so that they can readily and logically explain them to their soldiers.

It is a time-proven fact that American soldiers perform better when they *understand* the decisions that affect them. This is not to say that they must agree with them, but they do need to comprehend the rationale behind them. Few things anger soldiers more than a belief that they are laboring for stupid reasons or, worse, no reasons at all. You will never have time to explain every decision to every soldier, especially in combat. Precisely because of this, you *must* explain—or see to it that your subordinates explain—the rationale behind as many of your decisions as possible in peacetime, so that your soldiers will implicitly have confidence in your decisions when you do not have time to explain them.

The best way to develop confidence in your innate fairness and good rationale for decisions is to seek and consider the input of your principal subordinates. Your subsequent decisions—even if they do not always align with all their recommendations (and they will not)—will be both better informed and more readily and cheerfully accepted by your soldiers. Incidentally, this is another reason why candor and courage are important, too—without them, you can never be sure that you are getting the straight poop.

A well-trained unit in which soldiers lack confidence in their leaders will fall apart under the stresses of combat or even of difficult training, because the soldiers' hearts will not be in their work. An organization in which soldiers believe in their leaders but have not been trained as a unit will fail from the sheer frustration born of the inability to perform required tasks. Commanders must never allow either situation to occur.

The Importance of Discipline. One difference between a fine military unit and a mere rabble is the degree of obedience to the will of the leader. The combat value of units is determined by their training, experience, morale, and will to fight. It may be illuminating to approach the subject through a negative: "I say to this man go, and he goeth," is not a proof or even a test of discipline; he may start briskly at the go, and later turn off into a green or alluring pasture, instead of plunging onward through the morass of jungle and marsh beyond which lies his mission. A continual responsibility of military leaders, especially of those officers in direct command of soldiers, such as company and platoon commanders, is intelligent, willing, and cheerful achievement of assigned missions or compliance with orders. This is discipline. Fine discipline is the cement of a good organization. Where discipline is weak, leadership is faulty.

The way to obtain a disciplined command, as a habit of individual and group conduct, is to make certain of two things:

1. *The leader must be careful that orders are militarily correct and capable of execution by subordinates.*
2. *The leader must ensure by observation that orders are meticulously complied with by each individual.*

Don't be fooled by superficialities. Discipline goes deep and is the result of many mission achievements, of completed tasks, of compliance with orders, and of attaining little objectives as well as great ones.

Apply this reasoning to continuing responsibilities, such as in day-to-day training that encourages soldiers to meet the desired high standard; the maintenance of high standards of individual neatness of dress and personal appearance; the regular observance of military courtesies; always being at the prescribed place at the stated time. Apply your doctrine of good discipline to new and specific tasks that occur daily, even in combat: in the attack, to seize an area starting from a prescribed place, at a definite time, following a planned route; to go on patrol to accomplish a definite mission; to repair a truck; or to do any other necessary job. The habit of obedience, or achievement of mission, is the proof of discipline. The leader obtains it by example in meeting the goals established by his or her own commander, and by requiring compliance with orders or mission.

The officer has strong powers to exact obedience. But their use should be graduated from the mere statement of a shortcoming to show that it was observed to mild admonition, rebuke, denial of privilege, official reprimand, administrative reduction, and, as a last resort or for genuinely serious offenses, trial by court-martial. Consider always the soldier and his or her past record, the intent, and the gravity of the offense or failure. Act objectively and calmly. Choose always the lesser punishment until convinced that it will be ineffective. Never resort to scorn or ridicule. Get all the facts before action of any kind. Assume, as an example, failure of a group under an NCO to arrive at a distant point at the time prescribed. Quietly get from the NCO a statement of the reason. Assume these answers: "The bridge was out at Blankville and we were obliged to make a 30-mile detour." The lateness may be dropped. "I took a wrong road and lost the way." A check of the NCO's map-reading ability, or instruction, or caution, or mild admonition should be the action because it was simple carelessness or ignorance. "The truck driver let the gas tank run dry and we had to send back 10 miles for gasoline." This is buck passing and unacceptable; the NCO must be instructed firmly as to his or her responsibilities. "I spent the night in town and no one told the troops to be ready to move out." This is lack of appreciation of an NCO's responsibilities and, if indicative of habit or unreliability generally, consideration should be given to extensive training, or reduction. The point is, the leader must detect transgressions, determine the cause, and apply sound corrective action. If the leader habitually overlooks transgressions or lightly passes them by, he or she is lost; when the big test comes, the unit will fail to take the hill, and soldiers will die who should have lived.

The Power to Punish. Soldiers strive to please the leader they trust and admire and avoid acts that would lower their standing in his or her eyes. Even so, as long as most soldiers are young, and as long as people are people, there will continue to be human transgressions that require the application of the military leader's power to punish. The commander is authorized to use a carefully regulated power to enforce obedience and discipline, according to the severity of the offense and the past record of the offender. Considering the reduced size of our military forces and the more selective standards for retention, however, the leader must exercise that authority very wisely. This is not to say that the commander should withhold corrective disciplinary action from soldiers who need it, but rather that the commander must realize all the effects of the actions he or she has available. Some of the options that seem on the surface to be minor and possibly even lenient can in fact have far-reaching consequences that make the punishment totally inappropriate to the offense. Even the commonly used Article 15 punishments can be devastating to the future of a young soldier. All officers would do well to thoroughly understand the use of Article 15 and its less deadly cousin, Article 15 Summarized Proceedings. Both are administered by the commander. The Article 15 Summarized Proceedings stay with the soldier's record only while assigned to the unit, whereas the Article 15 becomes a permanent part of the soldier's record. Extensive instruction on the use of these procedures is provided in service schools and in unit instruction.

A commander may take one of five possible actions when an offense has been committed. First, a leader may counsel the errant soldier, to point out the error of his or her ways and provide motivation to err no further. Make sure that you ask why the soldier performed the offensive act and explain why the behavior was unacceptable. Then carefully explain what you expect in the future, and express your confidence that there will be no more problems of this sort. Also, explain the consequences of future repetitions of unacceptable behavior, including administrative actions such as letters of reprimand or admonition, nonjudicial punishment, administrative separation, and trial by court-martial. *Never* promise a consequence that you cannot personally enact. For example, as a company commander, never tell an NCO, "Next time, I'll take a stripe." Only the battalion commander can do that, so it is an empty threat and will only earn you scorn.

It is a good idea to record these counselings on the appropriate form and to get the soldier to sign it. (Whether the soldier signs or not, the counseling record is valid, however.) Later, such forms can be used to refresh your own memory of the soldier's actions, as well as trace the development (or lack thereof) of the soldier's character or duty performance.

Second, a soldier may be given extra attention and appropriate extra training without resort to recorded action. This is especially appropriate for trivial or minor offenses and first offenders. Most young soldiers intend to behave correctly, so this is usually an effective method to reinforce their discipline.

Third, Article 15 summarized proceedings may be imposed by the commander. This action allows for some elements of formal punishment, but it is a locally completed action and does not follow the soldier through his or her career. After this action, the offender should be counseled—not scolded, not threatened, but advised and guided as to the correct course of action for the future.

Fourth, the commander may impose formal punishment on a soldier or officer under Article 15. This punishment is nonjudicial, but it is a formal procedure as described in the Uniform Code of Military Justice. It is used for more serious minor offenses, and it becomes a permanent part of the soldier's or officer's record. Thus, the commander should realize that this action can have the effect of limiting the career of an individual on whom it is imposed. If this result is appropriate to the offense committed, then the commander should exercise this procedure, but it should be used only after careful consideration.

The fifth and most drastic disciplinary action available to a commander is to press charges to be heard before a court-martial. Such action is reserved for major offenses or repeat offenders who have proved to be deserving of more significant punishment. The court-martial is a judicial action that can result in a federal conviction. It is a grave matter indeed, and commanders should recommend such action only after careful and objective consideration.

Building Men and Women. Most soldiers are young men and women in their late teens or early twenties. As to their habits of obedience, behavior, diligence at work, ambition for advancement, and understanding of the obligations of the citizen to perform military service, they are individuals who have been produced by our family life, our schools, our churches, and our national environment. Don't sell them short or be deceived by critics who belittle our modern young citizens; when properly led they will be equal to any test, as their predecessors have proved, and the great majority of them will have all that it takes, and more, to provide for the national security. In any case, you, as the commander, must take those assigned to your unit and work out your destiny with them. You must resolve to lift each soldier entrusted to your care to that individual's highest level of capability. You must provide a high standard of leadership and sound discipline, and, when necessary, you must resort to a proper measure of punishment.

We are talking of building men and women. Good leaders are careful to issue clear instructions and to make certain they are understood. They avoid trivial, irritating restrictions. They never show favoritism or threaten or belittle. They are careful to note good work by a soldier and comment about it. "Good work, soldier" (but know and use his or her name) heads off many little disobediences. When a soldier does something wrong, skilled leaders tell or show the person how to do it right, and follow up to see that the lesson is learned. When necessary, they caution, or admonish, in private. When punishment is administered, it is done impersonally, objectively, without rancor. The goal is to convince transgressors that they have everything to gain by doing their duty and being good soldiers.

Occasionally a soldier is encountered who is so determined to avoid military service and to obtain a discharge that he or she chooses a most damaging course of action. The individual may deliberately commit a series of offenses in order to receive punishment under Article 15 or by court-martial, hoping to be separated from the service under AR 635-200 for incompatibility with service life or discipline. Here is a severe challenge to leadership. The temptation may be strong to "throw the book" at the offender. Such separations are a serious and permanent blot on an individual's life as well as his or her military record. The individual is thereafter considered to be unfit for further service and could not serve our country during war. There should be a sincere attempt at instruction and rehabilitation of the offender before permitting completion of such a disastrous course of action. As long as the leader feels that there is something good in an offender that can be brought out and developed, the officer should be loath to accept failure by having the soldier discharged.

Experienced officers know that many thousands of soldiers who have started off with the wrong concepts of military service, or sour on service in war or peace, have been brought into healthful understanding and have gone on to build commendable records of honorable service. Of course, there are others who cannot be influenced constructively and whose discharge is necessary. But first a sincere effort for rehabilitation should be made.

Finally, it can be verified by observation that few of our more effective Army commanders—starting at the platoon level—encounter serious problems of discipline or of offenses requiring severe punishment. Such extremes occur, but they are rare. Most soldiers choose to serve their country with honor and with pride.

TAKING CARE OF YOUR SOLDIERS

> Last in the chow line, but first out of his foxhole.
>
> —J. Kevin Hastings, veteran of the U.S. 324th Infantry Regiment, referring to the officers he most respected in combat during World War II

Why has there always been such heavy emphasis on the duty of commanding officers to take the best care of their soldiers? The commander provides as best he or she can for the physical, mental, and recreational welfare of the unit's personnel to make certain that they are receptive to training, able and willing to perform their individual missions capably, proud of their unit, and willing to perform military service in accordance with their enlistment contract.

The commander is provided with strong authority, under federal law, by virtue of his or her commission and specific assignment to command a particular unit. Authority is power. The commander has far broader power than is extended to any civilian chief. Obedience to the commander's orders in battle may result in forfeiture of life under conditions where refusal to obey a proper order could result in trial with sentence of death. This is power, indeed. But with this great power goes a heavy responsibility. The unit personnel are responsible to the commander, who, in turn, is responsible for them and all that they do in the performance of their duty. Authority and responsibility must go together

The commander must train his or her soldiers, and this is part of the job of taking care of them. Unless they are thoroughly and properly trained, the chance is increased that the unit will fall short of its mission. Individuals who are improperly or inadequately trained are far more likely than others to become battle casualties.

As an illustration, here are some of the things that are routinely done by a captain commanding a company upon completion of a long march before withdrawing to his or her own tent or quarters: assigns unit bivouac areas or quarters; designates locations for kitchen, vehicle park, and latrines; checks that arrangements are made for sick call or the medical attention of any soldiers who are injured or sick; provides for foot inspection; makes certain that arrangements are made for emergency issue of equipment. You must not look to your own personal comfort first or permit yourself to be entangled with some single facet of the task of getting the company settled. You must see to the whole job. Necessarily, you must see that the subordinate leaders follow the same course. When these tasks are finished, along with others as circumstances require, you may go to your own quarters. The needs of your troops come first. "Take care of your soldiers, and they will take care of you" is an old Army axiom to remember and to apply, for it is as true today as in earlier times.

The leader of a company or smaller unit must know a great deal about each individual. He or she must learn their names quickly, call them by name, and learn their specialties, their strengths, and their weaknesses. This creates a personal bond between the individual and the commander. "My captain knows me," the soldier will think, "and he knows of my hopes. He also knows the things I do wrong. My captain is a fair man and I am glad to serve in his company." The captain may think, "Jones is a lazy cuss. But if I am in a tight spot where I need someone I can depend on, I would want him with me." A bond of mutual understanding is essential to success in a command. Soldiers cannot be fooled. They will not mistake a poor commander for a good one. The relationship is one of daily contacts and is too continuous, too varied, and too close to allow for this kind of deception. If they are satisfied that their commander is taking good care of them, there is no question that they will do their best to accomplish an assigned mission, however they may regard it, and look out for the commander's interests in doing it.

The Officer's Relationship with His or Her Troops. Command is a very personal relationship. The smaller the unit, the more personal it becomes. How then should the officer conduct himself or herself with and before the unit personnel?

Some officers are always official, rather stern, unbending—not arrogant, not heartless, not unjust, but official, stern, unbending. Others have a warm and friendly relationship up to a point but are careful to maintain a dignity or reserve. They smile, make light of hardships, take things in stride. Still others carry this trait to the extreme, seem to strive to shed responsibility, to be "one of the boys." Which is correct? Certainly not the latter example, for this is the path of weakness and failure. But of the first two, we can say only this: Each officer must follow the course that seems most natural and that gives the best results. For that officer, that way is best. Be yourself. Be mature.

This too can be said: No officer can be on terms of personal intimacy with a few soldiers while holding others at a distance, without developing the conviction among the less favored that he or she is playing favorites. You must be objective, fair, and impersonal and do to each soldier what duty requires. You are the commander. You cannot display partiality or favoritism or preference and still do your duty.

Never use ridicule. The worst violation of common sense and human decency is to pour scorn or ridicule upon a soldier or permit others to do so. Never refer to a soldier as an "eight-ball" or call a poor marksman on the range a "bolo." To do so is to encourage others in the organization to use the same terms and apply them broadly. Pride is destroyed. Hatred may arise instead of tolerance.

Never talk down to your soldiers. Ours is the Army of a democratic republic, not a caste system or aristocracy. Soldiers will pick up on your tone immediately and will deeply resent condescension or patronization. You are their superior in rank only; they have earned theirs, as you have earned yours. Few things will inspire contempt more quickly and turn off a subordinate more permanently.

In any group of men and women there are variations in individual intelligence, mechanical aptitude, educational background, character, and all other human attributes. The military leader must take the individuals assigned to his or her unit, determine their potentials, assign them, train them for the most appropriate place in the organization, and do the best possible with them. The leader must get the most from each individual assigned to the unit, however far up or down the scale that "most" may be placed. Ridicule and scorn will reduce or destroy the capacity of any individual. Let it be clear that you recognize and value highly the talents of your soldiers and that you expect each one to achieve highly.

Health and Physical Welfare. The line commander has a definite responsibility for the health and physical welfare of his or her troops. Our splendid medical service will care for individuals who are sick, wounded, or injured; through checks, surveys, and examinations of all kinds our medical service will avoid or reduce all practicable hazards to the health of the command; they will inspect and advise commanders as to their health and sanitation situation. It is a joint job for the medical service and the organization commander.

What are some of the definite responsibilities if you are a line commander? You avoid unhealthy conditions in garrison or in the field as conditions allow. You insist on organization cleanliness through adequate cleanup and police measures. You insist on individual cleanliness as to person and clothing; this often requires that you provide measures for bathing, individual washing of clothes, or provisions to get clothing to a laundry and back again. In the organization dining facility you must provide at all times for adequate refrigeration and a high standard of sanitation, including facilities for washing cooking utensils and dishes and the cleanliness of all food handlers. There must be adequate lavatory and latrine facilities maintained in a sanitary, orderly condition. These are daily responsibilities of the line commander. They are illness-prevention measures in the eyes of the doctor.

There are the safety measures to avoid accidents. The military arts are especially hazardous. Great care must be used in the handling of weapons and explosives. Power equipment in the hands of a poorly trained individual or a novice may be lethal. Accidents in the driving of motor vehicles of all kinds are a continual threat. Proper maintenance and care of equipment reduces accident hazards by ensuring proper operation and reinforcing operator training. Weigh the risks involved with any operation and take appropriate action to reduce them as much as possible.

Civilian Education. Part of the reality of the modern era is that many soldiers come into the Army without all the academic skills they need to succeed in their military occupational specialties. Also, promotion boards for senior NCOs are placing increasing value on civilian educational achievements. It is essential, then, that commanders allow soldiers the opportunities to improve their personal academic skills and achieve the educational goals the Army demands of them.

Typically, the best soldiers will work the hardest and will have the least time to pursue outside educational interests or needs. When being considered for promotion to sergeant first class and above, they may be severely handicapped by a lack of at least an associate's degree. Moreover, unless they had an excellent civilian education prior to entry on active duty, certain of their academic skills (especially written and verbal communications skills) may be less than fully adequate for the performance of their duties at the senior NCO level.

Although training and deployment schedules are hectic and demanding, it remains encumbent on commanders to make educational opportunities available to soldiers who deserve and need them. Most education center counselors will work with commanders to build courses around unit schedules—support cycles, with their more predictable events, are a prime time for this sort of endeavor. Try to make the time available, and encourage your most promising enlisted soldiers to take advantage of these opportunities for academic self-improvement. The investment will bear rich dividends in terms of soldier confidence in the chain of command and will posture the best soldiers for success in all future Army service. It's part of the loyalty thing.

Receiving the New Soldier. The way a new soldier is received into an organization has a profound effect on his or her immediate impressions. It may have a lasting effect. Combat organizations of the Army have the ever-present problem of assimilating replacements (a poor word that some wise person should improve; let it never be forgotten that the replacement of today is the veteran of tomorrow). There is a place at once for the replacement's energy and skill, zeal in the cause, and determination to win.

Take those steps that will enhance the new soldier's immediate value, and avoid those things that will delay or destroy it. As the first step, take care of the new soldier's creature necessities and comforts, while at the same time letting him or her know that everyone in the organization is glad that he or she has arrived. See that the new arrival is fed, is given a place to stow personal gear and a place to sleep, and is made to feel a member of an up-and-coming organization that knows how to care for its members and intends to do it.

At the first opportunity, the unit commander must meet and interview the new arrival. Who is he or she? Learn about the person's training, service, age, home community, education, family, and service aspirations. Let the new arrival know that the commander wants to know his or her soldiers, is interested in them, and means to use them wisely. It is absolutely vital to avoid the image of being interested in and having time for only those soldiers who need extra training or punishment.

Repeat the process with interviews held by junior officers and appropriate noncommissioned officers. Decide promptly on the individual's squad assignment. Introduce the soldier to members of the squad, and encourage them to extend the hand of comradeship.

If these measures are done thoughtfully, newness will wear off promptly; shyness, wishful thinking of former comrades, and other deterrents will disappear and be replaced by satisfaction of assignment, a feeling of belonging, and determination to get on with the job, whatever it may be.

Coddling. Coddling is overlooking things soldiers do that are wrong. Avoiding night training, long marches, and training in rain or cold to spare physical discomfort are examples. It is to fail to send out a patrol when there is grave need for information because it is a dangerous mission, only to have many soldiers killed by surprise action of the enemy that might have been detected. Now is a good time to return to a definition of leadership in an earlier edition that starts, "The art of imposing one's will." The leader must be a determined person in matters concerning his or her duty and mission, holding subordinates squarely to the mark but at the same time providing in every way for their training, feeding and supply, administration, medical attention, and recreation, along with the human touch of welfare as it is needed. Coddling is weak and wrong, but full provision for the contentment of unit personnel is eternally right.

TRAINING

The other element of cohesion in units is tough, challenging, realistic training. As a commander, you are responsible for ensuring that your unit is ready to execute all assigned missions, in every environment from high-intensity war through peacetime engagement. War (or, more precisely, combat, since war is an act between states, whereas combat is action between units and people) is the hardest situation for which to prepare a unit, so use the tasks, conditions, and standards essential for victory in war as your "mark on the wall." Generally, if you prepare the unit for war, your soldiers will be psychologically, technically, and tactically prepared for success in less stressful operational environments. For example, there may be certain tasks, skills, and even attitudinal adjustments that must be made in an infantry battalion before it embarks on peacekeeping duties, but these can all be handled without a great deal of difficulty if the unit was prepared for war. It won't work the other way around.

You should seek opportunities for your subordinates to attend service schools that provide training or advanced training in their specialties. You must also train them on the job, using the already trained and experienced personnel in your unit. Finally, you must be watchful that personnel trained for a definite position, or MOS, are not erroneously or carelessly assigned to duties of lesser importance. Correct assignment of personnel is a command responsibility and is an essential element in training.

"Train and Maintain." When a unit is not in combat, the mission is to "train and maintain." Follow the training schedules with precision, using all available lesson plans, field manuals, and training aids. FM 7-0, *Training*, and FM 25-101, *Battle-Focused Training*, are recommended essential reading to fully understand the Army's training programs. Inject realism. Use and test your equipment. Conduct the training with vigor and enthusiasm, and it will rub off on your soldiers. Maintenance skills are essential combat survival skills. Maintenance, therefore, is an integral part of training.

Equipment. The unit must have its authorized equipment on hand (including spares and repair parts). It should be stored in equipment storerooms or on the proper vehicles as prescribed, ready for inspection or for movement. The junior officer has a major responsibility for the maintenance of unit equipment. Maintenance procedures are prescribed in technical manuals and bulletins. The end-of-day maintenance periods should be organized and performed with the same seriousness as the training periods earlier in the day, because learning and practicing proper maintenance is mission-essential training. Operational maintenance must be supervised by the officer and the NCOs; never leave the unit area until you, as the responsible officer, are certain that the equipment is combat ready.

The Reward of Thoroughness. The Army's splendid units of this period have proved the value of following procedures over and over again. Many Army units have passed from training situations directly into combat without delay or confusion, confident of their own combat readiness. The junior officer in a unit has a heavy responsibility. He or she must learn the standards of unit readiness and adhere to the book. The "book" is the accumulation of experience. Lean on the knowledge and the experience of the old hands. Be a tough inspector, fair and thorough, but never be chicken. Require your soldiers to prove their own capabilities and the readiness of their equipment. These painstaking steps can mean the difference between combat success and combat failure and, for some of your soldiers, the difference between life and death. The reward of thoroughness is confidence—and a clear conscience.

TRAINING JUNIOR LEADERS

One of the most important duties of a commander is to train junior leaders, officer and noncommissioned officer alike. It is a duty deferred or ignored by some commanders who convince themselves that it is easier and more convenient to do a job themselves, or who resort continually to precise, detailed instructions to be followed by rote. There is work in preparing a well-planned, well-conducted instructional program to develop self-reliant, confident junior leaders who can proceed effectively under mission-type orders. Hesitant commanders may rationalize that it is better for their subordinates to do a job right—that is, the way they would do it—than to chance a blunder. Such a course may seem to solve the requirements of the moment, but in the long run the performance of the entire unit will suffer. There must come a day when the commander is absent and an emergency arises; at that instant the lack of trained junior leaders to step in and do the job with confidence and skill will be painfully apparent. Besides, you will often be surprised by your subordinates' initiative and creativity.

Training includes more than studying and learning the contents of the applicable field manuals, regulations, and other publications. It must include a chance to practice what has been learned from books. This opportunity to practice is voided if you tell your subordinates not only what to do but also how to do it. It is natural for a senior officer to feel that he or she knows precisely how to tackle a particular problem and solve it in the minimum amount of time, with a minimum of effort, and with a minimum expenditure of materials. Knowing the correct procedure is one of the reasons you are the senior officer. However, when you impart this knowledge to subordinates in the form of step-by-step instructions, you rob them of a chance to use their own initiative, to think a problem through, to try their hand at arriving at the correct procedure. In so doing, you will have ignored a vital requirement of the training process.

In the performance of your leadership duties, see to it that you do not rob your subordinates of a chance to display their own initiative and capabilities. Battalion commanders should command only one battalion, not four companies. Company commanders should command only one company and not attempt to lead four or five platoons. Issue mission-type orders—that is, define precisely what is to be accomplished and furnish information about equipment and materials available, the time by which the assignment is to be completed, and any other pertinent information that may help define the boundaries of the problem. But then let your subordinates make their own plans as to the step-by-step accomplishment of the mission. They will make mistakes. Expect them—and be ready to deal with them. However, as time passes, the mistakes will be fewer and your own job will be made easier, for you truly will have trained your junior leaders. And who knows, in the process you may even learn better ways to accomplish a job for which you once knew the "right" procedure.

SPECIAL NOTE REGARDING NONCOMMISSIONED OFFICERS

NCOs are "the backbone of the Army." This is no idle phrase. Good NCOs are tremendously important to the Army. With them, a unit functions like a smooth-running machine; without them, even the best of unit officers will lead a hectic existence and will probably see poor unit performance. The NCO is a vital link in the chain of command.

In addition to being in the formal chain of command, NCOs also function in an NCO support channel that parallels the chain of command. The support channel begins with the command sergeant major and extends through subordinate unit command sergeants major to unit first sergeants and then to other NCOs and enlisted personnel of the units. This NCO support channel supplements the chain of command. Through it, the senior NCOs maintain a watchful oversight of many matters that affect the performance of the command. Although the commander ultimately is responsible for everything in the unit, matters properly within the purview of this support channel include development of NCOs, the setting and maintaining of performance standards for NCOs, supervision of unit operations within established guidelines, care of individual soldiers and their families, proper wearing of the uniform, appearance and courtesy of enlisted personnel, care of arms and equipment, care of living quarters, area maintenance tasks, and operation of recreational and other facilities. The list could go on and on. Operating within this support channel, however, in addition to the chain of command, the NCOs ensure the smooth functioning of the unit and lighten the load of the commander.

Good NCOs are made in much the same way that good commissioned officers are made. The preceding paragraphs apply equally to both commissioned and noncommissioned officers. But there is more. That the NCO holds his or her position without a commission is indicative only of relative rank and perhaps also of background and training. Each NCO should have specific duties and responsibilities assigned and should be delegated sufficient authority to enable accomplishment of these assigned tasks. NCOs spend most of their time among the troops in a unit. They are the ones who actually supervise the details involved in the accomplishment of the mission, but to do their jobs properly, they must have the respect of the troops. It is here that the attitudes of their superiors are important. *To accomplish their duties properly, NCOs must have the respect and support of their superiors.*

Accord your NCOs the same respect that you feel your superior officers should give you. Listen to their thoughts and ideas as you would expect fellow professionals to listen to you. Support your NCOs as you would expect to be supported by your company commander or your battalion commander. Ensure that your NCOs are properly trained, and include opportunities for them to exercise their own initiative and judgment. Expect and require that they carry their share of the inherent and assigned load of the unit. Do all these things, while not neglecting your own responsibilities regarding supervision and inspection, and you will be pleasantly surprised at how smoothly your unit functions. Your assigned tasks will be made lighter. You will have trained a true backbone for the unit—good NCOs.

UNIT ADMINISTRATION

The company, battery, and troop are administrative units in the sense that their commanders are required to prepare and forward to battalion or similar headquarters prescribed records and essential data. Supervision of these administrative, or paperwork, tasks is a continual responsibility of the commander. He or she may delegate to others the authority to prepare reports, check accounts and records, verify inventories, and submit whatever data are required, but it remains the commander's responsibility.

Unless data are correctly supplied as to each individual's records, all manner of complications can follow with respect to pay accounts, later claims for disability under veterans' benefits, and the like. Troop morale will suffer from poor maintenance of individual records or shoddy handling of personnel and pay actions. Strength reports originating in the company form the basis for all personnel accounting, and if they are incorrect, all is wrong. Individual and unit property records must be accurate. Strive for the *zero error* objective of all administrative managers.

Some officers have a serious misconception of this matter. Because they abhor paperwork, they shun the responsibility, or content themselves with slipshod results. Other officers devote so much of their own time to the task, instead of a proper decentralization with supervision, that they have inadequate time for other responsibilities, such as training. Both concepts are wrong. Here are some tests: Are there complaints about reports being submitted after the date or hour due? Do they "bounce" because of inaccuracies? Have outside checks of supply records or individual records shown an abnormally high number of errors? Is it necessary for the commander to personally prepare detailed reports that should be done by others? Or in nearly all cases is it necessary only that he or she sign the reports? The point is this: Unit administration requires sound leadership, just as there must be leadership in training, in food service management, in supply management, and in other responsibilities of command. The officer who neglects administrative tasks is riding for a hard fall, as is the officer who devotes so much time to the task that other responsibilities are neglected.

Leadership in Administration. Unless you are a capable administrator, you are unlikely to succeed as a commander. The first requirement is to learn, with regard to each responsibility within the organization, the records and reports that are required to be prepared; exactly what they include; how they are kept; and when they must be recorded or submitted.

The next step in management is to assign to appropriate personnel the specific missions of preparation and maintenance of records. This should be precise and include the what, how, when, and who.

As in all other tasks, there is an important element of training. Few individuals with clerical, typing, bookkeeping, and such skills reach the small units. Most of these trained individuals are screened out and assigned elsewhere. More than likely, the company or battery commander must find and train soldiers to perform these tasks.

This is not so difficult. The records to be kept are not complex. It is only that there are many of them, and each must be correct. Select individuals for the tasks who are intelligent, thorough, and reliable and who have some aptitude for such work. See that they have correct reference sources, or models, on which to base their work. Require the supervisory leaders to make careful checks, point out mistakes, conduct training periods, and report progress. As records must be legible, hold fast to the requirement for easily read, neat writing. Hammer on accuracy, completeness, and timeliness.

Once the separate administrative tasks are identified and allotted to specific individuals for execution, with a supervisory system for checking plus time for developing individual proficiency, the situation should be in hand. Thereafter the commander should be able in most instances to read a proposed paper or report, accept as facts the statements therein, and sign, if he or she agrees, with a minimum consumption of time. It is another form of leadership.

Military Correspondence. AR 25-50, *Preparing and Managing Correspondence*, provides detailed explanations and sample formats for nearly any kind of letter, memorandum, note, or endorsement that might be required. This regulation provides detailed guidance that should remove all doubt in the minds of officers and their clerks as to the correct forms of correspondence. Ensure that the regulation is available to and is used by your clerical personnel.

AR 25-50 also provides some commonsense rules to be applied to all correspondence to ensure that it is short, simple, strong, and sincere. You are urged to study these guidelines carefully and then restudy them periodically before you sign your name to a letter or report that has strayed from the objective of clear, simple writing.

Official Signature. An official signature consists of the name, grade, branch of the Army, organization, and title.

RICHARD D. AMES	K. K. KELLY
Major General, USA	Major, Infantry
Commanding	Commanding

FOOD SERVICE

It is true that the consolidation of many support activities has lifted much of the responsibility for food service out of the company and battery levels, and with it the routine appointment of a company food service officer to act for the commander in dining facility supervision. Even so, there remains a definite command responsibility whether in garrison or in the field, because what soldiers are provided for food influences their morale, their willingness to serve, and their efficiency far beyond the power of any printed words. Officers assigned to duty with troops must be mindful of the proper food service of their soldiers. The serving of a well-chosen menu, with well-prepared food items, in attractive surroundings is a matter of the first magnitude in garrison; in the field, the goal must be chow delivered on time, and hot when the mission allows.

Commanders must pay particular attention to daily operations to ensure that the Army Food Program is being carried out properly. The Army Food Program covers the personnel, procedures, and resources involved in feeding troops worldwide. The program was developed by experts of the highest standing and experience, and it is directed by the U.S. Army Troop Support Agency. Everything is included, from research and development of a food item through the cooking and serving process. The purpose of the program is to provide the best-tasting, most nutritious, and most wholesome meals possible within the basic daily food allowance (BDFA).

Dining Facility. In garrison, with a consolidated dining facility, what should a company commander or a company officer supervise? Local policies will establish the actual responsibility of unit officers. But in any case, a considerable interest by unit officers should be displayed in the quality of the food service. You must assure yourself that your troops are satisfied, or that you make appropriate recommendations to the responsible officials. You should know about the choice of menu components, the quality of food preparation, the attractiveness of the food service and the surroundings, the orderliness of the kitchen and personnel, and sanitation.

Field Feeding. The real challenge is the effective operation of a satisfactory food service under combat or field conditions. The food may be prepared in the area of the battalion trains, or the members of units in campaign or combat may dine on combat rations under the control of company officers. Clearly, there are important responsibilities for the company commander and his or her officers that require knowledge and leadership.

There are a number of important considerations that should receive your attention both before going to the field and in the actual operations of a field kitchen. Be certain that your unit has its authorized TOE field cooking equipment when organized as such, that it is in good working order, and that your food service personnel know how to operate it properly and safely. Provide training time for practice in setting up and taking down kitchen tents and in setting up and preparing the kitchen equipment for movement. Your unit may also be involved in the use of Kitchen Company Level Field Feeding (KCLFF) equipment used for preparing T-rations. If so, training time on this equipment should be included.

Ensure that your food service personnel know and practice field food service sanitation principles. And don't neglect basics, such as emplacement and camouflage techniques. Meals may have to be delivered to points designated by map coordinates, and the kitchen may have to be concealed from enemy observation. Finally, ensure that the needs of your food service personnel are properly accounted for in your unit loading plan. Provisions must be made for the field equipment and food, of course, but be certain that provisions are also made for a water trailer, water cans, gas cans, and any other support items required.

In the field, ensure that the kitchen is set up in an area with good drainage and easy access, and preferably with built-in concealment. Be certain that adequate waste disposal facilities are provided and that trash and garbage are properly disposed. Do not neglect camouflage as necessary, and precautions against CBRNE contamination. By all means, keep your food service sergeant informed of your plans so that meals can be properly prepared at the right times and delivered to the correct sites in the appropriate quantities. If you have done your job well while in garrison, you can rely on your food service sergeant and the other food service personnel to run the field kitchen without your detailed attention.

Under the assumption that a company or equivalent unit has a food service responsibility under field or combat conditions, the following items will assist an officer who may be assigned that responsibility.

Check Food Service Personnel and Be Certain of Their Qualifications. Go another step, and assure yourself of a workable division of duties and responsibilities as to working procedures. If the unit is to have good meals served at the right place, at the right time, under suitable conditions of sanitation considering all the circumstances, it will require good personnel, hard work, and effective leadership.

Check Food Supplies and Storage. Be certain that the food supply is adequate, stored properly to prevent spoilage, and secure from pilferage.

Check Kitchen Equipment. Assure yourself that food service personnel have the needed or authorized tools and equipment and that these are serviceable and properly cleaned.

Field Inspections. In campaign or field conditions, careful inspection of the food service is especially important. It is harder for the staff to maintain a high standard of cleanliness and to operate at the necessary high standard. Whenever a unit operates its own food service under field conditions, it becomes a first responsibility.

Kitchen equipment must be checked frequently to be certain that the required items are on hand and serviceable. In the field, an item broken or lost through carelessness or pilferage may cause real difficulties.

Sanitation is extremely important. It requires continual observation. Clean dishes used by the troops, or in food preparation, must be free from grease. Make spot checks, as often as circumstances may require; feel the dish surface, don't just look at it. If any grease is found, the item is unclean and may result in serious illness. You must make this check when it lies within your responsibility and be certain that adequate facilities are available to obtain cleanliness, which include a generous supply of hot water and soap or detergents.

Check the disposal of garbage and wastes. This can become a frightful nuisance, a source of intelligence information to the enemy, and a threat to health. An ever-watchful eye is needed.

LOGISTICS

"Forget logistics and you lose." So said Lt. Gen. Frederick Franks in the aftermath of Operation Desert Storm. All competent and successful commanders have understood this.

The United States spends an enormous amount of expense equipping and maintaining our armed forces at the highest level. Most individuals understand this clearly when they consider the amount of materiel and equipment present in any unit. The country provides this willingly but expects soldiers to use and maintain that equipment with the greatest efficiency and economy. The job of commanders is to obtain the items authorized for their units, see that they are in the hands of the appropriate soldiers, and that they are maintained in a first-rate condition.

Supply. Supply of unit equipment and the supply and resupply of consumable items is a continuing challenge for the unit commander and should always be an item of his or her attention and concern. The commander who understands the Army supply system and who builds a rapport with the quartermaster and ordnance units who supply and maintain equipment for his or her unit is usually a well-supported commander.

Supply in garrison is often a routine and orderly process. As soon as a unit enters field or combat conditions, however, different sorts of problems occur. In combat, consumption of supplies and wear and tear on equipment expand exponentially. Some items are lost or destroyed in battle. Property is pilfered if not carefully secured. Many items are in short supply. Resupply systems can be interdicted. The problems of supply of food, fuel, clothing, and other high consumption items are with the commander day and night. Units unaccustomed to frugality and who waste supplies pay for this in inconvenience, shortages, potential mission degradation, and sometimes lives.

Maintenance. In combat, any failure to maintain equipment in a safe and operable condition may be disastrous. A gun that fails to fire may result in a combat casualty. A tank, howitzer, radio, computer, truck, or even a field kitchen that fails at a critical moment will hinder operations and may prevent victory. Adequate maintenance will reduce the necessity for replacements and the down time associated with battlefield repair. Unit commanders are in direct charge of the equipment and the people who use it; they must understand the Army supply and maintenance system, be vigilant that equipment is being used and maintained rigorously, and that corrective action is taken. Continual training and inspection are necessary. Maintenance may mean the difference between success and failure on the battlefield.

COMMAND MANAGEMENT PRINCIPLES

Management in Leadership. Expert execution of command responsibility is never an accident. It is always the result of clear purpose, earnest effort, intelligent direction, and skillful execution. It is thoughtfully directed hard work.

Good management is an essential tool of the commander in the discharge of responsibilities. It is the judicious use of the available means to accomplish a mission.

Management often includes improvisations to make the best use of what you have to get what you want. Here, "what you have" includes your available resources in personnel, equipment, funds, and time. In the Army, you will rarely have all you want of anything; you must strive to succeed with what you have. The following steps have broad application to the approach of any problem.

Understand the Mission, Objective, or Job. Know exactly what is to be done. The mission may be a continuing command responsibility, such as discussed in this chapter. Or it may be a precise, detailed order. Or it may be a mission-type order that requires the officer to determine intermediate objectives.

Objectives of a commander are wisely divided into two categories. First is the long-range objective, which is to be attained in six months or a year and may involve accumulation of funds, construction, special training. Second is the short-range objective, which can be undertaken at once and completed quickly.

Develop a Detailed Plan. Next comes programming or scheduling, which is a part of planning. After deciding what is to be done there must follow how it will be done and who will do it. Part of this step is the issuance of orders or instructions, with whatever incidental training, discussion, or explanation is advisable or necessary.

Subordinate leaders and technicians must know their own responsibilities, as they must also know the soldiers and equipment available to them for the work directed. This requires the decentralization to subordinates of a stated degree of authority and responsibility. When this is done, subordinates need not stumble about in uncertainty but can go ahead with confidence and give their boss exactly what is wanted.

Announce a Definite Time for Completion of Mission. Timing is essential to coordination. It is especially true when two or more commands or groups must mesh together in teamwork. There are some variants. "Task to be completed by 1500 hours today." Or, "Not later than 3 November." Or, "Without delay." The last means to begin at once and bring it to successful completion with effectiveness and dispatch. Scheduling provides for coordination and ensures completion of the task at the projected time.

Let us now consider two necessary cautions. A good and necessary reason for scheduling is to keep appropriate pressure on subordinates. But this may be abused and poor results obtained. Some officers demand speed to the point of absurdity. Unless an officer is willing to accept the half-done and slipshod, he or she must allow sufficient time for the capable, willing worker to do the job well. Avoid becoming an officer who demands regularly that tasks be completed "yesterday."

The second caution is almost in conflict with the first. In Army life there are rare occasions when the leader must ask the almost impossible in life-or-death, victory-or-defeat situations. Ours differs from the humdrum vocations. Under some of these circumstances, brave soldiers will try, even at the risk of death, and sometimes the finest among them will succeed. They are the winners of the Medal of Honor: those who landed first in Normandy and pushed forward; those who drove on to the Yalu; the brave, superbly trained soldiers who executed search-and-destroy missions through the torrid, steamy jungles of Vietnam. In earlier days, they were the soldiers who stuck it out at Valley Forge, and one such officer carried the message to Garcia.

Complete Work—Clean as You Go. The phrase "clean as you go" means that jobs started are finished. It means order and thoroughness as a matter of course. It means policing an area to keep it clean and tidy, in contrast to a periodic, hurried cleanup to make it momentarily fit to be seen. It means each soldier on top of his or her job, all the time, and proud of it. It means confidence and pride in doing a worthwhile job well. Beyond all this, it means pride in organization and pride in military service itself, not a grudging minimum of unwelcomed service. The wise commander will make it a command practice and advance the principle: *clean as you go.*

Give Generously of Command Interest and Control. Management is not just planning; it is carefully supervising execution to ensure compliance with instructions and to adjust plans, if necessary. This includes minor on-the-spot adjustments, a word of praise or encouragement here, some prodding there. The commander must leave behind an improved and strengthened clarity of purpose and renewed determination to get on with the task and do it well. Many a well-planned mission has failed because of the commander's failure to properly supervise the execution phase.

Sizing Up the Task. The first step for the new commander is to evaluate or size up the assigned task based on the status of the organization and the individuals who are its members. Is it an old unit partially or wholly trained? Or is it newly activated with an experienced cadre and untrained members?

A New Unit. You must inform yourself about the total mission; the time available to complete it; the capabilities of the officers, noncommissioned officers, and technicians available to assist you; and the physical facilities available. Next you must gain knowledge about each member you are to train. Study the records to determine the education, civilian vocation or training, physical condition, aptitudes, age, and former military training of each. Interview each member; talk about these things; learn individual interests and aspirations; seek the personal understanding of the task and the cooperation to attain it from each. This approach is much better than the too-common assumption that all soldiers to be trained are without usable knowledge, equally able to absorb knowledge, and equally interested in doing well. From this first contact you must be seeking the exceptional individual who can be raised quickly to become a leader or a highly skilled technician; you must also identify the personnel who learn slowly or who may require patient handling and additional instruction to keep them abreast of the bulk of unit members. Along with these slow learners are others who join late, who have missed instruction because of sickness, or who have been absent from a period of instruction. This is a summary of the way good leaders learn about subordinates and guide their individual and collective progress to attain the necessary results.

An Old Unit. You must know the mission and the time, personnel, and activities available, as for any other task. But you must determine quickly such things as the following: Is it a fine, well-trained unit of high esprit? Or is it below standard in some specific way? Is it weak in discipline? Is it behind identical units in training or below their standards of training accomplishments? Has it failed in battle? Why? These are a few of the special situations that may face an officer in command of an established unit.

Once these matters have been identified and evaluated, the course to follow is similar to that described above. Study the records; study the individuals; plan a course that fits them and fits the goal. But in the case of an old unit, find the best way to remove the cause of the unit's difficulties or to exploit its strengths. Perhaps that way is the replacement of ineffective officers or noncommissioned officers. It may include improvement in things that have caused discontent, such as poor food service, poor living conditions, or poor policies regarding leaves and passes. These are only examples. You must determine all the reasons that the organization has been considered substandard, if that is the case, and then apply all measures within your control and power to correct them. Use the things that are done well as a base to build on. Accentuate the positive and go from there to eliminate the negative aspects.

Exploiting Acquired Skills. The most important asset of any unit is its soldiers' usable skills. This knowledge may have been acquired prior to entry into the Army, in school or college, or as a result of former employment and on-the-job training. The Army has an excellent classification system that identifies and records the special skills or knowledge of new personnel. After this initial classification, personnel are assigned to units or to training centers where opportunities will occur to use this experience. When the skill possessed by an individual is one for which the Army has a need, the classification and assignment procedure operates smoothly. Personnel should be placed where they can do the best work for the Army and the nation.

But there are some factors that need understanding. Civilian life has no specific counterpart for the infantryman or the artilleryman, for example, and soldiers trained to meet these needs must be produced in large numbers. More personnel may arrive with training in a civilian vocation than the Army needs on such duties; the surplus will be assigned to other duties. There are other civilian vocations for which there is no Army requirement whatever. These readily understood truths are stated because of the frequent charge of misuse of civilian talents, usually objecting to assignment to the infantry. They need understanding.

Acquired knowledge and acquired skills must be identified quickly, and to the extent such individuals are needed, they must be properly and promptly assigned. The tables of organization list most of the skills needed, each of which is identified as a military occupational specialty (MOS). You must obtain explicit information as to the unit's need for trained personnel. You must then consult the individual records to ascertain the personnel trained in the missions required. Then by interviewing and testing, as well as by observation, you decide what you have and balance it against your known needs.

One way of evaluating the Army training requirement is filling in the gap between the skills brought into the Army, or possessed by its members, and the training needed to meet the complete mission of the unit. If truck drivers are needed, the first place to search is among your troops to determine whether you have them but they are on other duties. The examples are infinite. The training load is magnified enormously unless the classification and assignment procedure is effective and properly used. Take for granted that every member desires to perform service of the maximum value to the Army and the nation. Make the best possible use of each soldier, whether officer or enlisted.

Our classification and assignment procedures are good. Observe and understand them. But in so doing, remember that assignment of individuals to specific duties is a command responsibility, and do your part as a responsible officer in placing your personnel wisely to meet Army needs.

Summary. This is an indication of the way you must apply your knowledge of leadership. You must size up your tasks and learn all about any special situation that confronts you. You must learn as much as possible about each assistant and each subordinate to be trained. You must form a team to address the mission before you, for a commander is only as strong as the collective strength of the unit. Then with your feet planted firmly on the ground and a feeling of confidence in your mind, go ahead and apply the fundamentals of sound leadership to your specific tasks day to day.

OFF-DUTY ACTIVITIES

It is Army policy to make life on posts and stations so attractive that all personnel will find activities to consume their time and interest. People vary in their interests. Many like to take part in athletics; some like to read or study; others prefer to remain in barracks with the activities immediately at hand. The point is this: There must be opportunities on a large scale for off-duty enjoyment, and there must be wide diversification in the activities available.

You cannot expect soldiers to maintain a positive attitude toward their unit if all they mentally associate with it is arduous training and demanding work. There is a great deal of truth to the old adage, "The unit that plays together stays together."

An effective off-duty program must be multifaceted. First, it should include group activities, such as picnics, parties, and even trips. A day trip (or overnight, if missions and funds allow) to a location of professional interest (a battlefield or museum, for example) is always useful, for it expands the professional horizons of all involved and does so in an environment that promotes group cohesion. If combined with a recreational opportunity, so much the better. Unit visits to long-term care units of local veterans' hospitals, (nursing home-type facilities for ex-soldiers and retirees) can also be powerful group experiences that serve the community and provide soldiers with great insights into the meaning of their profession.

Second, the commander should ensure that soldiers take advantage of off-duty opportunities made available by the garrison commander and his or her staff. Broad-based athletic and religious programs and even cultural events are regularly conducted and provide close, low-cost (or no-cost) opportunities for group participation.

Third, commanders should ensure that soldiers are aware of activities in nearby communities, such as services of the USO, civilian and civic organizations, churches, lodges, and the like. When the soldier leaves the post, these facilities provide a place to go and things to do. A sound off-duty program is the responsibility of the station commander, aided by his or her staff, in coordination with civilian leaders.

Company Recreational Program. Leadership is needed within a company if it is to have an attractive off-duty recreational program. Many commanders assign an officer or noncommissioned officer to supervise this program in addition to other duties.

Two programs have direct application to the company in barracks: the day room, or recreation room, and the organization athletic program. The latter may consist of informal athletics of many kinds for wide participation or the forming of company teams for scheduled games in a battalion, brigade, or station league.

Apply first attention to the day room. This is where soldiers will assemble when off duty, in fair weather or foul, more regularly than elsewhere. A proper day room is equipped to be interesting, well lighted, clean, and attractive in appearance. The equipment should include a number of comfortable lounge chairs; a larger number of straight chairs, because they require less space; a large table; small tables as appropriate; and floor or wall lamps. All must be strongly built. Part of the day room furnishings may be obtained by issue through the Morale, Welfare, and Recreation (MWR) programs or property book officers of the station. The remainder can be purchased from the unit fund, with the approval of the company commander. All or most of the following should be provided: basic supply of newspapers and magazines, television, pool table, and table tennis.

Strive to arrange facilities so as to provide reading (and studying) and hobby rooms away from the vociferous day room activity. Many soldiers are taking courses for various goals and should be assisted by access to a quiet area while off duty.

A good company athletic program is a matter of facilities, equipment, and guidance. The goal here is for everyone to participate. The officer in charge of company off-duty activities should survey what equipment is on hand and what is unserviceable and can be repaired. By conference with NCOs, he or she can determine what is wanted. Some athletic equipment may be obtained from the station MWR officer. Anything additional must be purchased with organization nonappropriated funds. A desired equipment list, with detailed costs, must be presented to the company commander for approval or amendment.

The Morale, Welfare, and Recreation (MWR) Officer. The MWR officer of posts and stations and of large units heads all recreational services including voluntary athletics. The closest coordination and support should be developed between commanders, especially small unit commanders, and the MWR officer.

The Chaplain. Religious freedom is one of the basic foundations of the American form of government. Military chaplains are provided in the U.S. Army to ensure availability of a free choice in the exercise of religion. Ordained clergy perform as an extension of their denomination in conducting worship services, administering rites and sacraments, and providing an active religious education program. The chaplain is a technical expert serving on the personal staff of the commander and assisting the commander in fulfilling command responsibility concerning matters of religion, morals, and morale as affected by religion.

The religious and spiritual welfare of the members of a command is an important factor in the development of individual pride, morale, and self-respect—essentials in a military organization. Although Kipling may have overemphasized the sordid side of the life led by some soldiers in saying that single men in barracks do not grow into plaster saints, it is a fact that the environment in some instances leaves a little to be desired. We are a religious people, and our soldiers are subject to a wholesome religious influence. There is a relation between morals and morale, just as there is a relation between fair and just treatment and morale, or cleanliness and morale. The organization commander who is mindful of the religious and spiritual environment of his or her soldiers will be able to make a strong appeal to the better instincts of all, which may be impossible to attain for those who neglect it.

The United Service Organization (USO). The facilities provided by the USO for off-post recreation are of high value and deserve strong support from commanders. The Army itself has made notable progress in improving the recreation and athletic facilities within military stations through its strongly supported Morale, Welfare, and Recreation program. The USO provides its assistance in nearby civilian communities through club and service facilities. A clean and attractive gathering place in cities visited by large numbers of servicemembers is essential. Their activities of many kinds provide needed recreation. The USO also sends shows to overseas commands to entertain our servicemembers.

Funds for support of the USO are raised through the annual Community Chest and United Givers Fund drives, all by volunteer giving.

Unit commanders can assist the USO program by knowing where the USO facilities are established, having a close acquaintance with the local leaders, informing their personnel about the USO, and regularly announcing current activities. A frequent visit to the facilities by commanders of all grades is helpful to demonstrate a sustained interest in the work of the organization. They can assist in the fund-raising program by telling civilian leaders of the need for USO in cities and towns near Army stations.

SPECIAL SITUATIONS

Leaders will encounter special problems in all categories of units. Several of these recurring problems, with suggestions as to their handling, are discussed below. There are others, of course, and some may have a "newness" that will startle the commander. In each such case, the commander must find a way to proceed quietly and objectively to determine the facts, weigh them, and decide what action, if any, should be taken for the benefit of the individual concerned or the unit, or for the broader good of the Army and the nation. The officer's commission is a public trust; he or she is expected to choose a wise course of action and have the courage to follow it.

Absence Without Leave. Commanders of companies will certainly be faced with the problem of absence without leave. It is the military version of "cutting classes" or staying away from work, which the civilian chief calls "absenteeism." But in a military command, it is far more serious. Skulking in battle is an aggravated form of being AWOL. Like other problems, it can be reduced greatly, if not entirely eliminated, by a sound approach.

Many times, soldiers are absent without leave when they could have received permission to be away if they had only asked for it. The first essential is that all personnel understand clearly how and when they may obtain passes for short absences and leaves for longer ones. As duty permits, there should be readiness to grant this authority within the scope of regulations. Then, when duty requires all to be on duty continuously, individuals will be better prepared mentally to accept the situation.

Instruction constitutes the preventive for a large portion of the unit. Good soldiers cherish their standing in the eyes of their superiors and of their fellows. They must understand that such an absence is prejudicial to promotion or assignment to a responsible position. They should be made aware of the effect on the standing of the organization itself, because senior commanders carefully watch the comparative standings of their units in AWOL and other transgressions. They must understand about loss of pay for the offense. The fact that essential instruction is missed will influence some soldiers. The problem can be reduced by being certain that sound, positive instruction is given about the gravity of the offense and its individual effects, although it may not be eliminated by this method.

After an individual returns from an unauthorized absence, there must be an interview with the commander. What was the true reason for the absence? Did the soldier understand the policy for leave or pass? Was he or she denied authority for absence? Was the absence an important one due to personal affairs, or trivial, or an example of character weakness? The commander must strive to learn the cause and, having done so, apply the corrective action indicated.

A series of such interviews is almost certain to disclose that the soldiers most prone to offend in this matter are those of lowest intelligence, education, or ambition or weakest in character. Patience in the instruction of such individuals may be necessary.

Strong action must be taken against repeat offenders who continue the practice despite instruction, sound appeals to pride, and milder punishments. Among such soldiers are those most prone to skulking on the battlefield or to self-inflicted wounds. They are the ones most likely to be asleep when they should be alert to detect enemy action or to turn back from a patrol. In such cases, the only remaining action may be to let trials by court-martial run their course.

The commander must face this problem with wisdom and sound action.

Debt. Debt beyond immediate capacity to pay is not a new problem of a soldier (or officer). "All men are good soldiers—when they are broke" is an ancient adage, but it has a vestige of truth.

There are merchants within the United States and abroad who go to extremes to get a soldier or an officer to place his or her name on an installment contract. It is one of our less admirable national traits that we want to have at once all the enjoyments of our contemporaries. It is the "travel now—pay later" philosophy extended. The legal assistance counselors, who see the seamy and greedy side of it, say, "Keep your pen in your pocket!"

What is the responsibility of the commander? He or she has a duty to instruct, because some soldiers, even senior NCOs and officers, contract obligations beyond their capacity to pay. The commander can be certain that the majority of temptations are disclosed. He or she can also prevent salespeople from encroaching on training time or even unit premises, unless the salesperson has an authorization in hand from an appropriate authority.

When the damage is done, the commander may assist the person in financial difficulty to get a consolidation loan, save interest, and make repayment on a reasonable basis. In some instances, such loans may be obtained from the Red Cross or Army Emergency Relief. This procedure includes the opportunity to instruct in order to prevent such problems from happening again.

An associated administrative problem is the barrage of letters from creditors demanding collection assistance. Commanders are required to assist bona fide creditors in collecting amounts properly payable to them; and there is the accompanying duty of a commander to protect his or her soldiers against fraudulent or incorrect claims. Some creditors resort to threats against the commander as well as the debtor. What course should a commander follow? He or she should require the creditor to provide a copy of the contract or written agreement or record to substantiate the claim, with a statement of all payments to date; consult the soldier to determine whether there is an agreement between creditor and debtor; and determine the soldier's version of the indebtedness. Some cases may require guidance from the legal assistance officer. When the commander is satisfied that there is an indebtedness, and payment is in arrears, it becomes his or her duty to require or try to arrange a suitable method of payment. Garnishment of pay is authorized when a creditor has obtained a legal, civil judgment against the debtor. A soldier (or officer) may be punished for failure to pay just debts.

This is an unpleasant duty that falls to commanders of companies and even to battalion commanders. Expect to encounter the problem, and regard it as a challenge to leadership.

Complaints. The commander must be accessible to members of the unit who wish to state a complaint. Some of the complaints will be petty. Others may be deliberate attempts to injure the reputation of another with a charge that is without foundation. There are other occasions, however, even in the best organizations, where genuine cause for dissatisfaction may occur. This is information that the commander must obtain, lest the morale of the unit be seriously impaired. Members must know that they may state a cause for complaint to their commander with the knowledge that he or she will give them a hearing and correct the grievance if convinced of its truth. The Army has an open-door policy that calls for easy access to the commander through appropriate subordinate commanders unable to resolve the complaint. Good soldiers will seek to avoid making a complaint. But when ideas, suggestions, or reports are stated to the commander that affect the welfare, the efficiency, or the morale of the unit or any individual therein, they must be heard sympathetically. If the condition reported can be corrected, or deserves correction, the action should be taken at once.

Delivery of Mail/Availability of Internet. Homesickness may well be the occupational disease of the American soldier. The student of soldier psychology might place communication with loved ones at home next in importance to food. Indeed, for the soldier with a wife or husband, a son or daughter, a sweetheart, or a beloved father or mother, contact from home by electronic or snail mail may be of far greater and lasting importance than food. Unit commanders must pay particular attention to this need. They must see to it that soldiers have internet connectivity when practicable, so that they can stay in touch by any of the various means that social media presents. They must also ensure that conventional mail reaches the soldier quickly and safely. They must make certain that none is lost and that the chance of malicious opening and violation of privacy is eliminated. Contact from home, be it by Zoom or by mail, is of the highest importance to the soldier, as it unites him or her for a moment with loved ones. Often this contact eases the soldier's apprehensions and dispels the worries that beset those deployed far from their homes. Knowledge that all is well at home secures a peace of mind so strong that it justifies every effort that good mail service and internet connectivity requires.

The commander must work to secure internet connectivity whenever possible or to send soldiers to the nearest area that allows for use of the internet whenever possible.

The commander must also check the method of receiving, guarding, and distributing conventional mail as it reaches the organization. Are the individuals charged with handling it completely trustworthy? Or are there unsolved complaints of rifled or stolen letters? Is the mail guarded scrupulously from the time of receipt until it is handed to the soldier? Or is it allowed to lie about, subject to scrutiny and mishandling? Or is it handed to the individual to whom addressed? Is it held for long hours after receipt to be distributed at the whim of some individual? Or is the mail delivered as promptly as circumstances permit after receipt? Are registered and special-delivery pieces handled with due regard to postal regulations? Are receipts for these classes of mail maintained exactly as prescribed? The soldier treasures packages from home. Are packages zealously guarded so that petty pilfering is prevented? These matters are very important in the daily life of the soldier.

Personal Problems. All soldiers have left interests or roots or problems behind them that may require their attention or action while in the military service. At home, the soldier would turn for advice to a parent, a friend, a lawyer, a minister or priest, or other person in whom he or she has trust and confidence. In the Army, the soldier usually turns to the company commander.

These occasions provide a fine opportunity for the commander to show a deep interest in the welfare of his or her soldiers. The commander should adopt an impersonal and kindly attitude in hearing these problems. If the matter is confidential, it must never be divulged improperly. When it is proper to do so, the commander should give the counsel he or she knows to be correct, or obtain the necessary information, or direct the soldier to an authority who can supply the information.

In many cases, after hearing the soldier's problems, the officer will need to refer the soldier to another authority: the legal assistance officer for legal matters; the American Red Cross field director for investigation and action about family conditions at home; the chaplain on matters of religion, marriage, and human relationships; the surgeon if the problem concerns health.

A good leader must have a genuine understanding of human relations. His or her tools are men and women; therefore, the leader must be able to deal with people. The necessary warmth of military leadership can be demonstrated when soldiers carry their baffling personal problems to their commander for advice or solution.

Letters or E-Mails to Parents. Many commanders make it a practice to write letters or e-mails to parents of their soldiers when important personal events have occurred in which they may take pride. The people of the United States are tremendously and vitally concerned about the progress of the Army to which they have given the services of their sons and daughters. Their impressions are formed by the reports they receive from their children and their neighbors and friends. No amount of big-name announcements as to Army morale and conditions will offset the local effect of an unfavorable report from a personal acquaintance. The interest and satisfaction that can be developed by contact, even by letters, with the folks at home are well worth the effort required.

Many company, battery, or platoon leaders write these letters. "Dear Mrs. Brown," they may write, "I am pleased to tell you that I have recommended your son for promotion to the grade of corporal. He has worked hard here and his record is splendid. He is a good soldier." Or this: "As you know, your daughter has been confined to the station hospital. I have visited her several times and have had frequent reports about her condition from her medical officer, the chaplain, and others. I can tell you now that she is well on the road to recovery, and her return to her organization will occur very soon. She is performing a service to her nation of which you and she, too, will always be proud." The American soldier is a young man or woman who is usually away from home for the first time. The soldier's parents are anxious that he or she perform credibly.

Have you been present when a letter is received from a young soldier? Do you know how it is discussed and passed around or read aloud to others? Can you not imagine the effect of a letter from a soldier's commander? Form the habit of writing personal letters to the parents of your soldiers on proper occasions. It places a human touch on a relationship that is impersonal and detached.

Hometown Press Releases. Although some soldiers will not want it for various reasons, the vast majority of young soldiers—officers and enlisted—will be pleased by positive publicity given to their accomplishments in their hometown newspapers. Achievements such as promotions, decorations, graduation from significant Army Schools, and so forth are newsworthy and can bolster your soldiers' morale. Ask the command's public affairs officer (PAO) how to go about this. It is usually a matter of filling out a brief form and mailing it.

14

Staff Assignments and Procedures

Any military organization has a commander who alone is responsible for all that the unit does or fails to do. All policies, basic decisions, and plans must be authorized by the commander before they are put into effect. All orders from a higher unit to a lower unit are given to the commander thereof and are issued by or for the commander of the larger unit. Each individual in the Army is accustomed to look to his or her immediate superior for orders and instructions. By this means, authority and responsibility are definitely fixed, and the channels of command are definitely established.

WHY A STAFF?

It should be apparent, however, that myriad details are involved with the day-to-day operation of any organization. As the size of the organization increases, the number and variety of the details increase. The commander cannot devote personal attention to all of them.

Therefore, a staff is provided as an aid to command. It serves to relieve the commander of details by providing basic information and technical advice by which he or she may arrive at decisions; by developing the basic decision into adequate plans, translating plans into orders, and transmitting them to subordinate leaders; by ensuring compliance with these orders through constructive inspection and observation for the commander; by keeping the commander informed of everything he or she ought to know; by anticipating future needs and drafting tentative plans to meet them; by supplementing the commander's effort to secure unity of action throughout the command; and by learning the commander's intent and working within it. In short, a properly functioning staff is an extension of the eyes, ears, and will of the commander.

Value of a Staff Assignment. An assignment to staff duties provides an opportunity for increased knowledge and capability for an officer of any grade.

The staff officer learns the detailed organization and missions of each component of the command and how the commanding officer and the subordinate commanders solve the problems that confront them. The staff officer works with each staff section and gains a knowledge of their responsibilities and methods. It is splendid training for a future assignment to command.

The realities of Army organization and officer professional development dictate that most of your service will be spent as a staff officer. In fact, the overwhelming majority of officers will only command at the company level for a total of less than eighteen months in twenty or more years of service. Thus, while command remains an officer's highest responsibility, and certainly the one to which all officers should aspire, staff duties are the ones that will be executed most often throughout the service of most officers.

Even though most officers will serve relatively little time actually in command positions, they are always in a position to *lead.* If command is the *highest* responsibility of an Army officer, the *first* responsibility is to lead. Whether by leading an entire staff as an executive officer or chief of a staff section, by making decisions under authority delegated by the commander, or even by forthrightly providing considered, candid professional opinions and recommendations, officers always lead as well as follow, in the way staff officers must. Keep this in mind as you learn about staff assignments and execute the responsibilities they entail.

STAFF ORGANIZATION

The organization of the staff of any military unit is prescribed in the tables of organization and is based on the duties and responsibilities of the commander it serves. The battalion is the smallest unit that has a staff, although even in the company there are officers and noncommissioned officers whose duties parallel those of staff officers. These include such duties as food service officer, supply officer, motor officer, and the training NCO and information and education NCO. At the company level, such duties generally are assigned in addition to the primary duties of the company officers or noncommissioned officers.

Staffs vary widely in their composition and structure, and staff officers' duties vary accordingly. Army staff officers' duties and responsibilities are outlined in great detail in FM 5-0, *Planning and Orders Production.* Of course, the exact details of each staff officer's specific duties vary with the particulars of each unit's missions and each commander's preferences. For the purpose of this book, however, the "generic" infantry battalion staff is used as a framework for discussing the subject.

Functional Areas. The staff assists the commander in the performance of four functional areas of responsibility:

1. Personnel and administration
2. Military intelligence
3. Operations and training
4. Logistics
5. Signal

In tactical units up to division level, these staff officers are designated S-1, S-2, S-3, and S-4, respectively. In divisions, corps, and (when we have them) field armies and army groups, the staff sections use the prefix "G," for "general" staff. In joint task forces and other joint staff positions, the prefix "J" is used, and in some combined commands, the letter "C" is used as the prefix to the number of each staff section.

General Staff. The four functional areas constitute the areas of responsibility of the general staff. The size of the staff section provided in each case is determined by the work load and complexity of operations encountered. The general staff operates under the supervision of the *chief of staff*, who is the principal assistant to the commander. The chief of staff transmits the will of the commander to those who act in the commander's name and is the principal coordinating agent to ensure efficient functioning of the staff and of all troops in the command. At lower levels, the *executive officer* fills the function of the chief of staff.

Special Staff. Beyond the overall framework of the four functional areas, there are certain specific functions that warrant the attention of a specialist. These specific functions vary with the size and the mission of the organization, but may include such matters as engineering, transportation, communications, ordnance, and medical support. To fill these specific needs, a special staff is provided, consisting of such staff officers as the engineer, the transportation officer, the signal officer, the ordnance officer, or the surgeon. These special staff officers usually operate under the direction of the general staff, reporting to such general staff officer as may be appropriate. For example, on matters pertaining to construction or maintenance of roads, cantonment areas, or similar facilities, the engineer deals with the G-4, while on matters pertaining to operations or training, he or she deals with the G-3. Similarly, the engineer deals with the G-2 on engineer technical intelligence matters.

Personal Staff. In addition to the general staff and the special staff, commanding generals are authorized a personal staff. This personal staff consists of authorized aides, or aides-de-camp, and other assistants. The purpose of the personal staff is to relieve the general from time-consuming personal matters so that he or she may devote full attention to matters of command and discharge heavy responsibilities with efficiency and continuity. The duties of an aide to a general officer are discussed in the final section of this chapter.

FUNCTIONS OF A BATTALION STAFF

The staff of a battalion includes the executive officer, the S-1, S-2, S-3, S-4, S-6, and special staff officers. The commander may either follow the TOE and assign staff officers to duties consistent with AR 611-1, which details their job descriptions, or "tailor" the staff in accordance with the mission and their experience, capabilities, and grades.

The commander may divide the staff's functions into administration and operations and, to reduce the span of control while enhancing supervision, assign the executive officer to manage administration and logistics and the S-3 to manage operations and training.

The Executive Officer. At battalion and brigade level, the executive officer usually serves as the second-in-command and as the principal assistant to the commanding officer. The XO usually directs, coordinates, and supervises the activities of the staff sections. The XO is often appointed the matericl readiness officer. It is the best possible training toward becoming a battalion commander.

The Adjutant (S-1). The S-1 is responsible for officer personnel management (the command sergeant major is usually charged with enlisted personnel management), unit strength management, legal administration, morale activities, and myriad administrative tasks, such as the management of the sponsorship program, support to the unit family support group, and so forth. The adjutant is usually also charged with maintaining the commander's official social schedule, as well as preparing and circulating his or her correspondence.

The Intelligence Officer (S-2). The S-2 is responsible for the production and dissemination of combat intelligence and counterintelligence matters. He or she assists the commanding officer and other staff officers in physical security matters, including safes, filing, clearances, intelligence training, and related duties. To fulfill the primary responsibility of producing combat intelligence, the S-2 collects, collates, evaluates, and interprets information on the enemy, weather, and terrain, which may influence the accomplishment of the unit mission. Of equal importance is the duty of disseminating this information to the commanding officer, other staff officers, subordinate commanders, and adjacent units.

The Operations and Training Officer (S-3). The S-3 has staff responsibility for planning the successive operations, organization, and training as directed by the commanding officer. In his or her field are operational directives, plans, tactical orders, command post exercises (CPXs), field training exercises (FTXs), the preparation of training schedules, training aids, ammunition requirements, school allocations and quotas, and a host of related duties. The S-3 prepares estimates and recommends to the commander actions or decisions for the accomplishment of the mission. It is a vital mission in which he or she works in close coordination with the executive officer and the commanding officer.

The S-3 has staff responsibility for the unit readiness of the command. Especially in this function, the S-3 works closely with the executive officer, who is usually appointed the materiel readiness officer.

The Logistics Officer (S-4). The S-4 is the battalion logistics officer and has staff responsibility for the logistics services and facilities available to the battalion. These are supply, transportation, maintenance, logistics plans and records, and other matters in the field of logistical support. The S-4 prepares logistical plans and appropriate portions of published plans and orders. During operations, he or she is responsible for the location and operation of the battalion trains. The S-4 is responsible for local security measures within the trains' area and for coordination with higher units and adjacent units in this responsibility.

The Signal Staff Officer (S-6). The battalion S-6 is the principal staff officer for all matters concerning network operations. The signal staff officer provides network transport, information services, and information management, conducts network operations to operate and defend the network, enables knowledge management (see FM 3-96), manages LandWarNet and combat net radios assets in area of operation, and performs spectrum management operations

The Special Staff. There are certain specific functions that warrant the assignment of a specialist. These special functions vary with the size, mission, and organization of a battalion or larger unit. Special staff activities are coordinated by some part of the unit staff. In some units, the members of the special staff command the related support unit, which increases their responsibilities. Because of their special technical skills, warrant officers often are assigned as special staff officers.

BASIC FUNCTIONS OF A STAFF OFFICER

The Army has standardized many of the basic staff techniques and procedures. The primary references are FM 5-0, *Planning and Orders Production*, and FM 6-0, *Commander and Staff Organization and Operations.* In addition, branch service schools have developed excellent texts; a useful example is the *Operations and Training Handbook*, U.S. Army Infantry School, Fort Benning, Georgia. Effective staff procedures are essential in the effective performance of mission by the entire unit; a fine staff accomplishes completed staff action with coordination and timeliness.

Informing. Exchange of information is the first key of good staff procedures. It is the first step in ensuring a common basis of understanding by staff officers and their commander in making estimates and decisions. No staff officer can operate effectively as a loner or in a vacuum. He or she must provide information to other staff officers, as well as receive information from them. This goal is attainable only by good working relationships within a staff and with other staffs, with a talent for good human relationships leading to fine coordination and teamwork.

Estimating. Estimates of the situation are formulated through the logical analysis of a problem by which one arrives at the workable choices of mission accomplishment and the selection of the one to be adopted. It is continuous, systematic, and as complete a process of reasoning as time permits. The accompanying list shows the basic steps in the process.

The sequence established by the standard estimate of the situation is applicable to all staff problems, not only the classic one of a combat decision. The staff uses the estimate procedure to recommend a course of action, and the commander uses it to choose or decide what he or she will order.

1. Mission (problem)
2. Situation and courses of action:
   a. Considerations affecting possible courses of action
   b. Opposing conditions
   c. Own courses of action
3. Analysis of opposing courses of action
4. Comparison of own courses of action
5. Decision

6. Analysis of results (to be completed after action has been taken on the decision)

Recommending. Staff officers make recommendations on their own initiative or when requested to do so. Such recommendations result from a careful estimate of the situation and, when offered, must be clearly and candidly presented, with neither equivocation nor ambiguity. If another staff agency is involved, which is usually the case, the recommendation must be coordinated with that agency. Should a failure to agree develop, the divergent view must be presented fairly and objectively. The staff officer is not settling a debate but is presenting facts and views and making a recommendation, on which the commander will make the final decision.

Never bring a problem to the commander without a cogent recommendation. Although it is true that bad news does not get better with age, make sure that you know what you are talking about when you apprise the commander (or XO) of a problem or negative development, and always have at least a tentative recommendation for its solution. Tactical and technical knowledge, coupled with creativity and initiative, go a long way toward helping a staff officer make recommendations; candor, coolness under stress, and the courage born of convictions do the rest.

Carrying out Decisions. When the commander chooses a course of action different from, or opposite to, the staff officer's recommendation, as happens in all headquarters, the staff officer applies his or her maximum talents to make certain that the commander's decision is executed precisely. No pique, no bruised feelings, no silent resolution that "next time she will get what I think she wants." The staff is an arm or a tool of the commander to assist in the discharge of heavy responsibility involving many people, many problems, and many techniques. But the commander is the responsible official, and the one who decides. The staff officer shows an understanding of this essential principle of carrying out the commander's decision to his or her highest capability, never indicating that he or she had recommended a different course of action.

Preparing Plans and Orders. Once the commander has stated a decision, the staff prepares the detailed orders for the entire command, if the circumstances require written orders and if time permits their preparation. Routine matters coming under previously approved policy are generally covered by SOPs and are handled by the staff without repeated visits to the commander; however, in the event of important matters or action that is unusual, the staff informs the commander at the first convenient opportunity. Orders involving missions for subordinate commanders, particularly tactical ones, are prepared with more consultation and approval by the commander.

Estimating the value of good plans and orders for achieving success in battle is hazardous at best; some authorities have stated that they are 90 percent of the battle and implementation 10 percent. Others insist that few plans survive the first shots of a battle. But it is certainly true that the commander must direct clearly *what* is to be done, and *where* and *when* it is to be done, with a specific mission for each element involved (the *who* factor); in some instances, a statement of the *why* is also helpful. Good plans and good orders are clear, concise, and precise. The staff officer should have in mind the ancient truth that "any order that *can* be misunderstood *will* be misunderstood."

The following quotation from *George C. Marshall: Education of a General*, by Forrest Pogue (Viking Press, 1963), both describes General Marshall and points out a major factor in his success as a commander: "The order must be comprehensive, yet not involved. It must appear clear when read in poor light, in the mud and rain. That was Marshall's job, and he performed it 100 percent. The troops which maneuvered under his plans always won."

PLAN FOR AN OPERATION ORDER

An accompanying illustration shows the plan format for an operation order, in the standard five-paragraph field order form. It is more than a mere form—it is a logical process of thought that should be used as a matter of course in the issue of simple instructions as well as detailed, written operations orders. It is helpful for staff officers as well as commanders to memorize the sequence and develop skill in its use.

Supervision. The execution or implementation of plans and resultant orders must have both command and staff supervision. The beginning staff procedure calls for being informed, which involves seeing, visiting, and inspecting by direct contact. Desk-bound staff officers are unlikely to serve the commander ably. Going out to see and check is the better way to keep informed. On the part of the staff officer, it requires tact, for he or she is not the commander. If erroneous action is discovered and the time factor requires it, the staff officer may issue corrective orders on the spot, but only in the name of the commanding officer: "The Commanding Officer directs. . . ." In the usual case the staff officer informs the commander at the earliest opportunity of his or her action. In inspections and reports to the commander, the staff officer is not a talebearer; most successful staff officers inform the subordinate commanders of their finding, discuss it with them, and inform them of the exact nature of the report to be made to the commander.

A staff visit has the essential purpose of coordination, teamwork, and mutual understanding between the commander and subordinate commanders, between staff divisions, and between the staff and subordinate commanders. Such visits may uncover misunderstandings and provide correction before becoming serious and requiring command action. The written word is not always clear, however carefully it is prepared, just as the oral word may also be misunderstood. Here is ever-present opportunity for the staff officer to serve the commander well and help in an important manner the achievement of the mission.

LIMITATION OF STAFF AUTHORITY

A staff officer, as such, has no authority to command. He or she does not prescribe policies, basic decisions, or plans, for that responsibility rests with the commander.

When it becomes necessary for a staff officer to issue an order in the name of the commander, responsibility for such an order remains with the commander

(CLASSIFICATION)

(Changes from verbal orders)

HEADING

Copy No.
Issuing Headquarters
Location of CP (Coord)
Date/Time Group
Message Reference No.

OPORD/OPLAN (Serial Number/Code Name)

References:

Time Zone:

BODY

Task Organization:

1. SITUATION:
   a. Enemy Forces
   b. Civilians/Noncombatants
   c. NGOs/PVOs, and Neutral Nation Observers, as applicable
   d. Friendly Forces
   e. Attachments/Detachments

2. MISSION:
Written in the directive form, using the third person voice.
   Example: "2-274th Infantry attacks commencing 05 0625 JAN in zone seize Objective WOLF (LV 815200)."

3. EXECUTION:
Commander's intent *(Succinct definition of what must be done with respect to the enemy and the terrain, and the desired end state).*
   a. Concept of the Operation *(Summarizes what is to be done to achieve success)*
   b. Subunit Missions
   c. Coordinating Instructions

4. SERVICE AND SUPPORT:
   a. Rations and Water
   b. Equipment/Supplies/Fuel
   c. Maintenance Plan
   d. Casualty Procedures
   e. Evacuation of Captured Personnel and Equipment

5. COMMAND AND SIGNAL:
   a. Command
      (1) Chain of command
      (2) Location of key leaders and radiotelephone operators
   b. Signal

ENDING

Acknowledgment Instructions:

Commander's Last Name
Commander's Grade

Annexes:

Distribution:

Authentication:

(CLASSIFICATION)

Generic Operation Order/Plan Format

even though he or she may not have seen the order as actually written or heard it if given orally.

Staff officers who exercise supervision over any phase of operations must restrict their control to the commander's announced intent, as well as specific directives. When circumstances arise that in their opinion may make advisable a deviation from established intent, even in the most minor degree, the situation should be presented to the commander for decision.

When a commander has given specific instructions to a staff officer, the actual issue of the necessary orders or instructions to members of the command are properly given in the name of the commander by the staff officer, thus "The Commanding Officer directs" or, if an adjutant or other specifically authorized staff officer is signing the instructions, "For the Commander." The orders may be given verbally or over the signature of the staff officer.

When a commander has decided on a policy to be followed and has indicated that policy to a staff officer, all future questions that fall completely under that policy should be handled without further reference to the commander. Routine matters are handled according to written or uniformly observed SOPs.

Some commands and staffs have a formal policy file for nonroutine but special matters of concern to the commander. All experienced staff officers will establish an informal personal policy file if a formal one does not exist. An established policy provides the limits within which the staff can function without continual reference to the commander.

A staff officer must never usurp the prerogatives of command. In the event of an unforeseen emergency when immediate action is imperative and the commander cannot be consulted, the staff officer should be prepared to state to the senior line officer with whom he or she is able to get in touch the action he or she believes the commander would desire. The decision then becomes the responsibility of the senior officer consulted and is not the responsibility of the staff officer.

Some commanders announce broad policies and desire members of the staff to proceed with confidence in the execution of tasks with little consultation on the details; others wish to give personal approval to at least the more important phases encountered. The staff must adjust itself to the method of operation desired by the commander.

COMPLETED STAFF WORK

Completed staff work is Army doctrine. The staff officer who is skilled will think through each problem assigned for staff action and then plan courses of action. The staff officer must determine the information needed and where to seek it and must list the individuals who have an interest in the problem, or knowledge of the problem, and plan to see them. He or she will determine all the angles and consider the varying viewpoints on important matters. "Legwork" is the requirement. A staff officer who neglects these essentials is unlikely to succeed.

The completed staff work doctrine means more work for the staff officer, but it results in more freedom for the commander to do and see the things that are

DECORATIONS, AWARDS, AND SERVICE MEDALS

U.S. ARMY AND DEPARTMENT OF DEFENSE MILITARY DECORATIONS

Medal of Honor
(Army)

Distinguished Service Cross (Army)

Defense Distinguished Service Medal

Distinguished Service Medal (Army)

Silver Star

Defense Superior Service Medal

Legion of Merit

Distinguished Flying Cross

Soldier's Medal (Army)

Bronze Star Medal

Purple Heart

Defense Meritorious Service Medal

Meritorious Service Medal

Air Medal

Joint Service Commendation Medal

Army Commendation Medal

Joint Service Achievement Medal

Army Achievement Medal

Prisoner of War Medal

Good Conduct Medal (Army)

Army Reserve Components Achievement Medal

National Defense Service Medal

Antarctica Service Medal

Armed Forces Expeditionary Medal

Vietnam Service Medal

Southwest Asia Service Medal

Kosovo Campaign Medal

Afghanistan Campaign Medal

Iraq Campaign Medal

Global War on Terrorism Expeditionary Medal

Global War on Terrorism Service Medal

Korean Defense Service Medal

Armed Forces Service Medal

Humanitarian Service Medal

Military Outstanding Volunteer Service Medal

Armed Forces Reserve Medal

U.S. ARMY SERVICE AND TRAINING RIBBONS

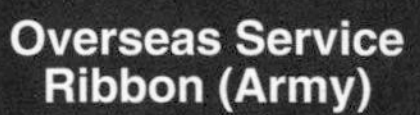

NCO Professional Development Ribbon

Army Service Ribbon

Overseas Service Ribbon (Army)

Army Reserve Components Overseas Training Ribbon

U.S. ARMY AND DEPARTMENT OF DEFENSE UNIT AWARDS

Presidential Unit Citation (Army)

Joint Meritorious Unit Award

Valorous Unit Award

Meritorious Unit Commendation (Army)

Army Superior Unit Award

ARMY SERVICE UNIFORM

Class A Uniform

Class B Uniform

NON-U.S. SERVICE MEDALS

United Nations
Medal

NATO Medal

Multinational
Force and Observers
Medal

Republic of Vietnam
Campaign Medal

Kuwait Liberation Medal
(Kingdom of *Saudi Arabia*)

Kuwait Liberation Medal
(Government of
Kuwait)

U.S. ARMY BADGES AND TABS

Combat and Special Skill Badges

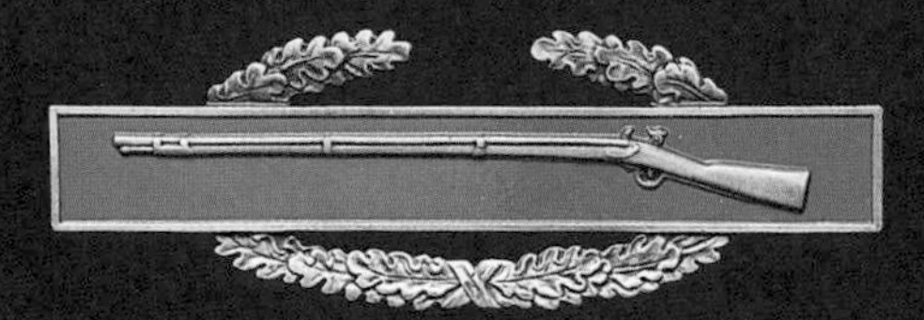

Combat Infantryman Badge
1st Award

Combat Medical Badge
1st Award

Combat Infantryman Badge
2nd Award

Combat Medical Badge
2nd Award

Combat Infantryman Badge
3rd Award

Combat Medical Badge
3rd Award

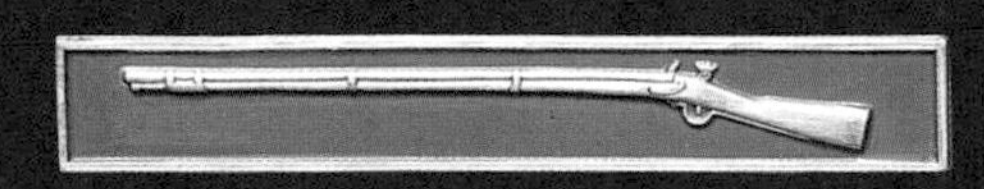

Expert Infantryman Badge

Expert Field Medical Badge

Combat Action Badge

Master Aviator
Badge

Basic Aviator
Badge

Senior Aviator
Badge

Master
Flight Surgeon
Badge

Basic
Flight Surgeon
Badge

Senior
Flight Surgeon
Badge

Master
Aircraft Crewman
Badge

Basic
Aircraft Crewman
Badge

Senior
Aircraft Crewman
Badge

Basic
Military Free Fall
Parachutist Badge

Jumpmaster
Military Free Fall
Parachutist Badge

Basic Parachutist
Badge

Senior Parachutist
Badge

Master Parachutist
Badge

Combat Parachutist
Badge (1 Jump)

Combat Parachutist
Badge (2 Jumps)

Combat Parachutist
Badge (3 Jumps)

Combat Parachutist
Badge (4 Jumps)

Combat Parachutist
Badge (5 Jumps)

Air Assault Badge

Glider Badge

Special Forces Tab
(Metal Replica)

Ranger Tab
(Metal Replica)

Sapper Tab

Presidents Hundred
Tab

Pathfinder
Badge

Salvage Diver
Badge

Second Class Diver
Badge

First Class Diver
Badge

Master Diver
Badge

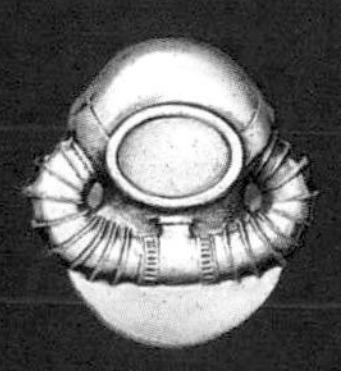

Scuba Diver
Badge

Special Operations
Diver Badge

Special Operations
Diving Supervisor
Badge

Basic Explosive
Ordnance Disposal
Badge

Master Explosive
Ordnance Disposal
Badge

Senior Explosive
Ordnance Disposal
Badge

Nuclear Reactor Operator Badge (Basic)

Nuclear Reactor Operator Badge (Second Class)

Nuclear Reactor Operator Badge (First Class)

Nuclear Reactor Operator Badge (Shift Supervisor)

Parachute Rigger Badge

Driver and Mechanic Badge

Marksmanship Badges

Marksman

Sharpshooter

Expert

Identification Badges

Presidential Service

Vice-Presidential Service

Secretary of Defense

Joint Chiefs of Staff

Army Staff

Guard,
Tomb of the Unknown Soldier

Drill Sergeant

U.S. Army Recruiter, Basic
(Active Army)

U.S. Army Recruiter, Basic
(Army National Guard)

U.S. Army Recruiter
(U.S. Army Reserve)

U.S. Army Instructor, Basic

COMPLETED STAFF WORK

How to Get It	How to Do It
Assignment of a problem and a request for a solution in such a way that completed staff work is readily possible.	Study of a problem and presentation of its solution in such form that only approval or disapproval of the completed action is required.
1. Know the problem.	1. Work out all details completely.
2. Make one individual responsible to you for the solution.	2. Consult other staff officers.
3. State the problem clearly, precisely; explain reasons, background; limit the area to be studied.	3. Study, write, restudy, and rewrite.
4. Give the individual the advantage of your knowledge and experience in this problem.	4. Present a single, coordinated, proposed action. Do not equivocate.
5. Set a time limit; or request assignee to estimate completion date.	5. Do not present long memorandums or explanations. Correct solutions are usually recognizable.
6. Ensure that you are available for discussion as work progresses.	6. Advise the boss what to do. Do not ask him or her.
Adequate guidance eliminates wasted effort and makes for completed staff work.	If you were the boss, would you sign the paper you have prepared and thus stake your professional reputation on its being right? If not, take it back and work it over; it is not yet completed staff work.

essential to the discharge of his or her own responsibilities. It also spares the commander from half-baked ideas, immature oral guesses, or voluminous memorandums that he or she has no time to study.

The final test of complete staff work is this: *If you were the boss, would you be willing to sign the paper you have prepared, and stake your professional reputation on its being right?* If not, take it back and do it over, for it is not completed staff work.

There is a special dividend that accrues to skilled staff officers. The young staff officer of today may be assigned to command much sooner than he or she has anticipated. The requirements of a fast-moving Army, on the vast missions of this period in our history, sometimes make unexpected demands. The officer who becomes an excellent staff officer, thoroughly trained in staff responsibilities, limitations, and procedures, will find on becoming commander that he or she knows what to expect from a staff and how to handle one. It is a major attribute for a successful commander.

A sometimes baffling situation confronts the staff officer regarding which problems or communications should be presented to the commanding officer, and which ones should be handled by the staff without reference to the commander. The commander must be informed of matters of importance, surely, and must be spared from trivia. But where to draw the line? There is no single answer, or policy, because commanding officers differ in their wishes, just as staff officers differ. Indeed, the confidence of the commander in the individual members of the staff bears upon both the problem and its solution.

Reproduced below is a memorandum for the staff, written many years ago, which was included in early editions of *Army Officer's Guide.* The author was Maj. Gen. Frank S. Cocheu, now deceased. It is far from universally applicable, of course. But it may point the way to the choice of a workable policy in the busy headquarters of today.

MEMORANDUM: For the Staff.

1. The following will be brought without delay to the attention of the Commanding General:
   a. Subjects of importance which require prompt action and are not covered by existing policies and instructions.
   b. Disapprovals from higher authority.
   c. Errors, deficiencies or irregularities alleged by higher authority.
   d. Communications that allege neglect or dereliction on the part of commissioned personnel.
   e. Correspondence or proposed correspondence conveying even a suggestion of censure.
   f. Appeals from subordinates from decisions made at this headquarters.
   g. Subjects which affect the good name or reputation of an officer or organization.
   h. Subjects involving financial or property irregularities.
   i. Serious accidents involving personnel of the command.
2. The following will be presented to the Commanding General for final action:
   a. Requests and recommendations to be made to higher authority.

b. Suggested disapprovals.
c. Communications that contain a suspicion of censure.
d. Communications that involve the good name of an officer or organization.
e. Reports of financial and property irregularities.
f. Letters to civil authorities in high positions.
g. Endorsements on efficiency reports.
h. Correspondence concerning war plans.
i. Communications of exceptional information.

3. A copy of these instructions will be kept exposed at all times upon the desk of each staff officer of this headquarters.

SELECTED STAFF PROCEDURES

The basic reference of staff officers in the understanding of command and staff relationships, responsibilities, functions, and procedures is FM 5-0, *Planning and Orders Production.* In addition, any well-organized headquarters has a staff manual or administrative procedures manual that contains the organization chart of the headquarters and details the responsibilities of each staff section.

Obtaining Background on a Staff Assignment. After studying the headquarters staff manual, the newly assigned staff officer should become familiar with all important staff actions of his or her particular section during some reasonable past period, such as the past six months or the last training cycle. Staff sections of tactical headquarters generally are obliged to keep journals wherein such staff actions are recorded. Individual staff officers maintain files of correspondence, "stay-backs," on which or together with which is a "Memorandum for Record" that summarizes the events leading up to the particular action in question, with a statement of concurrences or other information that might be useful for future reference. Normally, a unit or section SOP is also available that contains many items of useful information. From these sources, the newcomer can quickly get oriented. He or she will learn that the major problems confronting the staff section are few, although the details and ramifications may be myriad. All the problems quite likely have roots; and after they have been identified, the handling of details or new developments of an old problem will be less difficult.

The Staff Officer's Records. The commander relies on staff officers for the maintenance of required and useful charts, files, and records, as well as for the initiation of routine reports. This is an important responsibility for the staff. In the field, required records are prescribed in field manuals. In garrison, other records and correspondence must be available when needed. A staff officer should become completely familiar with the administrative requirements of his or her office, train the personnel charged with preparation of records or filing, and frequently check to ensure that they are maintained as desired.

However, the staff officer should not become "chart happy," covering office walls with useless status charts that look impressive but require major efforts by lower echelons to provide information and by staff assistants to prepare and

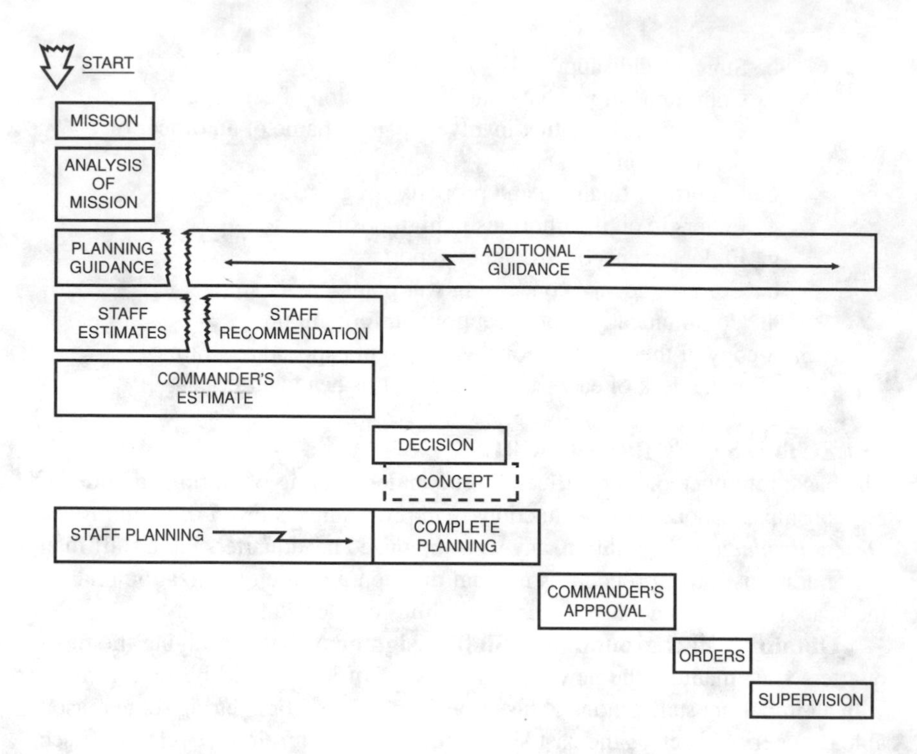

Sequence of Staff Actions

maintain. Similarly, beware of burdening subordinate units with endless reports of little consequence; this is a real temptation in the Information Age.

Mission Analysis and Staff Studies. Although the use of formalized procedures for mission analysis and preparation of written staff studies pertains more to the headquarters of major commands and the Department of the Army staff in Washington, the principles are generally applicable. In any case, the actual use of a staff study, and the analysis that leads to its preparation, depends on the complexity of the problem and the time available for its consideration. It is a good principle that formal, written staff studies should be avoided unless they are truly needed, but a staff officer at any level of organization may need to make such an analysis and write such a study.

The accompanying diagram, Sequence of Staff Actions, provides the basis for mission analysis and the staff estimate. Noteworthy is the planning guidance, which is often supplied by the commander; this guidance does not suggest the answer but may provide factors to consider, people to consult, the time to complete the study, or other matters the commander considers essential for staff understanding.

The staff study is a logical analysis that encompasses consideration of the specific problem, the facts and assumptions that bear upon the problem, information to set forth the facts, the reaching of conclusions, and the action the staff officer recommends. It should be both complete and brief. A good staff study presents a problem so the commander can reach an independent decision as well as exercise the option of accepting the recommendation submitted by the staff officer. The accompanying illustration, A Staff Study Format, should be followed, whether the study is prepared in twelve months or twelve hours, for it is time-tested and generally used. The condensed essential points should be presented on a single page, although some commanders will accept two pages. If detail is needed, attach annexes and appendices, as required, but keep the basic paper brief. Busy commanders depend on their staffs to save their time, and one way to do it is to prepare clear, concise, brief staff studies—when a staff study is needed.

Relations with Higher Headquarters. It is important for the commander to enjoy good relations with higher headquarters. His or her own standing and capacity for getting results will be enhanced when a favorable situation is created and harmed seriously when bad relations exist. Therefore, each staff officer must become personally acquainted with his or her counterpart on the next senior staff, for the same reasons that he or she must know opposite numbers in subordinate staffs. It is all a matter of acquaintance, understanding, and teamwork.

Relations Between the Staff and the Troops. The staff serves the troops as well as the commander. It has been said that the staff that serves the troops best serves its commander best. Good staff work requires the staff to know and appreciate fully the situation of the troops, their morale, their state of training, the state of their equipment and supply, and all other conditions affecting their efficiency.

As a rule, when a staff officer visits troop units, he or she first calls upon the commander. It is desirable that the staff officer state at once the purpose, if there is any in particular, of the visit. A cordial relationship between staff officers and subordinate commanders must be developed. It is desirable that before leaving, the result of the staff officer's observations be disclosed to the unit commander and then disclosed in exactly the same way to his or her commanding officer. Distrust is easily created and is difficult to dissipate. A mutual feeling of trust and confidence must be built up and confirmed by each recurring contact.

Coordination and Concurrences. Proper coordination among staff officers is absolutely essential. Without it, the commander easily could be placed in the position of issuing conflicting orders or directives.

During the course of preparation of any staff action, be sure to contact all other staff officers or staff sections that may have an interest in the problem or its solution. Obtain the views of other staff members as to the recommended solution, and take these views into account when completing your action. When your action is completed, but before presentation to your chief, take the completed package back to the interested sections to obtain their concurrence.

The purpose of this coordination is not to share responsibility with others—it is to ensure that your action has not compromised the position of other staff sections and that it is not in conflict with previous actions.

A STAFF STUDY FORMAT

Hq
Place
Date

Office Symbol Subject:

1. *PROBLEM:* Concise mission statement of what is to be accomplished by the study and/or the problem.
2. *ASSUMPTION(S):* Influencing factors that are assumed to be true. May be nonfactual or incapable of being proved.
3. *FACTS BEARING ON THE PROBLEM:* Essential facts considered in the analysis and in arriving at the conclusions.
4. *DISCUSSION:* An analysis and evaluation of the facts as influenced by the assumptions, with sufficient detail to support the conclusions.
5. *CONCLUSION(S):* Statement of the findings and the implications of an analysis and evaluation.
6. *ACTION RECOMMENDED:* Concise statement of what is to be done to solve the problem and who should do it. Tasks and milestones for their completion are often also included.

SIGNATURE BLOCK

ATTACHMENTS: (Supporting detail on each of the above sections.)

CONCURRENCES/NONCONCURRENCES:

(SECTION/AGENCY) Concur_________ Nonconcur_________

(SECTION/AGENCY) Concur_________ Nonconcur_________

(Each officer involved shows C/NonC by initials, followed by rank, name, position title, and telephone number. The reason for a NonC must be stated here or on a separate page.)

CONSIDERATION OF NONCONCURRENCES:

(The author of the staff study states the results of the consideration of any NonC. He or she then signs or initials the NonC.)

The concurrence of all interested staff sections is desirable before the completed action is submitted for the approval of the commander; however, it is not mandatory. It sometimes happens that different staff sections have conflicting viewpoints on the proper solution of a problem. In such a case, the views of the differing section should be so noted and considered in the action; however, if you are convinced that your solution is the correct one, stick to your guns. Just don't be bullheaded about it (and remember that you will have to continue working with the other section tomorrow and next week and next month). Usually the executive officer or the chief of staff will decide between conflicting points of view or recommendations. If he or she cannot, the commander must and will.

Maintaining Perspective. It is easy for an officer immersed in details at a desk to lose perspective. He or she may become desk-tied in viewpoint and fail to understand the impact that his or her recommendations, if approved, may have in the field. Some call it an occupational disease. It is a good reason for limiting the length of assignments. When it happens, the value of a staff officer shrinks swiftly. Never put aside your own appreciation of the impact of official communications that reach the field.

It is a wise course to visit subordinate units as opportunity arises. Go and see for yourself how projects of your interest are working. Talk with the officers concerned. Urge them to state their views. See if you can work out better ways to accomplish the ends sought with a higher standard, more quickly, at reduced cost. Get the feel of the field with all its enthusiasms and zeal, its hardships and human frailties. Then go back to your desk with perspective renewed. You will do a better job.

AIDES TO GENERAL OFFICERS

General officers commanding large units and other general officers occupying positions designated specifically by the Department of the Army are authorized aides, or aides-de-camp, as a personal staff. Duties of aides are not prescribed in training manuals but are such as the general officer may prescribe.

When a general officer is authorized two or more aides, the senior in rank generally is designated unofficially as the senior aide and in this capacity coordinates the activities for which he or she may be charged. This may include the issue of instructions to the other aide or aides, the secretarial staff that serves the general, orderlies, driver, and others. When a general officer uses two aides, one may be used to advantage as operations aide and one as administrative aide. Normally the maximum continuous tour of duty as an aide-de-camp is two years, although there is no limitation on length of tour.

Duties of Aides. As stated earlier in this chapter, there is great need for a personal staff to free the general from time-consuming personal arrangements so that he or she may perform heavy responsibilities with efficiency and continuity. A good aide sees to it that the tools and facilities needed by the general are on hand. Examples of duties that a general officer may wish an aide to perform are discussed below. The establishment of a new command post may involve an aide. The aide learns from the chief of staff the location of the area to be used by the general. Measures are taken at once to move all equipment and personnel in the routine use or service of the general to the new site and resume operations.

While the general is in the office or at the command post, an aide often is assigned to keep the general's appointment schedule to facilitate the transaction of business. In the performance of this task, the aide should consult the chief of staff and be guided always by his or her desires, for the aide is in no sense a link in the chain of command. It is likely that all generals expect their aides to maintain a list of future appointments, including social engagements, and remind them in ample time so that they may be kept. The general's spouse should be kept informed of the general's social engagements, especially those to which the spouse is invited, as well as those that require the general's presence outside of office hours.

Many generals wish their aides to assume charge of their personal and confidential files.

On change-of-station or other moves, the aide should take full charge of all arrangements as directed by the general so the general may work with full efficiency up to the time of departure and resume operations at once upon arrival.

Careful thought should be given by aides to arrangements for trips of whatever length or duration. The aide should make all arrangements for tickets, hotel reservations, movement of baggage, and all other matters. If appropriate, notification of expected time of arrival, security clearance (if required), and the number of people in the general's party should be sent to the proper officials.

The aide should keep a record of expenses incurred that are reimbursable by submission of the necessary voucher after completion of the journey. Many general officers provide the aide with a personal fund from which to make expenditures for the general for minor items, keeping an informal record as they are paid.

Welcoming committees or individuals meeting or receiving a general officer for some official or unofficial occasion may assume expenses on behalf of the visiting general from their own resources. This is a courtesy, of course, but general officers wish to avoid accepting the payment of expenses that are covered by their official travel allowances or that should be paid from their own funds. In such instances, the aide should ascertain the expenses and provide reimbursement with expression of appreciation for the thoughtfulness and the courtesy.

Aides are often used to transmit highly personal, important, or secret messages.

It is common practice for general officers to assign aides as assistants to staff officers, particularly in general staff sections of the headquarters. This is particularly the case when the general is to be absent and unaccompanied by aides, or during times of stress when additional staff assistants are needed in the staff section. Accordingly, a good aide will become schooled in the duties and techniques of each staff section to make himself or herself a useful member of the group, even if the assignment is temporary. While so assigned, the aide is responsible to the chief of section in the same manner as other officers.

Visiting dignitaries may be accompanied by an officer to serve them temporarily as an aide. Senior officers may be detailed to serve in this capacity for high officials, such as very senior officials of a friendly foreign power.

On social occasions, aides may assist in preparing the invitation list, or they may make the invitations orally for their general. When this is the case, they must be careful to provide the date, hour, place, nature of the event, and other appropriate information. If spouses are to be present, the aides will save much time by including information about dress, and whether the general's spouse will wear formal or informal dress. The nature of the occasion must be made clear, too; if it is to be a dinner, or a cocktail gathering that does not include a dinner, the invitation should not be susceptible of misunderstanding. For a social event being held at a club or hotel, the aide may be charged with making all arrangements after being instructed by the general. At a formal reception the aide is often asked to introduce the guests; in this case, the aide's position is near the head of the receiving line, where he or she greets all guests, ascertains their names, and introduces them clearly to the individual heading the receiving line. This action and position may cause the aide some concern until familiar with the experience. The aide too is a guest. Certainly he or she must not behave like a butler or servant, nor, for that matter, the host. Perhaps it is best described as the position of an elder son or daughter assisting parents in the pleasant task of entertaining friends.

For ceremonial occasions, aides must familiarize themselves with the appropriate regulations and the exact arrangements insofar as they can be foreseen.

Some aides prepare an SOP and a checklist for recurring activities, such as trips by automobile, air, and rail; movement of the command post or headquarters; and other events.

A general officer is obliged to meet a large number of people, military and civilian. Aides may render a most valuable service to their commander by maintaining a list of names of individuals with whom more than passing contact has been held or may be held. This file should contain, in addition to names, essential data about each individual, such as position or business connection, former military status if any, address, telephone number, current activities of military interest, and a statement of occasions when the general has been in touch with the individual. Before any occasional visitor, military or civilian, is permitted to fill an appointment with the general, the aide should brief the general on the visitor, including who the visitor is and what he or she wants, if the latter is known. When the general is to attend any large gathering, it will be helpful for the aide to ascertain the more important personages who will be present and inform the general in advance.

Care must be exercised by aides to avoid transgressing upon the fields of other staff officers. In this connection, many general officers require that their secretaries, chauffeurs, and orderlies work under personal control of the aides. When a pilot is provided, the aide and the pilot must work closely together, the pilot generally receiving all information about proposed trips from the aide.

General officers expect their aides to be models of military courtesy, tact, military appearance, and soldierly attitude and bearing. They also wish them to be unobtrusive and quiet, as well as ladies or gentlemen. Aides are cautioned that they are not commanders or assistant commanders. It is an honored, responsible position. Most general officers choose their aides based on their fine records, experience in extended combat, and decorations for valor, as well as for their personality and appearance. An officer chosen as aide to an Army general has an opportunity to acquire experience that will be extremely helpful as his or her responsibilities increase.

For lapel and other insignia worn by aides, see chapter 22, Uniforms and Appearance.

15

Warrant Officer Corps

Warrant officers are officers appointed by warrant by the Secretary of the Army, based on a sound level of technical and tactical competence. The warrant officer is a highly specialized expert and trainer who, by gaining progressive levels of expertise and leadership, operates, maintains, administers, and manages the Army's equipment, support activities, or technical systems.

The Army Warrant Officer Corps comprises over 20,000 men and women of the active Army and reserve components. Warrant officers are technical experts who manage and maintain the Army's increasingly complex, high-technology battlefield systems. They enhance the Army's ability to perform its missions. Warrant officers remain single-specialty officers whose career tracks are oriented toward progressing within their career fields, rather than focusing on increased levels of command and staff responsibilities.

Warrant officers are commissioned by the President with the consent of Congress and have the same legal status as their traditional commissioned-officer counterparts. Candidates who successfully complete Warrant Officer Candidate School are appointed in the grade of warrant officer one (WO1). Normally, after twenty-four months, they are promoted to chief warrant officer (CW2).

Competitive promotion to chief warrant officer three (CW3), chief warrant officer four (CW4), and chief warrant officer five (CW5) occurs at approximately six-year intervals thereafter. Updated information about the warrant officer corps may be found at the warrant officer website: www.hrc.army.mil.

HISTORY OF THE WARRANT OFFICER CORPS

The warrant officer designation has long been recognized by various navies of the world. In naval organizations, the warrant officer traditionally is a technical specialist whose skills and knowledge are essential for the proper operation of ships. The warrant officer grade in one form or another has been in continuous use in the U.S. Navy since that service was established. In the U.S. Army, the warrant officer lineage can be traced back to the headquarters clerks of 1896, later designated

Army field clerks. However, the recognized birthdate of the Army's Warrant Officer Corps is 9 July 1918.

On that date, an act of Congress established the Army Mine Planter Service as part of the Coast Artillery Corps and appointed warrant officers to serve as masters, mates, chief engineers, and assistant engineers of each vessel. An act of 1920 expanded the use of warrant officers, authorizing their appointment in clerical, administrative, and band-leading activities. In effect, the act of 1920 designated the warrant officer grade as a reward for enlisted personnel of long service and as a haven for former commissioned officers of World War I who lacked either the education or other eligibility requirements to retain their commissions after that war.

Between 1922 and 1935, no warrant officer appointments were made except for a few band leaders and Army Mine Planter Service personnel. In 1936, competitive examinations were held to replenish lists of eligible personnel, and some appointments began being made again. Warrant officers who were qualified pilots were declared eligible for appointment as lieutenants in the Army Air Corps in 1939. By 1940, warrant officer appointments began to occur in significant numbers for the first time since 1922, although the total strength of the Warrant Officer Corps decreased until 1942 because of the large numbers of warrant officers who were transferred to commissioned status during that period.

The second truly important piece of legislation affecting Army warrant officers was passed in 1941. An act of August 1941, amplified by an executive order in November of that year, provided that warrant officers could be assigned duties as prescribed by the Secretary of the Army and that when such duties necessarily included those normally performed by a commissioned officer, the warrant officer would be vested with all the powers usually exercised by commissioned officers in the performance of such duties. The act of 1941 also established two warrant officer grades, chief warrant officer and warrant officer junior grade, and authorized flight pay for those whose duties involved aerial flight.

Warrant officer appointments were made by major commanders during World War II, and warrant officers served in some forty occupational areas during that war. In January 1944, the appointment of women as warrant officers was authorized, and by the end of the war there were forty-two female warrant officers on active duty.

After World War II, the concept of using the warrant rank as an incentive rather than a reward was instituted. It was to be a capstone rank into which enlisted personnel could advance. This use of the warrant officer grade, combined with the earlier concept of using the grade as a reward for long and faithful service, resulted in mixed utilization so that, in practice, warrant officers became largely interchangeable with junior commissioned officers or senior enlisted personnel.

The Career Compensation Act of 1949 provided two new pay rates for warrant officers. The designations of warrant officer junior grade and chief warrant officer were retained, but the grade of chief warrant officer was provided with pay rates W-2, W-3, and W-4. In the Warrant Officer Personnel Act of 1954, these three pay rates became grades, and the warrant officer junior grade became warrant officer.

Warrant officers were used extensively during the Korean War, but by 1953 it had become apparent that use of the warrant officer grade as either a reward or an incentive was inadequate. Needed as a basis for continuation of the Warrant Officer Corps was a new concept consistent with functional Army requirements. From 1953 until 1957, the Department of the Army conducted an analysis to determine whether the warrant officer program should be continued and, if so, in what form and for what purpose.

In January 1957, as a result of the Department of the Army study, a new warrant officer concept was announced that affirmed the need for the warrant officer and the continuation of the Warrant Officer Corps. It stipulated that the warrant officer grade would not be considered as either a reward or an incentive for enlisted men or former commissioned officers, and it defined a warrant officer as "a highly skilled technician who is provided to fill those positions above the enlisted level that are too specialized in scope to permit the effective development and continued utilization of broadly trained, branch-qualified, commissioned officers."

In 1966, as part of a continuing effort to improve operation of the new concept, a study group was formed at DA with a mission to develop a formal warrant officer career program that would be responsive to future Army requirements while offering enough career opportunities to attract high-quality personnel. The study group examined all aspects of the Warrant Officer Corps and made a number of recommendations in areas such as pay, promotion, utilization, and education. Provisions for below-the-zone selection for promotion to grades CW3 and CW4 were implemented in 1967. The Regular Army program was reopened to warrant officer applicants in 1968 after having been closed for twenty years, and subsequent changes reduced service eligibility criteria and simplified application procedures. Since 1968, the military education available to warrant officers has been expanded. Before then, there was no formal progressive military schooling program for warrant officers. By the end of 1972, a trilevel education system had been established that provided formal training at the basic or entry level for warrant officers in fifty-nine occupational specialties, at the intermediate or mid-career level for fifty-three specialties, and at the advanced level for twenty-seven specialties. In 1973, the three levels of training were redesignated from "basic," "intermediate," and "advanced" to "entry," "advanced," and "senior." Simultaneously, as the result of successful testing of the concept, the Warrant Officer Senior Course (WOSC) was established to provide all warrant officers with access to the highest level of professional education.

In 1973, DA began to implement a plan to close the gaps in the warrant officer military education system by directing the expansion or modification of existing advanced (formerly intermediate) courses to accommodate all warrant officer specialties. Civil schooling opportunities were also increased during this period. The educational goal for warrant officers was upgraded from two-year college equivalency to attainment of an associate's degree, and warrant officers, for the first time, were authorized entry into fully funded civil school programs. As a means of aiding progression toward goal achievement, cooperative degree programs were established in colleges and universities near the installations conducting the warrant officer career courses. These programs were implemented to provide students in the military establishment the opportunity to complete requirements for MOS-related associate's degrees while in attendance at their career courses.

In consonance with increased educational opportunities, duty positions requiring warrant officers with bachelor's or master's degrees were validated for the first time by the Army Education Requirements Board (AERB), now the Army Education Requirements System (AERS). By the close of 1975, the Army's capability for professionally developing the Warrant Officer Corps had been significantly expanded, and warrant officers in the modern program were being offered developmental opportunities that their predecessors never had.

To satisfy the recognized need for qualified, highly trained individuals available to expand the active Warrant Officer Corps rapidly in time of emergency and to meet other Army requirements, reserve component warrant officers not on active duty and National Guard warrant officers are integrated into the Army's professional development program. Reserve component warrant officers are provided a balanced mix of training, experience, and career opportunities through periodic rotation at predetermined points in their careers between the Selected Reserve Troop Program Units (TPUs) and the Individual Ready Reserve (IRR), attendance at requisite military schools, and short periods of counterpart training with active component organizations.

The years following the inception of the Warrant Officer Career Program saw increasing warrant officer participation in the development of policies and programs. Increasingly, the warrant officer viewpoint was sought in the development of plans that had an effect on the Warrant Officer Corps. Since 1973, warrant officers have been authorized as voting members on various HQDA selection boards that consider warrant officers. Warrant officer positions were established in the Offices of the Deputy Chiefs of Staff for Personnel (G-1), Operations and Plans (G-3), and Logistics (G-4) to provide warrant officer perspective at the highest level of the Army.

In 1985, the Chief of Staff, Army, chartered the Total Warrant Officer Study Group to determine the future use, management, and professional development that would enable the Warrant Officer Corps to enhance the future combat readiness of the Army. The result of that effort was the establishment of the Total Warrant Officer System (TWOS) and a clear definition of the role of the warrant officer. In 1987, Congress approved the commissioning of warrant officers from the rank of CW2 through CW5. The warrant officer is expected to follow the same professional development guidelines as his or her commissioned officer counterpart.

The culmination of TWOS was the passage of the Warrant Officer Management Act (WOMA) as part of the FY 1992 and 1993 National Defense Authorization Act and approval of the Warrant Officer Leader Development Action Plan (WOLDAP) in 1992. On 5 December 1991, WOMA went into effect. WOMA is a major revision of Title 10, USC, and is the current basis for management of the active duty Warrant Officer Corps. Key provisions of the law include a single promotion system for Warrant Officers, tenure requirements based upon years of Warrant Officer service, the grade of CW5, and authorization for the Secretary of the Army to convene boards to recommend Warrant Officers for selective mandatory retirement.

WOLDAP was approved by the Chief of Staff, U.S. Army, on 27 February 1992. WOLDAP is a total Army plan designed to ensure both active and reserve Warrant Officers are appointed, trained, and utilized to a single standard. Key provisions of WOLDAP include an accession goal of eight years or less time in service for Warrant Officer Candidates, the establishment of a comprehensive Warrant Officer Education system, conditional appointment to WO1 upon successful completion of Warrant Officer candidate school, civilian education goals of an associate degree before eligibility for promotion to CW3 and a bachelors degree before eligibility for promotion to CW4, and the establishment of the Warrant Officer Career Center (WOCC).

The Army Training and Leader Development Panel (ATLDP) released on 22 August 2022, focused on training and leader development requirements for Warrant Officers as the Army transforms to the Future Force. This study (Phase III), the third conducted by the ATLDP, is part of the largest self-assessment ever done by the Army. The Warrant Officer study concludes the Army must make fundamental changes in the Warrant Officer cohort to support full spectrum operations. At the heart of the change is a complete integration of Warrant Officers into the larger officer corps, a process begun in the 1980s but never completed. Specifically, the study concludes that the Army needs to clarify the roles of Warrant Officers, then make changes to their professional development, training and education, and manning. The study recommended sixty-three changes to improve the training, manning and professional development of Warrant Officers. If fully implemented, these recommendations represent the most significant changes within the Army Warrant Corps since its inception in July of 1918.

On 9 July 2004, the eighty-sixth anniversary of the Army Warrant Officer Corps, all Warrant Officers began to wear the insignia appropriate to their branch. They no longer wear the Eagle Rising. The rank insignia of CW5 changed to one similar to other services, a silver bar with a single black band in the center. WO Division and Officer Division merged, and there was a consolidation of OES and WOES.

TOTAL WARRANT OFFICER SYSTEM

TWOS is the requirements-based, life cycle management system for warrant officers. It spans the period from initial recruitment to departure from Army service.

There are three distinct stages related to the schooling and experience of the warrant officer.

Warrant Officer (WO1/CW2). This stage extends from initial appointment to eight years of warrant officer service. Upon completion of Warrant Officer Career Candidate School (WOCCS), officers receive conditional appointments as WO1s. They attend the Warrant Officer Basic Course (WOBC) for technical and tactical training. Upon completion of WOBC, they receive their warrant officer MOS. An officer is commissioned and promoted to the rank of CW2 normally after twenty-four months of honorable and satisfactory service as WO1. A warrant officer's initial assignment should provide operational experience within the technical standards of his or her MOS.

After promotion to CW2, a mandatory nonresident Action Officer Development Course must be completed prior to attending Warrant Officer Advanced Common Core (DL) (WOACC-DL). It can be completed online via the Internet. It provides warrant officers with training in management techniques, communication skills, preparing and staffing documents, meeting and interview techniques, problem solving, writing, coordinating, briefings, and ethics. The course must be completed within one year of enrolment; however, CW2s have the flexibility to enroll at any time between twenty-four and forty-eight months of warrant officer service.

Senior Warrant Officer (CW3/CW4). This stage extends from about eight to nineteen years of warrant officer service. It includes attendance at Warrant Officer Intermediate Level Education (WOILE). The warrant officer serves in positions at battalion or higher, with increasing demands on his or her technical skills and leadership.

Master Warrant Officer (CW5). This stage extends from about nineteen to thirty years of warrant officer service. It includes attendance at the Warrant Officer Senior Service Educations (WOSSE) and assignments as master warrant officers.

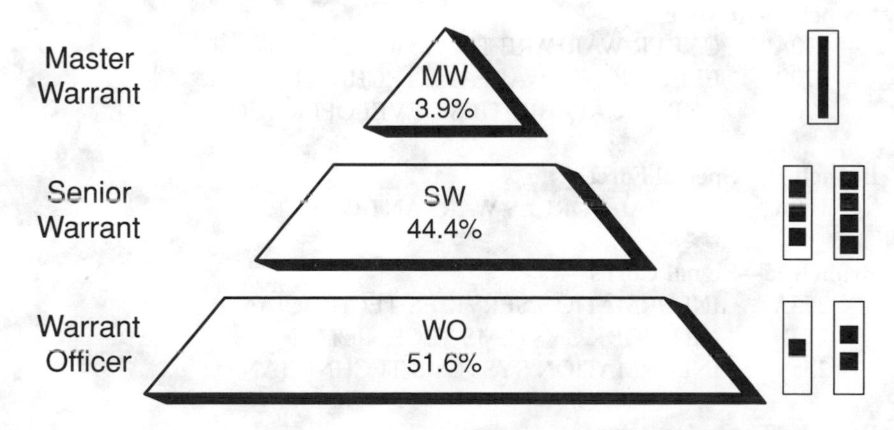

Warrant Officer Requirements by Rank Group

SERVICE AS A WARRANT OFFICER

Warrant officers serve at all levels of the Army and the Department of Defense, in an especially wide variety of positions. There are often multiple areas of concentration within each branch, and sometimes several military occupational specialties within each area. Here is a list of branches and their areas of concentration as of the time of printing.

Branch 12—Engineer

120A CONSTRUCTION ENGINEER TECHNICIAN
125D GEOSPATIAL ENGINEERING TECHNICIAN

Branch 13—Field Artillery

131A FIELD ARTILLERY TECHNICIAN

Branch 14—Air Defense Artillery

140A AIR AND MISSILE DEFENSE (AMD) SYSTEMS INTEGRATOR
140K AIR AND MISSILE DEFENSE (AMD) SYSTEMS TACTICIAN
140L AIR AND MISSILE DEFENSE (AMD) SYSTEMS TECHNICIAN

Branch 15—Aviation

150A AIR TRAFFIC CONTROL TECHNICIAN (RC)
150U TACTICAL UNMANNED AERIAL SYSTEMS (TUAS) OPERATIONS TECHNICIAN
151A AVIATION MAINTENANCE TECHNICIAN
153A ROTARY WING AVIATOR

Branch 17—Cyber

170A CYBER WARFARE TECHNICIAN
170B ELECTRONIC WARFARE TECHNICIAN
170D CYBER CAPABILITIES DEVELOPER TECHNICIAN

Branch 18—Special Forces

180A SPECIAL FORCES WARRANT OFFICER

Branch 25—Signal Corps

255A INFORMATION SERVICES TECHNICIAN
255N NETWORK SYSTEMS TECHNICIAN
255S INFORMATION SYSTEMS TECHNICIAN

Branch 27— Judge Advocate General's Corps

270A LEGAL ADMINISTRATOR

Branch 31—Military Police

311A CID SPECIAL AGENT

Branch 35—Military Intelligence

350F ALL SOURCE INTELLIGENCE TECHNICIAN
350G GEOINT IMAGERY TECHNICIAN
351L COUNTERINTELLIGENCE TECHNICIAN
351M HUMAN INTELLIGENCE COLLECTION TECHNICIAN
351Z ATTACHE INTELLIGENCE OPERATIONS TECHNICIAN
352N SIGINT ANALYSIS TECHNICIAN
352S SIGNALS COLLECTION TECHNICIAN
353T MILITARY INTELLIGENCE SYSTEMS MANAGEMENT/INTEGRATION TECHNICIAN

Branch 42—Adjutant General's Corps

420A HUMAN RESOURCES TECHNICIAN
420C BANDMASTER

Branch 64—Veterinary Corps

640A VETERINARY SERVICES FOOD SAFETY OFFICER

Branch 67—Medical Services Corps

670A HEALTH SERVICES MAINTENANCE TECHNICIAN

Branch 74—Chemical Corps

740A CHEMICAL, BIOLOGICAL, NUCLEAR AND RADIOLOGICAL (CBRN) TECHNICIAN

Branch 88—Transportation Corps

880A MARINE DECK OFFICER
881A MARINE ENGINEERING OFFICER
882A MOBILITY OFFICER

Branch 91—Ordnance

890A AMMUNITION WARRANT OFFICER
913A ARMAMENT SYSTEMS MAINTENANCE WARRANT OFFICER
914A ALLIED TRADES WARRANT OFFICER
915A AUTOMOTIVE MAINTENANCE WARRANT OFFICER
919A ENGINEER EQUIPMENT MAINTENANCE WARRANT OFFICER

Branch 92—Quartermaster Corps

920A	PROPERTY ACCOUNTING TECHNICIAN
920B	SUPPLY SYSTEMS TECHNICIAN
921A	AIRDROP SYSTEMS TECHNICIAN
922A	FOOD SERVICE TECHNICIAN
923A	PETROLEUM SYSTEMS TECHNICIAN
948B	ELECTRONIC SYSTEMS MAINTENANCE WARRANT OFFICER
948D	ELECTRONIC MISSILE SYSTEMS MAINTENANCE WARRANT OFFICER

16

Additional Duty Guide

The additional duty is the Army's way of assigning mission, administrative, housekeeping, and personnel-related responsibilities outside of primary military occupations. This chapter brings together the practical and detailed advice necessary to get started and develop a program to meet the needs of a particular assignment. It doesn't, of course, tell exactly how to do any particular job, rather it is a general guide.

A few of the mission-related duties, such as motor officer, food service officer, supply officer, training officer, and so on, require more time and effort than others. Many additional duties are assigned at battalion or higher level. Some duties at company level are assigned to noncommissioned officers.

The twenty-one jobs described herein were selected as the hard core of a seemingly unlimited number of additional duties. They are typical of those additional duties to be found, with minor variations, at almost every organizational level. What is said about each of them focuses somewhat on small-unit operations and shares advice about the most practical ways of carrying out the instructions, the intent, and the programs or procedures prescribed in the host of related official directives.

Early in any additional duty assignment you should get acquainted with the officially published material about that particular job. Unit reference libraries should have a copy of the most often used references. But how do you know they are complete and up-to-date? DA PAM 25-30 tells the story. The references listed for each of the additional duties described in this chapter were extracted from this pamphlet.

DA PAM 25-30, *Consolidated Index of Army Publications and Blank Forms*, is published in CD form or available online at http://armypubs.army.mil/2530.html. It is updated quarterly. This pamphlet contains a complete listing of all the Army's administrative publications (regulations, circulars, pamphlets, and so on); blank forms; doctrinal, training, and organization publications; supply catalogs and manuals; technical manuals and bulletins; lubrication orders; and modification work orders. In short, DA PAM 25-30 provides an index of all the information available about what has to be done in the Army, as well as how to do it. Sometimes reference copies of publications are kept in more than one library. The appropriate ones may be in the training office, maintenance shop, supply room, mess hall, or elsewhere. Be sure to check this possibility before concluding that a publication isn't available. If the unit library doesn't have the needed references, don't hesitate to order them; DA PAM 25-30 should be your guide.

If the unit library doesn't have the references you need, don't hesitate to order them using DA PAM 25-30 as your guide. More and more, the Army is accessing references online at www.usapa.army.mil. This is a fast and easy way to retrieve the information you need, and it saves the Army money.

The material in this chapter originally was prepared by Theodore J. Crackel and published in book form as *The Army Additional Duty Guide.* It has been updated as necessary and the format has been changed for inclusion in *Army Officer's Guide*, but it retains the flavor of the author and is incorporated here with his permission. The additional duties discussed are:

Ammunition officer
Army Emergency Relief officer
CBRNE officer
Claims officer
Class A agent (pay) officer
Courts-martial member
Duty officer
Food service officer
Income tax officer
Line-of-duty (LOD) investigating officer
Motor officer/maintenance officer
Postal officer
Range safety officer
Records management officer
Reenlistment officer
Safety officer
Savings officer
Supply officer
Training officer
Unit fund council, president/member
Voting officer

AMMUNITION OFFICER

Like most jobs, the effort and time that duty as ammunition officer takes depend on the type of unit to which one is assigned. Combat service support units may have little more than small-arms ammunition, which may be stored in the arms room. A combat unit may have a quantity and variety of ammunition that at first makes the job seem almost impossible. Regardless of the type of unit, the ammo officer's responsibilities (and problems) fall into three areas: storage, maintenance, and accounting.

No matter where or what type the ammo is, it needs protection. Proper storage eliminates most ammo trouble. First of all, be sure that any storage area is well policed of flammable materials of any kind. Pile ammo in neat stacks, separated by type, caliber, and lot number. Within the limitations imposed by the size of the storage area, locate the stacks far enough apart so that if one blows, others won't immediately go up too. In any case, keep the ammo off the ground. Have wooden strips laid between the tiers of cases to keep air circulating around the boxes to help keep them dry. Cover the stacks with a tarp or waterproof cover, but be sure to let the air circulate freely. Otherwise, moisture will be trapped and eventually cause damage. If there is no covering material available, stack the boxes in such a way that water will drain off. Repair and restencil damaged boxes immediately. (Unidentifiable ammo is automatically classified Grade 3 and is not fired.)

White phosphorus (WP or PWP) rounds need special attention. Segregate them to a clear area, and store them with projectiles nose-up (except 3.5-inch rockets, which should be stored nose-down). The white phosphorus filler can soften or melt in hot weather and could dislocate within the projectile if not properly stored. Segregate rockets of all types and point them all one way—either nose-down or toward a revetment or barricade.

When storing ammo in vehicles, check the TM and SOPs for proper storage locations and procedures. Be sure that the primers of large-caliber rounds are properly protected both in handling and in storage. Break the seal on small-arms ammo boxes only when absolutely necessary. Once the seal has been broken, however, remember to inspect and clean it periodically.

In many units, a portion of the basic ammo load is carried in trucks or trailers that belong to the ammo section of the support platoon. Local SOP establishes the degree of responsibility that you as the ammo officer will have for these vehicles, but from a practical—and tactical—point of view, you'll be very interested, since you depend on them to get needed ammo forward.

Check preventive maintenance (PM) on the truck. Even if ammo has been well cared for, it won't be of any use if the truck carrying it breaks down. This may take tact on your part, since vehicle maintenance is usually someone else's responsibility.

Here are some things to help keep this vehicle-loaded ammo in good shape:

- Place it on wooden floor racks and ensure that all drain plugs in the bed of the vehicle are open.
- Distribute the weight evenly over the entire bed of the vehicle.
- Brace the load to prevent shifting during movement.
- Place ridgepoles under the tarpaulin to prevent water pockets.
- Raise the tarpaulin periodically to air the load.

If part of the basic load must be stacked off the vehicle, stack it by vehicle load so that it won't have to be sorted when it's needed.

Ammunition stored in bunkers or other inside areas should also be on dunnage with stripping between layers. As usual, stack it by lot number with its nomenclature markings facing outward and readable. To allow for circulation, stack it at least 6 inches from walls and 18 inches from the ceiling.

There are special problems with ammo stored in tanks, APCs, and other combat vehicles. Keeping it dry is one. Open the ramp or hatches whenever possible and keep the air circulating. If ammo does get wet, wipe it off or, if possible, lay it out and let the air dry it. Any ammo that's unpacked will need a lot of care. Take care of the dirt on small-arms ammo by wiping it off with a clean rag. Tank and artillery ammo is a wholly different problem, but it, too, has to be cleaned and, if it's painted, touched up frequently. Like most things, there's a right and wrong way to go about this. The vehicle or weapon TM will give the lowdown on that particular ammo and outline any special problems.

Cartridges are usually either uncoated brass or steel with a varnish coat. Projectiles are generally either covered with enamel or a lacquer paint. Here are some general cleaning tips.

Brass Cartridges. Use a clean rag to wipe off the dirt, and use copper wool (not steel) to clean off corrosion. When it's clean, wipe it off with a rag dampened with solvent and let it dry.

Coated Cartridge Cases. Use steel wool to get the paint and foreign matter off corroded or rusty spots on cartridges and projectiles. Don't use steel wool on the rotating bands or fuses—for these *use only copper wool.* Clean the coated cartridge with solvent. *Caution:* Don't soak the cartridge in solvent. Unless one does a thorough job, it will have to be done over again in a short time. Finally, touch up the spots that are bare of varnish with the special epoxy varnish authorized for the job. (Never use any other type.) If the unit doesn't have or can't get this varnish, turn the ammo in to someone who can. It's important that all the exposed surfaces get proper protection from the atmosphere.

Projectiles. Clean corroded spots with steel wool, except the fuses. After that, use thinner to wipe it down. Again, be sure that it's wiped thoroughly. Touch up bare spots with the correct enamel to match the area concerned. Don't get the paint on too thick. If stenciled markings are painted over, they should be restored immediately.

While the ammo is being cleaned, you will have an excellent opportunity to inspect and inventory it and to check it for dents, bulges, or scratches. If a round seems badly damaged, turn it in to the ammunition support facility and let the experts make the final decision as to whether it is safe to fire. Not all dented or scratched rounds are bad. If dents and scratches are the only problems and if the rounds will seat properly in weapons, they are OK to fire. Try to wobble the projectile to see if it is loose. Check the eyebolt lifting plugs of separate-loading projectiles to see that they are not cracked. While inspecting small-arms ammo, look for short rounds (where the bullet has been pushed back into the case); loose bullets or long rounds; dents, burrs, and cracks in cases; and corrosion or dirt on the cartridge. On finding a bad round, turn it in. If you feel that you need technical assistance in matters of serviceability, routine maintenance, or supply, you should contact your ammunition support unit for a liaison or technical assistance team.

There are two aspects of ammunition supply—training and combat. Combat resupply is requested from the S-4 in accordance with the unit SOP and is moved forward on battalion ammunition trucks. Generally, the smaller-unit ammunition officer isn't involved at all. With training ammunition, however, it's a different story. Then the ammo officer is responsible for requesting it. Don't allow surplus ammunition to accumulate.

Requirements should be realistically determined, and requisitions should be only for that needed—not necessarily for the total quantities authorized. After training is finished, turn in the excess ammunition and the spent cartridge cases. Inspect small-arms brass thoroughly to ensure that no live rounds are included. Check with the S-4 or the ammunition supply point for the turn-in procedure.

You must work closely with the training officer to be sure that you have adequate ammunition for scheduled training. In addition, you can assist others by procuring special training ammunition or demolitions for instructional use.

Ammunition malfunctions except nuclear, biological, and chemical are reported in accordance with AR 75-1. You must make sure that both you and your personnel are familiar with the information in this regulation. Accidents or incidents involving nuclear, biological, and chemical ammunition are reported in accordance with AR 385-40.

Safety is always a concern around ammunition. All types merit special attention and careful handling. The FM and TM for each weapon spell out precautions. Conduct training in the safe handling and maintenance of ammunition often enough to instruct new personnel in the unit and to refresh the old hands.

The storage and maintenance of special munitions are spelled out in appropriate TMs and local regulations or SOPs. Study these carefully. Also seek out the expertise and advice of the Battalion Supply Officer (S-4).

References

AR 385-63, *Range Safety*
AR 385-64, *Ammunition and Explosives Safety Program*
AR 710-2, *Supply Policy Below the National Level*
PAM 700-16, *Ammunition Management*
AR 700-28, *Ammunition Management*
FM 4-30, *Ordnance Operations*

ARMY EMERGENCY RELIEF OFFICER

Army Emergency Relief (AER), a private nonprofit organization, came into being in 1942 to provide assistance to members of the rapidly expanding Army and their dependents who were faced with financial problems with which they were unable to cope and for which no appropriated funds were available. The Army Relief Society (ARS), now assimilated into AER, was established in 1900 to assist needy widows and orphans of Regular Army personnel. One of the big jobs of the AER officer is to see to it that all members of the command and their dependents know that they may receive financial assistance when emergencies arise that are beyond their ability to handle. Generally, AER officers are appointed only at installations with AER sections, which are Army-wide.

As the AER officer, you must familiarize yourself with the policies and procedures governing emergency financial assistance contained in appropriate regulations. You should confer with local American Red Cross representatives to understand the operating relationships of these two organizations.

As the AER officer, you are responsible for receiving, safeguarding, disbursing, and accounting for all the funds. You also prepare and maintain the required financial records and reports. You interview applicants for assistance, make necessary investigations, furnish counsel, and help them as necessary in accordance with AR 930-4. You may also make loan collections, although these are usually accomplished through AER allotments. AR 930-4 contains the details of section operation, including financial, accounting, and reporting requirements.

All members of the Army, active and retired, and their dependents, including spouses and orphans of deceased Army members, are eligible to receive emergency financial assistance from AER. Assistance is generally extended in the form of a loan (without interest), since Army people usually want and are able to repay. Repayment (by AER allotment) is usually in monthly installments so as not to cause hardship. Occasionally, a combination of a loan and grant or an outright grant may be the best solution. Assistance to dependents of deceased personnel is almost invariably a grant. AER also provides educational assistance, on a limited basis as a secondary mission, for dependent children of Army members pursuing undergraduate studies.

Government funds are not appropriated to provide emergency financial assistance to military personnel and their dependents. AER therefore relies on voluntary contributions from members of the Army family, repayments of loans, and income from investments to finance current operations. AER makes no appeal for funds outside the Army. Unsolicited gifts and legacies are accepted.

The Army conducts an annual fund campaign for AER to raise funds to sustain AER's capability to meet the emergency financial needs of Army members and their dependents.

Campaign material is furnished by National Headquarters, AER. As an overall guide, a dollar goal is established by installation, but neither units nor individuals should be assigned a dollar quota (see AR 1-10).

AER is organized to be an important and effective instrument of morale and exists only to help the Army take care of its own. The governing policies are intentionally broad to allow flexible utilization of AER by local commanders and the chain of command. As the AER officer, you will be most effective if you are aware of the view and policies of your commander and if you keep him or her regularly informed of the type and amount of assistance being rendered. It is also recommended that you seek out the expertise and advice of the Battalion Personnel Officer (S-1).

References

AR 1-10, *Fund Raising within Department of the Army*
AR 930-4, *Army Emergency Relief*

CBRNE OFFICER

As an additional duty, this assignment presents a critical challenge. It is this officer's responsibility to become proficient in ways to survive, exist, and function after a chemical, biological, radiological, nuclear-explosives (CBRNE) attack—and train the officer's unit to do the same.

First, it should be relatively easy for both the CBRNE officer and NCO to receive training at one of the Army's CBRNE schools. Find out from the unit training section what's available and apply right away. CBRNE training is a recurring subject, and repeated classroom hours are a drudgery for instructor and instructed alike. Practical hands-on training is without a doubt the best way, but it takes more planning, more material, and more instructors. Nevertheless, it's well worth the effort. The CBRNE officer must *plan ahead.* Plan for the time you'll need to prepare the instruction, get the supplies and training aids, and rehearse with assistant instructors. There are a number of training aids available for making training more realistic.

Proficiency tests and exercises are among the most effective training vehicles. Integrate CBRNE training whenever possible with unit tactical training. After all, the object is to train the unit to continue its mission in a CBRNE environment. The necessity of continuing the mission applies to all units, combat and support. Through experience and knowledge, you can devise other training situations to meet your unit needs.

You should seek professional advice and assistance from the staff chemical officer at brigade, division, or other higher headquarters. He or she will advise in planning unit exercises and can give technical assistance when necessary.

One of the most vital aspects of CBRNE defense is the accurate and timely reporting of enemy attacks. You will have to use your imagination and experience to devise effective training media in this area—it has to be done. Merely writing the data on a slip of paper and handing it to a soldier to be put into the proper format and transmitted doesn't accomplish much.

Once you have devised an effective training aid or system, you shouldn't keep it a secret. If you have set up a training course and trained assistant instructors to operate it, you can use it to train 500 people almost as easily as 50 or 100. You should expect to share your facilities with other units in the area. It's a two-way street and can be advantageous to all. Many units find it profitable (even necessary) to pool people and talents for CBRNE training.

The CBRNE officer's unit is required to have both survey and monitoring teams whose members are specially trained. You will have to set up these teams and train them. You will have to learn about area damage control. In rear areas, this is directed primarily toward minimizing the impairment of combat service support after mass destruction or mass casualty attack. In forward areas, this is directed toward minimizing interference with tactical operations and loss of combat power. The mission and organization of the particular unit determine to a large degree the mission of these teams, but in any case, their makeup and procedures should be covered in the unit SOP. Since these teams are sent into hazard areas to assist individuals and units that have been subjected to a CBRNE attack, they need additional training on first aid, evacuation, and decontamination techniques. Special effort here could pay large dividends.

In addition to Department of the Army CBRNE literature, almost every level of command publishes regulations, directives, pamphlets, or SOPs on the subject. These should be studied for pertinent information about local problems and special requirements.

Remember that, as the CBRNE officer, your effort should be aimed at keeping your unit operationally effective within a CBRNE environment. Also remember that although soldiers can learn the principles of CBRNE from lectures, films, and demonstrations, they can learn best by direct, firsthand experience with toxic and simulated agents.

References

AR 385-10, *Army Safety Program*

FM 3-11, *Chemical, Biological, Radiological, and Nuclear Operations*

AR 700-146, *Individual Chemical Equipment Management Program*

FM 8-285, *Treatment of Chemical Agent Casualties and Conventional Military Chemical Injuries*

FM 7-0, *Training*

TM 3-11.91, *Chemical, Biological, Radiological, and Nuclear Threats and Hazards*

CLAIMS OFFICER

The duties of a claims officer are primarily investigative. The gathering and accurate reporting of basic facts surrounding an incident are prerequisites to and necessary steps toward claims settlement. Investigations fall into two categories: those to determine the facts when a claim has been made, and those that report facts about an incident that may later give rise to a claim. The latter is usually referred to as a "potential claim" investigation.

Claims also are divided into two categories: small claims, those that may be settled for $750 or less, and large—more substantial—claims. Officers investigating small claims should remember the following:

- The amount of investigative effort is less in smaller claims.
- Evidence about small claims may be gathered by telephone or personal interview, from incident reports and other hearsay reports. Written statements of witnesses, estimates of repairs, and so on are not required.
- Use DA Form 1668 (Small Claim Certificate) for recording investigations, with brief summaries of the evidence attached.

- Be objective and fair during all phases of investigation, protecting the rights of both the claimant and the government.
- Evidence must establish that the amount claimed or agreed to is reasonable, that the claimant is the proper person, and that the government is liable for damage or injury.

The objective in handling any claim is to gather all possible evidence in the shortest practicable time and to stress facts and events that help to answer when, where, who, what, and how. In the full investigation required for large claims, this evidence should include statements of all available witnesses; accident reports, including those in civilian police files; photographs; maps; sketches; and so on. The investigator should visit the scene of the incident and make a physical inspection of any damage. If the claim involves injury, determine with the help of the claims judge advocate whether a medical examination is required and, if so, make the necessary arrangements. AR 27-20 provides some detailed guidance on investigations of specific incidents and the evidence required for dealing with any claim.

In completing the large-claim investigation, use DA Form 1208 (Report of Claims Officer) for the report. Make specific judgments about what the evidence says and recommendations as to the extent of the liability and the amount of compensation. Submit the report to the appointing authority, who comments if he or she desires and forwards it to the appropriate approving or settlement authority.

As the claims officer, you should assist persons who indicate a desire to file a claim. They should be given general instructions concerning the procedure to follow, necessary forms, and assistance in completing them. You may also assist them in assembling evidence; however, you may not disclose information that may become the basis of a claim or any evidence that you may have collected unless you have the permission of the claims judge advocate. Your opinions and recommendations will not be disclosed to a claimant. In addition, you may not represent a claimant in any way and must not accept any gratuity for your assistance. These claims should be presented to the commanding officer of the unit involved, or to the nearest Army post or other military establishment convenient to the claimant. Evidence to substantiate the claim should also be submitted with the claim.

One special situation worthy of mention is maneuver damage resulting from field exercises. In this instance anything that can be done on the spot for both the government and the claimant in the way of collecting evidence and advising the claimant how to proceed will save much work later. Most such on-the-spot investigations come under the "potential claims" category but they should be accomplished as fully as possible while all individuals and property involved are available. Reports are marked "potential claim" and submitted like regular claims.

Apart from more routine investigation and report preparation, you may find it helpful to have an idea of how compensation is determined. In cases of property damage that can be economically repaired, the allowable compensation is the actual or estimated cost of restoring it to the same condition it was in immediately before the damage. Allowances may be made for the depreciation or appreciation after repairs. When property is destroyed or cannot be repaired economically, the measure of the damage is the value of the property immediately before the incident less any salvage value. Lost property is compensated for on the basis of value immediately before the loss.

In claims involving personal injury and death, allowable compensation may include reasonable medical, hospital, or burial expenses actually incurred, future medical expenses, loss of earnings and services, diminution of earning capacity, pain and suffering, physical disfigurement, and any other factor that local law recognizes as injury or damage for compensation purposes. No allowances are made in any claim for attorney fees, court costs, bail, interest, travel, inconvenience, or any other miscellaneous expense incurred in connection with submission of a claim.

References

AR 27-20, *Claims*
AR 27-40, *Litigation*
DA PAM 27-162, *Claims Procedures*
FM 7-0, *Training*

CLASS A AGENT (PAY) OFFICER

Though most of the Army's payrolls have been automated, it may still, under some circumstances, be necessary for an agent (or an accountable disbursing officer) to pay troops, exchange foreign currency for military payment certificates (MPC) or U.S. currency, or make payments for specified purchases or rentals. Appointing orders specify what payments the agent may make.

If you are appointed class A agent, you are personally responsible for the funds entrusted to you and for the vouchers that account for that money. At your request, the unit commander should provide adequate armed guards to protect these funds and vouchers. If you have to keep funds overnight, they should be secured in an adequate safe. Field safes and combination-lock file cabinets are not normally considered adequate. If these are all that are available, an armed guard should be kept posted. In addition, provision should be made for frequent checks of the secured area by a CQ or duty officer.

Here is a list of things that you should *not* do while a class A agent:

- Insure entrusted funds.
- Use funds except as stated in the appointing orders.
- Gamble while entrusted with funds—even with only personal money.
- Loan, use, or deposit in any bank any portion of the funds, except as specifically instructed by the finance officer.
- Mix the funds with personal monies or attempt to balance the funds by adding to or deducting from them.

- Entrust the funds or paid vouchers to any other person for any purpose, even for the purpose of returning them to the finance office. Unless otherwise directed, you must return the funds or vouchers personally to the finance officer. (Other possible means could be through a class B agent, by courier, or by registered mail.)

Generally, class A agent duties involve only the payment of troops. In recent years, cash payment of troops has not been done. Nevertheless, each officer would do well to understand the process in case it is needed in the future. Here is a typical sequence of payday actions that such an agent might take:

- After receiving your orders, watch the daily bulletin or other local medium of notification for the time and place to pick up the funds and vouchers.
- Arrange for transportation (POVs should be avoided), guards, and a suitable place for payment (such as a day room or dining room).
- On the day of cash pickup, inspect vehicle condition in advance. (It is also a good idea to vary the route and timing of trips periodically.)
- Upon arrival at the finance office, you will need a copy of your appointing orders and your armed forces identification.
- Before you leave the office, you should verify the cash count given you. The currency is generally $1, $5, $10, and $20 bills, packed in bundles of $100. This eases the task of counting, but you should count the bills in each bundle to ensure that they contain the proper amount. You will receive a change list showing the number and denomination of bills making up the payroll. For example, it may show that you should have 280 $20 bills, 120 $10 bills, 100 $5 bills, and 203 $1 bills. You should check your count of various denominations against this list. *This is the time for you to bring any discrepancy to the attention of the disbursing officer, not later.*
- Before departure, you must sign a receipt for payroll money received. You then proceed directly to the unit or activity to be paid.
- Before starting to pay, it is best to break the payroll down into individual payments. Errors are easiest to correct while you still have the money and vouchers. You should have the exact number and denomination of bills necessary to make up all the different payments. So that everything will come out right, each payment is made up of the largest denomination of bills possible. (For example, if the amount is $237, use eleven $20 bills, one $10 bill, one $5 bill, and two $1 bills.) Each individual payment can be placed in a plain mailing envelope or paper-clipped to the payee's copy of the voucher.

Your pay team should be made up of two or three individuals. You will *personally pass the money* to the person being paid. A responsible NCO should assist by obtaining signatures on the original (white) vouchers. You may have a clerk pass to the payee a copy of his or her voucher after payment has been made.

- The pay table should be set up in an area where troops cannot congregate. Payments should begin promptly at the designated time.

- At the proper time, a team NCO should call the name of the first individual to be paid.
- The payee moves to the table, salutes, and signs the receipt for payment. As pay agent, you do not return the salute. The payee's signature, as part of the pay process, is checked with his or her identification card and the name on the voucher.
- If the signature is correct, the voucher is passed to you for payment. You then count out the payment and ask the payee to verify the amount. If a correction has been made by the finance office to the amount-paid entry on the voucher (lined out and a new amount entered), the payee is required to initial thc corrected amount when signing his or her name or receiving the cash.
- After verifying the amount paid, the payee receives his or her copy of the pay voucher.
- This procedure is repeated until all persons present are paid. Soldiers not present for the regular pay call may be paid as soon as possible thereafter. Hospitalized personnel on the payroll should be paid next if you can reasonably travel to their location. When considering the reasonableness of traveling to pay a soldier away from home station, you should keep in mind the hardship that not being paid on time may cause the payee. The mere fact that you must arrange special transportation or stay overnight does not in itself make the travel unreasonable. The payroll shouldn't include persons AWOL or in confinement, but if it does, they will *not* be paid; their pay is returned with other funds remaining and paid vouchers upon completion of the unit payment.
- Regulations allow the class A agent 24 hours after payment for the return of paid vouchers and cash, but you should return the funds as soon as possible.
- When the material is returned, a clerk will verify the vouchers and furnish a receipt for both the return of funds and the paid vouchers.
- The amount indicated on the receipt should agree with the amount of cash returned.
- You deliver the vouchers and receipt to the cashier and turn in all cash not paid to individuals. After verification, the receipt is signed by the finance officer and a copy returned to you. You are then properly relieved of responsibility for the funds that were entrusted to you. Such receipts, of course, should be retained at least a year in case any question should arise.

When you are paying soldiers that you don't know on sight, or when you pay commercial vendors, your biggest problem is establishing positive identification. In the case of paying soldiers, the problem is fairly simple. Since the vouchers are prepared from official records, the name on the voucher should correspond exactly to the name on the individual's armed forces identification card. The signature obtained must also match the signature on the identification card.

Occasionally a payee will sign his or her name in an incorrect form—that is, not the "payroll" signature. If the signature is not in the same form as the name shown on the voucher, the incorrect signature must be lined out and the payee required to sign it again so that it does agree with the voucher.

If for some reason the payee can't sign his or her name (e.g., a man with a broken arm), the payee may authorize some other individual to sign for him or her, and that signature will be regarded the same as if signed by the payee. This signing, however, must be certified to by two witnesses and the certificate attached to the voucher. The mark (X) of an individual unable to write must be witnessed by a disinterested person whose signature and address are placed adjacent to the mark (X).

Receipts for payment to commercial firms must be signed by a duly authorized officer or agent of the company. The receipt must be signed with the company name, followed by the autograph signature of the officer or agent together with his or her title. In addition to ensuring positive identification of the company representative, you must assure yourself that the representative is authorized by the company to receive payment.

When in doubt, consult the regulation or seek out expertise and advice from your servicing Finance Office.

Reference

AR 637-1, *Army Compensation and Entitlements Policy*

COURTS-MARTIAL MEMBER

The member of a military court is essentially a member of a jury. The president of the court is, in the same context, the foreman. Member duties, as prescribed in the *Manual for Courts-Martial*, 2019, are like those of a juror in that each member hears the evidence and arrives at his or her own determination of the guilt or innocence of the accused. In addition, if the accused is found guilty, courts-martial members determine the penalty. Each member, regardless of rank or position, has an equal voice and vote.

Most courts-martial are assigned a military judge who gives the members their instructions. In the absence of a military judge, the president of a special court-martial instructs the members of the court generally on their duties. The instructions are important and worth repeating here.

> As court members, it is your duty to hear the evidence and determine the guilt or innocence of the accused and if you find him guilty, to adjudge an appropriate sentence. Under the law, the accused is presumed to be innocent of the offense. The government has the burden of proving the accused's guilt beyond a reasonable doubt. The fact that charges have been preferred against this accused and referred to this court for trial does not permit any inference of guilt.

You must make your determination of the guilt or innocence of the accused based solely upon the evidence presented here in court and the instructions which I will give you. Since you cannot properly make that determination until you've heard all the evidence and received instructions, it's of vital importance that you retain an open mind until all of the evidence has been presented and the instructions have been given.

You must impartially hear the evidence, the instructions on the law, and only when you are in your closed session deliberations may you properly make a determination as to the guilt or innocence of an accused. Furthermore, with regard to sentencing, if it should become necessary, you may not have any preconceived idea or formula as to either the type or amount of punishment which should be imposed if the accused were to be convicted of this offense. You must first hear the evidence in extenuation and mitigation, as well as that in aggravation, if any; the law with regard to sentencing; and only when you are in your closed session deliberations may you properly make a determination as to an appropriate sentence after considering all of the alternative punishments of which I will later advise you.

It is the duty of the trial counsel to represent the government in the prosecution of this case, and it is the duty of the defense counsel to represent the accused.

Counsel are given an opportunity to put questions to any witnesses that are called. When counsel have finished, if you feel there are substantial questions that should be asked, you are given an opportunity to do so. The way we handle that is to write out the question, indicate at the top the witness to whom you'd like to have the question put, and sign at the bottom. This method gives counsel for both sides, the accused, and myself, an opportunity to review the question before it is asked, since your questions, like the questions of counsel, are subject to objection. Do not allow any other member of the panel to see your question. Whether or not your question is asked, it will be attached to the record as an appellate exhibit. There are a couple of things you need to keep in mind with regard to questioning. First, you cannot attempt by your questions to help either the government or the defense. Second, counsel have interviewed the witnesses and know more about the case than we do. Very often they do not ask what may appear to us to be an obvious question because they are aware this particular witness has no knowledge on that subject.

During any recess or adjournment, you may not discuss the case with anyone nor may you discuss the case among yourselves. You must not listen to or read any account of the trial or consult any source written or otherwise, as to matters involved in this case. You must hold your discussion of the case until you are all together in

> your closed session deliberations so that all of the panel members have the benefit of your discussion. If anyone attempts to discuss the case in your presence during any recess or adjournment, you must immediately tell them to stop and report the occurrence to me at the next session.

For any one of several reasons, an individual may be ineligible to sit as a member of a court-martial in certain cases. Whatever the reasons, they are usually referred to as "grounds for challenge." A number of grounds for challenge are listed in RCM 912(f), MCM, 2019. If you don't believe that you should sit on the court in a particular case, you should bring this to the attention of the convening authority before the court is formally opened. Usually, you will be relieved. If the grounds for challenge first come to your attention at the trial, the trial counsel will ask that you relate to them the "ultimate" ground for the challenge. You must be careful not to relate facts that, if heard by other members, might also prejudice or disqualify them.

Unlawful influence on any court member's decisions by a senior officer should never become a problem; however, members should bear in mind that the commanding officer, convening authority, or senior member of the court may not unlawfully attempt to influence their independent judgment. This doesn't mean that members should ignore the opinion of their seniors in court deliberations, but rather that they should not let position or rank sway their own judgment.

In order to reduce the influence that ranking members of the court might have, discussion and oral voting (when required) usually begin with the junior in rank. Votes by members of a general or special court-martial on the findings and on the sentence, and by members of a special court-martial without a military judge upon questions of challenge, are by secret written ballot. The junior member of the court usually collects and counts the ballots. This count is checked and announced by the court president.

A court has no power to punish its members. Nevertheless, members are expected to conduct themselves in a dignified and attentive manner and misconduct as a member of a court may be a military offense. No member should ever become a champion of either the prosecution or the defense. Such partisan behavior would cast substantial doubt on the fairness of a trial. You must be particularly careful in your contact with counsel or other members of the court when the court is not in session. You may, however, without referring to any case pending or then being tried, carry on a normal official and social relationship with other prospective or appointed members of the court. It is proper to ask such administrative questions as the scheduled date for the court to meet, location of the trial, physical arrangements for the trial, and other matters that have no bearing on the issues of a case.

Generally, there is no objection to making notes during the trial, and these can also be taken into closed session so long as they are purely for the court member's individual use.

The guiding principle to follow at all times is that you should perform your duties without being subjected to any out-of-court influence, direct, indirect, or covert. The legal rights of an accused demand no less.

You should be well acquainted with the *Manual for Courts-Martial*, 2019. Reading it will provide a good orientation for court duty, even without having had the experience of participating in a trial. A prospective but inexperienced court appointee may also take advantage of opportunities to watch actual trials, particularly those conducted by general courts-martial. When appointed to serve on a court, you should resolve to do your best, as conscientiously as you can.

References

AR 27-10, *Military Justice*
DA PAM 27-7, *Guide for Summary Court-Martial Trial Procedure*
DA PAM 27-173, *Trial Procedure*
MCM 2000, *Manual for Courts-Martial,* 2002 edition

DUTY OFFICER

One of the most frequently recurring duties for junior officers is battalion or brigade staff duty officer (SDO). Field-grade officers also find themselves being assigned as unit or installation field officer of the day. While most of the specific tasks associated with these duty requirements are covered in detail in applicable local SOPs or duty instructions, there are several general points common to all.

As the commander's representative during non-duty hours, the duty officer is responsible for maintaining selected aspects of the command's health, welfare, training, morale, and discipline. During a tour as a duty officer, officers are usually required to remain in the headquarters when not conducting inspection tours. On weekdays, duty officer tours typically begin shortly before retreat (or close of business) and continue until a specified time on the following morning. On weekends, tours usually begin around 0800–0900 and continue for the next 24 hours.

Some duty officer positions allow the officer to sleep in quarters on post when not performing required tasks, but many, especially including those during assignment to units with short-notice deployment missions, require the officer to remain alert in the headquarters throughout the tour. Also, except for duty rotations conducted on a Friday–Saturday night or Saturday–Sunday night, officers are usually required to return to their regular duties immediately after completion of their staff duty tour, with a short break for personal hygiene.

A typical duty officer tour begins with reporting to the duty supervisor (the unit adjutant or equivalent officer). Officers who have not previously performed SDO duty in the particular unit should report early to become familiar with staff duty instructions. Duty officers will then be informed of any special orders or other important matters, such as impending alerts, ongoing missions, or emergencies that may require special reporting procedures. As always, taking detailed notes will aid retention of the knowledge imparted. Duty officers must ensure that they thoroughly understand their special orders as well as the routine requirements of their duties before the departure of the duty supervisor at the end of his or her duty day.

As the designated representatives of the commander, duty officers should always present especially sharp appearances. Save a fresh uniform for the tour and put on highly polished boots. For men, a shave before reporting is appropriate.

Typically, the duty detail will include a staff duty NCO, a junior enlisted runner, and possibly a driver for a staff car or other vehicle used to make inspection rounds. At the outset of a duty tour, if you are not acquainted with the enlisted soldiers who will share the tour with you, you should introduce yourself and learn a little about their backgrounds, their units of assignment, military occupational specialties, and other useful bits of knowledge that might help you perform the duties more effectively. Discuss the routine and special requirements of the upcoming tour with them, and be sure that everyone understands his or her role. Usually, enlisted soldiers of the duty detail are required to be alert at all times and are usually allowed compensatory time immediately following the tour. Make sure the soldiers understand their duties thoroughly, as well as your own.

Inspection tours of important facilities are common requirements during duty tours. For example, ammunition storage points may be inspected for security. On-post entertainment facilities and guest quarters may be inspected to ensure security and to verify the appropriateness of what is going on there.

Although changes in policies over the last decade have, in many cases, resulted in a reduction or elimination of nonduty hours inspections of troop billets in garrison, such inspections—for the maintenance of good order, healthy living conditions, and soldier morale—are still common in training units and in forward-deployed areas. During the inspections, take special care to avoid officiousness, but maintain a serious demeanor consonant with your duty requirements. Unless otherwise directed, do not seek fault *per se*, but observe carefully and check specifically all items required by the duty instructions you received.

Sometimes, duty officers are required to fill out checklists while inspecting dining facilities, motor pools, and so forth; enter required information objectively. You are the commander's designated representative, and it is your duty to inspect specified facilities and activities, maintain good order and discipline, and report accurately what you find. Commanders are responsible for everything their units do and fail to do, so effective commanders always appreciate detailed, candid, complete reports from their duty officers about activities that occur when they are away from the unit area.

Use common sense. When you discover a situation that can be best remedied by a subunit commander or staff officer, the appropriate action is often to contact him or her and report what you have found. They should be informed of serious or important issues immediately and follow up with you as appropriate. If you encounter possible violations of the Uniform Code of Military Justice or other directives, deal with them immediately. If possible, take suspects into custody and advise them of their rights under Article 31, UCMJ (it is always a good idea to have handy the "Rights Warning Card"). Contact their commander or designated representative (e.g., duty officer or NCO) and record the details as an affidavit for inclusion in the duty log. If necessary for situations beyond your capability to control, request assistance from the post law enforcement unit.

If you have doubts about what should be reported officially, ask the duty supervisor before the beginning of your tour. Adjutants and secretaries to general staffs are acutely aware of the commander's preferences and can usually tell you quite easily what should be officially reported and what can be handled in other fashions. Your duty instructions will also usually include specific information about how to report discrepancies with established policies or standards. Follow them closely. Tours as a duty officer are *never* opportunities to "get even" or to enact "gotchas" against anyone; they are, however, situations requiring astute, thorough, especially conscientious performance of actions required by the commander in his absence. Your performance as a duty officer should reflect the commander's philosophy as well as his or her specific instructions and intent.

Maintain a detailed and complete duty log. In some cases, maintenance of the log is the responsibility of the staff duty NCO, but as the duty officer, you must ensure that the log includes all pertinent entries. Enter all of your significant activities, including the times and results of inspection tours and significant communications. One of the most common tasks required of duty officers is to relay important information—news of personal emergencies, requests for Red Cross assistance, notification of alert or impending deployment, and important operational developments. Record in the log detailed information about the receipt and relay of such information and ensure that the information is effectively and quickly passed to the appropriate authorities. Thorough familiarity with unit alert procedures is a must.

Duty officers' tours typically end at the beginning of the following duty day. Review your log and prepare a brief, but concise, verbal report. Again, present a sharp appearance when reporting to the duty supervisor for your outbrief. Appear alert; men should have a fresh shave. Be prepared to submit your duty log and any required checklists. Apprise the duty supervisor of any especially significant events that occurred during the night, and answer his or her questions succinctly and directly.

During weekend duty tours, brief your successor as you would the duty supervisor; posture him or her for success by delivering a thorough report.

Before departing, thank the enlisted duty detail. A note to their commander, first sergeant, or command sergeant major commending their performance and attitude is always appreciated and appropriate when warranted. Similarly, in the case of substandard duty performance, counsel the soldiers concerned and advise their supervisors of unsatisfactory duty performance or attitude.

FOOD SERVICE OFFICER

There is little doubt that the quality of Army food, both its preparation and serving, contributes to morale. It is one of the main—and could be the chief—

motivating factors contributing to overall unit performance. Because of this, the food service officer's job is one that must be taken seriously, not only from the point of view of unit effectiveness but also personally, because of the challenges it presents to your leadership abilities (see chapter 13). The days of jokes about Army chow and mess halls are past. In today's Army, food is prepared by professionally trained sergeants and cooks who take great pride in their work and accomplishments.

When you are assigned as a food service officer, you are the commander's direct representative in the dining facility. You are also the direct supervisor of the food service sergeant. His or her attitude will make a big difference in how hard or how easy your additional duty will be as food service officer. The size of the dining facility for which you are responsible will determine the size of your job. A dining facility for just your company may feed only 100 to 200 meals per day or less. A large, consolidated dining facility may feed up to 7,500 meals per day. It is also possible that you will be a company dining facility officer but that your company eats in a consolidated facility that is the responsibility of another unit. In this case, you may have to do some coordination with the food service officer who actually runs the facility, but you will not be directly responsible for feeding until your unit goes to the field with its own field kitchen.

When you are appointed as a food service officer, one of your first tasks is to set up a meeting with your food service sergeant. You need to assure him or her, without question, that you will be fulfilling your duties, but at the same time recognizing the food service sergeant's professional authority to handle the day-to-day details. Before talking with the food service sergeant for the first time, you should familiarize yourself with AR 30-22, *Army Food Program.* This details your responsibilities, which are considerable, and gives you a feel for how a dining facility operates. The food service sergeant should then be able to give you a complete rundown on the following:

1. Dining facility administration
   — Current regulations and SOPs
   — Scheduling for cooks
   — Reports
   — Sanitation
   — Security
2. Training
   — OJT for MOS 94B cooks
   — Cross-training programs
   — Training for field operations
3. Food preparation, cooking, and serving
   — Menu planning
   — Production worksheets
   — Standard operating procedures
4. Accounting
   — Requesting rations

— Inventories
— Head count data and cash control
— Required records needing your signature

As part of this meeting, you also need to let the food service sergeant know how you intend to operate. For example:

- You will be checking food preparation, serving, and sanitation periodically. Also, you will be paying particular attention to the production worksheet.
- You take seriously your responsibility to sign certain reports.
- You want the food service sergeant to discuss with you problems he or she is having and improvements he or she wants to make.
- You are available at any time, day or night.

How successful you are as a food service officer depends on how well you follow through with the stage you have set. Let's face it, the dining facility sergeant may wish to see as little of you as possible. It is only natural, since the sergeant views the dining facility as his or her domain. This is what makes leadership interesting and challenging.

How you conduct inspections of the dining facility will set your chances for success or failure. Remember that daily activities must go on in the dining facility even during your inspections. Make allowances for this, but do not let it be an excuse for a poor operation. Arrange your schedule around the dining facility schedule. A good checklist for your inspections is contained in TM 4-41.12, *Food Program Operations.* You can also get a checklist from your headquarters or installation food adviser.

The food adviser is a good person for you to get to know. Not only is he or she an experienced food service warrant officer, but the adviser is familiar with many facilities and can help you compare your facility with the others on your post. The food adviser is also your primary source for training about your responsibility as a food service officer.

AR 30-22 is the primary regulation you will use as a food service officer. It contains information on your specific responsibilities and the responsibilities of those above and below you in the dining facility chain of command. Among other responsibilities detailed in AR 30-22, you must do the following:

- Make a monthly review of requisitioning procedures in your dining facility.
- Review receiving procedures twice each month on an unannounced basis.
- Update the commander at least quarterly on actions needed to improve security in your dining facility.
- Track the accountability of at least four high-dollar food items each month and report the results.
- Conduct a monthly formal review of production schedules.
- Make a review of the disposition of leftovers each week.
- Compute the dining facility inventory each month.

When your unit goes to the field, food service becomes even more important because the soldier looks forward to mealtime even more in the field than in garrison. It is also harder for you to carry out your responsibilities as dining facility officer, because you are working harder than usual at your regular job when your unit is in the field. About the only way to ensure good dining facility performance in the field is to have excellent performance in garrison and let it carry over into the field because of good work habits. You are working with different equipment in the field: field ranges and immersion heaters, water trailers, and ice chests. Training needs to be done on this equipment in garrison, and it needs to be well maintained before it is brought to the field.

Remember, as the food service officer, you represent the unit commander in the dining facility. He or she expects the best-quality food for the troops. You have been entrusted to ensure that this happens. It is a job that will test your leadership skills but also offers lots of personal reward every time a soldier gets a good meal. Your dining facility sergeant and his or her cooks have a strong professional pride in what they do. Use this to your benefit.

References

AR 30-22, *Army Food Program*

PAM 30-22, *Operating Procedures for the Army Food Program*

TM 4-41.12, *Food Program Operations*

ATP 4-41, *Army Field Feeding and Class I Operations*

TC 4-02.3, *Field Hygiene and Sanitation*

TB MED 531, *Facility Sanitation Controls and Inspections*

A complete list of dining facility reference materials is contained in AR 30-22.

INCOME TAX OFFICER

This additional duty assignment is made so that members of a command have the opportunity to get counsel and answers to questions about income tax. As the income tax officer, you are not expected to be a full-fledged tax expert, but you are expected to know how the law applies to soldiers. The staff judge advocate, if available, is the place to send persons for assistance in complicated situations, and that office may even offer some periodic formal training for all tax officers. Unit income tax officers should be sure to take advantage of it.

The questions that income tax officers are most often asked deal with filling out returns and claiming legitimate deductions. In addition to the instructions received with each tax form, it behooves the income tax officer to get a copy of *Your Federal Income Tax*. It's revised each year to reflect changes and interpretations of the laws and provides easy-to-understand examples of how they are applied in specific cases. Many items of interest to the military person are indexed under Armed Forces. The nearest IRS office should be able to supply a copy. Another excellent source of tax information is Publication 3, Armed

Forces/Tax Guide. This IRS publication is published annually and is available at www.irs.gov.

These annual references are a good bet for helping the income tax officer stay up-to-date. However, here are some general guides:

- Military personnel should always include their permanent home address on their return to help establish that they are in the armed forces.
- Military personnel have an automatic extension when in a combat zone. They must file within 180 days after leaving the area or after being discharged from a hospital outside CONUS. This rule applies to a joint return but not to the taxpayer's spouse if filing separately. If a soldier decides to take advantage of the extension, he or she indicates COMBAT ZONE on the return when it is finally filed. If his or her spouse files their joint return while the military member is still in a combat zone, the spouse also marks the form COMBAT ZONE. (It is then unnecessary for the military member to sign the return.)
- Military personnel who are prisoners of war who have been detained in a foreign country against their will have ninety days after release to file a tax return.
- All military personnel on duty outside the United States and Puerto Rico are allowed until 15 June to file. However, anyone using the extension must pay interest on the amount of tax from the due date and explain the reason for the delay.

Here's a guide to some of the taxable and nontaxable pay and allowances.

Taxable income:

Basic pay

Reserve training basic pay

Dislocation allowance

Lump-sum payments such as separation pay

Nontaxable income:

Forfeited pay (but not fines)

Mobile home moving allowances (actual expense)

Subsistence, or the value of subsistence

Uniform allowances

Quarters allowances or the value of quarters furnished

Housing and cost-of-living allowances received to defray the cost of quarters and subsistence at a permanent duty station outside the United States

Payments made to beneficiaries of soldiers who died in active service

Pay received by enlisted personnel or warrant officers for any month or part served in a combat zone. For officers, the first $500 of such monthly compensation is nontaxable.

That part of a dependency allotment contributed by the government

Sick pay is nontaxable when one is in the hospital or not duty-assigned for thirty days or longer.

Reservists not on active duty may deduct the cost and maintenance of uniforms over and above allowances received, and if reserve drills are in a location away from the reservist's *place of business*, he or she may deduct the cost of the round-trip transportation.

A quick trial set of calculations usually indicates that service personnel should take the standard deduction. Exceptions might be those buying a house or other property, or those with substantial loans who pay a significant amount of interest.

Military personnel are also required to file state income tax returns. Since the requirements vary from state to state, it may be necessary to advise the individuals to contact their states for copies of any material available.

References

Armed Forces/Tax Guide (Internal Revenue Service Pub. 3)
Your Federal Income Tax (Internal Revenue Service Pub. 17)
Instructions for Preparing Your Federal Income Tax Return (IRS)

LINE-OF-DUTY (LOD) INVESTIGATING OFFICER

"Line of duty" and "misconduct determinations" are phrases used when investigating whether an individual's disease or injury was incurred while the person was conducting himself or herself properly *in his or her role as a member of the Army.* Those assigned to investigate and make these determinations must do so carefully, since they will have a great effect on the concerned individuals and their dependents. Investigations are made primarily to provide data for the administration of federal statutes affecting the rights, benefits, and obligations of members of the armed forces.

Under specific circumstances, a finding could cause a person to be separated from the service without entitlement to severance pay. In the case of death, these investigations could lead to findings that would make some persons' dependents ineligible for many or all normal benefits. Or an LOD investigating officer could confirm that those concerned are indeed entitled to all benefits under the law. The importance of conducting a thorough and impartial investigation in accordance with the requirements of applicable regulations cannot be overemphasized.

The job of the LOD investigating officer is generally to investigate, record, evaluate, make findings, and report these findings. The details of how to accomplish this are found in AR 638-8 (*Army Casualty Program*). Here is a brief rundown of what you, as the LOD investigating officer, should be doing and some guidelines you can follow.

The Investigation. You must notify the individual concerned of the impending investigation. If practicable, he or she should be permitted to be present at the examination of witnesses if the investigation is to go beyond the examination of documentary evidence. If not present, the individual will be permitted to respond to adverse allegations.

You should visit the scene of the incident as early as possible. You should record all the relevant evidence, such as statements of witnesses, photographs, diagrams, letters, results of laboratory tests, observations and reports of local officials, extracts of local laws and regulations, descriptions, weather reports, the date and exact time of the incident, and so on. This documentation should reflect every fact and circumstance that you will consider in making your findings and report. Before the testimony of witnesses is taken, they must be advised of their rights under Article 31 of the Uniform Code of Military Justice or the Fifth Amendment to the Constitution.

The subject of the line-of-duty investigation is permitted to submit evidence or statements, sworn or unsworn. Before his or her statement is taken, the subject must be advised of legal rights under Article 31, Uniform Code of Military Justice, and the purpose of the investigation. If a statement from the individual is not obtained, you must state the reason in your report.

After you have acquired the evidence, you must evaluate it and determine, in your judgment, the exact circumstances under which the injury, disease, or death occurred. You make a summary of your findings in the "Remarks" section of the report of investigation. Findings as to line of duty and misconduct must be arrived at by the investigating officer in all actions except death, in which case Headquarters, Department of the Army, makes the final determination.

In every formal investigation, the purpose is to find out whether there is evidence of intentional misconduct or willful negligence that is substantial enough to rebut the presumption of "in line of duty." To arrive at such decisions, several basic rules can be applied to various situations. The specific rules of misconduct are as follows:

Rule 1. Injury or disease proximately caused by intentional misconduct or willful negligence is not in line of duty. It is due to misconduct. This is a general rule and must be considered in every case in which misconduct or willful negligence appears to be involved.

Rule 2. Mere violation of military regulations, orders, or instructions, or of civil or criminal laws, if there is no further sign of misconduct, is no more than simple negligence. Simple negligence is not misconduct.

Rule 3. Injury or disease that results in incapacitation because of the abuse of alcohol and other drugs is not in line of duty. It is due to misconduct.

Rule 4. Injury or disease that results in incapacitation because of the abuse of intoxicating liquor is not in line of duty. It is due to misconduct.

Rule 5. Injury incurred while knowingly resisting a lawful arrest, or while attempting to escape from a guard or other lawful custody, is incurred not in line of duty. It is due to misconduct.

Rule 6. Injury incurred while tampering with, attempting to ignite, or otherwise handling an explosive, firearm, or highly flammable liquid in disregard of its dangerous qualities is incurred not in line of duty. It is due to misconduct. This rule does not apply when a member is required by assigned duties or authorized by appropriate authority to handle the explosive, firearm, or liquid, and reasonable precautions have been taken.

Rule 7. Injury caused by wrongful aggression, or voluntarily taking part in a fight or like encounter, in which one is equally at fault in starting or continuing the fight, is not in line of duty. It is due to misconduct. The rule does not apply when a person is the victim of an unprovoked assault and sustains injuries in an attempt to defend himself or herself.

Rule 8. Injury caused by driving a vehicle when in an unfit condition, and the member knew or should have known about it, is not in line of duty. It is due to misconduct. A member involved in an automobile accident caused by falling asleep while driving is not guilty of willful negligence solely because he or she fell asleep.

Rule 9. Injury because of erratic or reckless conduct, or other deliberate conduct without regard for personal safety or the safety of others, is not in line of duty. It is due to misconduct.

Rule 10. A wound or other injury deliberately self-inflicted by a member who is mentally sound is not in line of duty. It is due to misconduct.

Rule 11. Intentional misconduct or willful negligence of another person is charged to a member if the latter has control over and is thus responsible for the former's conduct, or if the misconduct or neglect shows enough planned action to establish a joint enterprise.

Rule 12. The line of duty and misconduct status of a member injured or incurring disease while taking part in outside activities, such as business ventures, hobbies, contests, or professional or amateur athletic activities, is determinable as any other case under the applicable rules and facts presented. To determine whether an injury is due to willful negligence, the nature of the outside activity should be considered with the training and experience of the member.

After you have completed your investigation and evaluation and have arrived at your findings, you are ready to complete the report. These reports are submitted on Report of Investigation—Line of Duty and Misconduct Status (DD Form 261). You must be sure that all your conclusions are based solely on the evidence reflected in and attached to your report and that the source of all the evidence is properly reflected. The finished report is forwarded to the appointing authority.

References

AR 15-6, *Procedure for Administrative Investigations and Boards of Officers*
AR 638-8, *Army Casualty Program*
AR 635-40, *Disability Evaluation for Retention, Retirement, or Separation*

MOTOR OFFICER/MAINTENANCE OFFICER

As the new motor officer, your first questions are usually about the type and quantity of equipment you'll be responsible for. This varies from a few wheeled vehicles to more than a hundred, and from wheeled vehicles alone to a variety of wheeled and tracked vehicles, such as those found in a tank or mech-infantry company.

One of the keys to success of the maintenance mission is the motor sergeant, who most likely was selected for technical knowledge, mechanical ability, and, most important, aptitude for organization and supervision. He or she is responsible to you for the implementation of your policies and the enforcement of SOP and regulations governing the operation of the maintenance section. It's important that you work through and with him or her.

If you just joined the unit, you should get acquainted with the maintenance area right away and make some mental notes about the general conditions of the shop and the appearance of the mechanics:

- Is the shop adequately lighted?
- Are safety precautions, such as fire regulations, followed?
- Is the shop floor free of grease, oil, and dirt?
- Are blocks used under jacked-up vehicles, and are the wheels chocked?
- Are vehicles that are not being worked on removed from the shop?
- Do mechanics seem adequately supervised?
- Are mechanics using the TMs to perform or check their maintenance properly?
- Are mechanics using the equipment inspection and maintenance worksheet (DA Form 2404) in their work?
- Are mechanics or supervisors making proper entries in the equipment record folder and historical records?

The answers to these basic questions will help you get the feel for the adequacy of your maintenance organization. If you answered yes to all these questions, the organization is a real gem. If you answered no to one or more questions, you need to take a closer, more critical look at the operation.

Next, you should study the unit maintenance SOP. You should find out what the policy has been and not be in too big a hurry to change things. Check higher headquarters maintenance SOP and directives to see what can be expected in the way of support and what is required to receive it. Usually when unit work load exceeds the capabilities of a section, help can be obtained from battalion maintenance. (The same is often true of direct support maintenance assistance to separate companies.) One shouldn't ask for help unless it's needed, but when it is, neither foolish pride nor independence should be allowed to get in the way of ensuring that a unit's vehicles are ready to move.

Repair parts supply support is normally handled by the battalion maintenance shop. The repair parts stocked locally are determined by consolidated authorized organizational stockage lists (AOSL), organizational maintenance technical manuals and parts manuals on individual items of equipment, and documented demand experience of the unit. A list of the repair parts to be stocked locally is known as the prescribed load list (PLL). Necessary parts or assemblies other than those stocked are ordered by the battalion parts section from the direct support maintenance unit servicing it. Arrange to follow the paperwork on some repair parts supply actions to see how the system works. You will probably find that you spend a lot of time checking on parts that have been ordered and not received. *Proper follow-up is necessary.* After designated periods (dependent on the priority of parts requests) the parts clerk should initiate tracers to check the status of requisitions. You should be sure that he or she does and that you are informed of the answers received. Most repair parts are now ordered through automated systems.

When you become a maintenance officer, you usually must sign for some or all of the tools, equipment, and vehicles assigned to your section. Here, as everywhere else, there should be a complete joint inventory with the old hand-receipt holder or property book officer. Don't sign for anything that can't be seen or found. Somewhere, sometime, there has to be an accounting, so don't get stuck with the responsibility (and pecuniary liability) for something that was never there.

One of the big bugaboos in an automotive maintenance account are the tool sets. The new maintenance officer cannot be expected to know what is in these sets or even what all the items look like. *PS Magazine*, however, has come to the rescue and published complete lists of the tools with drawings to assist in their identification. Get these issues or reprints of them:

- Automotive Mechanics Tool Kit, Issue 156
- No. 1 Common, Issue 160
- No. 2 Common, Issue 176

Some separate units responsible for their own semiannual PM service may also get a No. 1 Supplemental Set. If that is the case, see *PS* Issue 172.

Reprints of these *PS Magazine* articles are often available in text or in the form of supplementary material for service schools. All these tool sets are also described in applicable supply publications.

In addition to these "common" tool sets, Tool Kit, Special Set A, consists of special tools peculiar to each different vehicle and not included in the organizational maintenance or general mechanic's tools. These special sets are described in the organizational maintenance parts manual (-20P) for the applicable vehicle.

Basic issue items (BII) that should accompany each vehicle are described and listed in the back of the applicable operator's technical manual (-10).

In touring the maintenance area, you undoubtedly will see a variety of miscellaneous equipment, such as generators, portable heating units, air compressors, and so on. Most of these require special operator training and licensing. You should find out who is responsible for the operation and maintenance of this equipment. One soldier and an alternate should be appointed by the motor sergeant to operate each item. It's also a good idea to know who supports you with organizational and higher maintenance for this equipment. Operator manuals or instructions and applicable equipment record forms should be available for inspection and daily use. Certain safety requirements apply, particularly to these pieces; for example, gasoline must not be stored in these items when they're kept in the shop.

Not all maintenance jobs can be accomplished at the organizational maintenance level. To find out exactly what one can and cannot do, see the maintenance allocation chart in appendix II of each -20 manual.

It's easy to see how important publications are to the maintenance effort. There's no substitute for the proper manuals. You should have the -10, -12, -20, and -20P (Parts Manual) for each item of equipment in your unit. Check AR 25-30 at armypubs.army.mil for the latest changes to technical manuals, supply manuals, and lubrication orders.

Today's Army requires more and more reliable information to plan replacement and procurement of the equipment needed to keep its operational readiness at the highest possible level. Data that provide DA with information about equipment age, reliability, and potential are recorded in the Army equipment record forms. The organizational level is the first step in the Army Management Maintenance System (TAMMS). DA PAM 750-8 is the guide and authority to this system. It contains a list of the necessary reportable equipment forms and instructions for completing and maintaining them.

Training organizational maintenance personnel is always a problem. In the average unit, the ideal solution is to requisition school-trained replacements, or to utilize available schools by sending the unit's own untrained replacements. The only disadvantage to the latter course is shortened retainability, but it does offer the advantage that you can personally select your maintenance people. The least acceptable but probably most often used method is supervised on-the-job training (SOJT). SOJT is most effective when the new individual is paired with a knowledgeable "old hand." This apprenticeship arrangement can be doubly rewarding, since the senior individual gets an opportunity to develop leadership abilities. Personal interest on the part of both the motor officer and the motor sergeant is necessary to the success of the SOJT program. Individuals work harder and better when they know their supervisors are interested in and appreciate their efforts. The lack of formal training can be offset somewhat by local instruction.

If, after observing the operation for a while, you are not satisfied with the performance of your maintenance section, get together with the motor sergeant and work out a new arrangement. Support units may find that a job-by-job assignment of mechanics is best, whereas in a tactical unit, one mechanic may be assigned the responsibility for providing a single platoon with organizational maintenance support. This allows the mechanic to become intimately familiar with the equipment he or she is responsible for. These are but two of many possible solutions. Adopt the one that's best for the unit.

Being a motor officer is one job that requires some time—time to learn what it's all about, time to learn the mechanics' capabilities, time to supervise. The commander depends directly on the motor officer and maintenance section to provide vehicles and equipment for training or combat. If you learn your job well, train your people, and apply your knowledge and experience, you'll provide timely and efficient support.

A word about command maintenance inspections: These inspections are scheduled to check on the status of vehicles and equipment. They are usually scheduled and announced in advance so extra effort can be exerted to have everything in the best possible shape. They are designed to help motor officers tighten up on their maintenance procedures. Properly viewed, they are a valuable aid in the maintenance effort.

References

AR 710-2, *Inventory Management Supply Policy Below the National Level*

AR 750-1, *Army Material Maintenance Policy*

DA PAM 750-8, *The Army Maintenance Management System (TAMMS) User Manual*

PS Magazine—all issues

TM 9-243, *Use and Care of Hand Tools and Measuring Tools*

TM 9-8000, *Principles of Automotive Vehicles*

POSTAL OFFICER

The postal officer has responsibility for the overall operation of unit mail service including active supervision of the mail clerk and daily checks and inspections of the mail room.

The mail room should be a separate and secure room, utilized for no other purpose. If registered or certified mail is handled, have a field safe or, at a minimum, a locked container that can be physically secured to prevent its removal. Secure official registered or official certified mail that is held overnight in accordance with AR 380-5.

Only one mail clerk may have the keys or combinations to the mail room and locked containers. The unit postal officer has the second key (there must be only two) or the only other copy of combinations. This second set of keys and/or combinations should be sealed in separate envelopes (PS Form 3977) marked to identify the contents and signed across the flap (the postal officer and clerk) to protect against tampering. The envelopes should be kept in a safe place, such as a company safe.

Mail clerks are appointed on DD Form 285 (Appointment of Military Postal Clerk, Unit Mail Clerk, or Mail Orderly). Enough copies should be made to allow distribution to the individual, the unit file, the serving APO or post office, and the battalion or consolidated mail room, if one is used. The forms must be validated by the serving postal facility.

Mailboxes or receptacles for outgoing mail must be strong enough to make deposited mail reasonably safe and must be physically secured to prevent removal.

Establish hours for collection and distribution of mail, based on the schedule of the serving postal unit. Record mail collection hours on a USPS Label 55 and post on all mail receptacles. Ensure that mail is picked up promptly as posted.

Incoming mail must be delivered personally to addressees and not, for example, left on bunks or footlockers. Hold mail for soldiers temporarily absent or en route for later delivery. Immediately forward or return to the sender mail for personnel no longer with your unit.

The mail room must maintain a unit directory of all personnel (other than dependents) who receive mail through the unit. Individuals joining or departing prepare a DA Form 3955, which is used in making or updating the directory. These forms are maintained in alphabetical order in one file, regardless of grade or status. Upon departure, locator cards are retained as prescribed in DoD 4525.6-m, Vol. II.

Undeliverable mail can be a problem. This is the type of mail that tends to pile up because it takes extra time to process. You should watch for this in your checks and inspections. DoD 4525.6-m, Vol. II, discusses the procedures and proper endorsements for returning undeliverable mail. A copy of this regulation and applicable changes must be kept in each mail room.

You should check daily that all registered, numbered insured, and certified mail is properly accounted for. You should verify the registers each day and retain them on file. Subordinate units, such as companies serviced by a battalion mail room, give a receipt for this; the original of this receipt remains at the battalion mail room. The duplicate is your record, and you can use it to check accountability and delivery of these pieces. Each addressee signs the duplicate upon delivery of his or her mail. All undeliverable mail is returned to the source from which it was received. A chain of receipts must be maintained on all accountable mail.

Here are things to watch out for: the obstruction of correspondence and the theft or receipt of stolen mail (18 USC 1701, 1702, 1708), the mailing of obscene or indecent matter (18 USC 1461), and removal of postage stamps from mail (18 USC 1720). The unit postal officer should report promptly any known or suspected postal offenses, including the loss, theft, destruction, or other mistreatment of mail, to the installation postal officer, the postal officer at the serving APO, or the local military investigative agency. Unit mail clerks suspected of mistreatment of mail should not be relieved of postal duties while under suspicion or investigation.

Free mailing privileges have been granted military personnel serving in specifically designated combat areas. Personal letters, postcards, and tape-recorded correspondence qualify for this service when they have the complete return address in the upper left corner and the word *free* handwritten in the upper right corner. Envelopes should be no larger than 5 inches by 11 inches and should not be endorsed "Air Mail."

Even when the free mailing privilege is not in force, any member of the armed forces may send letters without a stamp. Postage is collected from the addressee. This service is extended to handle emergency correspondence when stamps are not available and shouldn't be overused. The envelope is marked "Soldier's Mail" and signed by the unit commander.

Mail service is particularly important in the field as a morale factor, so unit postal officers should try to ensure that everyone receives his or her mail as quickly as possible, especially those individuals in units that are cross-attached. Make arrangements also for proper handling of mail for dependents overseas who remain at "home station" during field exercises.

There are times in the field, usually around payday, when the troops need money orders, stamps, and other postal items. The unit postal officer authorizes the mail clerk to purchase them for these individuals, using the unit mail clerk's receipt for funds and purchase record (DD Form 1118), which the mail clerk and the purchaser complete in duplicate. The original is kept by the unit, and the duplicate is given to the purchaser. The purchaser acknowledges receipt of the item by signing the original. The original copies are retained in the unit files.

Courses of instruction for postal clerks are generally given periodically by the serving postal facility. Every unit postal officer should take advantage of this training to refresh abilities of mail clerks and to train necessary replacements.

References

AR 25-51, *Official Mail and Distribution Management*
DoD 4525.6-m, Vol. II, *DoD Postal Manual*

RANGE SAFETY OFFICER

The designation of range safety officer conveys a special responsibility and obligation to see that conditions and procedures that are normally safe do not become unsafe, resulting in injury to personnel and damage to equipment. Specific safety requirements for each different weapon and category of ammunition are spelled out in the applicable technical and training publications. Before going to the range, the safety officer should become familiar with these requirements. The range safety officer works for the range officer-in-charge (OIC), who has overall responsibility for the conduct of firing. The latter should be contacted for instructions or considerations about the following:

- Maintenance and policing of range.
- Selection of competent and qualified range safety personnel assistants.
- Preparation of necessary maps.
- Posting range guards, barriers, and signals.
- Prescribing the wearing of steel helmets under certain conditions.
- Stationing of ambulances, emergency-type medical vehicles, and medical personnel.
- Arrangement for alternative means of medical evacuation, such as by air, and the applicable notification procedures, frequencies, signals, and so on.
- Measures to protect down-range personnel.
- Taking suitable precautions to prevent unauthorized trespass or presence on ranges.
- Any other duty or activities to ensure safe operation of the ranges.

As the range safety officer, you should contact your safety team as soon as its members are designated and brief them on what their jobs will be. It is important that they understand what is expected of them. The specific safety requirements that apply to the weapons and ammunition that will be on the range should be reviewed with them. Remember, however, that although you can delegate part of your authority in regard to safety, you cannot delegate your responsibilities.

One of your first duties at the range is to conduct a safety orientation for all personnel. You should do this prior to the opening of the range. You should also supervise the safe handling of ammunition. Both before and during firing, all ammunition and explosives or hazardous components must be handled and assembled in the manner prescribed by applicable safety regulations and appropriate technical manuals and field manuals.

- Place all ammunition at firing sites out of range of any weapon backblast, and store it so as to minimize the possibility of ignition, explosion, or detonation.
- Issue ammunition to troops only on the "ready" or firing line.
- Cover all ammunition to protect it from the elements and against direct rays of the sun. Provide enough air circulation around the ammunition to maintain uniform temperature.
- Transport and store boosters, rockets, fuses, detonators, chemical munitions, and so on separate from other ammunition and as specifically prescribed.
- Fuse ammunition only on the firing line and only as needed.
- Do not allow any round of ammunition, including practicc and blank ammunition, to be forced into the chamber of any type weapon.

See that weapons are handled safely. Be personally aware of all the special requirements of each weapon, such as the danger areas behind weapons such as rockets and recoilless rifles.

Are necessary warning signals and signs in place? Display range and danger flags and, when necessary, warning signs or flashing red lights at appropriate points. Proper warning devices are available from the office responsible for range maintenance and supervision.

Restrict all firing to designated firing points. No person should leave the firing line or remove material from it without permission from the range safety officer or the officer in charge of the range.

Individuals assigned as range safety officers supervise the handling of misfires, hangfires, and cookoffs. It's important that all personnel understand the nature of these malfunctions as well as the proper preventive and corrective procedures.

You must make the final determination *before firing* that settings placed on weapons and ammunition will impact the rounds within safety limits. This includes settings on fire-control equipment, fuse setting, and correct ammunition and charge. After firing, require all weapons to be clear and safe before they are removed from the firing line.

You position yourself where you can exercise maximum supervision over the safe conduct of firing. You should have no other assigned duties on the range while acting as range safety officer. Your job is to minimize the possibility of accidents. At the same time, you keep your assistants organized so that you can accomplish your mission without unduly interfering with the smooth progress of training.

References

AR 75-15, *Policies for Explosive Ordnance Disposal*
AR 385-63, *Range Safety*
AR 385-10, *The Army Safety Program*
FM 3-34.82, *Explosives and Demolitions*
AR 700-28, *Ammunition Management*
TM 9-1300-200, *Ammunition, General*
DA PAM, *Ammunition and Explosives Safety Standards*

RECORDS MANAGEMENT OFFICER

One of the things inspectors dig into first is a unit's records. A large part of what goes on in every unit is reflected in its files of actions, transactions, and training. A job as records management officer involves inspecting and supervising unit recordkeeping to ensure compliance with regulations and established procedures. In most units files are everywhere—the orderly room, supply room, training office, maintenance shops, mess hall. The records management officer must find them.

The Army has established a functional files system, a method for keeping records and reference material, and a guide to its disposition when no longer needed. Since your role as unit records management officer is mostly advisory, you need to know the system thoroughly. You should study the references listed below. Unit files, wherever they are, are the direct responsibility of those who maintain them. Nevertheless, as records management officer, you have the responsibility to correct any misuse of the system. This takes a real diplomat and leader.

The records management officer's job is one of constant inspection, which should concentrate in two areas—proper *categorization* and proper *labeling.*

Many papers are difficult to classify for filing. However, there is always one major subject that serves as a basis for filing. For most units, these major subjects—or functions—are spelled out in AR 25-400-2.

Because of the nature of the particular unit, there may be a need for special files not described in AR 25-400-2, but provided for in AR 340-18. For example, an engineer unit may have a need for mapping and geodetic files. Special files are authorized, as needed, below division level.

As you dig into the mechanics of the system you'll see why labeling of file folders is important. The label tells what is in the folder, how long it will be kept in the active file, and what to do with it at the end of the active period.

You have an indirect responsibility for all the files maintained by your unit. It's your job to help those with direct responsibility by detecting errors and helping correct them. Therefore, the more you know about the system, the better you can manage the attendant records.

Reference

AR 25-400-2, *The Army Records Information Management System* (ARIMS)

REENLISTMENT OFFICER

Officers with this primary duty are found in all major commands and in other commands or installations where the enlisted strength exceeds 5,000. Installations and organizations not authorized to have career counseling personnel on a primary-duty basis appoint an officer and a noncommissioned officer to carry out the reenlistment functions on an additional-duty basis. The reenlistment officer's job is to aid the commander in the reenlistment effort and to provide guidance and assistance to NCO career counselors.

It is a cardinal rule that every reenlistment officer keep his or her commander informed on all matters pertaining to the reenlistment program. This includes changes in the qualifications or procedures used in processing applicants for reenlistment. These are outlined in AR 601-210.

As the reenlistment officer, you should be alert for changes in this and other pertinent documents by maintaining contact with reenlistment and personnel sections of higher headquarters. You should kccp up-to-date on the availability and requirements for reenlistment options and other specialized career options.

The reenlistment officer should make sure that orientations are conducted for newly assigned officers and enlisted personnel in grades E-5 and above, informing them of the policies, procedures, responsibilities, and objectives of the reenlistment program. The program needs to be a continuing thing, but in the long run its effectiveness is a reflection of the prevalent attitude of the unit.

There are two films for use as part of the orientation of new officers and NCOs. *The Company We Keep* offers suggestions and presents a philosophy for effective reenlistment programs. *The One That Got Away* (MF 12-9323) may be shown to officers and NCOs at the discretion of the commander. If used, it should precede *The Company We Keep*. These films are available through the audiovisual communication center serving the unit.

New commanders want to be apprised of their responsibilities in relation to the reenlistment program. They are as follows:

- Counseling and interviewing eligible individuals (see AR 601-210) eight to ten months prior to the expiration of term of service (ETS) and forwarding the name and mailing address of each individual recommended for reenlistment to the career counselor, who includes him or her in a direct-mail reenlistment campaign. This campaign consists of five reenlistment information folders, mailed one per month by DA.

- Requiring all personnel completing their first tour (or with four years or less for pay purposes at ETS) to attend a showing of the film *Something to Build On* approximately four months before ETS.
- Informing recommended individuals of the reenlistment opportunities that will be available to them at the time of separation or within three months thereafter.
- Reenlisting individuals desiring unbroken service the day following the date of discharge, even when that day is a nonduty day.
- Establishing procedures to bar untrainable or unsuitable individuals from reenlisting (see para. 8c, AR 635-200).

Occasionally a person is encountered who is felt to be exceptionally worthy of retention in the active Army but, for some reason, is not qualified for reenlistment. (The reason could be lost time, over-age, conviction of a minor offense, medical, or other.) In this case, you may request a waiver of reenlistment disqualification in accordance with AR 601-210. Waivers are also possible for personnel who want to attend an Army service school but do not meet the minimum prerequisites for the desired course. These waiver requests are submitted as directed in paragraph 4-26, AR 600-200.

If the unit has room for a separate reenlistment office, it will do the most good if it is located where it is conspicuous enough to draw attention, yet private enough to allow an informal, friendly atmosphere for interviews. In any case, each unit should have an effective display of reenlistment literature. In the unit area, this should be a self-service display stocked with current reenlistment information.

Use originality in the conduct of reenlistment ceremonies. The national flag should always be in the immediate vicinity, but this doesn't mean that reenlistments need to take place in an office. You can use the PIO facilities available to publicize enlistments and give your program a boost.

There is one important clerical area in the field of reenlistment—the Reenlistment Data Card (DA Form 1315). It is designed as an aid to the reenlistment program. The form is initially completed at the Army Reception Station and forwarded with the individual's 201 file. If not, the unit personnel officer prepares one. When a person is transferred or reassigned prior to ETS, the losing unit commander makes an appropriate entry in the enlistment status section of the form. Review of the facts on the card will help prepare the officer for the interview. From it one can learn age, dependency status, level of civilian education, civilian occupation, GT score, the top scores in the individual's qualifications battery, and whether the soldier is qualified for a reenlistment option. Prior interviews and viewing of a reenlistment film are recorded on the DA Form 1315. Stereotypical remarks such as "will not reenlist" or "does not like Army" are *not* to be used. Chances are that you or someone else will have to interview the individual again, and some pertinent information from the record about the previous interviews will help jog thinking or serve as a guide to the new interviewer.

The reenlistment officer is required to maintain sufficient statistics to indicate the reenlistment efforts of each company-size unit. Local regulations usually spell out the nature and form of the required statistics.

In talking to a soldier about reenlistment, it's not surprising to hear gripes about KP, guard, bed check, or the billets. You should remember, however, that you are not reenlisting the soldier for a career of these things. What the Army wants are skilled technicians and leaders. You can point to such soldiers in your unit as examples that any young soldier should strive to emulate. You must try to make reenlistment the "in" thing to do, and gear your program to the individual.

In terms of broad objectives, the most important goal of any reenlistment effort is to select and retain those qualified soldiers who have shown the potential for greater service to themselves and the Army. You assist in evaluating the capabilities and attitudes of every person in the command. You must remember that the soldiers you reenlist will serve as trained replacements under you or some other commander. So you should never shortchange anyone for the sake of good-looking reenlistment statistics.

References

Enlisted Ranks Personnel Handbook, which contains the following:
AR 635-200, *Active Duty Enlisted Administrative Separations*
AR 601-2, *Army Recruiting Support Programs*
AR 601-280, *Army Retention Program*
AR 614-200, *Enlisted Assignments and Utilization Management*

SAFETY OFFICER

Briefly, the safety officer's job is to develop a sustained safety education and accident-prevention program. Normally, much of this effort is necessarily directed toward creating an interest in safety on the part of all personnel in the command. Here are some guidelines for the newly appointed safety officer:

- Become familiar with the Army and subordinate command safety regulations (385-series), DA PAM 385-1 (*Small Unit Safety Officer/Noncommissioned Officer Guide*), and local SOPs.
- Hold periodic briefings to keep supervisors, platoon leaders, and NCOs alert to safety requirements and programs.
- Promote original campaigns to keep individuals constantly aware of their responsibilities for accident prevention, both on and off the job.
- Be sure that directives, policies, plans, and procedures on safety are realistic in terms of primary unit mission.
- Review accident statistics or reports to identify trouble areas and apply practical corrective measures.
- Investigate and report each accident accurately and promptly, irrespective of its severity, degree of injury, or damage cost.

- Organize a unit safety council to make recommendations and suggestions to improve the accident-prevention program.
- Conduct safety inspections, recommend action to remove or control hazards, and determine the need for safety training. Inspections help identify unsafe conditions or persons *before* accidents occur. This fact should be used to advantage.

Design safety programs to reduce or eliminate accidents—in essence, to reduce hazards and to develop safe behavior among unit personnel. DA PAM 385-1 (*Small Unit Safety Officer/Noncommissioned Officer Guide*) is an excellent guide and reference for planning a unit safety program. Here is a simple program checklist based on the requirements of AR 385-10. This can be modified to meet any specific unit situation.

- Do SOPs include provisions for safe practices and procedures?
- Does the commander personally review the accident experience of the command periodically?
- Does the commander include safety as a topic in staff meetings?
- Has the safety officer been appointed on orders?
- Does the command receive and display safety publications, posters, and material?
- Does the safety officer conduct training and prepare safety material for presentation by others?
- Does the safety officer conduct safety inspections and surveys?
- Are summaries of accident data periodically assembled and reviewed?
- Are reports, records, and other accident information safeguarded as prescribed by current regulations (AR 385-10)?
- Do all unit SOPs contain clear and concise instructions on the reporting of accidents?
- Are accident investigations thorough and timely?
- Are all accident reports being reviewed carefully for completeness and accuracy?
- Is remedial training for drivers involved in traffic violations required?
- Is disciplinary action initiated when traffic violations are the primary cause of an accident?
- Are the safe driving rules for winter driving brought to the attention of the entire command?
- Is private-motor-vehicle accident prevention emphasized?
- Does the command participate in local safe-driving campaigns?
- Are off-duty pass or leave personnel required to comply with directives regarding safe operation of private motor vehicles?
- Does the command have an adequate safety awards program?
- Is the program effective in stimulating interest in the reduction of accidental injury and/or property damage?
- Are individual safety awards being used and presented properly?

The best accident is one that has just been prevented. The time-proven methods that help keep accidents to a minimum are the "three E's of safety": engineering, education, and enforcement. *Engineering* is identifying, locating, and eliminating hazards; compensating for those that cannot be removed; and avoiding the creation of hazards in new designs or operations.

Education and training have three aspects: the development of positive safety attitudes, the knowledge necessary for safe performance, and the skill level necessary for safe performance.

Engineering and education can prevent most accidents, but there are some people who just won't be careful. For them, strict *enforcement* of safety practices, backed by prompt corrective action, is necessary. Punishment should not be for having an accident, but rather for violation of an order or procedure in effect to prevent such an accident.

Keep in mind the objectives of the program: to reduce hazards and develop safe behavior. Engineering and inspections help achieve the former, and education and enforcement contribute to the latter.

Safety inspections are generally of the "continuing" type, conducted to discover accident-causing conditions or procedures throughout a unit area. As part of the program, the safety officer should invite periodic inspections of specific areas by specialized teams. The installation safety director can help arrange these. Every unit safety inspection should cover all the activities of the unit in as much detail as possible.

Accident records and reports from previous inspections will indicate areas that may need particular attention. For each such inspection, a checklist should be used and a record of deficiencies kept. Appendix G of DA PAM 385-1 contains a suggested safety inspection list. Additional safety criteria and information may be obtained from publications (ARs, TMs, FMs, TBs, *PS Magazine*) appropriate to the unit and equipment.

The safety officer should serve as recorder for the safety council (not as its chair). The recorder prepares a detailed agenda before each council meeting, including enough detail to show the extent of problems to be discussed and the need for doing something about them. The ideas and suggestions that come out of the meeting are also recorded. They can be used as a basis for developing safety-promotion campaigns. To add emphasis, council members should be selected by the unit commander. The deciding factors when selecting individuals should be their interest in safety problems and leadership ability. Here are some items that could be considered by the council:

- Accident experience and trends.
- Accident reports, including cause analysis and corrective action.
- Review of new equipment or procedures to determine any potential hazards and appropriate corrective action or SOP changes.
- Safety programs and recommended solutions.
- Evaluation of safety suggestions.
- Implementation of Army safety policy regulations and programs.

- Planning and implementation of safety contests, demonstrations, and orientation of personnel.

The safety officer, by virtue of his or her interest in safety and the experience gained, is normally the best-qualified person to act as accident investigator. A detailed discussion of accident investigation, reporting, and analysis is contained in DA PAM 385-1. This material should be studied in depth before attempting to investigate an accident.

Remember that one major purpose of accident investigation is to provide information that will be useful in preventing similar occurrences. It is essential that the investigator go beyond the superficial causes and determine the *why* of the accident, seeking reasons, not alibis.

A good safety officer will take advantage of the safety management extension courses and subcourses presented by the U.S. Army Safety Center. Information on these and other nonresident courses can be obtained by visiting www.safety.army.

In planning unit safety programs, remember that in the end, safety is a result of each individual's interest. Once this is developed sufficiently, accident statistics will begin to improve.

References

AR 95-1, *Flight Regulations*

AR 385-10, *Army Safety Program*

AR 600-55, *The Army Driver and Operator Standardization Program (Selection, Training, Testing, and Licensing)*

DA PAM 385-1, *Small Unit Safety Officer/NCO Guide*

TM 5-682, *Facilities Engineering: Electrical Facilities Safety*

SAVINGS OFFICER

The savings officer is primarily a "general sales and promotion manager," whose objective is to plan and conduct a continuous educational program that will encourage regular and systematic savings. He or she is responsible for the savings bond program. The whole idea is to encourage voluntary savings. It is accomplished through the payroll savings plan for the purchase of savings bonds.

Savings bonds have been sold continuously by the Treasury since 1935. The payroll savings plan, initiated in 1941, has proved to be an easy and safe way for the average soldier to accumulate capital. The decision to participate in the program is entirely up to the individual. Savings officers must never use coercion, reprisals, or threats of reprisal to induce personnel to enroll in the program. Instead, the case for bonds should be presented on its own merits, together with information on how to enroll in the payroll savings plan—nothing more. The Army is very explicit that *this is all the savings officer can do.* How best to go about doing this is the problem.

Each year the Army conducts a savings bond campaign during the month of May. The savings officer should arrange for a person-to-person canvass to explain the advantages of the savings program and to solicit participants. This covers explaining the benefits to both the government and the individual. If the soldier is already buying bonds, increasing his or her present deduction can be suggested. During the canvass, each soldier should be provided with an SF 1192 form (Authorization for Purchase and Request for Change, United States Series EE Savings Bonds). The completed forms are sent to the finance section serving the unit. The savings officer may follow up periodically but cannot harass or coerce. A good time to follow up any prospect—even those already taking a bond—is at the time of a pay raise or promotion when a payroll deduction would be most painless.

The Assistant Comptroller of the Army for Finance and Accounting furnishes material and guidance for the annual savings bond campaign.

In view of the many benefits, the savings officer should try to make the presentation of the program continuous and dynamic. Explain the program, sell its merits and advantages, make enrollment as easy and painless as possible—and *that is all.* Nothing else is required; nothing else is expected; nothing else is wanted; nothing else is allowed.

Reference

AR 690-500, *Pay and Allowances Administration*

SUPPLY OFFICER

Duties of the unit supply officer vary from unit to unit. It may simply require responsibility for the operation of a small-unit supply room that does little more than dispense expendable items and provide storage for the unit equipment. Or it may involve a large supply operation complete with requisitioning and accounting responsibilities.

In most company-size units, a supply sergeant (under the supply officer's supervision) runs the supply room. As a new supply officer, your first move should be to meet the supply sergeant and visit the supply room. There you should determine what property the unit is authorized. There should be an up-to-date file of TOEs, MTOEs, FMs, SCs, TMs, and so on that apply to the supply activities of the unit.

Organizational property, such as weapons, vehicles, and other items, is included in modified tables of organization and equipment (MTOEs). Other items, such as helmets, canteens, and ponchos, are authorized by CTA 50-900. Additionally, field and garrison furnishings and equipment authorizations, such as tentage, data room equipment, and office furniture, are authorized by CTA 50-909. (This equipment can be accounted for as organization or installation property. The latter does not normally accompany a unit to the field or on a change of station.)

Personal clothing is repaired and replaced with the cash maintenance allowance paid monthly to each enlisted soldier. The regulation also outlines certain conditions for gratuitous issue and repair of personal clothing and describes the handling and disposition of clothing upon discharge or absence of the owner.

Morale, Welfare, and Recreation Service property, such as athletic, welfare, and recreation supplies, has no basic publication. Its expendability is established by the Army Master Data File (AMDF).

All nonexpendable unit property, except components of sets, kits, or outfits (SKOs), is listed on hand receipts in the unit supply files. These files are a logical next step in the new supply officer's introduction to the supply room. (Nonexpendable and durable components of major end items must be recorded on hand receipt annexes. Nonexpendable and durable components of SKOs must be recorded on component hand receipts.)

In accordance with AR 25-400-2, unit supply files are established for supply control within the organization and are not records of accountability. Every unit has them. They should include as a minimum the following:

Hospital and Absence without Leave File. Personal clothing of individuals in each of these categories is inventoried, and a record of the inventory is kept in this file.

Gratuitous Issue File. This file contains documents initiated in connection with gratuitous issue or repair of personal clothing.

Work Order File. This file contains work order requests for the repair of unit property. Responsibility for the items submitted for repair is temporarily transferred to the repairing agency. These work requests and job orders assist in supply control.

Laundry Files. These assist in the control and supervision of individual and organizational clothing sent to the post laundry. Most posts offer three laundry services: (1) monthly payroll deduction rates, (2) cash-per-bundle rates, and (3) piece rates. The monthly payroll deduction rate plan has a maximum bundle limitation, and a piece authorization is used. The deduction is made by means of a roster. The per-bundle rates are collected when the laundry is turned in. The cash and signed vouchers (DA Form 3136) accompany the bundles. Individual piece rate is handled as determined by the officials responsible for the local laundry service (DA Form 2741).

Two accountability records may be maintained in the unit supply files, at the discretion of the property book officer. They are the Organizational Clothing and Individual Equipment Record (DA Forms 3645 and 3645-1), which record individual draw and turn-in of CTA 50-900 property; and personal clothing records, which are initially prepared upon entry into the service and forwarded to each new unit. Posting is accomplished by the supply sergeant (for example, as when uniform authorizations change).

You are now ready to take a critical look around. Is the supply room secure? Are "No Smoking" signs posted (AR 700-15) and are fire extinguishers present, filled, clean, and in operating order? Are the last inspections of the firefighting equipment entered on the inspection tags in accordance with local fire regulations? The general appearance and organization of the supply room are clues to the quality of supply management one can expect to find.

Next check the storage of equipment. Metal tools should be clean, free of rust, and oiled. Cutting edges should be protected. Wooden handles should be free of paint and treated with linseed oil (handles fitted to hammers, axes, and other like tools should be wedged to ensure secure mounting). Blankets and wool items should be clean and adequately mothproofed. Mattresses should be stored off the floor in mattress covers. They should be stored flat, no more than three high, and shouldn't have anything piled on top of them. Excess supplies, salvage, and unserviceable items should be turned in promptly. Don't be afraid to dig around in the bottoms of bins to locate excess salvage items. Every supply room has them, and they should be gotten back into the system—someone else may need them.

Canvas items should be dried and cleaned before storage. Canvas should always be stored off the floor on dry, clean dunnage. This allows air to circulate freely around it. Poles and stakes should never be rolled with tents for storage. Tents should be pitched periodically and inspected. They should be clean and dry before storing. They should be tagged when stored with nomenclature, National Stock Number (NSN), date of storage, and date last aired. Canvas repair kits (authorized at company/detachment size unit) can be used to repair small holes and rips and to replace missing grommets.

Equipment that uses flammable fuels is stored in accordance with fire regulations. A mop string partially inserted into the fuel tank acts as a wick and ensures that tanks are dry and safe.

Sometimes material for unit projects is unavailable through normal supply channels. On occasion it can be found in the salvage yard of the supply installation serving the unit. As a rule, the items can be issued as long as they are not used for their originally intended purposes. See AR 735-5 for details.

Supply control in most unit supply rooms is a pretty simple thing. Items that are lost, damaged, or no longer needed are reported or turned in to the commander or property book officer. The loss, damage, or destruction without fault or neglect is reported to the commander or property book officer. For damaged property, he or she prepares a letter explaining the circumstances and forwards it to the next higher commander for concurrence, with a request for relief of accountability. If granted, the items are removed from the hand receipts and property book.

When fault or neglect is involved and pecuniary liability for the loss is admitted, collection is made in cash on a cash collection voucher (DD Form 1131), listing the property and value and including the statement "Used in lieu of a report of survey, AR 735-5," or it is made in the form of payroll deductions by issuing a statement of charges (DD Form 362). Receipted copies of either act as relief from responsibility. When pecuniary liability for the loss, damage, or destruction of property is not admitted, a report of survey (DA Form 200) must be initiated by the hand receipt holder, the accountable officer, or the person most knowledgeable.

Nonexpendable components of an SKO, such as hand tools, that are worn out though fair wear and tear are turned in, and replacement is requested. The chain of accountability is maintained by using a turn-in tag (shoe tag), DA Form 2402. The lower portion of the tag is a receipt for the item and is later exchanged for the new replacement item.

Study in detail the discussion of accountability and supply procedures contained in AR 735-5.

Check the qualifications of supply personnel next. Even if they are experts in garrison supply procedures (which is rare), there is a lot that can be done to ensure adequate support in the field. For example, the supply section usually operates in rear areas, making daily trips between its unit and supply base. Map reading is an often neglected but necessary subject. The same is true of CBRNE training, vehicle maintenance, and all other basic military skills. The supply section is often overlooked or allowed to miss unit training on these subjects. Guard against this. The training (or lack of it) that the supply section receives could someday be the deciding factor in some critical combat decision. Chances are that most local supply personnel could use some training, so you may be able to arrange larger classes and assist in presenting them.

Responsibilities for supply and property are serious. Laxness in this area can be expensive in a personal way, not to mention its impact on the unit mission. It's wise for a new supply officer to immediately become familiar with the regulations, orders, instructions, and SOPs that apply to supply and property accountability.

References

Unit Supply Update
AR 25-400-2, *The Army Records Management Program*
AR 40-5, *Army Public Health Program*
AR 190-11, *Physical Security of Arms, Ammunition, and Explosives*
AR 700-15, *Packaging of Materiel*
AR 700-84, *Issue and Sale of Personal Clothing*
AR 710-2, *Supply Policy Below the National Level*
AR 735-5, *Property Accountability Policies*
DA PAM 710-2-1, *Using Unit Supply System (Manual Procedures)*
DA PAM 710-2-2, *Supply Support Activity Supply System (Manual Procedures)*
ATP 4-42, *Material Management, Supply, and Field Services Operations*

TRAINING OFFICER

Your duties as a training officer can vary considerably from unit to unit. In one, you might have a fairly free hand in planning and conducting your unit's training. In another, training may be directed almost completely from above. In either case, your responsibilities are considerable.

In all units, depending on what portion of the Army training program (ATP) is being conducted, certain training is mandatory. Whether your unit is undergoing individual or unit training, you should study FM 7-0 and FM 7-10 to get a good overall picture of training as the Army conducts it. For most combat-ready units, the Army subject schedules that pertain to the particular unit contain a detailed plan and program of training. You should be sure that you have an applicable copy and have studied the requirements outlined in it. There are also training notes, sequence charts, and lesson outlines included that will help you. Next you should check the training directive of higher headquarters for a definition of training policies and particular requirements or objectives to be accomplished.

One of your first steps is to find out who actually plans the day-to-day training. You may be expected to do it, or it may be directed by higher headquarters. In any case, you will want to know how far ahead it is necessary to plan. This depends to a great extent on the availability of training areas, facilities, and training aids. It takes a certain amount of time to procure training aids, but this shouldn't be much of a problem. Training areas, however, are another matter. Unless you are at a most unusual post, they must be requested far ahead at periodic range conferences. These conferences are usually held semiannually or quarterly, with more frequent updating sessions. Planning should therefore be done far enough in advance to register unit requirements at the "3" shop before these main meetings. If all the needed facilities are not available at first, they may become available at the interim meetings. This, however, requires that you remain flexible and alert to short-notice opportunities. Major activities, such as armor-unit tank gunnery programs, may be scheduled by division. Other activities, such as brigade or battalion field exercises, will be scheduled by those headquarters. You must fit your unit training into the schedule to complement the plans of higher headquarters. For example, you should have platoon and company tactical training before a battalion field exercise rather than after it.

Combat service support units have special problems when it comes to field training. The supported units don't just vanish when the servicing unit needs field training. A solution is to send one element of the supporting company to the field at a time to operate during field exercises of the maneuver units that depend on its service. Even this may not work for all organizations. You have to apply your imagination. Regardless of the organization, there is never enough time, facilities, or money to do all the training desirable. You will have to plan ahead and in detail, using knowledge, experience, and ingenuity to get the job done right.

Time is an important element in training. Obviously, time must be included in the unit's schedule for training, but it is not so obvious in the planning phase that sufficient time must be made available to instructors to study, prepare, and rehearse their classes. No class should be given without a rehearsal, preferably several. Someone should critique each lesson sufficiently in advance to allow time for instructors to rewrite or revise weak or unsatisfactory parts. The time required is well spent.

Stick to the essentials in training. When writing or reviewing a lesson, always ask, What does the soldier need to know to accomplish the mission? Be sure that soldiers get all they need to know and that they learn it. Don't waste training time—the instructors' or the students'—on unnecessary or irrelevant information.

As much training as possible should be "hands-on" practical work. Even simple tasks are learned best when done rather than described. Next best is a demonstration with actual equipment. For example, camouflage training, basic to every unit, is sometimes difficult to accomplish, since there are few places that allow the tree cutting necessary to do an effective job. Here is where a demonstration of one vehicle or position properly camouflaged can do the trick. At the position, a soldier can see what's expected, and by moving to an enemy viewpoint, both proper and improper examples can be illustrated. Night-light discipline is another example. One can talk about it for hours, but a simple nighttime demonstration of a few typical sources of light is easy and effective. Don't forget to include a demonstration of how to solve the problems brought up in discussion—in this case, how to shield necessary light.

For specialized training and training beyond the scope of unit capabilities there are service schools in almost all theaters. DA PAM 351-4 with its current changes lists these schools. School quotas are requested through the battalion or post S-3 (G-3) office.

Training paperwork generally falls into two areas: training schedules and individual training records. Both are important. Local training directives usually spell out the requirements and the form of both.

Training schedules must be timely; they must be prepared early enough to allow the instructors reaction time to prepare their classes, to schedule classrooms or training facilities, and to obtain necessary training materials.

Individual training records, when properly maintained, are important to the unit program. See that they are kept up-to-date. In addition to revealing how much each soldier is progressing toward fulfilling the training requirements, they also tell which classes need to be rescheduled or which individuals need to be sent to another unit's class for makeup work. Here is an area where close cooperation and coordination between training officers can pay dividends. One makeup class could suffice for an entire battalion (or more). If training is staggered among units, some soldiers who need makeup work can often be sent to another unit giving the same instruction as a part of its regular program. To do this, you have to stay on top of your own requirements and the training being conducted by other units around you. You should be prepared to offer the same help to other units that you may want for your own.

The standards of training in your unit will largely reflect the attitude, interest, and ingenuity you bring to your job. You should seek help and ideas from your subordinates as well as your superiors. There is always more than one solution to a problem, and you must always be prepared to pick or recommend the best.

References

AR 350-1, *Army Training and Education*

ATP 350.3, *Multi-Service Tactics, Techniques, and Procedures for Survival, Evasion, and Recovery*

AR 385-10, *The Army Safety Program*

FM 7-0, *Training*

TC 3-21.75, *The Warrior Ethos and Soldier Combat Skills*

UNIT FUND COUNCIL, PRESIDENT/MEMBER

Unit fund councils are composed of at least one commissioned officer or warrant officer and two enlisted members of any pay grade. The number of members should be limited but must include at least these three individuals. At company level, the commissioned officer is usually the unit commander. At higher headquarters, the commissioned officer is the unit commander or one of his or her staff. The enlisted representatives are members of the unit. The council president and members of the council are designated by the unit commander.

The council must meet at least once each quarter (more frequently when necessary) at the call of its president. Every council member has a duty to do the following:

- Ascertain and ensure that the unit fund is being properly administered and safeguarded as provided in AR 215-1, DoD 7000.14-R, Vol. 13, and local regulations.
- Determine that all income has been received in full, properly recorded in the book of accounts, and accurately reflected in the financial statements.
- Approve the amounts and purposes of all expenditures of the fund. Such approvals may be of a general nature, such as total expenditures authorized for running a contest, fund administration, or recurring-type program expenses, or they may be of a specific nature (such as expenditures for a particular purchase of supplies, equipment, or awards).
- Review the fund financial statements and other fund records as required to ensure that all expenditures are made in accordance with approved council actions and within the purpose for which the fund was established.
- Assure the accountability of all fund-owned property and the conduct of physical inventories of such property, and recommend disposition of surplus.
- Assure that audits are scheduled and conducted as prescribed, and review reports of audits and inspections and take appropriate action thereon.
- After the council has examined monthly accounts, found them correct, and approved the expenditures, both the recorder/custodian and the president sign the Cash, Property and Reconciliation Report (CPRR). See DoD 7000.14-R, Vol. 13.

References

AR 215-1, *Military Morale, Welfare, and Recreation Programs and Nonappropriated Fund Instrumentalities*

DoD 7000.14-R, *Financial Management Regulation, Vol. 13, Nonappropriated Funds Policy and Procedures*

VOTING OFFICER

The voting officer's job is to provide general voting information and to assist in the procedures of registering and requesting absentee ballots. He or she must also provide state election information, to include election date, officials to be elected, and constitutional amendments and other proposals to be voted on.

The Army Voting Assistance Program Unit Voting Assistance Officer Handbook is published each federal election year to help voting officers. Other informational aids, such as posters, are made available well in advance of general election dates.

Although much of the information in this pamphlet is applicable to any absentee voter, it is particularly directed to members of the armed forces and Merchant Marine and their spouses and dependents, citizens temporarily residing outside the United States, and overseas civilians who may or may not intend to return to the United States. As specified in the Federal Voting Assistance Act, these persons are as follows:

- Members of the armed forces while in active service, and their spouses and dependents.
- Members of the Merchant Marine of the United States, and their spouses and dependents.
- Citizens of the United States temporarily residing outside the territorial limits of the United States and District of Columbia, and their spouses and dependents when residing with or accompanying them. As specified in the Overseas Citizens Voting Rights Act, these persons are:

  Citizens of the United States who are outside the United States and may not now qualify as a resident of a state, but who were last domiciled in such state immediately prior to departure from the United States and are not registered to vote and are not voting in another state.

The Federal Voting Assistance Act sets up both mandatory and recommended procedures for absentee voting by specified categories of people as guidance for the states. But each state makes its own voting laws. It's important, therefore, that voting officers consult the summaries of the state laws in question before attempting to counsel persons on how to apply for registration or absentee ballots.

A special application form is printed and distributed for persons covered under the act, called the Post Card Registration and Absentee Ballot Request

(Standard Form 76, revised 1981), commonly referred to as the federal postcard application (FPCA).

The Department of Defense has directed the services to issue FPCAs directly to all eligible personnel for general elections taking place at two-year intervals (see AR 608-20).

The FPCA is used to apply for an absentee ballot and registration if the state or territory so authorizes. Standards of acceptance and procedures vary from state to state. Filling out an FPCA and sending it to the proper officials of a person's home state does not always entitle that person to absentee registration or voting privileges. In some states, it does; in others, the FPCA serves as a request for the state's own forms, which must be filled out and returned before final action is taken on the request.

In a few states, one FPCA serves for all elections in that calendar year. But one FPCA can never be used for more than one person. For instance, a spouse who is authorized by a state to use the FPCA must submit a separate form with his or her own signature.

In addition to abiding by the state's individual requirements for using the FPCA, voting officers should advise their personnel to follow these general rules:

- Print by hand or use a typewriter to fill in the form.
- Be sure that all requested information is supplied, and be sure that it is printed clearly and legibly.
- Show the name of the applicant twice—once printed or typed and once in the applicant's own handwriting. Anyone may fill out the card, but only the person who is to receive the ballot may write his or her name on line 15 (signature of person requesting ballot), unless the state specifies otherwise.
- Include street and number, rural route, or place of residence on the FPCA. It is also essential that an applicant include the name of his or her home county. This helps state officials speed action on the application when the form is not sent directly to the home county.
- Clearly print or type the military address so that no letter or number will be misread. Military addresses, particularly in abbreviated forms, are often confusing to civilians.
- Indicate the applicant's legal voting residence, which must be in a place where he or she actually lived—not just a residence of record. No more than one such address may be given. If the applicant has had more than one such address in a state, give only the legal residence address.
- If required, have the FPCA certified by a commissioned officer, unless the state specifies that a noncommissioned or warrant officer's attestation will be accepted. Civilians not attached to the military should have the FPCA certified by a notary public or other person authorized to make attestations.
- Before addressing the FPCA, check the state's mailing instructions. In some cases, the card is to be addressed to the secretary of state (who then

sends it to the proper local official); in other cases, it is to be addressed to a local official, such as the county clerk or auditor, or to an election board.

- Mail the FPCA as early as the state permits. No postage is required if mailed within the U.S. postal system or the APO/FPO system.

If FPCAs are not available, use a letter as an application for a state absentee ballot or registration. Provide the same information as the FPCA and mail it in the same way you would the FPCA.

The state, city, and county (or township) in which a person lived before entering the military usually is considered a legal residence for voting purposes unless he or she establishes residence elsewhere.

All states permit persons in the armed forces to acquire a new voting residence within their jurisdictions. When this is accomplished, voting rights in the old state of residence are lost.

Persons desiring to acquire a new voting residence must meet the new state's legal requirements. They must have lived within the state for the required length of time and presently must intend to make the new state their permanent home.

Time spent in military or federal service counts in meeting the total residence requirements. When there is any question pertaining to voting residence, seek an answer through the legal affairs officer.

Many states permit registration by absentee process, and some will register a qualified voter when they accept a voted absentee ballot. In others, a voter must be registered before applying for a ballot. Procedures vary and must be understood and followed exactly on a state-by-state basis.

Application for registration should always be made as early as the state permits, especially when registration must be completed before applications can be made for absentee ballots.

In some states, registration is permanent. Where such permanent registration laws are in effect, a person is not required to reregister for each election so long as certain requirements are met. In general, the requirements are that the applicant vote regularly and not legally change his or her name or move away from the area (such as precinct or district) where registered.

Most states permit minors to apply for registration if they will be of legal voting age by the date of the election.

When a ballot is received from a state, the envelope containing the ballot should not be opened until instructions on the envelope have been read. This is important, because some states require that the envelope be opened in the presence of a commissioned officer, notary public, or other authorized person. If there are no instructions on the outside of the envelope, it may be opened as any other mail.

States usually include full instructions inside the ballot envelope with the ballot form as a guide for persons voting by absentee process. Voting officers should help personnel follow these instructions or advise them when no instructions have been sent by the state.

Polls or straw votes are prohibited in relation to elections or voting choices. In addition, no commissioned, warrant, or noncommissioned officer may attempt in any way to influence any person's choice of candidate. The actual marking of the ballot—the voting—must be done secretly. It's required by law.

When possible, the voting officer should provide a place where ballots can be marked in secret. A fabricated voting booth, however crude, will not only meet the requirement but also offer an opportunity to publicize the voting effort.

Voting officers who need more information should not contact state or local officials but should write to the Adjutant General, ATTN: DAAG-DPS, Department of the Army, Alexandria, Virginia, 22331.

References

AR 608-20, *Army Voting Assistance Program*

The Army Voting Assistance Program Unit Voting Assistance Officer Handbook

PART III

Social and Family Matters

17

The Social Side of Army Life

A feeling of belonging—born of unity of purpose—naturally pervades Army communities. It is enhanced by the shared challenges, hopes, expectations, and even fears among the families of the soldiers at the installation. This community of interest is a noteworthy contrast to the experience of newcomers in most civilian communities, wherein neighbors may know one another but often live in quite different social worlds. Officers and their families are accepted as immediate, full-fledged members of the military community upon their arrival at a new post.

This chapter is an extension of chapter 1, because Army social and family culture springs from the same traditions and time-honored practices we call Army customs and courtesies. The subjects presented can assist in creating an atmosphere and an environment for a pleasant, gracious, and rewarding life. This is important for those in the Army because of the official missions assigned, the frequent changes of station, and the challenges related to living conditions.

Some social activities in which officers and their families participate are similar to those engaged in by other professions. Some are unique. As they come from all states and regions of our country, officers and their families tend to adopt activities that have been enjoyed elsewhere. Army families have the great advantage of being bound together by the common interest of the Army mission and association in joint undertakings. For these reasons, community interests are likely to be extensive and participation quite general. As in social and civil life elsewhere, including religions, fraternal, and similar activities, each individual receives in benefit and enjoyment about the same as he or she contributes.

ARMY SOCIAL CUSTOMS

Just as with other Army traditions, Army social customs can clearly be traced to historical practices, many of which have their origin in necessity or practicality. As the Army has evolved, many of what were once practical necessities have now become traditions. Most of these nevertheless remain quite enjoyable, although others may seem amusingly archaic. It is therefore useful to understand their origins, as such a comprehension can enrich your understanding of Army life; it will also give meaning to what otherwise might *seem* to be artificial.

Over the years, the officer will experience the social life of the Army station as well as that of the civilian community. The newcomer can be confident that the social practices learned in the Army will stand him or her in good stead within the civilian community. There are infrequent formal dinners and receptions that are held for necessary purposes, such as to honor visiting senior officers or for the reception of a high government official. Take them all in stride. This chapter has been prepared to be helpful. Army social life consists of pleasant human relationships, adding to the enjoyment of service associations, and should be approached with enthusiasm.

Increasingly, Army social functions originate at the small-unit (company or battalion) level, thanks to the evolution of family readiness groups. As such, these are necessarily joint enlisted-officer affairs. Given that most officers and many enlisted soldiers in today's Army come from essentially similar socioeconomic backgrounds, this is part of the natural evolution of Army social life. Embrace this change, which brings leaders and those they lead closer together. Cohesion in a meritocracy such as the Army is definitely enhanced by joint socializing, but always fractured by fraternizing. You and your family should always strive to be friends *to* your troops and theirs, but avoid being friends *with* them, so long as there is the chance that one day you may have to make difficult decisions affecting their lives.

A Special Message for Newcomers. The newly arriving officer, or officer and family, will get a sincere welcome from the members of the military community. Set aside any concern you may have about "newness." Unlike civilian communities, where residents may reside for years and years, the assignment of officers to a station or duty may continue for only three or four years, and when it is a command assignment, the tenure may be even less. So in fact, all members of an Army garrison, from the most senior to the most junior, are "new" in the sense of length of residence.

No person entering an Army community for the first time need be disturbed about the customs or the social practices. In the first place, Army people are understanding of the special problems of the newcomer and, if given the opportunity, are usually pleasantly helpful. Let people know who you are and that you are pleased to be among them. Seek acquaintances among individuals who share your special interests, backgrounds, or activities. You need not be the life of the party or the best-dressed person in attendance. Just be pleasant, and you will do well.

As a junior officer, you will not be expected to do a great deal of formal entertaining, which initially would mean the return of obligations, in due time, and in a manner entirely appropriate to your means. Participate in "Dutch treat" gatherings, informal picnics, and such other normal social events that come your way. Later, as you learn your way around and have developed acquaintances, you may even organize inexpensive share-the-cost gatherings. Eventually, you will even help some newcomer newer than yourself get off to a pleasant start.

Manage your personal financial affairs capably and firmly. You are aware of your pay and allowances, as well as the obligations you have assumed. Live within your means. Practically all your peers have the same challenges you do. Face them and adjust to them. Do a top-notch official job for the Army, and then have such good times with pleasant associates as circumstances permit.

The Officers' Community Club. The officers' club frequently sponsors social activities for officers and their families. There are facilities for dining and for holding private dinner parties, plus a program of dances, card parties, and other forms of group entertainment desired by the members.

It is likely that you will wish to participate in those activities at the club or elsewhere that are sponsored by the unit to which you are assigned. An officer with permanent station at a post having a club should become a member at once. To fail to do so will cause the officer and the adult members of his or her family to miss much of the available social life of the post. Most major units hold periodic membership drives to encourage officers to join the club for the benefit of all.

Cultural Opportunities. There are many cultural opportunities available at an established military station. They provide opportunities for the enjoyment of acquired interests and the development of new ones. Concerts, fairs, and, in foreign stations, joint cultural events, such as Oktoberfests in Germany, are staged periodically for the enjoyment and edification of all. Although their resources have in many cases diminished over the last decade, many post libraries are well equipped, and new books and magazines are acquired regularly. Libraries are under the direction of professional librarians who may be relied on to render assistance. In many cases, interlibrary loans are available.

Opportunities for Community Service. Military stations and adjoining civilian communities provide opportunities for community service in many worthy causes. Members of Army families may find rewarding activities that pertain to problems of the garrison or join with civilian organizations in work beneficial to the community. There are a variety of organizations on every post that welcome volunteers. The Red Cross sponsors grey ladies and nurses' aides. The United Services Organization (USO) needs helpers and hostesses. Chaplains (and off-post clergy) need teachers and other lay support. Schools are supported by the Parent-Teacher Association. Scouting and sports activities, such as Little League baseball, need the support of parents. The Army Community Service (ACS) also provides many opportunities for volunteer work.

Social and Recreational Opportunities for Young People. An Army station is a healthy and interesting environment for young people because of the many active organizations that thrive on most posts. Youth sports leagues and scouting organizations have been established at most Army installations where families are present. The chaplain sponsors religious activities for young people. Adults are always needed to serve as leaders in these activities, and those who do will find the experience rewarding and interesting. Officers and their spouses should plan to be active at least in those activities in which their children participate.

BUILDING SOCIAL GOODWILL

This process is the development of respect and esteem. Essentially, it is the sum of an indefinite number of impressions made on others in the community. This discussion deals with the fundamentals.

Strive to Be on Good Terms with All. Military family members have likes and preferences in developing friendships and friendly associations, just as other people do. However, it is wise to avoid cliques. Avoid open expressions of criticism or dislike, which develop hard feelings. Do not always restrict your social associations to the same few individuals. Broaden your acquaintances. By so doing, you will reap the reward of building more friendships and finding more people of interest.

Upon Receiving a Social Invitation, Express Appreciation at Being Included. Accept at once if that is your wish. Decline at once if unable to accept, and extend a courteous word of appreciation. If there is uncertainty about being able to accept, state the reason for the uncertainty. It is better to accompany uncertainty with a declination so that the host can invite others. If invited to accept or decline at a later time, give the answer as soon as possible. Do not keep the host dangling in uncertainty longer than necessary. Remember in expressing appreciation that it is the hostess or host who is extending the courtesy—not the guest by accepting.

Pay Particular Attention to Written Invitations. On many, if not most, you will see the initials "R.S.V.P." or the words "Regrets Only" in the lower left corner. The former is an abbreviation of the French words *repondez s'il vous plait* ("the favor of a reply is requested"), asking that you acknowledge the invitation by either accepting or declining so that your host or hostess will be able to make firm plans regarding the number of attendees. "Regrets Only" assumes that you will be there, and requires a response only if you cannot attend. Do your host or hostess the courtesy of responding so that there will be no doubt regarding your intentions. *Hint:* Establish and maintain a social calendar and write in the occasion, date, and hour—and then keep track of your commitments. Few things spawn doubt about your personal—and even *professional*—sense of responsibility faster than a "blown" social engagement.

For any Social Affair, Arrive Promptly. There are problems of timing that apply with equal force to the simple home dinner and the formal dinner for a large gathering. It is rude to arrive late at a dinner unless the tardiness is indeed unavoidable; even then the host should be informed by telephone if possible. Prompt arrival is also essential at other social events, such as a cocktail party or a gathering for a specific purpose where the presence of all invited guests at a stated time is clearly desirable. When the invitation is for a stated period, such as "6–8," customs differ. Invited guests may arrive after 6 and depart before 8, but they customarily do not remain more than briefly after the hour stated because it may interfere with other plans of the host and hostess.

PRACTICAL TIPS

The Right Clothes. In these days of informality, selection of the proper clothes to wear on various occasions can be a problem. The first point to consider is whether the occasion is essentially a military function or a private affair. For military functions, the proper uniform may be specified, thus settling the problem immediately. Civilian dress standards and their military equivalents can be defined as follows.

Formal. Gentlemen wear tuxedos, and ladies wear evening gowns. The military equivalent is the blue mess or white mess uniform, or the Army blue uniform with black bow tie.

Informal. Gentlemen wear business suits, and ladies may wear suits or dressy cocktail attire. The military equivalent is the class A uniform, or Army blue uniform with four-in-hand tie.

Casual. This is a tricky one, because it has so many interpretations. Gentlemen are never wrong to wear a sport jacket and dress slacks (have a tie in your pocket, just in case), and ladies should wear a blouse and skirt or slacks or a simple dress. Most hosts who specify this dress standard today, however, mean simply a collared sport shirt (golf shirt) and dressy trousers for men. If you are unsure, ask someone you trust who has attended a similar event put on by the same host. *Hint:* At unit social events at the battalion level and below, blue jeans and athletic shoes are often OK, but beware of assuming that the same is true of any event above that level.

Athletic Attire. Increasingly, many unit social functions involve some sort of sporting or athletic activity. When this standard is prescribed, wear whatever is normally worn for that sport, but bring casual shoes for pre- and post-event wear; don't forget a sweater, sweatshirt, or warm-up suit if the event will extend into a cool evening or is being conducted on a cool day. You will undoubtedly be sitting or standing around for most of the affair.

When in doubt, it is better to overdress than to underdress for the occasion.

The Right Words. At any social gathering, guests should strive to make the occasion pleasant for all. Conversation is important. Visit with all or many of the other guests. It is better to be a good listener than a good talker, but a good listener needs to be adept at starting subjects of conversation. Suitable subjects can include mutual but absent friends, the other guest's recent leaves or other interesting experiences, mutual family interests, current sporting events, and so forth. Think of a subject above the commonplace, and start the conversation. Then listen. Avoid controversial subjects such as politics and religion. Try to avoid shop talk—you deal with it all day long, often seven days a week.

The whole point of such social affairs is to allow officers and their spouses to become better acquainted in a convivial atmosphere. Avoid being a wet blanket. Do not take over the party. Use the opportunity of your host's hospitality to broaden your social horizons and strengthen friendships. Social gatherings of military people in many cases include officers of various ranks, experience, and age. Most senior officers enjoy association with younger officers and their wives or husbands. Make a point of offering a pleasant greeting and having at least a brief conversation with the senior officers and spouses present. The older or senior guests will do their part in making the younger ones feel at ease. Make a special point of greeting your boss and his or her spouse. Meet them as social equals, for such is your status as an officer or officer's spouse. Be polite and respectful, but never subservient or obsequious. If guests of honor or house guests are present, make a special effort soon after arrival, and engage them in sincere conversation.

The length of stay at a social function depends on the program of the host. Unless a reason exists to the contrary, a good rule to observe is to depart no more than an hour after the service of dinner is completed. In these busy days within military circles, the custom is for early departure. When the commanding officer is present as a guest, especially at a formal event, he or she is first to depart and has the special obligation of departing at the appropriate time. For purposes of general illustration, if dinner guests arrive at 8:00 P.M., goodnights are often appropriate by 10:30. At informal affairs of any kind, another good guidepost is to depart before the host and hostess can possibly begin to wonder about the departure intentions of their guests. In making a departure, do not dawdle. There are few things more irritating than a guest who prepares to depart and then stands in the hall or doorway, narrating little nothings while hosts stand patiently by. Arise, express your appreciation for the occasion, obtain outer wraps, say your gracious farewells, and leave.

Social Calls. The matter of social calls presents a special category for consideration. In the years before World War II, social calling was an established Army custom. Social calls were useful to welcome newcomers, to broaden acquaintances, and to bid pleasant farewells prior to departure. The custom was largely set aside during the war years, and it was never really revived.

In recent years, the practice of social calling has been largely replaced by periodic unit or station receptions at which newcomers are welcomed by and introduced to unit or station personnel. Similar affairs are often used as a means of bidding farewell to departing officers and their spouses. These gatherings are usually called "hails and farewells," and all officers are encouraged to attend when one is held by their unit.

Some Tips for Hosts and Hostesses. At the dusty, dreary posts of the western frontier, the Philippines, or Panama—where so much Army tradition was born—unit or post social events were the only social activities available to officers and their spouses. Frequent, or even regular, dances, dinners, or other affairs were the sole relief from the hard living and tough missions inherent to assignment on the fringes of the American empire. Since World War II, however, these situations have largely changed. Keeping the following corollaries to these changes in mind will help hosts and hostesses schedule and arrange truly appreciated social events, enjoyed by all.

Off-Post Living. Many officers live off post. Schedule events to allow time for officers to leave duty at a reasonable hour, return home, change clothes, and travel to the event's location by the designated start time. End events in time for them to return to their quarters at a reasonable hour, especially if the following day is a duty day.

Unmarried Officers. An increasingly large proportion of the officer corps is single. Although some will bring dates to Army social events, not all will. Be sensitive to the implications of this social dynamic, for they are several and important. First, do not schedule *too many* social events—one of the reasons why so many officers are single is that the operational tempo of today's smaller, yet more frequently deployed Army limits their opportunities to meet potential mates. Next, ensure that any games or other planned activities will be enjoyable for unescorted officers as well as for couples. Too frequent—or too awkward and irrelevant—social events will be seen by single, unescorted officers as "mandatory fun sessions" and will either unduly restrict them during their precious free time or cause them to not attend.

Two-Career Couples. Many officers are married to civilian professionals or other career persons (including other officers) who have social schedules of their own. Be aware of the legitimate competing social requirements facing many officers today, and schedule social events accordingly. Ultimately, accommodating everyone is impossible, but do not chastise or think ill of the officer who must send his or her regrets to attend a conflicting affair.

Remember the Children. Getting good babysitters is difficult in many parts of the United States. American society is evolving, and in some areas, it is becoming increasingly difficult to find dependable, mature child care at reasonable prices. Plan events well in advance, to allow plenty of time for arranging child care. Schedule some of them during the day on weekends so that children may attend. End them at times that will allow babysitters to be relieved at a proper hour—or before the kids reach "tracer burnout." All these considerations will add to the enjoyment and density of attendance at your social events.

Calling Cards. Tied to the custom of social calling, the use of calling cards has also largely faded into the past. When paying social calls, calling cards were left in prescribed numbers by both the officer and his wife for adult members of the visited officer's family. (They were usually left in a silver tray by the door.) The use of calling cards was helpful for remembering names, and they were also sometimes used to check attendance.

There is no *official* military counterpart to the civilian business card. Still, in recent years, many officers have found it convenient to use such cards in their contacts with other officers, civilian officials, and private business concerns. They can easily convey the myriad communications addresses so often possessed in this Information Age (official and unofficial e-mail, commercial and DSN voice and fax numbers, addresses, and so forth). Today, with the software available for use on personal computers and the proliferation of commercial office and business centers equipped with color printers, officers can pretty much design their cards exactly as they please. There is no longer a protocol, as there once was with calling cards. Although it is a matter of individual taste, try not to be garish or extravagant with your design. Distinctive unit insignia ("regimental crests"), unit patches, or other organizational insignia are fine; so are rank insignia (although, presumably, these would benefit only those who cannot read your rank as it is written). Avoid the temptation of cards with rampaging tanks, giant jump wings, enormous ranger tabs, fluorescent lightning bolts, or—even worse—skulls, daggers, or other fantasy art. The only people who will be impressed with such things are probably not worth giving a card to anyway. More importantly, most people, civilians and military alike, will gain a poor impression of you and the Army in general from such adolescent expressions of uncontrolled clip-art mania. Remember, almost as much as personal contact, your card makes and leaves a lasting statement about you: Informative is good, imaginative is better, but tasteless is bad.

Introductions. Here is a social skill that is worthy of cultivation and maintenance. Introductions are a key part of establishing communication with a stranger. Making introductions in a socially consistent format allows the persons being introduced to concentrate on the names, not on the manner of introduction. Here are the simple guidelines:

Introducing your wife to any man (except chiefs of state and some very high-ranking church dignitaries): "Mary, this is Colonel Mustard."

Introducing your husband to another man: "Colonel Brown, may I present my husband, Jack," or "Jack, this is Lieutenant Victor." (The correct order is to introduce the junior to the senior or the younger to the older, as noted below.)

Introducing one lady to another: "Mrs. Jones, may I present Mrs. Green," or "Mary, this is Mrs. Green."

Introducing one officer to another: "Major Smith, this is Captain Miller."

Introducing yourself to an officer senior to you: "Sir, may I introduce myself? I'm Captain Jones." (Wait for the other to extend his hand.)

Introducing yourself to an officer of equal or subordinate rank: "I'm Captain Jack Jones." Extend your hand. If you are in uniform, there is no need to include your rank when introducing yourself to peers, as such usage can easily appear pompous or phony. Never introduce yourself as "Captain (P) Someone," or "Major Promotable Someone." Not only is such usage improper (it's not your rank), but it is indicative of insufferable egotism. The administrative decisions of a promotion board are of no relevance in an introduction, as they make no difference regarding the conduct of the person to whom you are being introduced.

In recent years, a similarly inappropriate practice has arisen in which some officers knowingly introduce themselves to subordinates by their first and last names only: "Hi! I'm Bob Someone!" This is both confusing (if the officer is not in uniform, the subordinate won't know what rank he or she is) and impolite. Army rank is not an indicator of social caste; it is, rather, a legitimate professional title that carries certain requisite decorum on the part of all involved in the introduction process. Don't be embarrassed to use it, *in any company.* Denying knowledge of your rank to newly met persons deprives them of recognition of how to address you, puts them ill at ease ("Why is this officer being so familiar?"), and may even signal disdain for the Army and its traditions. You've earned your rank; certainly don't abuse its use, but definitely be proud of it.

When introducing children or teenagers to adults, say: "Sergeant Jones, this is Jane Smith," or "This is my daughter Jane."

Respond to an introduction by saying: "How do you do, Colonel Green?" or "Sir, I'm pleased to meet you."

To thank a host and hostess on departing a social function, say: "Thank you for a delightful/enjoyable/fun evening."

If you must leave a function noticeably early, say: "Mrs. (Hostess), I'm so sorry I/we must leave early." (Then briefly state your valid reason.)

Here are some guidelines for using titles with names:

Major General Jones: "General Jones."
Brigadier General Smith: "General Smith."
Colonel Green: "Colonel Green."
Lieutenant Colonel Stone: "Colonel Stone."
First Lieutenant Whitman: "Lieutenant Whitman."
Second Lieutenant McDonald: "Lieutenant McDonald."
Chief Warrant Officer (CW2–CW5) Coffee: "Mister Coffee."
Warrant Officer Lake (WO1): "Mister Lake."
Command Sergeant Major Ryan: "Sergeant Major Ryan."
First Sergeant Topp: "First Sergeant Topp."
All other noncommissioned officers: "Sergeant Lastname."

Gentlemen are introduced or presented to ladies, not the reverse. This holds even though the gentleman may be very distinguished and the lady very young. *Exceptions:* the President of the United States, a royal personage, or a high-ranking church dignitary.

The most common way to make introductions, always in good taste, is to state the names in proper sequence: the lady, senior, more distinguished, or more elderly first. "General Smith, Captain Jones"; "Miss Youthful, Colonel Adams"; "Mrs. Gracious, General Stone." Use a rising inflection for the first name pronounced ("General Whitman? Cadet Johnson"). The more formal method can also be used: "General Smith, may I present Captain Jones?"

Acknowledgment of an introduction by saying "How do you do?" is always appropriate. When men are introduced, they shake hands, standing, without reaching across another person, if possible. They may say nothing, just look pleasant or smile, or say a courteous, "It is nice to meet you" or "How do you do?"

When women are introduced to each other, with one sitting and one standing, the seated one rises to greet her hostess, or a distinguished lady, as an act of respect. This would apply, for example, to the wife of a very senior officer. In the usual case, the seated lady does not rise. The reply to an introduction may be a simple, "How do you do?"

When a man is introduced to a lady, he does not offer his hand unless the lady proffers hers. In Europe, men are taught to take the initiative in handshaking. A lady does not refuse a proffered hand.

At a social occasion, host and hostess should shake hands with guests in greeting and upon their departure.

Remembering Names. The Army *is* people, and there is no sweeter sound to most people's ears than their own names, pronounced properly and spoken with respect. Soldiers and their spouses meet such a large number of people that remembering names is a true challenge. Nevertheless, it is an important skill. To remember names at all, you must first understand them clearly. At the time of introduction, be sure to hear and understand the name correctly. Repeat it aloud to ensure correctness and to aid your memory; your interest will please and flatter many. During the occasion, strive to use the name in conversation, and fix the name to the face of the person. Be very careful when making introductions to state names correctly and distinctly.

It is probably a sign of the evolving state of American culture, but it has become accepted in some quarters of the Army to mispronounce long or unusual names, or to use nicknames even in a formal environment. Few actions more quickly reveal the shallowness of intellect or the insincerity of the person who introduces someone as "Lieutenant Alphabet" or "Sergeant Z."

Upon arrival at a new station, it is not a bad idea to use a notebook in which to record the names of people you should remember. List their complete names and middle initials. List names of husband and wife. For family friends, list names of children and birthdays, if known. After the passage of months or years, such a record may be of much interest and assistance.

Social Precedence and Protocol. An understanding of basic principles of social precedence and protocol is essential to Army leaders. A simple definition of *protocol* is "the code of international politeness."

Precedence becomes a factor for host and hostess to consider in deciding who sits where at a dinner or in deciding places at a speaker's table or positions in a receiving line or other formal occasion. The place of honor is on the right or in the lead. *Examples:* The superior walks on the right of the subordinate. Subordinates step aside for the superior to be first to enter a door. The superior is last to enter an automobile but first to leave, and the right rear seat is reserved for him or her. The superior returns the salute of the subordinate. An officer of lower rank is presented to the officer of higher rank. Age is a decided consideration in social precedence, especially in a group of mixed professions or vocations.

Formal Receptions and Receiving Lines. For understandable reasons, the formal reception is probably used more within military stations than by other professional groups. This is likely to be true because of the frequency of official visits by military and civilian dignitaries. They are convenient for other special events such as a wedding reception honoring a newly married bride and groom, or the introduction at a social occasion of a newly arrived commander and his or her spouse. Invitations to official receptions are accepted as first priority unless duty precludes attendance.

An officer wears the uniform prescribed, or as local custom makes appropriate. Spouses dress appropriately for the occasion. In the evening, a formal evening dress or tuxedo would be customary for civilian spouses. In the afternoon, a cocktail dress or coat and tie would be in good taste. In these days of swift changes and high prices, a civilian lady dressed neatly in a nice street dress or a civilian gentleman in a business suit would cause no lifted eyebrows at a social event on an Army post.

Strict protocol is observed at formal receptions. At an official function, the host ranks first, then the hostess, and then the honorees. For example, the commanding officer of the unit holding the reception is on the right of the receiving line, with his or her spouse on his or her left; next is the ranking honored guest, with his or her spouse on his or her left; other officers and their spouses extend the line in the same manner. If civilian officials are members of the receiving line, their place is indicated by the host, who should be guided by good judgment.

It is customary upon arrival at a reception for the guests to pass through the receiving line. An aide, adjutant, or protocol officer is often used to announce the names of the guests to the host. Discreetly and clearly announce your rank and name, and your escort's name: "Major and Mrs. Jones." If you are single and are escorting a date, be sure to make clear your escort's last name: "Captain Green and Miss Hudson." Except at the White House, or a diplomatic reception, the lady precedes the gentleman through a receiving line. At the White House the gentleman precedes and is presented to the President of the United States, then he presents his wife to the President. Each person in a receiving line usually introduces the guests to the person next on his or her left. A simple, cordial greeting, using correct names, is in order.

Keep in mind the social customs of today's Army. If in the increasingly unlikely event that a lady is wearing gloves, she removes them for gentle but firm handclasps. It is both awkward and rude to have a drink in your hand when you proceed through the reception line. Having a cigarette or cigar in your hand (don't even *think* about between your lips) will definitely make a lasting impression that your reputation is unlikely to recover from.

Some occasion—dinner, dance, cocktails, and so on—follows, after which you should make appropriate farewells to the hosts and the guests of honor. If the activity is large, such as a ball or dinner-dance sponsored by a major unit, and the attendance list includes several hundred officers and their spouses or guests, it is not improper to simply leave after the colors have been retired for the evening.

Return of Hospitality. It is a good and sensible rule that hospitality accepted should be returned. Those who accept all invitations that come their way and never return the hospitality are inconsiderate or, worse, moochers and will be regarded as such.

Here is another good and sensible rule: No person or family should entertain socially beyond their means. In the service, there is no occasion or precedent for "keeping up with the Joneses." Most officers are dependent on service pay. They have nothing to gain by trying to maintain a social standard beyond their bank accounts, and if they do so, they will receive more disapproval than favor. Still, hospitality accepted must be returned.

The saving grace is that hospitality accepted need not be returned with identical reciprocity, or even by the same means. For example, Colonel Oldtimer is in command of a brigade, and his wife is a gracious lady who has a well-equipped and fairly large home of which she is quite proud. Three new officers join the command. Soon, all three with their spouses or dates are invited to dinner. The Oldtimers go to some trouble to entertain them well, with their best china, linen, and silver, which they have accumulated during twenty-four years of service in the United States and abroad. They serve lavishly prepared food and exotic drinks, recipes for which they acquired in their extensive travels. They have done this with no thought of artificially impressing the newcomers, nor to emphasize their more senior position (and larger paycheck). They wish to show the newcomers a good time, that they are welcome and respected as valuable members of the organization. Also, they have accumulated nice things that they enjoy using.

No quality officer would expect his subordinates to return such hospitality by identical means. Instead, a simple home-cooked dinner would more than suffice, as would an outdoor picnic or dinner at the officers' or community club. To further reduce expenses, two or more officers may combine resources as joint hosts and hostesses. In fact, the most appropriate response to the Colonel and Mrs. Oldtimer's graciousness might be dinner prepared and attended by all three lieutenants and their spouses or dates. No excuses need be offered, and to do so would express poor taste. All officers have faced the same financial hurdles. They know the costs of raising a family and the expenses of frequent changes of station. They have the same unfavorable views of those who elect to live beyond their means or to show off their independent wealth. Don't do it.

Bachelor officers should follow the same course. If bachelors do not have quarters suitable for entertaining, there are a number of solutions. A dinner or cocktail party at the officers' club or a picnic are examples of appropriate returns of hospitality.

Here is a social program that will meet the requirements and lie within the financial capabilities of nearly all officers: Belong to the officers' club and attend its social events if desired. Participate in functions arranged by your unit to the extent practicable. Attend all official receptions because you are expected to attend them, if invited; in general, they establish no obligation. Accept invitations from your commanding officers, staff officers, and close associates and from those other families whom you know and like or would like to get to know better. With graciousness and politeness, decline the others; the "bus to Abilene" never gets there, and you can't please everyone.

Being a Good House Guest. In the course of your service, you will find yourself the guest in friends' or acquaintances' homes on many occasions. Thoughtlessness or ignorance can play havoc with even the most cherished friendships. Conversely, thoughtfulness and grace will strengthen them. As in all other human relationships, social manners are the result of a proper regard for the rights and needs of others.

The host and hostess should be informed as accurately as practicable of the exact hour of planned arrival and departure. This will enable them to plan your visit and to resume their normal activities when you have left. Particularly annoying is the guest who accepts an invitation and states only that he will arrive, "during the day." Such vagueness is extremely frustrating for the host and hostess and places unnecessary limitations on their activities. Equally troublesome is the guest who neglects to announce a definite time of departure. The hosts wish to extend all possible consideration to the guest, but upon his departure will wish to resume other activities. They may wish to keep or cancel important appointments or obligations, or may need to complete necessary shopping and errands. Uncertainty in these routine arrangements is displeasing. Of course, the guest who arrives to stay for a few days and fails to inform the hosts about plans for departure—until descending with packed bags to say farewell—exhibits the ultimate lack of manners.

Some ways to inform your hosts are by email or telephone call, which might state: "Thanks for your kind invitation. We accept with pleasure. We plan to arrive at your home soon after 4:00 P.M. Friday afternoon and must start our return before 9:00 A.M. on Monday." Having established these days and times bend heaven and earth to keep them.

A guest must adjust to the conditions of the household. Keep your own things picked up and the room tidy. Leave the bathroom in the same condition as you find it, or better. On the day of departure, the bed should be stripped and the used bed linens folded and stacked at the foot of the bed. A guest should share in the household work to the extent that is welcome or acceptable. In this way, hosts and guests have more uncrowded hours in which to enjoy one another, and the hosts are spared excessive strains.

If plans have been made or suggested, the guest must show pleasure in sharing them. Adaptability is the order of the day. Both host and hostess will require *some* time to be by themselves to take care of personal responsibilities. Make it easy for them to do so. Take a walk, write a letter, read a book, go running. This permits the hosts to perform routine household tasks. An invitation by a guest to take the family of the host out to dinner is often a welcomed courtesy, but it must not be pressed if once proffered and declined.

Prior to departure, be certain that no personal belongings are left behind for the host to package and mail. A remembrance to the hosts in the form of flowers, candy, or a book, given either before or after departure, is appropriate and generally most appreciated.

Guests who fail to do these things and others that suggest themselves prove the old saying about making the hosts twice glad—glad when the guest arrives, and glad when the guest departs. Being a good house guest may result in future invitations. The essence is regard for others.

Expressing Gratitude. In the Age of Information, with all manner of convenient communications media, there is no substitute for a hand-written "thank-you" note—an email simply will not suffice. Even if you brought an appropriate house gift or intend to return the hospitality in the near future, this is a matter of basic manners. Whether the host arranged a cocktail party, prepared dinner, or provided accommodations, a significant effort was made in providing hospitality for you. You should send a thank-you note within 48 hours of your return to your quarters. It doesn't have to be much; engraved stationery or a store-bought card is *not* required. Simply pen a note reiterating your sincere appreciation, and mail. If your hosts are husband and wife, the note goes to the wife, the hostess. If a bachelor officer, write it yourself; if male and married, your spouse should do it. Never send it to your host, especially if he is superior in rank—such acts reek of bootlicking, unless he or she, too, is unmarried. (Even then, be careful to word the note without a trace of sycophancy, but rather sincere and respectful appreciation for hospitality rendered a fellow professional.) *Never* send it through the message center or drop it in your boss's in-box.

STANDARDS

It is natural for a newcomer to the military community to wonder about the moral standards that will be encountered. Many, particularly spouses, have little or no concept of what to expect in the Army environment. Since the advent of the all-volunteer Army in 1973, it is no exaggeration to say that a huge percentage of Americans have never known *anyone* on active duty. As anyone who has been in the Army knows, the portrayal of Army life in the popular media is often inaccurate and misleading. Rarely do Army or Navy JAG officers helicopter in to work, and Army wives do not lead the lives seen on television. Here is some reassurance for those about to enter the greater Army community.

Soldiers are drawn into the Army from the population. For the most part, officers are men and women of above-average education and moral fiber. Many enlisted soldiers are also quite well educated, and the vast majority of them are of impeccable character as well. They could not remain long in the Army if this were not so. Army spouses are generally people who share their spouses' values and, in many cases, their origins. They bring into the service the customs, standards, and expectations common in the communities in which they have lived. When they enter the Army, they are joining a society that is committed to a common, selfless cause. Soldiers are subject to military discipline and to the Uniform Code of Military Justice (UCMJ). They may be punished, even dismissed, for breaches of this restrictive and demanding code. In the case of officers, they may be punished for "conduct unbecoming an officer," an article of the UCMJ that defines as illegal many acts for which civilians cannot be prosecuted, and even some that are widely tolerated.

In such a closely knit community, it is natural that each person should wish to deserve the goodwill of associates rather than their scorn. For these reasons, the moral and conduct standards encountered in the military service are at least as high as, and almost certainly higher than, those encountered in almost all other walks of American life.

Soldiers and their wives or husbands are people. Among them will be those whose frailties show. All the problems of men and women have not been solved by the military, but remember, the Army consists of American soldiers who have *volunteered* to put their lives on the line to support and defend the Constitution against all enemies, foreign and domestic, and to bear true faith and allegiance to the same. They have no hope of extravagant pay; they work outrageous hours at difficult and sometimes exhausting or even potentially deadly tasks; they have forsworn more level paths and chosen many harder "rights" over myriad easier "wrongs." Of what other large community in this country can the same be said? Where should you feel safer or more at ease? Of whose acquaintance should you feel more proud?

LIVING OFF POST

A large portion of Army families live off post because there are too few sets of family quarters to meet the need. Many officers have duty assignments within a post or station but reside with their families in a nearby civilian community. This condition presents challenges that must be faced and solved, as well as opportunities to be embraced.

To the extent that you can, take care in the selection of the neighborhood in which you ultimately reside. Do some research, and choose an area with low crime rates and good schools. Check with the local police about areas to avoid; check with the state board of education to see how the schools in the general area are rated. Consider the roads and weather if you are considering living a significant distance from the post; if you are going to be assigned to a rapid-deployment unit with frequent geographical or temporal restrictions and short response requirements, don't consider living too far away, or in an area likely to get snowed in or cut off by flooding.

Most officers and family members wish to enjoy congenial relations with the residents of their new community. You and your family may wish to participate in some local educational, religious, fraternal, or recreational facilities. Generally, these facilities are made available to newcomers upon application or upon invitation. A little tact is appropriate. Remember that the good things enjoyed by a community were obtained by the vision and work of the residents and paid for by taxes, dues, or donated funds. It is likely that the residents will be entirely willing to share many of these things with the transient military family, but they will do it more willingly when it is obviously appreciated.

There are some negative considerations. If you think your own hometown or your last station is preferable to the present location, keep it to yourself. If you think a particular merchant or landlord is overcharging, don't denounce *all* local merchants and landlords as cheats and gougers; just take your patronage elsewhere. Avoid doing or saying things that irritate and annoy the longer-term residents of the area. They may determine whether you receive the hand of fellowship or the cold shoulder.

It is wise to meet and make yourself known to the businesspeople and others with whom you have contacts. This means the banker, the grocer, the mechanic servicing your car, the clergyperson at your church, the principal at your child's school. If you will need credit arrangements, make them in a businesslike manner. Establish yourself as a new and desirable member of the community.

You will have the same privileges regarding facilities and activities on post as you would have if living in on-post quarters. That is, you will have full privileges at the officers' open mess and post organizations and facilities of all sorts. Judicious choices must be made between the activities in town and those on post. The primary interest may properly lie on the military installation. Divide your activities between post and town on the basis of convenience, and use both as desired.

As time passes, associations may ripen into friendships, and life itself may be enriched by these local contacts.

18

Family Affairs

In many important ways, the Army itself is a family; in fact, it is a family *of* families. Good units take on a family atmosphere in which soldiers bond to one another through shared hardships and good times, are inspired by unit traditions and heritage, and are nurtured by concerned and fair leadership. Unit veterans—especially combat veterans—should also be a significant part of the unit "family" and can add richness and cohesiveness to the overall experience of belonging to a good unit.

In addition to the family nature of the units themselves, there are other "family affairs" about which officers must be concerned. These affairs can essentially be categorized as *personal*—that is, each officer's individual family needs and obligations—and *extended organizations.* In an expeditionary Army such as ours, in which units are frequently deployed away from home station on lengthy training or operational missions, it is not only a moral but also a practical imperative that you look after your soldiers' families' welfare during these stressful separations.

PERSONAL

Considerations for Married Officers. As discussed earlier, an American Army officer's profession is a *calling*, not merely a job. By your oath, by the requirements of your duty position, and by the obligations of Army tradition, you will frequently have to sacrifice certain interests of your personal family for the good of the organizational one. "Selflessness" does not mean "when it is convenient." As a leader in your unit and extended organizational families, you will quite often have to place the preferences of your personal family behind the needs of the other two. If you are married and haven't yet discussed this with your spouse and children, do so soon. This required prioritization can be damaging to personal family relationships if the basis is not understood. Putting the unit's requirements before your personal family's preferences does not mean that you do not love and cherish your family; this is the path you have chosen, and there are heavy responsibilities involved.

This does not mean that professional needs *always* take priority over family needs. Far from it. As a spouse and parent, it is your responsibility to properly look after and support your family's emotional, financial, educational, recreational, and other needs to the best of your ability. You must strike a fine balance between the two, and every good Army boss is not only fully aware of this but also heartily supports it. There is a big difference between devotion to duty and neglect of family, just as there is a distinction between seeing your family's needs and conjuring excuses to avoid long hours or difficult duty. You should never fabricate artificial "crises" to justify staying at work when you should be home. The intricate, sensitive details related to good spousal and parental relations are outside the scope of this book. If you need advice, the unit chaplain is an excellent and fully legitimate source from which to start seeking sincere, effective counsel.

Considerations for Single Officers. If you are not married, seriously consider the impact of your profession on any relationship. Army life is *not* for everyone. In many cases, potential spouses are not willing to make the extraordinary sacrifices required by the exigencies of the Army calling. In some cases, they are not emotionally *capable* of doing so. Pursuing a demanding career while an Army spouse is extremely difficult. Before you enter a serious relationship with someone, look carefully and candidly at his or her own aspirations and desires.

Although it is not wise to overstate them (or talk about them *too* early in a relationship), be direct and candid about the realities of Army life. On the one hand, the sacrifices that must be made—frequent dislocations, long separations, heavy work loads—must be understood and their import frankly evaluated. On the positive side, a joint investment into the security of the Republic is a rewarding and even exciting adventure that should appeal to most patriotic Americans. Opportunities to travel also appeal to many. Don't forget to remind him or her that you will not be in the Army *forever* and that after you retire you may be willing and able to shift priorities dramatically. To avoid these issues is to obscure reality and will only result in problems down the line.

THE EXTENDED ORGANIZATIONAL FAMILY

Since the 1980s, the Army has made enormous strides in formalizing its system for looking after the needs of family members. From the beginning of World War II through the Vietnam era, the Army was overwhelmingly populated by conscripted enlisted soldiers whose families tended to remain in their home areas while their spouses deployed to war. During neither the Korean nor the Vietnam Wars did Army units deploy and stay until it was over. Instead, they deployed as units in the early phases of each conflict, and the personnel rotated home individually after a year, while the unit stayed in the combatant theater. As a result, the families of long-term professionals (NCOs and officers) either went back to their hometowns until their soldiers returned or stayed at the home station, supported by units that had not deployed. There were no combined training centers, relatively few lengthy training events away from home station, and rarely expeditionary deployments.

The volunteer Army changed all that. As a result of the end of the draft in 1973 and steadily climbing military pay, an increasing number of junior enlisted soldiers brought in or started families. Along with deployments to lengthy joint training exercises and the Jungle Operations Training Center (JOTC), during the Army training revolution of the late 1970s and 1980s, units began frequent, even lengthier rotations through the various combined training centers. Operation Urgent Fury in 1983 saw the dawn of an epoch in which many Army units became truly rapidly deployable, as the United States embarked on strategies to roll back Communism and more aggressively defend its citizens, interests, and allies. The period from 1994 to 2012 saw an increase in operational tempo of 300 percent as Army units deployed all over the world in pursuit of the goals and objectives inherent to the United States' role as the remaining superpower. From 2001 to the present, operations in Afghanistan, Iraq, and other regions have increased a demand for a comprehensive, formal system for the care of Army families during the many periods of protracted, and often indefinite, separation from their soldiers.

Family Readiness Groups. Family readiness groups (FRGs) are designed to assist family members during periods of deployment or emergency. They also provide a means of information flow between the military and family members about family-related and, in some cases, operational issues. Rear detachment commanders (RDCs) and FRGs provide primary support for family members within units during deployment. During periods in garrison, FRG activities are directed at developing a sense of community among families in partnership with their unit—the extended organizational family.

Training for members forms the basis for all successful programs. The FRG link with commanders is critical. Without commanders' active support, FRGs could not exist. FRG legitimacy flows primarily from the commander's approval and from chain of command support. FRG volunteers will be prepared only if family support volunteer training and organization have command emphasis. Officers must understand, support, and provide the training needed to produce quality care for family members.

Family separation creates psychological stress for both the deploying soldier and the family members left behind. Soldiers who feel that their families can manage without them are better able to concentrate on performing their duties. Commanders demonstrate genuine concern for their soldiers when they actively support FRGs as part of unit predeployment readiness. During deployments, FRGs assist the rear detachment in sustaining families of deployed soldiers by exchanging support and transmitting accurate information between families and the RDC.

FRG Volunteer Training. FRG training initiatives, such as the Army Family Team Building (AFTB) program, are intended to provide support, outreach, and information to family members while at the duty station and during periods of family separation. In many cases, the G-1/S-1 is the command structure link with the mission of supporting and coordinating FRG training initiatives. These officers coordinate periodic "sustainment seminars" to ensure that all units have several trained volunteers who can work collectively with unit family support programs. FRG volunteers are often the first contact points for family members who need help. Due to their unique functions, FRG volunteers are often in a position to best evaluate family coping abilities given differing situations. Moreover, the volunteers can arrange early assistance to prevent a crisis situation from occurring. Well-run FRGs can act as a buffer between family problems and the soldiers downrange. The more resilient a family can be, the more the soldier can focus on the mission at hand. Training—in relevant subjects such as problem solving, crisis intervention, newsletter preparation, and so forth—should be conducted routinely. Additionally, training should periodically be reevaluated based on situations encountered before and during deployment. Some FRG volunteers need special training. Refresher courses may benefit others with past experience.

Organizing FRGs. Commanders and other leaders must be deeply involved in the organization and activities of FRGs. The *Army Family Readiness Group Leader's Handbook* is an excellent reference. FRGs can be formed quickly during mobilization or deployment preparations, but they work much better if they are ongoing activities established prior to deployment Leaders should participate in the process of establishing FRGs, but the ultimate effectiveness of a group lies in the strengths of the families themselves, their training, the imagination of FRG leaders, and the initiative of FRG volunteers.

FRGs are most effective at battalion and company levels. They work best when NCO, junior enlisted, or junior officer spouses with energy, interest, and natural leadership abilities are motivated to serve in leadership roles. FRGs are more successful with willing volunteers. These spouses are usually perceived as more readily able to understand the economic and lifestyle challenges of soldiers and their spouses. During major deployments, especially those involving activation of reserve component forces, there is a need for family support for those who do not fit the normal FRG criteria. These include guardians of children whose parents have deployed and active and reserve families of all services who have temporarily moved from military installations and rear detachments. Family support for this category should accommodate the needs of all military families, regardless of active or reserve status or service affiliation.

FRG Composition. FRGs are self-help organizations made up primarily of unit spouses but may include guardians of dependent children, parents and "significant others," and soldier volunteers. FRGs are open to families and guardians of all members of the command, regardless of a soldier's rank. Participation is voluntary.

Volunteer spouses in FRGs are sometimes associated with military rank or duty position. Leaders can be selected by the organizers or group members. There are many instances of spouses of junior soldiers serving in leadership positions. The main requirements are a caring attitude and a willingness to work. FRG success may depend on the degree of ownership volunteers feel toward the group.

FRG volunteer leaders may find that their homes become communications centers. It is important for FRG leaders to remain calm, be reassuring, and recognize that no leader can please everybody. FRG leaders can call on the rear detachment, family advocacy agencies, and chaplains for support.

It is important that FRG leaders take care of their families and themselves. FRG volunteer leaders tend to "burn out" when they provide direct services to unit families or substitute for the RDC in providing official assistance to spouses. Those who help others must have help.

Commanders and FRGs should make a concerted effort to include non-command-sponsored families in FRG support activities, social events, and the distribution of information.

FRG Operations and Functions. In effect FRGs, the operations and functions are anticipated in advance of unit deployments. Potential problems that may arise during the unit's absence have been addressed, and general solutions established.

Commanders of deploying units and rear detachments should clearly define the operational boundaries of FRG activities for FRG leaders. FRGs without boundaries may experience conflicts among volunteers and loss of mutual support within the group.

Disseminating correct and timely information is very important. FRGs should assist or participate with the rear detachment in conducting periodic information briefings. FRGs need help from RDCs in preparing newsletters and paying postage. A reading file for families to review during the week is helpful if they are unable to attend scheduled information briefings. The data in the reading file must be kept current.

Rosters and email lists are important tools for FRGs. They are more accurate when the information is gathered as part of unit in-processing. In addition to names, addresses, and phone numbers, rosters may contain information on language proficiency (e.g., does the spouse speak, read, write English; does a family member have a unique physical condition or special education requirement?). This information is useful in making sure the FRG is appropriately helping.

Rosters and email lists should be transformed into useful "telephone trees" and used to leverage social media to provide verbal support to FRG members or to transmit valuable information rapidly. These products form positive support relationships by bonding families of servicemembers in the same company or platoon.

FRGs perform an important function by helping rear detachments and military communities keep track of dependents. During deployments, some dependents leave the area to live with family while others visit relatives or friends and then return to their quarters. This makes 100 percent accountability of dependents very difficult for rear detachment and community commanders. In overseas areas, this impacts on noncombatant evacuation operations planning.

Family Care Plans. A family care plan pertains to the care of dependents when a single parent is deployed to the field or assigned to an unaccompanied tour of duty overseas. In two-parent families, it is presumed the non-deploying parent or spouse will look after dependents. Single parents must have a dependent care plan that can be immediately activated. The plan is filed with the company commander. The adequacy of family care plans over time is the responsibility of the soldier. Unit commanders are responsible for inspecting family care plans and judging their adequacy (see AR 600-20).

Family care plans must be carefully screened by unit commanders, and hard questions asked, such as:

- Can an elderly grandmother care for dependent children?
- Will the designated guardian want to leave home to live at Fort X for a year with little or no notice?
- Does the guardian drive?
- Does the guardian speak English?
- What is the soldier's plan if the guardian refuses to come when needed or leaves after three months?

RDCs should also monitor family care plans, include guardians in the distribution of information, and provide support as appropriate. If a family care plan fails after deployment, the RDC must coordinate with the deployed soldier and his or her unit on actions to resolve the situation. Deployed soldiers whose family care plans fail are not able to concentrate on their duties until the situation is rectified.

Family Finances and Household Budgets. Family finances is an important area that can spawn problems for families during deployment. Soldiers and spouses must decide who will pay the family bills during deployment. If the spouse is to pay them, the spouse must understand what amount of money will be available and how it will be received. The soldier may elect to have a specific amount of money sent to the spouse in the form of an allotment or have his or her military pay sent to a checking account accessible to the spouse.

Family budgets are sometimes disrupted due to the loss of income from a soldier's second job. If monthly financial obligations depend on this supplemental income, the soldier needs to coordinate with creditors so that payments are adjusted to put them in line with the actual income available during the deployment. Single parents or dual military parents whose children will be cared for by a relative or guardian should ensure that an allotment is in place to provide for their children's needs during the deployment.

Assistance in managing family finances is available before deployment. For instance, the Army Community Services (ACS) agency is staffed to provide help in planning household budgets.

Preparing Children for Parental Deployment. Soldiers need time to devote genuine attention to their children. The disruption of the family during deployment can have continuing effects on how children see themselves and the world in which they live. It can affect their performance in school and their response to authority. The following actions may help children cope with the absence of a deployed parent.

- Talk with each child individually.
- Explain what's going to happen. Be honest and keep it simple.
- Talk about where you are going, what you will be doing, and why you are going. A map or globe is helpful.
- Talk about when you expect to be home in terms of special events.

Children should be included in family goodbyes at the point of departure.

Family Emergencies Requiring Return of Soldiers. Some family emergencies warrant return of the soldier, if existing military operations permit. The field commander determines whether mission conditions will allow the soldier's return.

- Emergencies involving the death, critical illness, or injury of a member of the immediate family. Immediate family members include the soldier's spouse, children, siblings, and parents, or guardians who raised the soldier in place of parents.
- Critical illness or injury means the possibility of death or permanent disability.
- Illnesses such as the flu and injuries such as a broken arm, although not minor, are not considered emergencies.

Prior to your departure, educate spouses and guardians on the early return policy and notification process. The spouse or guardian must contact the Red Cross, representatives of which will verify the nature of the emergency. The rear detachment or the FRG can assist the family in contacting the Red Cross, if necessary.

The soldier's commanding officer must be notified by Red Cross message verifying the nature of the emergency before the commander can make a decision to return the soldier.

Disposition of Pets During Deployments. Pets become important members of a family, especially to children. Pets of single-parent soldiers or of dual military families also need a place to go, especially if the family relocates while the parent is deployed. The disposition and care of family pets must be considered before deployment. It is inhumane to go off and leave a pet to fend for itself. If the soldier intends to retain ownership of the pet through and after the deployment, arrangements should be made with a nonmilitary neighbor or friend. If the soldier does not plan to keep the pet, it should be offered for adoption before turning it over to an animal shelter. Make this part of the family care plan.

Rear Detachments. When units deploy, their commanders designate rear detachments to administer, command, and control all nondeploying personnel, but also to interface with and support the FRG. Hence, the selection of an RDC is an important and sensitive one; the position should not be allowed to automatically devolve to the least competent officer in the unit. The RDC is responsible for remaining personnel and equipment and for assistance to families of deployed soldiers. He or she is the focal point for all family support matters that require official actions or approval. Hence, RDCs must be mature officers with excellent initiative; they must be sensitive to and work well with family members.

The RDC works closely with FRGs to stay abreast of matters requiring expeditious handling by the rear detachment or the FAC. Rapid and effective resolution of families' problems has a significant impact on the morale of soldiers and their families.

RDC Family-Related Administrative Functions. RDCs must be knowledgeable of all support agencies and methods of contacting them. They also do the following:

- Provide explanations to forward commanders concerning requests for emergency leave.
- Work closely with the FRG.
- Provide assistance that requires official action (e.g., pay matters, official travel training, AER loans).
- Ensure that all family contacts are recorded in an official logbook. Maintaining a record is essential for proper evaluation and follow-up.
- Publish telephone numbers for and other means of contacting the RDC.
- Forward mail for deployed soldiers. The RDC can serve as a postal supervisor prior to unit deployment. (A soldier in the rear detachment should possess a valid mail clerk appointment card.)
- Distribute or forward the Leave and Earnings Statements of deployed soldiers to authorized recipients.
- Control storage, security, and inventory of POVs and personal items of deployed soldiers who reside in the barracks prior to deployment.
- Provide briefings to families to pass accurate information and reduce rumors.
- Refer certain problems to an appropriate agency of the FAC if the problem is beyond the scope of the rear detachment.
- Gather information on all immediate family members, including:
  - — Location of children's schools.
  - — Addresses, email, and phone numbers of soldier's and spouse's next-of-kin or other relatives with whom the spouse might reside in the soldier's absence.
  - — Spouse's work addresses and phone numbers (especially important for those families of soldiers attached for deployment).
  - — Spouse's ability to speak English.
  - — Spouse's ability to translate foreign language into English.
  - — Spouse's accessibility to savings or checking accounts.
  - — Physical, mental, or financial conditions that might require rear detachment assistance to family members.

The Army Family Advocacy Program. Although child abuse and spouse abuse have always been part of the darker side of the human condition, they have reached unprecedented proportions since the 1960s.

The Family Advocacy Program is the Army's principal weapon in this fight. As outlined in AR 608-18, *The Army Family Advocacy Program*, its objectives are to prevent spouse and child abuse, to encourage the reporting of all instances of such abuse, to ensure the prompt assessment and investigation of all abuse cases, to protect victims of abuse, and to treat all family members affected or involved in abuse.

The program involves a wide variety of medical, legal, and social professionals from many installation agencies—as well as chain of command representatives—who participate in various committees that monitor both the program and the status of individual cases of abuse that are reported or discovered by its members. Commanders should embrace the potential of this program and request instruction, available through Army Community Services, that will enhance prevention of this potentially devastating phenomenon. Every installation commander has established a twenty-four hour means of reporting and responding to child or spouse abuse.

19

Personal and Financial Planning

Forewarned, forearmed, to be prepared is half the victory.

—Cervantes

A prudent officer will have a financial plan to ensure personal security. Bachelor officers should plan for security in later life and for the reasonable possibility of leaving the ranks of the unwed. Married officers face the greatest challenge, for their plans must include security for their families. The foundation for the officer's financial planning is what the government provides in return for faithful and often hazardous service. Government benefits are a solid base. Objective, realistic foresight should identify the basic requirements for each person, while timely action will start the mission. As the years pass, the initial program must be adjusted as changing circumstances warrant. The right start at the right time is important.

The Army officer accepts definite hazards. Short tours of foreign service may require separation from your family. Each decade beginning with the 1940s has found our country involved in a war or combat short of war, with intervening threats to the nation's security. Most officers of today's Army have been in combat in southwest Asia. In planning for the future, you should include the possibility that part of your future active service may be in combat. It is a harsh reality that combat may result in death or disability, but you should recognize the possibility. Further, the very nature of the Army mission requires that you engage in activities that are hazardous or that you work with hazardous equipment, either of which also poses definite risks. You should identify as accurately as possible your future responsibilities and goals and decide what action is necessary to attain them.

This chapter is provided to help all officers in their financial planning for personal and family security. Other sources of study and reference are cited for detailed guidance. The chapter's goal is to provide a concise, informative introduction to the subject. Thought and foresight are needed. Since the Army wife or husband may need to manage these affairs during the absence of the servicemember, many officers see the wisdom of sharing with spouses the entire process of deciding and acting on program development.

If the task or the responsibility seems heavy, be encouraged by the truth that most career officers lead interesting, rewarding lives and at the same time provide for the uncertainties and potential requirements of their future years.

FINANCIAL SECURITY

Financial security is a continuous requirement while one is on active service and after one is retired. A security program provides for both emergencies and planned events. Army officers should identify their family goals through retirement by planning, budgeting, saving, and otherwise providing for such events as college education, buying a home, paying a major dental bill, and going into retirement. Even without the hazards of combat, parents must face with realism the possibility of the untimely death of either husband or wife. Having foreseen their obligations and responsibilities, they must have taken the security actions that are reasonable, prudent, and adequate for their situation.

The above motivation introduces the need for financial planning and should urge you to implement a planned security program and review its adequacy and currency at such critical stages as going overseas, the birth of a child, a promotion, retirement, and others. A partial review is in order each year when you compute income taxes. Such planning and reviewing calls for the following steps, which will be detailed in subsequent sections:

- Learn the elements of security planning.
- Review your current family and financial status.
- Choose your short- and long-range goals.
- Identify your service benefits.
- Establish a reasonable insurance program.
- Supplement your estate with investments.
- Be aware of the many assistance societies.
- Take care of vital personal administration.

FUNDAMENTALS OF FINANCIAL SECURITY

Security planning consists of evaluating your financial situation, comparing it with your requirements and goals, and deciding what must be done to create a program that meets today's needs while providing security for the future. Estate analysis is a similar term usually employed by insurance salesmen who are oriented toward security through insurance programs. As used here, *security planning* involves your total financial program, including salaries, property, savings, insurance, governmental benefits, investments, and other monetary assets. It involves the integration of these finances into a formal plan. Finally, it takes into consideration today's and the immediate future's needs as well as those of the far future after retirement. Overcommitment of your salary to security programs can deny you and your family the opportunities of building a living estate of wonderful, memorable, and educational recollections of vacations, entertainment, hobbies, and those other events and circumstances that might not constitute absolute necessities. A term often encountered is *insurance poor* for one who buys too much insurance, but the term applies equally to other types of estate building. Thus, security planning is a management science to some extent, yet it could also be classified as an art because of its intimate relation to personal philosophies.

Reference Material. There are many useful sources of information to assist you in estate planning, such as the following:

Guide to Personal Financial Planning for the Armed Forces, Stackpole Books, Mechanicsburg, Pennsylvania.

Estate Planning for Military Personnel, published by Dyke F. Meyer, Col., USAF Ret., C.L.U., San Antonio, Texas.

Financial Planning Guide for Military Personnel, published by the Armed Forces Benefit Association.

The United Services Automobile Association (USAA) and the Military Officers Association of America (MOAA) also provide informative pamphlets on topics of interest.

STATUS, GOALS, AND REQUIREMENTS

The next steps of security planning involve determining your financial situation, establishing your goals, and deciding on the actions required to meet these goals. The first two, status and goals, are discussed in this section along with the broad plan of action for establishing security. Detailed sections then follow on how to implement the broad plan of action.

Your Financial Status. Your financial status is based on your salary and any other income. To analyze your situation further, it is necessary for you to consider your service benefits, insurance programs, savings, investments, Social Security credits, real estate equities, and any other assets. This is the point where you realize what you are worth financially should you live, be incapacitated, or die. It is generally useful to study and review other personal affairs records at this time along with your financial data.

Your Financial Goals. Next, you have to determine personal and family goals. Officers who are fulfilling an obligation for a few years and those who are contemplating careers of twenty to thirty years will have different goals. However, since this is an officer's guide, an officer's career is assumed to be the planning base. Conservative financial planning could be based upon achieving the grade of major and retiring after twenty years' service. The length of service is most influential on security planning. For example, an officer who is involuntarily separated with at least six but less than twenty years of service is entitled to a separation payment equal to 10 percent of annual base pay times years of service. Officers may retire voluntarily with twenty years of service, and only a few will be allowed to serve more than thirty years before retirement. The amount of retired pay you can expect to receive is a function of your final base pay, or the average of the three years of highest base pay, your total years of service, and the date you entered the Army. Three different retirement systems are now provided by law, depending on when you entered the Army. See chapter 24, Retirement, for details of the retirement system that applies to you.

Next comes your personal plans as to married life. One of the reasons marriage is often called a partnership is the financial relationship of the spouse. The spouse's attitudes, philosophies, financial responsibility, monetary assets, age, health, and other characteristics all influence to some degree an officer's planning. Then there are the children. How many? College education? Establish in business? What is the influence of their personality, intelligence, hopes, educational plans, and other factors that affect the family's way of life and long-range program? These general and philosophical planning components serve to assist in the identification of financial goals and requirements so that subsequent planning to meet them can be realistic, practical, and logically derived. Once an officer-spouse team has considered all these matters with objectivity, the decisions are made clearer and the courses of action are more readily discerned.

You may be assisted by considering your future lives as comprising a few main periods, and then estimate the conditions that are to be expected within each one, such as the following examples:

- The period of active duty, which should continue until ________ (year) at the earliest, or until ________ (year) at the latest.
- The years in which children are growing up, receiving education, and becoming self-supporting, now foreseen until ________ (year).
- Family wishes for the period immediately following military retirement and reestablishment in a civilian community of their choice.
- The years between military retirement date and eligibility to receive Social Security payments in ________ (year).

Also, consider the situation and requirements that would follow the death of either husband or wife during each of the above periods.

A wise start requires candid and searching answers to personal questions, such as those that follow.

How Are the Spouse and Children to Be Protected? The reasonable person recognizes that the duration of life is uncertain. He or she proceeds without undue morbidity or overemphasis to face facts. Proceed as you would in making a military estimate. Is your spouse qualified to be self-supporting if suddenly faced with the necessity to do so? Or would he or she need training to resume a former vocation or to undertake a new one? Consider your spouse's present age, health, potential, wishes, and inclinations. At what age should it be considered that he or she would be unable to provide a part of the required income, or would not wish to do so? When there are children, at what year will each be prepared for life and be self-supporting? Are there physical or mental frailties to be considered and provided for? How much income per month will be needed, and for how many years, for the support and education of the children? Be realistic and inclusive. Later in your estimate you will balance potential outgo against potential income to see whether there is a gap to bridge. Right now, just determine the facts.

What Is Your Present Life Insurance Program? The family insurance program stands beside family net worth as a sound foundation to family security. It is true that military service establishes important benefits for family protection far above those of most vocations. But it is equally true that the hazards are greater, and the possibilities of estate accrual are more difficult. The termination of some benefits upon retirement, with the reduction or uncertainties of others, requires careful identification for each family situation. There are gaps to be filled in service benefits during your active career. There are more and wider ones to face upon retirement. Life insurance of the appropriate type, in carefully determined amounts, is one proven way of providing for the potential needs of the future.

Do You Have an Investment Program? Many military families have entered upon such a program with farsighted wisdom. They may have bought a home and are increasing their equity through rental. Others are buying securities, such as government bonds, mutual funds, and common stock, or have opened an individual retirement account. Some have developed an organized program involving a combination of these possibilities. Others simply rely on the Thrift Savings Plans.

Do You Have Reasonable Expectancy of an Inheritance? If so, include it as a possibility. But don't place extensive reliance on it; old age, ill health, and many other hazards, plus taxes, can reduce an estate. Just weigh the possibilities and go ahead.

To What Extent Do Present Military Survivor Benefits Meet the Total Needs? Basic information in this chapter can be studied as a start in this investigation of potential resources. You should make a detailed study of departmental publications and consult experts about some phases of the subject.

You will see at once that for all officers the benefits are valuable. The officer on active duty and the officer's family are better protected than ever before. Upon your retirement, there are important benefits, but there are also potential gaps in protection that must be identified. Study is needed to apply these laws to your situation and the requirements of your family.

What Sort of Program for the Future Can You and Your Spouse Afford Today? This chapter provides much information about future security, so let us seek for a moment to establish a balance. Each family has its own way of life, with its daily needs and its own dreams for the future. However, a family's immediate and continuing happiness—such as children growing up with proper food, clothes, books, family outings, dinners at the club—is important, too. Prudence is needed for family security, of course, and its importance must not be minimized; but prudence is also required in avoiding overcommitting the family income for estate building. Once an officer-spouse team has considered all these matters with objectivity and realism, the best decisions are usually clear. For most families, expenditures of today can be controlled so that some part of income can be diverted to the family program of the future.

Should Your Family Have a Budget? Officers and their spouses usually manage their finances with either a formal budget or a working budget. With a formal budget, a detailed system of planned expenses and records serves to guide the expenditures for the various parts of the living and the security programs. In the working budget, the officer-spouse team establishes habits that allow them to accomplish the basic parts of their programs while not exceeding their income or depleting their cash reserve. A cash reserve is maintained in both cases for emergencies and for planned spending on major events (or things) in the near future. A two months' salary cash reserve is recommended. Savings beyond this should be invested if the other elements of the security program are met. Advised is an interest-bearing checking account for day-to-day use, and a money market mutual fund with check-writing privileges that you can use to accumulate funds for larger expenses such as a new appliance, a trip, clothes for college, and so on. Those who maintain a formal budget usually do better at managing their finances, but some people do not enjoy the detailed accounting required. Having a program, meeting it, and staying within one's income are also practical ways to security without detailed budgeting. The choice is yours.

How Much to Rely on Credit? In this matter of budgeting, the use of credit deserves your special attention. In our society today, the use of credit is a normal occurrence. Credit is readily available in the form of charge cards such as VISA, MasterCard, and many others. Using these credit cards can be a handy way to do your shopping, including shopping at the post exchange and commissary, without the necessity of carrying large amounts of cash or writing checks for the purchases. The card-issuing companies accumulate the charges over a monthly period and present you with a consolidated bill at the end of the month. Problems can arise when the size of the monthly bill exceeds your ability to pay, since any unpaid amounts are carried forward to the next month's bill at interest rates that are generally in the range of 1.5 percent or more per month. Without careful attention on your part, it is easy to accumulate large charge-card balances on which you are paying high interest rates. What this means is that over a period of time you can wind up paying 20 percent more for your purchases than would have been the case had you used cash. It also indicates that you are living beyond your means; that is, your outgo exceeds your income. It is a path that can lead to severe financial problems.

If you do use charge cards, it is far better to schedule your purchases so that you can pay the bills in full each month. This may even be better than using cash, since you have the advantage of a "float" of your funds during the month. That is, you use the card-issuing company's funds during the month, while your funds may be drawing interest at the bank until the time you must write a check covering the month's accumulated bills. Do not assume, however, that it will be any easier to pay your bills next month than it is now. You will find that each month has its own special needs that you had not forecast, and that this month's unpaid bills will be at least 1.5 percent greater next month. If you find yourself falling behind in paying off credit purchases, it is suggested that you put the credit cards aside and return to using cash for your purchases. In any event, do not fall into the trap of accumulating large, unpaid credit card balances.

Your Financial Plan. Once you have faced the basic questions of financial status and goals, estate planners say that you are now ready to make your basic financial plan. Here is what you must do:

- Establish a solvent financial situation.
- Build a cash reserve.
- Establish a sound insurance program.
- Invest your remaining income.

You, however, are an officer and cannot complete your security planning until you first consider the benefits of being in the service and a citizen of the United States.

BENEFITS ACQUIRED THROUGH MILITARY SERVICE

There are many benefits acquired through military service. They are provided by the government in recognition of the hazards of military service as well as to supplement your pay and to provide an added inducement for you to select military service as a career. You should be aware, however, that benefits provided by law can also be altered or withdrawn by changes to the law. Over a period of years, Congress and successive administrations do change the benefits. In general, such changes are aimed at improving the benefit provided, but in some recent instances, benefits previously counted on have been withdrawn or curtailed. This section attempts to clarify your earned benefits, but you are cautioned to study the regulations, to stay up-to-date with the latest changes as reported in the various service journals, and to seek the advice of experts in the field to help you with your plans.

Military Health Services System

The Department of Defense operates one of the nation's largest health care systems. Approximately 10.26 million individuals are eligible to receive care through the Military Health Services System (MHSS).

Since the end of the Cold War, the U.S. military has dealt with new challenges to its organization and mission. Fewer men and women are on active duty. Along with fewer combat forces, there have been reductions in support forces, including physicians and other medical professionals. In fact, the number of doctors, nurses, and medical technicians in military service has declined over 50 percent in some locations.

Two more factors led to changes in the MHSS: the rising cost of health care, and the continuing requirement to maintain a trained and ready medical corps to support our troops, in peace or combat. Costs for medical care are rising in both civilian and military communities due to many complex factors. Some of the major causes are:

- Improved technology—new diagnostic procedures, new machines, new treatments.
- Increased utilization—people are taking more responsibility for their health and seek health care at a greater rate than ever before.
- Aging of the population—applies mostly to growth in the number of people in the 65+ age group. Specifically, in the case of the military, a larger retired population versus active-duty population has put new demands on the MHSS.

Although providing health care during peacetime is an important mission, the number-one priority of the MHSS is to support emergency operations. A new approach was needed to meet peacetime demands for health care while preserving the capability within the active-duty medical corps to deploy and support military men and women on operational missions.

Tricare. In response to the challenge of maintaining medical combat readiness while providing the best health care for all eligible personnel, the Department of Defense introduced Tricare. Tricare is a regionally managed health care program for active-duty and retired members of the uniformed services, their families, and survivors. Tricare brings together the health care resources of the Army, Navy, and Air Force and supplements them with networks of civilian health care professionals to provide better access and high-quality service while maintaining the capability to support military operations. Tricare's four geographic regions are each organized around a military hospital that acts as a "lead agent" in coordinating health care for beneficiaries in that region.

Eligibility for Tricare. Tricare is the health benefits program for all the uniformed services: Army, Navy, Air Force, Marines, Coast Guard, and the Public Health Service. All active-duty members and their families, retirees and their families, and survivors may participate in at least one of the three Tricare options.

Until October 2001, most retirees sixty-five and over were shifted to the Medicare system and needed to buy "Medigap" insurance to fill in where Medicare left off. In the most sweeping improvements to the Department of Defense's health care system in nearly thirty years, as legislated by the 2001 National Defense Authorization Act, effective 1 October 2001, these retirees gained access to expanded medical coverage known as Tricare For Life. To participate, a retiree must be a uniformed-service beneficiary who has attained the age of sixty-five, is Medicare-eligible, and has purchased Medicare Part B. Tricare For Life is a permanent health care benefit. For more information about eligibility, benefits, enrollment requirements, and costs associated with Tricare For Life, contact a Tricare office or visit its website, www.tricareonline.com.

It is important that you know your medical benefits and, for your security and that of your family, understand where gaps could occur that must be covered by other measures. This section summarizes the more important provisions. However, you may have to consult the regulations for details as to your particular situation. In case of any doubt, consult the nearest health benefits adviser/health care finder.

Health Care Programs. One of the central features of Tricare is the choice of health care plans it offers. While all *active-duty men and women are automatically enrolled in an option called Tricare Prime,* other eligible individuals may choose among three programs: Tricare Prime, Tricare Standard, and Tricare Extra.

Tricare Prime is a managed-care program similar to a health maintenance organization (HMO). Participants must enroll for a period of twelve months, pay an annual enrollment fee (retirees only), and seek treatment exclusively from a preapproved network of military and civilian health care providers. Membership in Tricare Prime necessitates that you live within a convenient distance from a military medical facility. The benefits of Tricare Prime are lower out-of-pocket expenses and a preexisting network of medical professionals so that every move doesn't force you to find competent medical professionals in an unfamiliar community. Weigh these against giving up complete control of who your medical providers are.

Tricare Select replaced Tricare Standard and Extra in 2018. Tricare Select is a self-managed, preferred provider plan. Participants have unrestricted choice of health care providers and medical facilities but pay higher deductibles, cost shares, and excess physician charges. It is chosen most often by those who have established relationships that they wish to maintain with civilian physicians. Additionally, it may be used by beneficiaries who have other health insurance, where Tricare Standard is the second payer.

The programs are specifically designed to fit the needs of individuals based on their medical needs; geographic constraints; and personal comfort level regarding costs, choice of health care providers, and so on. Talk to the health benefits adviser/health care finder about your family's needs so that you can make the choice that is right for you and your family.

Who Is a Family Member? Before examining medical care provisions for an officer's family members, it is important to define the term. *Family member* means any person who bears to a member or retired member of a uniformed service, or to a person who died while a member of a uniformed service, any of the following relationships:

1. The lawful wife.
2. The unremarried widow.
3. The lawful husband.
4. The unremarried widower.
5. An unmarried legitimate child, including an adopted child or stepchild, who has not passed his or her twenty-first birthday, regardless of whether the child is dependent on the active-duty or retired member; or who has passed his or her twenty-first birthday but is incapable of self-support because of a mental or physical incapacity that existed before the twenty-first birthday and is, or was at the time of death of the active-duty or retired member, dependent on the member for over half of his or her support; or who has not passed his or her twenty-third birthday and is enrolled in a full-time course of study in an institution of higher learning approved by the Secretary of Defense or the Secretary of Health and Human Services, as the case may be, and is, or was at the time of death of the active-duty or retired member, dependent on the member for over half of his or her support.
6. A parent or parent-in-law who either is dependent on the active-duty or retired member for over half of his or her support and is residing in a dwelling place provided or maintained by the member or was at the time of death of the member dependent on him or her for over half of his or her support and was residing in a dwelling place provided or maintained by the member.

In addition, Tricare benefits are extended to divorced spouses of active, retired, and deceased servicemembers under certain conditions. The coverage is dependent on the date of divorce or annulment, the length of the marriage and its overlap with active-duty service, and the divorced spouse's status and coverage by an employer-sponsored health care plan. See the regulations for details.

Eligibility for Medical Care. When applying for any kind of medical care at a military or civilian facility, military personnel, active and retired, and their dependents are required to provide evidence of their eligibility by means of the Uniformed Services Identification and Privilege Card (ID card). Military hospitals also issue an identification card for use in connection with medical services. If a person uses or allows another to use these cards to obtain medical care to which he or she is not entitled, a fine of up to $10,000 and imprisonment for up to five years may result.

To minimize the possibility of fraudulent use of military health benefits by unauthorized persons, as well as to improve the control of military health care services, the Defense Enrollment Eligibility Reporting System (DEERS) was established. Using DEERS, a military health care facility can quickly check a person's eligibility to use the facility. You as a servicemember are automatically enrolled in DEERS, but you must take special action to enroll your family members. See your personnel officer. Enrollment of family members is necessary for them to obtain health care either at a military facility or at a civilian facility under the provisions of Tricare.

Tricare Prime

Membership. If you have selected the Tricare Prime option for your family, you will need to select a primary care manager (PCM), which is the single point of contact for nonemergency health care for enrollees in Tricare Prime. Depending on the enrollee's status, locale, and availability of medical professionals, he or she may select a PCM at a nearby military hospital or clinic (which is most likely for active-duty members), or he or she may request a civilian professional who is a member of the contracted prime network in a nearby community. A PCM is a doctor (or group of doctors) who administers your primary care, supported by a team of administrative personnel. PCMs provide and coordinate medical care, maintain health records, and recommend preventive and wellness service. The PCM also refers you to specialists or arranges for hospital admission when necessary.

Preventive Care. With an emphasis on keeping families healthy, Tricare Prime includes a variety of preventive and wellness services at no additional charge, including eye exams, immunizations, hearing screenings, mammography, Pap smears, well-baby checkups, prostate exams, and other cancer-prevention and early-diagnosis exams.

Out-of-Area Care. If you should need nonemergency medical care while away from home, it will be covered, provided you obtain prior approval from your PCM using a special toll-free telephone number. Care authorization is required for all routine medical care received out of the area or at another facility. If you need medical help after hours and cannot reach the PCM, be sure to follow up with the PCM the next duty day to ensure that your authorization is approved.

Prescriptions. All persons eligible for Tricare, as well as those retirees now on Medicare, currently may take prescriptions to any military pharmacy and have them filled, free of charge. Bear in mind that this opportunity is limited to medications carried by the military facility.

Dental Care

Active Duty. The Defense Department's Tricare Active-Duty Family Member Dental Plan offers basic preventive and restorative dental care to enrolled families of active-duty sponsors in the uniformed services. The care is provided by civilian dentists. Claims are filed, either by the dentists or by the families who received the dental care, with the civilian contractor that operates the dental plan for the services.

The plan covers persons living in the fifty states, the District of Columbia, Puerto Rico, Guam, the Virgin Islands, and Canada.

Retirees. Effective 1 October 1997, the Defense Department began offering a Tricare Retiree Dental Program to retired military members (including the U.S. Coast Guard), their eligible family members, and unremarried surviving spouses of deceased military retirees (no age limits on eligibility). The program features a variety of diagnostic, preventive, restorative, endodontic, periodontic, and oral surgery services from civilian providers, at specified levels of cost sharing.

The program is paid for by premiums collected through payroll deduction from those who receive retired pay. Those who don't receive retired pay are billed directly for premiums by the contractor that runs the program.

Initial enrollment in the program is for at least twenty-four months. After the first twenty-four-month period, enrollees may choose to stay enrolled on a month-to-month basis.

Dependency and Indemnity Compensation

The Survivors Dependency and Indemnity Compensation Act of 1974 provides for compensation to survivors for the loss of an officer or soldier whose death is attributable to military service. All *service-connected deaths* that occur in peace or wartime in the line of duty qualify the eligible survivors for Dependency and Indemnity Compensation (DIC). Retired officers' survivors may qualify also, provided the Department of Veterans Affairs (VA) rules the death to have been from service-connected causes. DIC payments are not authorized to the survivors of a retired officer whose death is not service-connected. (A small pension may be paid to the widow or widower, as discussed later, provided her or his income is below a stated minimum.)

Compensation may be increased for widows and widowers with children under age eighteen or with a child over eighteen who is incapable of self-support. An additional sum may be awarded for each child between eighteen and twenty-three who attends a VA-approved school. All such compensation is income tax–free. The officer's surviving children will receive compensation should the spouse die. Parents of a deceased soldier likewise qualify for compensation. Your finance officer or any VA office can assist you in computing the various possible compensations and pensions for your particular family situation. The website that discusses DIC and SBP is www.military.com/benefits/survivor-benefits/the-survivor-benefit-plan-explained.html.

Pay and Death Gratuity

At the time of a sevicemember's death, there will be pay due for a month or a fraction of a month. The spouse receives this arrears in pay upon claim to the nearest Finance Center, or by writing to the Defense Finance and Accounting Service—Cleveland Center, P.O. Box 99191, Cleveland, OH 44199-1126.

In addition to the arrears in pay, a death gratuity is payable to the survivors of a servicemember who dies while on active duty, active duty for training, or inactive duty training (weekly drills), and of a retired servicemember who dies of a service-connected disability within 120 days after separation from the active list. These payments should be included in the determination of assets and benefits in an individual's estate security plan.

Survivor Benefit Plan (SBP)

The Survivor Benefit Plan (SBP) allows members of the uniformed services to elect to receive a reduced retirement pay in order to provide annuities for their survivors, for another person with an insurable interest in the servicemember, or, under certain conditions, for an ex-spouse. Under SBP, the federal government pays a substantial portion of the overall cost. Participation in the SBP is open to all future retirees, including members of the reserve forces when they attain age 60 and become entitled to retired or retainer pay. Present retirees have previously been afforded the opportunity to participate.

Under SBP, future retirees may elect to leave up to 55 percent of their retired pay to their selected annuitants. Participation in SBP is automatic at the 55 percent rate for those still on active duty who have a spouse or children. Prior to retirement, the member may fill out an election form, DD Form 1883, electing a reduced coverage, or the member may decline to participate. However, the spouse must concur in the election of reduced or no coverage.

There are a number of points about SBP that deserve careful consideration. Most decisions are irrevocable. However, deductions under the plan cease during any month in which there is no eligible beneficiary. Further, the coverage may be switched to a new spouse if the retiree remarries. An important advantage of SBP is that the amount of coverage is tied to the consumer price index (CPI). Whenever retired pay is adjusted on the basis of the CPI, the amount selected by the retiree as an annuity base, the deduction from retired pay covering the adjusted base, and the annuity payable to selected beneficiaries are adjusted accordingly. The plan thus provides for automatic cost-of-living adjustments.

For members who retire after 1 October 1985, the law provides for a two-tiered system of payments under which the surviving spouse receives 55 percent of the base amount of retired pay until age 62. At age 62, at which time the survivor becomes eligible to receive Social Security payments based on the member's earnings, the amount payable under SBP drops to 35 percent of the base amount, continuing for the life of the survivor. Despite the reduction, total income to the survivor remains the same or increases, since the SBP payment combined with the Social Security payment generally exceeds the 55 percent of the base amount selected. Note, however, that the reduction is effective whether or not the surviving spouse actually is collecting Social Security payments based on the member's income. Retirees who choose the full amount of their retirement pay as the base amount also may select from one of four supplemental plans, under which the retiree may elect to have the SBP payments continue after the survivor reaches age 62 at the 55 percent rate, at a 50 percent rate, at 45 percent, or at 40 percent. The added cost for a supplemental plan is paid by deduction from retired pay based on the member's age when he or she signs up for SBP. The extra costs for the supplemental plans are commensurate with the added coverage provided.

Prior to 1 October 1985, the law provided that when the survivor reached age 62, the SBP payment would be reduced by the amount of the Social Security payment based solely on the servicemember's military pay, with a maximum reduction of 40 percent of the SBP payment. The law now provides that the SBP payments for survivors of pre-October 1985 retirees will be either that computed under the old system or the amount computed under the newer two-tiered system, whichever is greater.

The plan also provides that if a retiree dies of a service-connected cause, as a result of which his or her survivors are entitled to DIC payments (see the previous discussion earlier in this chapter), the SBP payment will be reduced so that the total of the two payments—DIC and SBP—will be equal to the full amount otherwise payable under SBP. Thus, the total income to the survivors from SBP, DIC, and Social Security can be equal or slightly higher than the SBP income alone, depending on the extent of the offset for Social Security. But since SBP (as well as Social Security) is tied to the CPI, the retired member also is assured that the spending power he or she has provided through SBP will remain relatively constant regardless of the effect of inflation.

It is apparent that participation in SBP has definite advantages and, perhaps, disadvantages that call for study, advice, and the use of care in deciding whether coverage is desired and in selecting the amount. SBP is not necessarily a panacea for all military retirees, but it deserves careful consideration as a benefit accruing to you as a result of military service, which is available for use in planning your estate.

Social Security

Military personnel accrue Social Security benefits for active service as an important element of their overall security program. Participation is mandatory. Under the Social Security program, your base pay is taxed at a prescribed rate. The maximum taxable earnings are increased as average wage levels rise. At the same time, the benefits payable also increase. The amount of the benefit received from Social Security depends on the average monthly earnings (AME) and the number of years of credit that have been accrued. The benefit of most general interest is the monthly income that an officer and eligible members of the officer's family attain when the officer reaches age sixty-five. (Reduced benefits may be selected for ages sixty-two, sixty-three, and sixty-four, as well.) Payments are provided for disabled officers and for their spouses and children. Monthly income is provided for an officer's widow or widower with children under eighteen years of age, or for children alone, or for the widow or widower at age sixty, or for dependent parents. These important benefits supplement other insurance-type programs. Application for benefits may be made at any Social Security office.

Note: It is important that you ensure that the Social Security Administration has credited you with the correct contributions. This may be accomplished by requesting the status of your account at approximately three-year intervals. Postcard-type forms are available at any Social Security office to make this request.

Scholarships and Education Loans

College education costs for children are of growing concern in view of the increased costs of tuition and board. In planning one's estate, settlement options of life insurance and educational endowment insurance are sometimes purchased to provide this coverage. Some provide this security by purchasing bonds and participating in other savings and investment programs. Should this be insufficient, there are a number of scholarships that can be won through good grades and successful competition through examinations. Many states have statutes and educational programs that provide scholarships and financial aid to dependents of disabled and deceased veterans.

The Junior G.I. Bill or War Orphan's Educational Assistance Act of 1964 provides benefits to sons and daughters of service personnel who die or become disabled as a result of a wartime or extra-hazardous peacetime disease or injury. Up to $376 per month for forty-five months may be received. One cannot predict one's disease or disabling injury, but this benefit should be known to servicemembers.

The National Defense Education Acts of 1958 and 1965 provide low-cost loans to college-age family members of service personnel. This too cannot be specifically included in an estate plan, but it is comforting to know that it is there to assist peacetime veterans and their family members.

Young married officers will be encouraged by some insurance salespeople to provide for the college education of children through insurance contracts. Such endowment insurance for education is expensive and may not be the best method for you. Competing with insurance for this purpose are the above programs, savings programs such as government bonds, and long-range investment programs. More information about DoD education efforts can be learned from www.dodea.edu/Offices/HR/employment/benefits.

Burial Rights and Benefits

The government provides burial and funeral allowances in the event of the death of active-duty, retired, or honorably discharged officers. The allowances vary from complete for an active-duty officer's death to a plot allowance for an honorably discharged officer. These allowances should be considered in your security program. You should acquaint your spouse or other responsible family members with your desires and with an understanding of what services and allowances are available. The government respects its departed servicemembers and desires that interment be proper and at as little expense to the members' survivors as possible.

Officers on Active Duty, Active Duty for Training, or Inactive Duty Training. The government will take charge of most of the arrangements if the surviving kin so desire. The standards of the services provide every proper consideration; it is usually best to leave arrangements to the military authorities at the place of death. The services and allowances provided include the following:

1. Preparation at the place of death (pickup, embalming and preservation, casket, and hearse service to the cemetery or shipping terminal) and funeral director's services.
2. Cremation and a suitable urn for the ashes (if requested in writing by an authorized survivor).
3. A flag for the casket.
4. Transportation of remains with a military escort.
5. An allowance for interment: $3,100 maximum for burial in a private cemetery; $2,000 maximum when remains are consigned to a funeral director and are interred in a national or post cemetery; or $110 maximum when remains are consigned directly to national or post cemetery officials. (The allowances are meant to defray the costs of such items as the coach, flowers, vault, church services, obituary notices, car for the family, services of the funeral director, opening and closing of the grave, and use of cemetery equipment.)

Retired and Honorably Discharged Officers. The Department of Veterans Affairs will pay an allowance of $300 as reimbursement for funeral expenses if death occurred in a VA hospital or if the veteran was entitled to VA disability compensation. In addition, if burial is in a private cemetery, an additional plot allowance, not to exceed $828, is authorized. Honorably discharged (not retired) officers are entitled only to the $150 plot allowance unless they were in receipt of VA pensions at the time of death. If the retiree died from service connected causes, an allowance not to exceed $2,000 is authorized in lieu of any other burial benefit.

Social Security Death Benefit. In addition to any allowance or reimbursement received from the VA or military service, survivors of retired officers also are entitled to a lump-sum death benefit payment of $255 upon application to the nearest Social Security office.

Family Members. When a family member dies while the member is on active duty (other than for training), transportation may be furnished for the remains at Army expense. Certain privileges related to national and post burial sites are authorized and are discussed below.

Control of Cemeteries. The control of all national cemeteries except Arlington and those operated by the service academies and the Soldiers' and Airmen's and the Naval Homes was transferred to the Veterans Administration (now the Department of Veterans Affairs) by PL 93-43. Post cemeteries remain under control of the Army.

Burial in a National or Post Cemetery. A deceased active-duty officer may be buried in any national cemetery in which grave space is available. The surviving spouse, minor children, and, in certain instances, unmarried adult children are also eligible for burial in the same cemetery. Only one grave site is authorized per family unit. Burial at a post cemetery is subject to determination and approval of the post commanding officer and in keeping with Army regulations. At post cemeteries, immediate members of the family beyond those authorized at national cemeteries may be buried on post approval. Certain posts will accept grave site reservations in writing for surviving spouses. Retired and honorably discharged servicemembers are also entitled to burial in a national cemetery, if space is available. Detailed additional information on these matters can be secured from a superintendent of a national cemetery or from the Office of Support Services, DA, Washington, D.C. 20315. Local veterans organizations will also provide much assistance and information.

Honors and Headstones. Military honors will be provided by the Army when requested for interment at a national or post cemetery. At certain national and at civilian cemeteries, these honors are rendered by a local reserve unit or a veterans organization upon request.

Headstones and markers that are proper, appropriate, and otherwise suitable are provided in memory and honor of the servicemember at little or no cost. Such provision is made by the service, the Department of Veterans Affairs, and/or the state. No servicemember's grave will be unmarked because of lack of funds. At national and post cemeteries, this honor is virtually automatic with no action, other than consent, required by the kin.

In summary, your government is grateful and proud of your service; in the grim instance of death, it is desired that your burial be in keeping with your dedication. Consequently, most of the expenses of burial are paid by or shared by the government as your deserved and earned benefit.

Advice for the Widow or Widower. Although the primary purpose of this chapter is to enable officers and their spouses to understand their rights and benefits and to plan for all eventualities in their future lives together, it also contains a great store of information for those who have already become widows or widowers. The Army and Air Force Mutual Aid Association will provide upon request a booklet *Notes for a New Widow.* Like a similar one titled *Help Your Surviving Spouse—Now!* provided by the Retired Officers Association, it has blanks to be filled in to summarize vital information and guidance for the new widow. The guidance is equally applicable to widowers. A thoughtful officer will, of course, keep the information up-to-date as financial plans and estates are altered.

The Army takes thoughtful, effective, and thorough care of the families of its deceased members. This extends to funeral arrangements and wise counsel as to the settlement of all residual affairs. The Installation Retirement Services Offices, at stations throughout CONUS, are available to provide information and assistance. The local ACS agency also can provide help. Veterans' organizations are eager to help all service widows or widowers with the complexity of administration of affairs. Contact all these plus the Social Security field office, your banker, and all insurance companies. Finally, make all contracts and sign all papers with care and counsel. Your spouse would want you to grieve, but be sensible with your emotions when it comes to your family's security. All administrative matters of long-range importance can be settled within reasonable periods of time and do not require your immediate attention while you are in mourning.

Veterans' Benefits

There are a number of important benefits provided by the government for veterans of service to their country during times of strife. These programs, along with the foregoing survivors' benefits, are administered primarily by the Department of Veterans Affairs, with assistance from other federal agencies. Veterans of Cold War conflict or peacetime service are not included in certain veterans' acts. The basic benefits are educational assistance, VA home loan, VA compensation, and VA pension, which assist active duty, discharged, and disabled veterans and provide financial help to survivors. For more detailed information, visit the VA website: www.va.gov.

Educational Assistance. There are several educational assistance programs now in existence for active-duty personnel. Eligibility is determined by dates of service.

Post–Vietnam Era. The Veterans Education and Employment Assistance Act of 1976 (PL 94-502) established a contributory system for persons entering military service on or after 1 January 1977. Participants could contribute from $25 to $100 monthly from their pay up to $2,700. The government matched the contributions on a two-for-one basis. Enrollment in the plan is prohibited for persons entering the military after 1 July 1985. (See Montgomery G.I. Bill, below.) Eligibility to use the benefits begins after completion of six years of active duty. The amount of the fund, up to $8,100 maximum, is divided by the number of months of entitlement to determine the monthly payment to cover educational expenses. This benefit must be used within 10 years of leaving the service. Unused contributions by the *individual* will be refunded.

Montgomery G.I. Bill. The Veterans Educational Assistance Act of 1984 (VEAP) established a new program for individuals entering the military service after 1 July 1985. The program originally was to operate during a three-year trial period, but it proved so popular that it was made permanent in 1987. The permanent legislation was renamed the Montgomery G.I. Bill Act in recognition of the preeminent role played by Representative G. V. (Sonny) Montgomery of Mississippi in helping the legislation achieve permanent status. Participation is automatic, with a $100 per month contribution from basic pay each month for the first 12 months of service, unless participation is specifically declined. Participation in the program is closed to graduates of one of the service academies or a senior ROTC program in which they were in receipt of ROTC scholarship funds. Three years of active service or two years of active duty plus four years in the selected reserve or National Guard entitle an individual to a post-service benefit of $404.88 per month for 36 months. The benefit is $328.98 per month for 36 months for a two-year active-duty enlistment. Supplemental benefits also are available to those who extend their active-duty service or fill critical shortage specialties. The current limit on supplemental benefits is $400 per month, although with the right combination of years of service and duty in critical skills areas, a soldier could qualify for a total benefit of $1,300 per month for the 36-month period. For reservists, no contribution is required. Those who enlist for six years can receive $192 per month for 36 months without having to wait until their obligation is completed.

Post 9/11 GI Bill. If you have at least ninety days of aggregate active-duty service after 10 September 2001 and are still on active duty, or if you are an honorably discharged veteran or were discharged with a service-connected disability after thirty days, you may be eligible for this VA-administered program. For details of this bill, which provides significant education benefits to you, including provision for coverage for your children, visit www.benefits.va.gov/gibill/post911_gibill.asp.

VA Home Loan. The home loan program under the VA provides the opportunity to buy a home with a low down payment. (FHA in-service home loans are also available to active-duty officers and should be investigated.) Virtually all veterans who have served on active duty since 16 September 1940, as well as present active-duty personnel who have served at least 181 days of continuous duty, qualify for the VA home loan program.

VA Compensation. Compensation is payable by the Department of Veterans Affairs to an officer who becomes disabled while on active duty or active duty for training, provided the disability was incurred in the line of duty. The amount of the compensation is based on the extent of the disability, from 10 percent to 100 percent, which is a legal assessment by the VA in consideration of the officer's future capability to earn a living. The compensation goes to the officer. The amounts vary with the percentage and kind of disability, when the disability occurred, and the number of dependents. If the officer dies of service-related disabilities either while on active duty or after leaving the service, compensation for survivors is payable under dependency and indemnity compensation, discussed earlier.

Disabled servicemembers with less than eight years' active duty may or may not receive service disability retirement or severance pay, depending on the circumstances. They could receive the VA disability compensation. Officers who are involuntarily separated with more than five but less than twenty years of service are entitled to a separation payment equal to 10 percent of annual base pay times years of service. These matters should be clarified before you leave the hospital and are separated.

Combat-Related Special Compensation (CRSC) and Concurrent Retirement and Disability Program (CRDP). CRSC and CRDP are parts of a congressional initiative to align military retirement compensation with that of other federal service disability and retirement programs. Until recently, military retirees with disabilities waived their regular retired pay to receive VA disability compensation. This system was unique to military retirement, since all other federal service retirees have been entitled to receive full regular retirement compensation and disability compensation.

The main differences between these programs are as follows: CRSC is for military retirees with combat-related disabilities. CRSC is not subject either to taxation or division with a former spouse. CRDP on the other hand, makes no requirement that disabilities be combat-related. CRDP payments are taxed and subject to division of retired pay with a former spouse.

Combat-related and service-connected disabilities are separate types of disabilities that are obtained under different circumstances. Combat-related disabilities are defined by DoD. Service-connected disabilities pertain to any verified disability or injury incurred while in the service. The disabilities may be combat or non-combat-related. Because CRDP and CRSC both seek to restore lost retired pay, they are mutually exclusive. In other words, retirees cannot receive CRDP and CRSC payment at the same time. For the same reason, the Special Compensation for the Severely Disabled (SCSD) program has been repealed and replaced with CRSC or CRDP. For more information, view the CRSC website at www.crsc.army.mil.

Life Insurance Programs

All estate planners advise that the first step to security after having a sound financial base is to establish an insurance program. But how much insurance is enough? There are two solutions to the problem. One is the "needs approach"; the other is to follow the advice of insurance counselors and recommended literature. The "needs approach" calls for an analysis of how much the survivors will require should the breadwinner die. This includes living expenses, housing, and education of children, among many other things. Consider your spouse's earning capability. Remarriage is a difficult planning factor, to say the least, but it should be considered. If in doubt, seek the advice of a reputable insurance salesperson or insurance estate planner, and match the advice received against that of the literature referenced at the beginning of this chapter.

One rule of thumb is that you should be covered by insurance in an amount up to five times your annual income. If your family also depends on income from your spouse, she or he should have similar coverage.

The need for expert counsel becomes more apparent when one recognizes that there are seven basic types of life insurance: reducing term, level term, ordinary (whole) life, limited pay life (twenty- or thirty-year pay), endowment, annuity (retirement income), and variable insurance. There is no less expensive way to provide for security; however, as more options are selected the costs increase. Other options are found within the basic kinds: war risk, aviation coverage, suicide, disability waivers, dividend options, double indemnity, settlements, endowments for education, surrender cash values, and loan values. These should be carefully checked in your contract; at least you should understand what you are paying for. Many military benefit and mutual associations offer packaged family insurance programs that provide coverage for the entire family at moderate rates. As in other financial programs, if you do not pay very much, you do not accrue much cash or loan value. In insurance programs, however, you do have security not provided by other investment-type programs.

Servicemember's Group Life Insurance (SGLI). This government-sponsored insurance program provides up to $400,000 of life insurance coverage for active-duty members and for drilling members of the National Guard and reserve. It also is available to members of the retired reserve under age sixty at slightly increased rates, depending on age. Upon separation from active duty, the coverage is continuable as low-cost term insurance for five years under the Veterans' Group Life Insurance (VGLI), after which you are guaranteed the right to convert to a permanent insurance program, whatever the status of your health. Since this conversion to permanent insurance would be at rates applicable to your attained age, it could be expensive. You should consider this as a bonus program and definitely not drop or forgo other insurance coverage.

Survivor Benefit Plan (SBP). The SBP was discussed in an earlier section as a qualified benefit. Actuarial studies indicate that the cost of coverage under SBP is significantly less than the cost for the same coverage by a commercial insurance company.

Other Kinds of Insurance. In addition to life insurance, other kinds of insurance are necessary for security. These include fire insurance on a home, insurance against loss or theft of personal property, personal liability, and automobile insurance. Some later-referenced associations package these kinds of insurance at lower costs to service personnel; your consideration and study of their programs are in order. Each will provide literature, advice, and cost and coverage quotations for your particular situation. As in the case of life insurance, studying referenced literature and seeking the advice of experienced officers and professional counselors can save you significant funds.

Investment Programs

By carefully budgeting your salary and after establishing the other priority components of a financial estate, you arrive at the opportunity for saving and investing. Actually, your potential retired pay and Social Security income can be considered investment programs, since both will provide what amounts to an annuity when you retire or attain age 65. Further, both Social Security and retirement income provide relatively fixed purchasing power, as opposed to fixed income, since both are adjusted to offset cost-of-living increases. You may not have realized that you (with the government's assistance) began two investment programs upon entering the service.

An investment program may consist of fixed dollar investments and variable dollar investments. Fixed dollar investments are those that have a fixed value and return a certain yield, such as bonds and savings accounts. Variable dollar securities grow or decline in value at the risk of the investor. These include common stocks and investment in real estate. There is increased risk in such purchases, since there is no guarantee of eventual sale at a stipulated price. The purchase of carefully selected common stocks, or carefully chosen real estate, is made with the goal of increasing values, combined with dividend or rental return. Most investment advisory companies recommend a suitable mix between fixed and variable dollar investments—highly rated bonds with highly rated common stocks, for example. There are also mutual funds that have investments wholly in common stocks or wholly in bonds, with many having a judicious mix of their own selections of each type of security.

Also do not forget to consider the Army's Thift Savings Plan in your investment program. Go to www.tsp.gov for more details.

Each servicemember also is entitled to establish an individual retirement account (IRA) to which he or she can contribute up to $3,000 per year, which may be wholly or partially deductible. In any case, the earnings in the IRA accumulate tax-free. However, the funds in the IRA cannot be used until age 59½ without incurring a penalty. When used, the proceeds from the account are taxed as ordinary income.

When considering investment programs, do not overlook your checking account. Many banks, savings and loan associations, and credit unions pay interest on checking accounts, provided a minimum balance is maintained in the account. Although the interest rates on these accounts may be relatively low, it is far better to be earning interest than to maintain a small balance in an account where you quite likely will be required to pay a monthly service charge, a charge for each check written, or both. The difference between earning interest and paying service charges can add up to significant savings.

How should you enter this interesting and potentially rewarding field? First, you should enter into investment planning as early in your career as possible. *The less you have to invest, the more important it is to start your investment program early and do your homework carefully.* There are good books on investment planning. There are good magazines devoted to business, finance, and investment. There are investment advisory services. There are investment counselors. There are brokers who can provide study material and detailed information, and many will assist in financial planning. As an initial step, go to your local libraries. See what they have available for your reading and study. When you have some investment money—make your start.

Assistance Facilities and Associations

Servicemen and women have many assistance and advisory associations that will help them and their family members in solving problems related to personal affairs and security. There are also many mutual insurance associations that provide all types of insurance at reasonable rates while also assisting in associated administrative matters. Knowing about them can affect your security planning. They also can and will assist your family members while you are away or should you die. You should determine your eligibility and study what each association can provide.

Army Community Service (ACS). The Army has consolidated a number of services to personnel and their families with the missions of providing information, assistance, and guidance in meeting personal and family problems beyond the scope of their own resources. The functions of Army Emergency Relief, Army Personnel Affairs, Survivors Assistance Officer, and Retired Affairs are integrated with ACS as local conditions permit. Information is provided on financial assistance, availability of housing, transportation, relocation, medical and dental care, legal assistance, orientation of new arrivals, care of handicapped children, recreation, and many other community matters. The theme is "self-help, service, and stability." The ACS is particularly devoted to the assistance of family members whose sponsors are away for whatever purpose. If you or your family members have a problem, the local ACS on post is a place to begin for information and assistance.

Department of Veterans Affairs (VA). The many services, rights, and benefits of veterans are administered by officials of the VA, assisted in many cases by members of the American Legion and Veterans of Foreign Wars. Most larger cities and county seats have offices where you can get assistance. The VA's website is www.va.gov.

American Red Cross. There is a broad program of directed and voluntary aid to military personnel by officials and volunteers of the American Red Cross. Their assistance in hospitals is well known. There is a director at each post to assist in emergency notifications, investigations, provision of financial aid, child welfare, and other related matters of emergency. Its website is www.redcross.org.

Army Emergency Relief. Although primarily an association to give financial loans and grants to enlisted personnel, the Army Emergency Relief officer at every post can also assist an officer or the officer's family if the situation warrants. Each case is determined on merit and need, usually in association with American Red Cross officials and those of the ACS program.

Social Security Field Offices. The many complex rules and administrative matters of the Social Security program can be clarified at its field offices located in most larger cities. Personnel there will determine eligibility and assist in obtaining the many benefits that this program provides. A toll-free number is available: (800) 772-1213. The Social Security website is www.ssa.gov.

Army and Air Force Mutual Aid Association. This mutual insurance association began in 1879. It offers low-cost life insurance to Regular Army and Air Force officers, warrant officers, and those reserve officers and warrant officers in the indefinite category who have completed three years of continuous active service. The association offices are at 102 Sheridan Avenue, Fort Myer, Arlington, Virginia, 22211. They provide free services for survivors in preparing survivor benefit and other claims on the government. They will also store your valuable papers in their vault. All officers are urged to investigate the association's programs. Its website is www.aafmaa.com.

Armed Forces Benefit Association. This worthwhile mutual association has similar eligibility rules to those of the Army and Air Force Mutual Aid Association and provides low-cost group life insurance. It is at 909 North Washington Street, Alexandria, Virginia, 22314. Its website is www.afba.com.

Association of the U.S. Army. Officers who have joined the Association of the U.S. Army are eligible for group life insurance at very low cost. Generally no physical examinations are required, and there are no exclusions. All risk protection and disability waiver of premium payments are included. You can get more information by writing to AUSA Members Life Insurance Plan, 1529 18th Street NW, Washington, D.C., 20036. AUSA's website is www.ausa.org.

United Services Automobile Association. USAA provides a wide range of insurance and financial services to active-duty, retired, National Guard, reserve, and formerly commissioned and warrant officers in all branches of the U.S. military. Officer candidates within twenty-four months of commissioning may also apply. USAA tailors its products to meet the needs of officers and, when financial results justify them, declares regular and special dividends that help to lower insurance costs.

In addition to auto insurance, USAA provides homeowner's, renter's, boatowner's, personal property, and personal liability coverages. Also, the USAA family of companies provides products and services including life and health insurance, no-load mutual funds, brokerage services, and travel, banking, and merchandising services.

Officers are invited to write to USAA, 9800 Fredericksburg Road, San Antonio, Texas, 78288, or call (800) 531-8080. Its website is www.usaa.com.

Armed Forces Insurance. This association, formerly the Armed Forces Cooperative Insuring Association, offers low-cost insurance of personal and household effects against fire, theft, and other like losses without regard to station or duty and at actual cost to members. Eligibility is like that for the Army and Air Force Mutual Aid Association. Members do the promoting and most of the assessing. The association's promptness and payment of reasonable claims without argument are well known throughout the service. Automobile insurance and personal liability insurance are also offered. Officers are invited to write to Armed Forces Insurance, P.O. Box G, Fort Leavenworth, Kansas, 66027, for complete information. AFI's website is www.afi.org.

Other Associations. There are other associations that insure and assist military personnel. Only a few of the best known and longest established societies have been mentioned. There are, in particular, a number of mutual life insurance associations that offer low-cost protection. Their names are excluded in the interests of brevity, but exclusion does not imply that such associations are more expensive, untrustworthy, or unreliable. The ones recommended can save you considerable sums of money should you insure with them or similar organizations.

PERSONAL ADMINISTRATION RELATED TO SECURITY PROGRAMMING

No security program review is complete without a study and inspection of one's personal administration. Such matters as records, state and federal taxes, estate and death taxes, wills, trusts, powers of attorney, and joint ownership of property are discussed below. You should review these records each year (at income tax time, for instance) and, in particular, prior to going overseas with or without your spouse. All matters should be clearly explained to your spouse and the places where your records are stored clearly identified.

Records. You should know where all personal records are and their status. The important ones should be filed in a safe deposit box, a safe, a fireproof vault (the Army and Air Force Mutual Aid Association stores life insurance policies and wills, for example), or some other secure location. Photostatic or otherwise certified copies of the original may then be kept at home as your working file. A partial list of important records would be birth certificates, marriage certificates, wills, powers of attorney, trusts, deeds, mortgages, active-duty and retirement orders, automobile titles, life insurance policies, bonds and stocks, baptism and confirmation papers, and other valuables and valuable papers. Less important records could be filed at home with a personal file of military records. Such a file would include financial records, promotion records, personal property records, bank and savings account records, other kinds of insurance policies, and so on; the more important ones can and should be filed in a strong, relatively fireproof, portable box.

State and Federal Taxes. All officers must file a federal and (in most cases) a state income tax return each year and pay these taxes as required. Your finance officer withholds federal pay for this purpose and advises you of the amounts withheld each year by means of the W-2 form. Under provisions of the Soldiers' and Sailors' Civil Relief Act of 1940, you are exempted from paying state income and personal property taxes to a state where you temporarily reside because of military duty, provided you can prove to that state that you are domiciled—legally reside—in another. You must pay federal income tax on your basic pay, dislocation allowances, and per diem pay in excess of needs, and all other sources of taxable income. All units and posts have personnel advisors who can assist you in most uncomplicated filing matters. Your spouse is not exempted from state income and property taxes if he or she earns an income and possesses property in that state. It is wise to establish a legal residence in a single state and pay your taxes there accordingly; however, there are states that have no income taxes, and some officers establish a legal residence in one of them in the interests of lower overall taxes. Such establishment should be done carefully, having in mind the future and not a saving of a few dollars through legal loopholes. Since the career family moves often, this problem will recur. Facing it with honesty, foresight, and advice from experts is in order.

Estate and Death Taxes. The federal government may tax your estate in the event of your death if the adjusted gross estate is sufficiently large. State death taxes take the form of an inheritance or estate tax. Each state has a different system of such taxation; therefore, consult with a nearby legal officer. Your will (and estate) must be examined, appropriately implemented, and taxed by a state official. The process known as probate is accomplished prior to execution of the will or disbursement of the estate. Bank officers will record the contents of safe deposit boxes and may not allow the papers therein to be taken or assumed until after the probate. It is possible to compute your potential estate tax while you are alive. The matter of estate taxes is complicated, however, and the advice of a specialist in estate planning is recommended. Total income, including insurance benefits, is the basis in many states for the inheritance tax. The issue for you is one of awareness that such taxes exist so that they can be considered and expected.

Wills. Without exception, all references and counselors advise an Army officer to execute a will and to keep it current. Your spouse should also have a will. If you die without one, the state will, in effect, make one for you. Unfortunately, regardless of how little you have, the state cannot make such a will and distribute your estate as well as you could have done by a will. On a draft, write what you believe you desire in a will, then go to the legal assistance officer for expert help.

There are many dos and don'ts for will making, a few of which follow: Don't do it by yourself; don't change it by pen and ink unless legally supervised; don't delay doing it; don't fail to reexamine it upon records review and especially when changing legal residence; don't specify amounts (in general, use percentages); don't sign more than one copy; do choose an executor with business sense; and do keep the original signed copy in a secure place. Specify names and property or assets only if you have to; many a family has had temporarily or permanently strained relationships develop out of contests for estate assets. Use relations and terms like "to the children, share and share alike." The important advice is reemphasized: Make a will with legal help, and keep it current.

Power of Attorney. A power of attorney is a legal instrument whereby one person may designate another person to act in his or her behalf in legal or personal matters. This authority may be granted by a husband to his wife or any other person of legal age and capacity. It can be made very general and unlimited in scope, or it can be restricted to specific functions, as desired by the grantor. Do not fail to understand that when a power of attorney is given to another person, that person has the legal right to act under it; and if the powers granted include the right to buy or sell or to give away the grantor's property, such acts are binding. *It has been called the most potentially dangerous document that man has ever devised.* Therefore, prudence demands that you entrust such powers only to an individual in whom you have complete trust and confidence.

The document has particular value to officers. Many officers provide their spouses with power of attorney. Thereafter, when the officer is traveling extensively or is assigned overseas with the family remaining behind, essential business or legal matters may be performed without the officer's presence or signature.

The authority given in a power of attorney, unlike that of a will, becomes invalid upon death of the grantor.

By all means, consult a legal assistance officer or attorney in the preparation of this document. Explain carefully the purpose in mind. It may be true that a restricted rather than a broad form may be best. Again, as with a will, there are variations among state laws. A special form must be used to authorize cashing of government checks. A power of attorney must be acknowledged before a notary public.

The use of a form to give to another these powers is discouraged. Consult a legal assistance officer or an attorney, and have it done correctly and thoughtfully.

Trusts. A trust is an agreement whereby an individual gives property to a second party, the trustee, for the benefit of a third party, the beneficiary or beneficiaries. Individuals, trust companies, or banks act as trustees. Trusts are special legal contracts. Trusts may be specified while you are alive, as testaments for financial income to specified individuals after your death, as revocable, and as irrevocable. As contracts, they require legal and business advice and should be entered only with knowledge and care.

Joint Ownership by Husband and Wife. For many years, this guide presented a discussion advocating a joint bank account for husband and wife and joint ownership with right of survivorship of other property such as real estate, stocks and bonds, and the family automobile. In view of sudden orders for TDY, foreign service unaccompanied tours, and other temporary family separations, the method has definite advantages. The wife or husband staying behind to manage the family and its affairs has fewer strictures. On the death of one of the joint owners, the property passes immediately to the other. It is not subject to the cumbersome processes of the probate courts, and there is the avoidance of potentially heavy legal fees. These are important advantages.

As the years have passed, experience has proved that the method has weaknesses. Some states do not recognize joint ownership with right of survivorship. Others are community property states in which spouses are presumed to own half the estate accumulated during the marriage. Difficulties arise when the joint owners are divorced or one becomes incompetent. In the absence of a will, difficult problems arise if the co-owners die in a common disaster. There are estate and inheritance taxes, federal as well as state, and if the estate becomes sizable, joint ownership presents additional problems. There are gift taxes that may need to be paid in the transfer of title to a spouse. These are all matters for expert guidance, including the special situation of the state in which assigned to Army duty.

Consult the legal assistance officer, or consult a civilian attorney. State all the facts with candor. Joint ownership is not condemned. Under the conditions of Army service for most officers, it may be the preferred method of ownership. Get legal advice so your decision can be made with prudence and wisdom.

Personal and Property Records. Your complete security program can be detailed on a form and changed with each review if necessary. Many associations provide such forms, and it is in your interest to secure them and to fill them in.

You will find as the years go by that such a record requires several pages that can be conveniently bound in a folder or a notebook. Property records are similarly inclined to grow and can be included in your folder. The need to keep the file up-to-date, in a relatively secure location where your spouse can find and review it, is obvious but worthy of noting.

STORAGE OF IMPORTANT DOCUMENTS

Military families should gather together important documents and safeguard them in a secure file so they are immediately available, if needed. It's important for the soldier and the spouse to jointly organize their important document file so that both understand the status and significance of each document.

Document	On Hand	Location
Marriage certificate		
Birth certificates		
Baptismal certificates		
Adoption papers		
Citizenship papers		
Passports		
Armed Forces ID cards (check expiration)		
Wills		
Family medical records		
Family dental records		
Immunization records		
Social Security cards/numbers		
Court orders (divorce/child custody)		
Copy of emergency data card		
Copy of SGLI election form		
Addresses/phone numbers of immediate family		
Powers of attorney		
Copies of TDY/PCS orders		
Insurance policies (life, auto, home, personal property)		
Leave and Earnings Statements		
Bank account numbers (checking/savings)		
Checkbook		
List of investments/bonds		
Deed/mortgage papers		
Copies of installment contracts		

Document	On Hand	Location
Credit card/club card	_____	____________________
Federal and state tax records	_____	____________________
Driver's license	_____	____________________
Car registration, title, and inspection certificate	_____	____________________
POV shipping documents (OCONUS)	_____	____________________
Warranties on car or appliances	_____	____________________
Inventory of household goods	_____	____________________
Pet health/vaccination records	_____	____________________
Ration book (OCONUS)	_____	____________________
Gas coupons (OCONUS)	_____	____________________
Extra keys (car, house, safe deposit box)	_____	____________________
Diplomas/school transcripts	_____	____________________
Spouse's employment resume and work experience information	_____	____________________
Family photo albums	_____	____________________
List of important phone numbers (FSG, RDC, FAC, emergency numbers)	_____	____________________
Dependent child care plan	_____	____________________

SUMMARY

Security or estate planning calls for step-by-step analysis of one's situation, goals, and benefits, and supplementing these with savings, insurance, and investments. The plan must be reviewed and altered as each change in your personal and financial situation occurs but, in particular, prior to departure for overseas and combat. There is no simple plan or checklist for you to follow because of the numerous financial programs and the personal variables introduced by each officer and his or her dependents. This chapter should have provided you with a base for such planning; by following its sections in order and checking off each point as it applies to you, you will be able to draw up your own plan. Then, by reading the references and seeking counsel, you will be able to ensure security for you and your family members.

PART IV

Quick Reference

20

Pay and Allowances

Army personnel are assured that their services are appreciated by our government. There have been times in the past when military pay rates lagged far behind the pay scales in private industry, and indeed even far behind the pay scales of federal civilian employees, but periodically the pay rates are adjusted to provide approximate comparability with the civilian sector.

BASIC PAY AND ALLOWANCES

How Pay Is Established. Federal pay rates are set using the Department of Labor's Employment Cost Index, which tracks the cost of all salary raises nationwide. Federal civilian pay rates are adjusted each year in relation to rates for comparable work in private industry. Originally, the adjustments to pay were to be made effective 1 October each year, but in recent years the adjustments have been made three months later on 1 January.

Although the comparability study applies strictly to federal civilian employees, the Federal Employees Pay Comparability Act of 1990 requires that military personnel receive the same average pay increase as civil servants. Thus, an increase in federal civilian pay automatically results in an equitable increase in military pay.

Military Pay. The Defense Finance and Accounting Service (DFAS) website displays the current military pay information and rate tables. Access it at https://mypay.dfas.mil/mypay.aspx.

Service Creditable for Basic Pay. As seen in the tables, pay increases in each grade with length of service. In computing the years of service for pay purposes, credit is given for all periods of active service in any regular or reserve component of any of the uniformed services. Credit may also be granted for service other than active duty. Officers are advised to consult the finance officer servicing their pay accounts with a statement of all their military service for consideration of credit that may be given under the laws.

Basic Allowance for Subsistence (BAS). The BAS is the same for all officers, regardless of grade. The amount is adjusted periodically as the base pay is adjusted. The BAS is not subject to income tax. The monthly rate for officers in effect 1 January 2022 was $280.29.

Basic Allowance for Housing (BAH). In 1998, BAH was initiated as the new housing allowance. BAH is based on geographic duty location, pay grade, and dependency status. The intent of BAH is to provide uniformed servicemembers accurate and equitable housing compensation based on housing costs in local civilian housing markets, and it is payable when government quarters are not provided.

DoD and the services developed BAH to improve and enhance housing allowances for all members, taking into account complaints about the old variable housing allowance (VHA) program. Eligible members receive one monthly dollar amount for BAH, in place of a separate VHA and basic allowance for quarters (BAQ). A grandfathering provision, known as rate protection, keeps individuals from experiencing reductions in housing allowances, as long as their status remains unchanged. Rate protection continues until the member incurs a change in status, defined as a PCS move, a decrease in grade, or a change in dependency status. Promotions are specifically excluded in the definition of change in status. Like BAQ, BAH distinguishes between with dependents and without dependents, but not the number of dependents.

Members will not see dramatic increases in housing allowances in any one year, because changes are being phased in over a multiyear period, based on the need to protect individuals from decreases (rate protection). Generally speaking, the new rates are increased in high-housing-cost locations, for junior members, and for without-dependent members.

A primary reason for the BAH was the awareness that the old VHA/BAQ system was unable to keep up with housing costs, and members were being forced to pay larger out-of-pocket costs than originally intended. With BAH, increases are indexed to housing cost growth instead of the pay raise, thus protecting members from any further erosion of housing benefits over time.

The BAH is designed to be inherently fair because the typical servicemember of a given grade and dependency status arriving at a new duty station will have the same monthly out-of-pocket expenses regardless of the location. For example, if the out-of-pocket cost for a typical O-3 with dependents is $100, the typical (median) O-3 with dependents can expect to pay $100 out-of-pocket for housing if assigned to Miami, New York, San Diego, Fort Hood, Camp Lejeune, Minot, or any other duty location in the United States. Once the member arrives, rate protection applies, and the member will receive any published increase but no decrease in housing allowances.

If any individual officer chooses a larger or more expensive residence than the median, that person will have greater out-of-pocket expenses. The opposite is true for an individual who chooses to occupy a smaller or less expensive residence. Only for the median member can we say that the out-of-pocket expense is the same for a given pay grade and dependency status at any location in the United States.

The BAH employs a civilian-based method of measuring comparable housing costs that is superior to the old VHA housing survey that measured members' spending on housing.

In computing BAH, local price data of rentals, average utilities, and insurance are included. Data are collected—in the spring and summer, when housing markets are most active—on apartments and town homes or duplexes, as well as on single-family rental units of various bedroom sizes. A multitiered screening process is used to ensure that the units and neighborhoods selected are appropriate, considering commuting distance and the type of neighborhood of comparably paid civilians.

CONUS Cost-of-Living Allowance (CONUS COLA). Since 1994, Congress has provided a cost-of-living allowance for members stationed in the continental United States when the cost of living other than housing exceeds the national average by 8 percent or more. However, DoD is using, at least initially, a figure of 9 percent over the national average to determine eligibility. The amount of this allowance depends on the rank, years of service, and dependent status of the member and is also tied to the portion of income that the member spends on basic needs, so that the allowance pays a larger percentage to the lower ranks. It differs from other allowances in that it is considered taxable income. The CONUS COLA is adjusted based on calculations of "spendable income" prepared by the Bureau of Labor Statistics. To determine the COLA for an area, go to www.defensetravel.dod.mil/site/conus.cfm.

Definition of Dependent for Quarters Allowance. Use of the term *dependent* has long since fallen into disfavor, supplanted by the more acceptable *family member* to describe the status of an officer's spouse and children. Still, *dependent* is used for legally defining who is authorized certain benefits and privileges. The law that authorizes the allowance for quarters for an officer with dependents provides that it includes at all times and in all places the lawful spouse and the legitimate unmarried children under twenty-one years of age. There are important additional provisions for other dependent family members, and for rulings in these instances, the best course to follow is to consult the unit personnel officer or the finance officer.

Reserve Officer Pay. Reservists in a Troop Program Unit (TPU) are paid one day's pay (prorated on the basis of a monthly salary at the appropriate pay grade) for each four-hour Unit Training Assembly (UTA) performed as part of a monthly Inactive Duty Training (IDT) period, usually on a weekend. Normally, a drill weekend consists of four UTAs; the reservist would draw four days' pay for that drill weekend. During the twelve to fourteen days of annual training, reservists are paid on a day-for-day basis.

OTHER PAY AND ALLOWANCES

In addition to basic pay and to allowances for subsistence and for quarters (for those officers not furnished government quarters), the following provisions for special pay or allowances are important to understand. This listing is necessarily concise. For more information, see your servicing Finance Office or refer to the DoD Financial Management Regulation (DoDFMR), Volume 7A: Military Pay Policy and Procedures—Active Duty and Reserve Pay, dated March 2008 and found online http://comptroller.defense.gov/Portals/45/documents/fmr/Volume_07a.pdf.

Hazardous Incentive Duty Pay. Several types of duty assignments are classed as hazardous. Incentive pay is authorized for performance, under competent orders, of these types of duty: parachute duty, demolition duty, exposure to toxic pesticides, and experimental stress duty. A member may receive hazardous duty pay for two types of such duty for the same period if qualified and required to perform multiple hazardous duties in order to carry out the mission of the unit. (An example would be to engage in regular and frequent parachute jumps and also to perform demolition duty when required to accomplish the mission of the unit.) The monthly rate of hazardous duty pay for all soldiers regardless of rank is $150.

Flight Pay. Flight pay, technically "Aviation Career Incentive Pay," ranges from $125 to $840 per month for rated commissioned officers or aviation warrant officers who meet certain "gate" requirements as determined by credits they receive for operational flying time.

Diving Duty Pay. Policies regarding diving pay were revised in August 2007. An officer on diving duty may receive diving pay between $110 and $240 per month, depending on his or her classification.

Career Sea Pay. This a taxable special pay based on pay grade and years of sea service. To qualify for this pay, officers, warrant officers, and enlisted soldiers with three or more years of documented sea service must be assigned and on orders for duty aboard a qualifying vessel. Career sea pay for officers ranges from $150 to $750 a month.

Foreign Language Proficiency Pay. The language pay program applies to two categories of servicemembers: career linguists who are proficient in a critical language and who hold a designated specialty or are assigned to a documented linguist position; and noncareer linguists who are proficient in language but not assigned to a designated position and are not required to be proficient in a language to maintain their specialty or branch code. DoD and the Army completed a major overhaul of FLPP in 2006. One major change made was that career and noncareer linguists now receive the same special pay, which can range from $100 to $500. See AR 611-6 (Army Linguist Management) and MILPER messages 06-233, 07-137, and 07-182 for the latest policy changes.

Special Pay for Officers Serving in Positions of Unusual Responsibility and of a Critical Nature. The Secretary of the Army may designate positions of unusual responsibility that are of a critical nature to an armed force under his jurisdiction and authorize special pay to officers performing the duties of such a position. Monthly rates of responsibility pay range from $50 for designated majors and below to $150 for designated colonels.

Engineering and Scientific Career Continuation Pay (ESCCP). Officers may be entitled to ESCCP if they are not receiving any other accession or career

MONTHLY ACTIVE COMPONENT BASIC PAY TABLE

(Effective January 1, 2023)

	<2	>2	3	4	6	8	10	12	14	16	18	20	22	24	26	28	30	32	34	36	38	40
Commissioned officers																						
O-10												17,675.10	17,675.10	17,675.10	17,675.10	17,675.10	17,675.10	17,675.10	17,675.10	17,675.10	17,675.10	17,675.10
O-9												17,201.40	17,449.80	17,675.10	17,675.10	17,675.10	17,675.10	17,675.10	17,675.10	17,675.10	17,675.10	17,675.10
O-8	12,170.70	12,570.00	12,834.30	12,908.10	13,238.40	13,789.50	13,918.20	14,441.70	14,592.60	15,043.50	15,696.60	16,298.10	16,700.10	16,700.10	16,700.10	16,700.10	17,118.30	17,118.30	17,545.80	17,545.80	17,545.80	17,545.80
O-7	10,113.00	10,582.80	10,800.30	10,973.40	11,286.00	11,595.30	11,952.60	12,308.70	12,666.60	13,789.50	14,737.80	14,737.80	14,737.80	14,737.80	14,813.70	14,813.70	15,110.10	15,110.10	15,110.10	15,110.10	15,110.10	15,110.10
O-6	7,669.20	8,425.20	8,978.10	8,978.10	9,012.60	9,398.70	9,450.00	9,450.00	9,987.00	10,936.20	11,493.60	12,050.40	12,367.50	12,688.80	13,310.70	13,310.70	13,576.50	13,576.50	13,576.50	13,576.50	13,576.50	13,576.50
O-5	6,393.30	7,202.10	7,700.40	7,794.30	8,105.70	8,291.40	8,700.60	9,001.80	9,389.70	9,982.80	10,265.40	10,544.70	10,861.80	10,861.80	10,861.80	10,361.80	10,861.80	10,861.80	10,861.80	10,861.80	10,861.80	10,861.80
O-4	5,516.40	6,385.20	6,812.10	6,906.30	7,301.70	7,726.20	8,254.80	8,665.50	8,951.10	9,115.50	9,210.30	9,210.30	9,210.30	9,210.30	9,210.30	9,210.30	9,210.30	9,210.30	9,210.30	9,210.30	9,210.30	9,210.30
O-3	4,849.80	5,497.80	5,933.40	6,469.80	6,780.30	7,120.50	7,340.10	7,701.60	7,890.60	7,890.60	7,890.60	7,890.60	7,890.60	7,890.60	7,890.60	7,890.60	7,890.60	7,890.60	7,890.60	7,890.60	7,890.60	7,890.60
O-2	4,190.70	4,772.70	5,496.90	5,682.60	5,799.30	5,799.30	5,799.30	5,799.30	5,799.30	5,799.30	5,799.30	5,799.30	5,799.30	5,799.30	5,799.30	5,799.30	5,799.30	5,799.30	5,799.30	5,799.30	5,799.30	5,799.30
O-1	3,637.20	3,786.00	4,576.80	4,576.80	4,576.80	4,576.80	4,576.80	4,576.80	4,576.80	4,576.80	4,576.80	4,576.80	4,576.80	4,576.80	4,576.80	4,576.80	4,576.80	4,576.80	4,576.80	4,576.80	4,576.80	4,576.80
Commisioned Officers with over 4 years' active duty service or more than 1,460 reserve points as an enlisted member or warrant officer																						
O-3E				6,469.80	6,780.30	7,120.50	7,340.10	7,701.60	8,007.00	8,182.50	8,421.00	8,421.00	8,421.00	8,421.00	8,421.00	8,421.00	8,421.00	8,421.00	8,421.00	8,421.00	8,421.00	8,421.00
O-2E				5,682.60	5,799.30	5,983.80	6,295.50	6,536.70	6,715.80	6,715.80	6,715.80	6,715.80	6,715.80	6,715.80	6,715.80	6,715.80	6,715.80	6,715.80	6,715.80	6,715.80	6,715.80	6,715.80
O-1E				4,576.80	4,887.00	5,067.90	5,252.70	5,433.90	5,682.60	5,682.60	5,682.60	5,682.60	5,682.60	5,682.60	5,682.60	5,682.60	5,682.60	5,682.60	5,682.60	5,682.60	5,682.60	5,682.60
Warrant officers																						
W-5												8,912.10	9,364.20	9,701.10	10,073.40	10,073.40	10,578.00	10,578.00	11,106.00	11,106.00	11,662.50	11,662.50
W-4	5,012.40	5,391.30	5,546.10	5,698.20	5,960.70	6,220.20	6,483.00	6,877.80	7,224.30	7,554.00	7,824.30	8,087.70	8,473.80	8,791.50	9,153.60	9,153.60	9,336.30	9,336.30	9,336.30	9,336.30	9,336.30	9,336.30
W-3	4,577.70	4,767.90	4,964.10	5,027.70	5,232.30	5,635.80	6,055.80	6,253.80	6,482.70	6,718.20	7,142.40	7,428.30	7,599.60	7,781.40	8,029.50	8,029.50	8,029.50	8,029.50	8,029.50	8,029.50	8,029.50	8,029.50
W-2	4,050.30	4,433.40	4,551.00	4,632.30	4,894.80	5,302.80	5,505.60	5,704.50	5,948.10	6,138.60	6,310.80	6,517.20	6,652.80	6,760.20	6,760.20	6,760.20	6,760.20	6,760.20	6,760.20	6,760.20	6,760.20	6,760.20
W-1	3,555.00	3,938.10	4,040.70	4,258.20	4,515.00	4,893.90	5,070.60	5,318.70	5,561.70	5,753.10	5,929.20	6,143.40	6,143.40	6,143.40	6,143.40	6,143.40	6,143.40	6,143.40	6,143.40	6,143.40	6,143.40	6,143.40

Enlisted members																						
E-9							6,055.50	6,192.90	6,365.70	6,568.80	6,774.90	7,102.80	7,381.50	7,673.70	8,121.60	8,121.60	8,526.90	8,526.90	8,953.80	8,953.80	9,402.30	9,402.30
E-8						4,957.20	5,176.50	5,312.10	5,474.70	5,650.80	5,968.80	6,130.20	6,404.40	6,556.50	6,930.90	6,930.90	7,069.80	7,069.80	7,069.80	7,069.80	7,069.80	7,069.80
E-7	3,445.80	3,760.80	3,905.10	4,095.30	4,244.70	4,500.60	4,644.90	4,900.50	5,113.50	5,258.70	5,413.50	5,473.20	5,674.50	5,782.50	6,193.50	6,193.50	6,193.50	6,193.50	6,193.50	6,193.50	6,193.50	6,193.50
E-6	2,980.50	3,279.90	3,424.80	3,565.50	3,711.90	4,042.20	4,170.90	4,419.90	4,496.10	4,551.30	4,616.40	4,616.40	4,616.40	4,616.40	4,616.40	4,616.40	4,616.40	4,616.40	4,616.40	4,616.40	4,616.40	4,616.40
E-3	2,730.30	2,914.20	3,055.20	3,199.20	3,423.90	3,658.50	3,851.70	3,874.80	3,874.80	3,874.80	3,874.80	3,874.80	3,874.80	3,874.80	3,874.80	3,874.80	3,874.80	3,874.80	3,874.80	3,874.80	3,874.80	3,874.80
E-4	2,503.50	2,631.60	2,774.10	2,914.80	3,039.30	3,039.30	3,039.30	3,039.30	3,039.30	3,039.30	3,039.30	3,039.30	3,039.30	3,039.30	3,039.30	3,039.30	3,039.30	3,039.30	3,039.30	3,039.30	3,039.30	3,039.30
E-3	2,259.90	2,402.10	2,547.60	2,547.60	2,547.60	2,547.60	2,547.60	2,547.60	2,547.60	2,547.60	2,547.60	2,547.60	2,547.60	2,547.60	2,547.60	2,547.60	2,547.60	2,547.60	2,547.60	2,547.60	2,547.60	2,547.60
E-2	2,149.20	2,149.20	2,149.20	2,149.20	2,149.20	2,149.20	2,149.20	2,149.20	2,149.20	2,149.20	2,149.20	2,149.20	2,149.20	2,149.20	2,149.20	2,149.20	2,149.20	2,149.20	2,149.20	2,149.20	2,149.20	2,149.20
E-1 >4 mos	1,917.60	1,917.60	1,917.60	1,917.60	1,917.60	1,917.60	1,917.60	1,917.60	1,917.60	1,917.60	1,917.60	1,917.60	1,917.60	1,917.60	1,917.60	1,917.60	1,917.60	1,917.60	1,917.60	1,917.60	1,917.60	1,917.60
E-1 <4 mos	1,773.00																					
Academy Cadets, Midshipmen and ROTC	1,273.20																					

NOTES:

1. Basic pay for an O-7 to O-10 is limited by Level II of the Executive Schedule in effect during Calendar Year 2023 which is: 17,675.10
   This includes officers serving as Chairman or Vice Chairman of the Joint Chiefs of Staff (JCS), Chief of Staff of the Army, Chief of Naval Operations, Chief of Staff of the Air Force, Commandant of the Marine Corps, Commandant of the Coast Guard, Chief of the National Guard Bureau, or commander of a unified or specified combatant command (as defined in 10 U.S.C. § 161(c)).
2. Basic pay for O-6 and below is limited by Level V of the Executive Schedule in effect during Calendar Year 2023 which is: 14,341.80
3. For the Sergeant Major of the Army Master Chief Petty Officer of the Navy, Chief Master Sergeant of the Air Force, or Sergeant Major of the Marine Corps, or Senior Enlisted Advisor of the JCS, basic pay for 2023 is: 9,786.00

MILITARY RESERVE COMPONENT DRILL PAY - 2023

(Effective January 1, 2023)

Pay Grade	2 or less	Over 2	Over 3	Over 4	Over 6	Over 8	Over 10	Over 12	Over 14	Over 16	Over 18	Over 20	Over 22	Over 24	Over 26	Over 28	Over 30	Over 32	Over 34	Over 36	Over 38	Over 40
Commissioned officers																						
O-7	10,113.00	10,582.8	10,800.30	11,286.00	11,952.60	12,666.6	13,789.5	14,737.8	14,737.8	14,737.8	14,737.8	14,813.7	14,813.7	15,110.1	15,110.1	15,110.1	15,110.1	15,110.1	15,110.1			
1 Drill	337.10	352.76	422.2	459.6	491.2	491.26	491.2	493.7	493.7	503.6	503.6	503.6	503.6	503.6	503.6							
4 Drills	1,348.40	1,411.04	1,688.8	1,838.6	1,965.0	1,965.04	1,965.0	1,975.1	1,975.1	2,014.6	2,014.6	2,014.6	2,014.6	2,014.6	2,014.6							
O-6	7,669.20	8,425.20	9,987.0	10,936.2	11,493.6	12,050.4	12,367.5	12,688.8	13,310.7	13,310.7	13,576.5	13,576.5	13,576.5	13,576.5	13,576.5	13,576.5						
1 Drill	255.64	280.84	332.9	364.5	383.1	401.68	422.9	443.6	443.6	452.5	452.5	452.5	452.5	452.5	452.5							
4 Drills	1,022.56	1,123.36	1,331.6	1,458.1	1,532.4	1,606.72	1,691.8	1,774.7	1,774.7	1,810.2	1,810.2	1,810.2	1,810.2	1,810.2	1,810.2							
O-5	6,393.30	7,202.10	9,389.7	9,982.8	10,265.4	10,544.7	10,861.8	10,861.8	10,861.8	10,861.8	10,861.8	10,861.8	10,861.8	10,861.8	10,861.8	10,861.8						
1 Drill	213.11	240.07	312.9	332.7	342.1	351.49	362.0	362.0	362.0	362.0	362.0	362.0	362.0	362.0	362.0							
4 Drills	852.44	960.28	1,251.9	1,331.0	1,368.7	1,405.96	1,448.2	1,448.2	1,448.2	1,448.2	1,448.2	1,448.2	1,448.2	1,448.2	1,448.2							
O-4	5,516.40	6,385.20	8,951.1	9,115.5	9,210.3	9,210.30	9,210.3	9,210.3	9,210.3	9,210.3	9,210.3	9,210.3	9,210.3	9,210.3	9,210.3							
1 Drill	183.88	212.84	298.3	303.8	307.0	307.01	307.0	307.0	307.0	307.0	307.0	307.0	307.0	307.0	307.0							
4 Drills	735.52	851.36	1,193.4	1,215.4	1,228.0	1,228.04	1,228.0	1,228.0	1,228.0	1,228.0	1,228.0	1,228.0	1,228.0	1,228.0	1,228.0							
O-3	4,849.80	5,497.80	7,890.6	7,890.6	7,890.6	7,890.60	7,890.6	7,890.6	7,890.6	7,890.6	7,890.6	7,890.6	7,890.6	7,890.6	7,890.6							
1 Drill	161.66	183.26	263.0	263.0	263.0	263.02	263.0	263.0	263.0	263.0	263.0	263.0	263.0	263.0	263.0							
4 Drills	646.64	733.04	1,052.0	1,052.0	1,052.0	1,052.08	1,052.0	1,052.0	1,052.0	1,052.0	1,052.0	1,052.0	1,052.0	1,052.0	1,052.0							
O-2	4,190.70	4,772.70	5,799.3	5,799.3	5,799.3	5,799.30	5,799.3	5,799.3	5,799.3	5,799.3	5,799.3	5,799.3	5,799.3	5,799.3	5,799.3							
1 Drill	139.69	159.09	193.3	193.3	193.3	193.31	193.3	193.3	193.3	193.3	193.3	193.3	193.3	193.3	193.3							
4 Drills	558.76	636.36	773.2	773.2	773.2	773.24	773.2	773.2	773.2	773.2	773.2	773.2	773.2	773.2	773.2							
O-1	3,637.20	3,786.00	4,576.8	4,576.8	4,576.8	4,576.80	4,576.8	4,576.8	4,576.8	4,576.8	4,576.8	4,576.8	4,576.8	4,576.8	4,576.8							
1 Drill	121.24	126.20	152.5	152.5	152.5	152.56	152.5	152.5	152.5	152.5	152.5	152.5	152.5	152.5	152.5							
4 Drills	484.96	504.80	610.2	610.2	610.2	610.24	610.2	610.2	610.2	610.2	610.2	610.2	610.2	610.2	610.2							
Commisioned Officers with over 4 years' active duty service or more than 1,460 reserve points as an enlisted member or warrant officer																						
O-3E				6,469.80	6,780.30	7,120.50	7,340.10	7,701.6	8,007.0	8,182.5	8,421.0	8,421.00	8,421.0	8,421.0	8,421.0	8,421.0	8,421.0	8,421.0	8,421.0	8,421.0	8,421.0	8,421.0
1 Drill				215.66	226.01	237.35	244.67	256.7	266.9	272.7	280.7	280.70	280.7	280.7	280.7	280.7	280.7	280.7	280.7	280.7	280.7	280.7
4 Drills				862.64	904.04	949.40	978.68	1,026.8	1,067.6	1,091.0	1,122.8	1,122.80	1,122.8	1,122.8	1,122.8	1,122.8	1,122.8	1,122.8	1,122.8	1,122.8	1,122.8	1,122.8
O-2E				5,682.60	5,799.30	5,983.80	6,295.50	6,536.7	6,715.8	6,715.8	6,715.8	6,715.80	6,715.8	6,715.8	6,715.8	6,715.8	6,715.8	6,715.8	6,715.8	6,715.8	6,715.8	6,715.8
1 Drill				189.42	193.31	199.46	209.85	217.8	223.8	223.8	223.8	223.86	223.8	223.8	223.8	223.8	223.8	223.8	223.8	223.8	223.8	223.8
4 Drills				757.68	773.24	797.84	839.40	871.5	895.4	895.4	895.4	895.44	895.4	895.4	895.4	895.4	895.4	895.4	895.4	895.4	895.4	895.4
O-1E				4,576.80	4,887.00	5,067.90	5,252.70	5,433.9	5,682.6	5,682.6	5,682.6	5,682.60	5,682.6	5,682.6	5,682.6	5,682.6	5,682.6	5,682.6	5,682.6	5,682.6	5,682.6	5,682.6
1 Drill				152.56	162.90	168.93	175.09	181.1	189.4	189.4	189.4	189.42	189.4	189.4	189.4	189.4	189.4	189.4	189.4	189.4	189.4	189.4
4 Drills				610.24	651.60	675.72	700.36	724.5	757.6	757.6	757.6	757.68	757.6	757.6	757.6	757.6	757.6	757.6	757.6	757.6	757.6	757.6

Warrant officers																					
W-5												8,912.10	9,364.2	9,701.1	10,073.4	10,073.4	10,578.0	10,578.0	11,106.0	11,106.0	11,662.5	11,662.5
1 Drill												297.07	312.1	323.3	335.7	335.7	352.6	352.6	370.2	370.2	388.7	388.7
4 Drills												1,188.28	1,248.5	1,293.4	1,343.1	1,343.1	1,410.4	1,410.4	1,480.8	1,480.8	1,555.0	1,555.0
W-4	5,012.40	5,391.30	5,546.10	5,698.20	5,960.70	6,220.20	6,483.00	6,877.8	7,224.3	7,554.0	7,824.3	8,087.70	8,473.8	8,791.5	9,153.6	9,153.6	9,336.3	9,336.3	9,336.3	9,336.3	9,336.3	9,336.3
1 Drill	167.08	179.71	184.87	189.94	198.69	207.34	216.10	229.2	240.8	251.8	260.8	269.59	282.4	293.0	305.1	305.1	311.2	311.2	311.2	311.2	311.2	311.2
4 Drills	668.32	718.84	739.48	759.76	794.76	829.36	864.40	917.0	963.2	1,007.2	1,043.2	1,078.36	1,129.8	1,172.2	1,220.4	1,220.4	1,244.8	1,244.8	1,244.8	1,244.3	1,244.8	1,244.8
W-3	4,577.70	4,767.90	4,964.10	5,027.70	5,232.30	5,635.80	6,055.80	6,253.8	6,482.7	6,718.2	7,142.4	7,428.30	7,599.6	7,781.4	8,029.5	8,029.5	8,029.5	8,029.5	8,029.5	8,029.5	8,029.5	8,029.5
1 Drill	152.59	158.93	165.47	167.59	174.41	187.86	201.86	208.4	216.0	223.9	238.0	247.61	253.3	259.3	267.6	267.6	267.6	267.6	267.6	267.5	267.6	267.6
4 Drills	610.36	635.72	661.88	670.36	697.64	751.44	807.44	833.8	864.3	895.7	952.3	990.44	1,013.2	1,037.5	1,070.6	1,070.6	1,070.6	1,070.6	1,070.6	1,070.5	1,070.6	1,070.6
W-2	4,050.30	4,433.40	4,551.00	4,632.30	4,894.80	5,302.80	5,505.60	5,704.5	5,948.1	6,138.6	6,310.8	6,517.20	6,652.8	6,760.2	6,760.2	6,760.2	6,760.2	6,760.2	6,760.2	6,760.2	6,760.2	6,760.2
1 Drill	135.01	147.78	151.70	154.41	163.16	176.76	183.52	190.1	198.2	204.6	210.3	217.24	221.7	225.3	225.3	225.3	225.3	225.3	225.3	225.3	225.3	225.3
4 Drills	540.04	591.12	606.80	617.64	652.64	707.04	734.08	760.6	793.0	818.4	841.4	868.96	887.0	901.3	901.3	901.3	901.3	901.3	901.3	901.3	901.3	901.3
W-1	3,555.00	3,938.10	4,040.70	4,258.20	4,515.00	4,893.90	5,070.60	5,318.7	5,561.7	5,753.1	5,929.2	6,143.40	6,143.4	6,143.4	6,143.4	6,143.4	6,143.4	6,143.4	6,143.4	6,143.4	6,143.4	6,143.4
1 Drill	118.50	131.27	134.69	141.94	150.50	163.13	169.02	177.2	185.3	191.7	197.6	204.78	204.7	204.7	204.7	204.7	204.7	204.7	204.7	204.7	204.7	204.7
4 Drills	474.00	525.08	538.76	567.76	602.00	652.52	676.08	709.1	741.5	767.0	790.5	819.12	819.1	819.1	819.1	819.1	819.1	819.1	819.1	819.1	819.1	819.1
Enlisted members																					
E-9							6,055.50	6,192.90	6,365.70	6,568.80	6,774.90	7,102.80	7,381.50	7,673.70	8,121.6	8,121.6	8,526.9	8,526.90	8,953.80	8,953.8	9,402.3	9,402.3
1 Drill							201.85	206.43	212.19	218.96	225.83	236.76	246.05	255.79	270.7	270.7	284.2	284.23	298.46	298.4	313.4	313.4
4 Drills							807.40	825.72	848.76	875.84	903.32	947.04	984.20	1,023.16	1,082.8	1,082.8	1,136.9	1,136.92	1,193.84	1,193.8	1,253.6	1,253.6
E-8						4,957.20	5,176.50	5,312.10	5,474.70	5,650.80	5,968.80	6,130.20	6,404.40	6,556.50	6,930.9	6,930.9	7,069.8	7,069.80	7,069.80	7,069.8	7,069.8	7,069.8
1 Drill						165.24	172.55	177.07	182.49	188.36	198.96	204.34	213.48	218.55	231.0	231.0	235.6	235.66	235.66	235.6	235.6	235.6
4 Drills						660.96	690.20	708.28	729.96	753.44	795.84	817.36	853.92	874.20	924.1	924.1	942.6	942.64	942.64	942.6	942.6	942.6
E-7	3,445.80	3,760.80	3,905.10	4,095.30	4,244.70	4,500.60	4,644.90	4,900.50	5,113.50	5,258.70	5,413.50	5,473.20	5,674.50	5,782.50	6,193.5	6,193.5	6,193.5	6,193.50	6,193.50	6,193.5	6,193.5	6,193.5
1 Drill	114.86	125.36	130.17	136.51	141.49	150.02	154.83	163.35	170.45	175.29	180.45	182.44	189.15	192.75	206.4	206.4	206.4	206.45	206.45	206.4	206.4	206.4
4 Drills	459.44	501.44	520.68	546.04	565.96	600.08	619.32	653.40	681.80	701.16	721.80	729.76	756.60	771.00	825.8	825.8	825.80	825.80	825.8	825.8	825.8	825.8
E-6	2,980.50	3,279.90	3,424.80	3,565.50	3,711.90	4,042.20	4,170.90	4,419.90	4,496.10	4,551.30	4,616.40	4,616.40	4,616.40	4,616.40	4,616.4	4,616.4	4,616.4	4,616.40	4,616.40	4,616.4	4,616.4	4,616.4
1 Drill	99.35	109.33	114.16	118.85	123.73	134.74	139.03	147.33	149.87	151.71	153.88	153.88	153.88	153.88	153.8	153.8	153.8	153.88	153.88	153.8	153.8	153.8
4 Drills	397.40	437.32	456.64	475.40	494.92	538.96	556.12	589.32	599.48	606.84	615.52	615.52	615.52	615.52	615.5	615.5	615.52	615.52	615.52	615.5	615.5	615.5
E-5	2,730.30	2,914.20	3,055.20	3,199.20	3,423.90	3,658.50	3,851.70	3,874.80	3,874.80	3,874.80	3,874.80	3,874.80	3,874.80	3,874.80	3,874.8	3,874.8	3,874.8	3,874.80	3,874.80	3,874.8	3,874.8	3,874.8
1 Drill	91.01	97.14	101.84	106.64	114.13	121.95	128.39	129.16	129.16	129.16	129.16	129.16	129.16	129.16	129.1	129.1	129.1	129.16	129.16	129.1	129.1	129.1
4 Drills	364.04	388.56	407.36	426.56	456.52	487.80	513.56	516.64	516.64	516.64	516.64	516.64	516.64	516.64	516.6	516.6	516.64	516.64	516.64	516.6	516.6	516.6
E-4	2,503.50	2,631.60	2,774.10	2,914.80	3,039.30	3,039.30	3,039.30	3,039.30	3,039.30	3,039.30	3,039.30	3,039.30	3,039.30	3,039.30	3,039.3	3,039.3	3,039.3	3,039.30	3,039.30	3,039.3	3,039.3	3,039.3
1 Drill	83.45	87.72	92.47	97.16	101.31	101.31	101.31	101.31	101.31	101.31	101.31	101.31	101.31	101.31	101.3	101.3	101.3	101.31	101.31	101.3	101.3	101.3
4 Drills	333.80	350.88	369.88	388.64	405.24	405.24	405.24	405.24	405.24	405.24	405.24	405.24	405.24	405.24	405.2	405.2	405.24	405.24	405.24	405.2	405.2	405.2
E-3	2,259.90	2,402.10	2,547.60	2,547.60	2,547.60	2,547.60	2,547.60	2,547.60	2,547.60	2,547.60	2,547.60	2,547.60	2,547.60	2,547.60	2,547.6	2,547.6	2,547.6	2,547.60	2,547.60	2,547.6	2,547.6	2,547.6
1 Drill	75.33	80.07	84.92	84.92	84.92	84.92	84.92	84.92	84.92	84.92	84.92	84.92	84.92	84.92	84.9	84.9	84.9	84.92	84.92	84.9	84.9	84.9
4 Drills	301.32	320.28	339.68	339.68	339.68	339.68	339.68	339.68	339.68	339.68	339.68	339.68	339.68	339.68	339.6	339.6	339.68	339.68	339.68	339.6	339.6	339.6
E-2	2,149.20	2,149.20	2,149.20	2,149.20	2,149.20	2,149.20	2,149.20	2,149.20	2,149.20	2,149.20	2,149.20	2,149.20	2,149.20	2,149.20	2,149.2	2,149.2	2,149.2	2,149.20	2,149.20	2,149.2	2,149.2	2,149.2
1 Drill	71.64	71.64	71.64	71.64	71.64	71.64	71.64	71.64	71.64	71.64	71.64	71.64	71.64	71.64	71.6	71.6	71.6	71.64	71.64	71.6	71.6	71.6
4 Drills	286.56	286.56	286.56	286.56	286.56	286.56	236.56	286.56	286.56	286.56	286.56	286.56	286.56	286.56	286.5	286.5	286.56	286.56	286.56	286.5	286.5	286.5
E-1	1,917.60																					
1 Drill	63.92																					
4 Drills	255.68																					
E-1<4 mo	1,773.00																					
1 Drill	59.10																					
4 Drills	236.40																					

continuation bonus; are below rank of brigadier general; hold a degree in engineering or science from an accredited college or university; have been certified by the Secretary of the Army as technically qualified for detail to engineering or scientific duty; have completed at least three but less than fourteen years' active engineering or scientific duty as a commissioned officer; are serving in or are selected for assignment to a critical engineering or scientific military specialty requiring an engineering or scientific degree and are in one of the armed forces that has a critical shortage; and execute a written agreement to remain on active duty for assignment to engineering or scientific duty for at least one year, but not more than four years. The maximum amount payable is $3,000 for each year of obligated service for which the officer has agreed to remain on active duty.

Medical and Dental Special Pays. Medical and dental officers on active duty for more than one year are entitled to several types of special pay designed to bridge the gap between military base pay and the income these military medical professionals might expect from civilian practice. These include a Variable Special Pay, ranging from $1,200 to $12,000 annually. They are entitled to annual Additional Special Pay if they complete their internship or initial residency training and agree to serve on active duty at least one year. Officers who are board certified within their particular specialty are entitled to Board-Certified Pay, an annual payment ranging from $2,500 to $6,000. A few highly trained medical and dental officers in critical specialties may also be eligible for an annual bonus as Incentive Special Pay. Finally, a Multiyear Special Payment is authorized for doctors who are in the grade of O-7 and below, who have completed residency requirements or specialty training, and who agree to remain on active duty for periods of from two to four years. Amounts of the Multiyear Special Payments vary by length of service commitment and medical/dental specialty. Veterinarians, optometrists, psychologists, pharmacy officers, and nonphysician health care providers are entitled to retention special pay in varying amounts. Nurses who qualify for appointment as Nurse Corps officers in one of the Military Departments and who during the period of 28 October 2004 to 31 December 2008 execute a written agreement to accept a commission and serve on active duty as a Nurse Corps officer for a period of not less than three years may be paid a lump sum accession bonus up to a maximum of $30,000. Certified Registered Nurse Anesthetists (CRNA) are also eligible for Incentive Special Pay.

Officers' Uniform and Equipment Allowance. Officers may become entitled to an initial uniform allowance and/or an additional active duty uniform allowance for the purchase of required uniforms and equipment. Officers are entitled to an initial uniform allowance upon first reporting for active duty for a period of more than ninety days; upon completing at least fourteen days of active duty or active duty for training as a member of a reserve component; upon completing fourteen days of inactive duty training as a member of the Ready Reserve; or upon reporting for the first period of active duty required of a member of the Armed Forces Health Professions Scholarship Program. Amount payable is $400 regardless of active or reserve component, branch of service, source of commission, or previous enlisted status. Officers of the reserve components, officers of the Army of the United States, those without component, and ROTC graduates appointed in the Regular Army may be entitled to an additional active duty uniform allowance in the amount of $200 per DoD Financial Management Regulation, Volume 7a, chapter 30, paragraph 3003.

Station Allowances, Overseas. Certain station allowances are authorized to help equalize the expenses of living at an overseas station, which includes Alaska and Hawaii.

Cost-of-Living Allowance (COLA). COLA is a supplement to regular pay that reflects the higher prices of goods and services at the overseas station. It equalizes purchasing between servicemembers overseas and their CONUS-based counterparts and varies by rank, location, and number of dependents. The website for Overseas COLA is www.defensetravel.dod.mil/site/colaCalc.cfm.

Overseas Housing Allowance (OHA). Soldiers living off base at an overseas duty station also receive OHA in addition to their regular BAH. OHA compensates servicemembers for the majority of their housing expenses. It has three components: rental ceiling, utility or recurring maintenance allowance, and move-in housing allowance (MIHA). Rental ceilings are computed using actual rents as reported through finance centers. Rental ceilings are set such that 80 percent of personnel with dependents have rents fully reimbursed. Unaccompanied members or personnel without dependents are entitled to 90 percent of the with-dependents rate. The utility or recurring maintenance allowance is paid monthly to defray expenses paid directly to utility companies; it is based on average expenses reported by personnel with dependents. The MIHA has three components: miscellaneous—a fixed-rate, lump-sum payment that reflects average expenditures to make dwellings habitable (e.g., supplemental heating equipment, wardrobes); rent—an actual dollar-for-dollar payment made in the field for customary or legally required rent—related expenses, such as rental agents, fees; and security—an actual expense component paid for security-related enhancements to physical dwellings when quarters must be modified to minimize exposure to terrorist or criminal threats in areas that the State Department considers to be subject to terrorist activities. Allowances are periodically updated based on new cost data and review of currency fluctuations. The website for overseas housing allowance is www.defensetravel.dod.mil/site/oha.cfm.

Also, ask your servicing Finance Office about Temporary Lodging Allowance, Interim Housing Allowance, and Advance of Housing Allowance (OHA or BAH). Each of these is described in the Joint Forces Travel Regulation, Volume I (reference (d), and Military Service procedural instructions for entitlement provisions, payment procedures, and systems requirements.

High-Deployment Per Diem. Effective October 2001, DoD announced a new entitlement called High-Deployment Per Diem that would be payable to soldiers who are deployed 401 days or more out of the preceding 730 days. Five categories of deployment were to be tracked and counted toward the established thresholds: operations (includes TDY), named exercise (includes TDY), unit training (includes TDY), home station training, and mission support TDY. Not regarded as deployment for the purposes of this entitlement are: study or training at a school; administrative, guard, or garrison detail duties at the member's permanent duty station; hospitalization at or near the soldier's permanent duty station; or absences due to disciplinary action. Soldiers who met the required number of days were to be paid $100 for each day on which the member is deployed. However, the enduring nature of the Global War on Terrorism and the necessity to redeploy units and individuals on multiple overseas tours compelled DoD to suspend this program indefinitely. High-Deployment pay would be taxable income except if it is earned while serving within a combat zone, making the per diem eligible for the Combat Zone Tax Exclusion.

Hardship Duty Pay—Location (HDP-L). This pay compensates soldiers assigned to areas with extraordinarily arduous quality of life conditions. Pay is based on duty location and ranges from $50 to $150.

Combat Zone Tax Exemption. All military pay earned in a combat zone is exempt from federal income tax, although some states tax soldiers' income. Officers, warrant officers, and enlisted soldiers in a combat zone during any part of a month are eligible to have all of that month's military pay exempted from taxation. Bonuses and special pays earned as a result of reenlistment and service extension contracts are also tax-free. Commissioned officers are now tax-exempt up to a level equal to the base pay of the Sergeant Major of the Army plus Imminent Danger Pay of $225 a month. Military income above these levels is fully taxable.

Hostile Fire/Imminent Danger Pay (HF/IDP). Soldiers qualify for Imminent Danger Pay of $225 a month when they serve outside the continental United States and are subject to physical harm or danger as a result of wartime conditions, terrorism, civil insurrection, or civil war. In some locations, HF/IDP is paid in addition to Hardship Duty Pay–Location.

Combat-Related Injury Rehabilitation Pay (CIP). This is a special pay for servicemembers who, while in the line of duty, incur a wound, injury, or illness in a combat operation or combat zone designated by the Secretary of Defense and are evacuated from the combat operation or from the combat zone for medical treatment. Such servicemembers are entitled to CIP for each month they are hospitalized for treatment of such wounds, injuries, or illnesses. The amount payable of CIP is $430 a month. CIP may be paid in addition to any other pay and allowances to which the servicemember is entitled or authorized, except Hostile Fire/Imminent Danger Pay (IDP). The monthly amount of CIP shall be reduced by any amount of HF/IDP received for the same month. CIP will be paid at the full monthly rate for any month in which the servicemember is eligible for the pay.

Assignment Incentive Pay. This voluntary program is intended to attract volunteers to serve extended tours in South Korea and formerly for deployed soldiers to extend their tours in Iraq, Afghanistan, and Kuwait. The program is open to officers, warrant officers, and enlisted soldiers. For more information about this program, go to https://militarypay.defense.gov/Pay/Special-and-Incentive.

Per Diem. This pay is for the reimbursement of subsistence expenses incurred during periods of official military travel within the continental United States. Per Diem is based on duty location and includes a lodging rate, local meals rate (when government mess is not available), proportional meals rate (when government mess is available for some but not all meals), as well as $5 per day for incidental expenses.

Family Separation Allowance (FSA). In addition to any Per Diem or other entitlements, servicemembers with dependents may be entitled to receive an additional $250 per month. To receive this tax-free allowance, the following criteria must be met: the servicemember must have dependents; must be on temporary duty away from the permanent duty station for a continuous period of more than thirty days; and the dependents do not live at or near the temporary duty station.

Dislocation Allowance. This allowance is paid when servicemembers make a permanent change of station (PCS) move from one duty location to another. The allowance varies based on pay grade and whether or not the servicemember has dependents.

DEPLOYED FINANCIAL BENEFITS

If you are deployed, here is a rollup of the entitlements and special pays that you may be eligible to receive: Per Diem, Family Separation Allowance, Basic Allowance for Subsistence, Basic Allowance for Housing (based on home station), Cost of Living Allowance (COLA, if eligible prior to deployment), Hardship Duty Pay-Location, Hostile Fire/Imminent Danger Pay, Combat Zone Tax Exclusion, Savings Deposit Program, Special Leave Accrual, Tax Filing Extension, Special Extension Entitlements (for extended periods of deployment). For more information about these entitlements and special pays, contact your servicing Finance Office or visit the DFAS website at www.dfas.mil/militarymembers/payentitlements/specialpay/.

RECEIPT OF PAYMENTS

Leave and Earnings Statement. A DFAS Form 702, Leave and Earnings Statement (LES) is prepared by the Defense Finance and Accounting Service (DFAS) twice monthly for every soldier. All items of entitlement or obligation are shown. Partial payment of earnings can be made twice a month or full payment can be made on the last duty day of the month. Be sure to examine your LES closely for any discrepancies. If you see something that doesn't look right, ask your servicing Finance Office about it.

Advance and Partial Pay. Prior to starting travel, an officer may draw an advance of pay. The approval of the unit commander is often required but is granted liberally. This is meant to assist the officer in purchasing transportation tickets or in paying automobile expenses. Should an officer need funds at another time, he or she may draw up to the amount accrued for that month as a partial payment of that month's salary. Partial payments are not encouraged routinely.

However, no officer is expected to borrow money at interest when that officer's own salary can provide for an emergency.

DEDUCTIONS FROM PAY

Federal Tax. Military base pay is subject to federal and state income tax. As in the civilian community, an amount is withheld from the pay each month, based on the number of exemptions claimed and the estimated total pay for the year. The amount withheld is itemized on the LES. About 1 February of each year, DFAS generates a W-2 form showing total taxable pay and the total deducted for federal income tax and state income tax. For more information, the IRS website is http://mypay.dfas.mil/mypay.aspx.

Quarters and subsistence allowances are not taxable; dislocation allowances and incentive pay are taxable.

Social Security Tax. DFAS is required to deduct Social Security tax (FICA) from each pay account. See chapter 19, Personal and Financial Planning.

In addition to taxes, there are several other common deductions, which are explained below.

Army and Air Force Exchange Services (AAFES). All unpaid deferred payment plan (DPP) charges that are overdue will be deducted from a soldier's pay.

Uniformed Soldiers' and Sailors' Home (USSH). All Regular Army enlisted soldiers and warrant officers have $0.50 automatically deducted from their pay to contribute to the upkeep of these homes.

Servicemember's Group Life Insurance (SGLI). Soldiers may elect to purchase up to $250,000 worth of life insurance. Premiums for this coverage are deducted monthly from their pay. The amount of coverage and cost of SGLI changes periodically.

Fines and Forfeitures. Commanders can take away some pay if soldiers get into trouble and get an Article 15 or a court-martial conviction.

Garnishments. Soldiers who are delinquent in alimony payments to a former spouse, child support payments, IRS payments, or commercial debts can have their military pay garnished to meet their financial obligations. After the court and legal actions are complete, the individual or agency sends a request to the DFAS in Cleveland to be processed, and the debts are collected from the soldier's pay.

Government Property Lost or Damaged (GPLD). Soldiers who lose or damage government property through their own negligence have to pay to fix or replace the property. When this happens, the Army automatically deducts the money from the soldier's paycheck.

Bad Checks. If soldiers write "bad checks" (i.e., checks that are returned due to insufficient funds in the checking account) at the post exchange, commissary, finance office, clothing store, club system, shopette, or any other government agency, the money due the government agency is automatically deducted from the soldiers' pay. This deduction also includes the returned check fee charged by the agency, which ranges from $25 to $35 per check.

Montgomery G.I. Bill. Active-duty soldiers who elect this option have a specific amount deducted from their pay for a specific period to pay for this educational benefit.

ALLOTMENTS

An allotment is a designated portion of a soldier's pay and allowances that is authorized to be paid to qualified allottees. A qualified allottee is any person or institution to whom the allotment is made payable. Allotments are a convenient way to meet financial obligations on a regular basis. They are sent even if soldiers deploy, are on leave or TDY, or in the hospital.

There are various types of allotments and limits as to the number of allotments soldiers may have at one time. Active-duty soldiers are authorized no more than six discretionary allotments at one time. Discretionary allotments are regulated by a soldier's personal judgment. Allotments used to make contributions to the Army Emergency Relief (AER), Combined Federal Campaign (CFC), or American Red Cross; to repay a debt to the government; and to buy savings bonds do not count toward the six discretionary allotments.

Soldiers should first visit the DFAS website at https://mypay.dfas.mil/mypay.*aspx* since many types of allotments can be initiated by soldiers themselves. Some types of allotments must be initiated by your servicing Finance Office. The list below gives examples of the different type of allotments and the purpose for using each type.

- Bank: payment of consumer loans, car note, deposits to a bank account.
- Voluntary support: support of legal dependent.
- Mortgage payment: payment of home mortgage or rent.
- Nonindividual allotments: payment to mutual fund and insurance companies.
- Charitable contributions: contribution to AER, CFC, and the like.
- Loan: AER or Red Cross loan repayment.
- Authorized purchases: purchase of U.S. savings bonds.
- The Thrift Savings Plan. See chapter 24, Retirement.
- Medical (TRICARE) (for retirees)
- Dental

21

Travel

This chapter provides information normally required by an officer performing travel under individual travel orders, moving family members on permanent change of station, and shipping household goods. Excluded are unusual situations and special cases of travel not generally encountered by the majority of individuals performing routine travel. (The Joint Federal Travel Regulations can be found at https://travel.dod.mil.)

TRAVEL OF OFFICERS

Travel Status Defined. Officers are entitled to travel and transportation allowances as authorized in accordance with existing regulations only while actually in travel status. They are in travel status while performing travel away from their permanent duty station, upon public business, pursuant to competent travel orders, including necessary delays en route incident to mode of travel and periods of necessary temporary duty or temporary additional duty. Travel status, whether travel is performed by land, air, or sea (except as a member of a ship's complement), commences with departure from the permanent duty station or ship and includes any of the following conditions:

- *Temporary duty or temporary additional duty:* Travel in connection with necessary temporary duty or temporary additional duty, including time spent at a temporary duty station or a temporary additional duty station, without regard to whether duty is required to be performed while traveling, and without regard to the length of time away from the permanent duty station. Temporary duty (TDY) assignments are normally limited to periods not in excess of six months.
- *Permanent change of station (PCS):* Travel from one permanent duty station to another permanent duty station.
- *Delay:* Delay incident to mode of travel, such as necessary delay while awaiting further transportation after travel status has commenced.
- *Travel to and from hospital:* Travel to or from a hospital for observation or treatment.

- *Travel by air:* Travel performed by military or commercial aircraft when proceeding from one duty station to another under orders of competent authority. Air travel includes one or more landings away from the starting point and the necessary delays incident to this mode of travel. Aircraft flights for crew training purposes come under this category.
- *Aerial training flights:* Aerial flights for training purposes made in the absence of travel orders when it is necessary to remain away overnight.
- *Special circumstances or conditions:* Special circumstances or conditions not heretofore defined that may be determined jointly in advance, contemporaneously, or subsequently by the secretaries of the uniformed services to constitute a travel status.

Traveling with Troops. An officer is traveling with troops when he or she is physically traveling as a member of, or on duty with, any body of troops that is subsisted en route from a mobile kitchen trailer, field range, ship's galley, or other comparable facilities for preparing complete cooked meals en route.

Members traveling with troops will not be paid mileage or reimbursed on a per diem basis for expenses incurred. Under no circumstances will members obtain meals on meal tickets, or meal receipts. Transportation and sleeping accommodations, if available and required, will be furnished in kind.

Group Travel. Group travel is a movement of three or more members traveling under one group order from the same point of origin to the same destination when a member is designated in the order as being in charge of the group.

Standard of Accommodations. When a member is entitled to transportation in kind and uses government transportation requests, such member shall be furnished first-class accommodations, except when travel is by commercial aircraft. Then, tourist-class accommodations normally are furnished.

Type of Carrier on Which to Travel. Within the continental United States, the individual may elect to travel by any mode of transportation at personal expense, subject to reimbursement upon completing the journey.

Termination of Travel Status. Travel status terminates upon return to the permanent duty station or upon reporting at a new permanent duty station ashore or afloat, except that travel status terminates when the member reaches the assigned port if the vessel to which he or she is reporting for duty is already in port.

Transportation in Kind. Transportation in kind includes travel by all modes of commercial transportation and military facilities, but does not include travel at personal expense. In all matters pertaining to transportation, consult your transportation officer *prior to commencing travel.*

Transportation Request. A transportation request (TR) is a requisition issued to a commercial carrier to furnish specified transportation services. The TR is issued by the transportation officer or other competent authority.

TRAVEL ORDERS

No reimbursement for travel is authorized unless orders by competent authority have been issued therefor. Reimbursement for travel is not authorized when the travel is performed in anticipation of or prior to receipt of orders.

Travel orders issued under unusual conditions that are not originated by competent authority must be approved by competent authority to allow reimbursement for travel expenses.

Permanent Change of Station. The term *permanent change of station*, unless otherwise qualified, means the transfer or assignment of a member of the uniformed services from one permanent station to another. This includes the change from home or from the place from which ordered to active duty to first station upon appointment or call to active duty, and from last duty station to home or to the place from which ordered to active duty upon separation from the service, placement upon the temporary disability retired list, release from active duty, or retirement.

Temporary Duty. The term *temporary duty* means duty at a location other than a permanent station to which a member of the uniformed services is ordered for 179 days or less under orders that provide for further assignment to a new permanent station or for return to the old permanent station.

Blanket or Repeated Travel. Blanket travel orders are issued to members who regularly and frequently make trips away from their permanent duty stations within certain geographical limits in performance of regular assigned duties. Travel must not be solely between place of duty and place of lodging.

TEMPORARY DUTY TRAVEL

Travel at Personal Expense. When authorized travel is performed by commercial transportation at personal expense, you will be reimbursed for the actual costs incurred. Note that first-class travel by air is not reimbursable. Be sure to retain receipts to be able to support your claim.

Travel by Privately Owned Vehicle (POV). For TDY actually performed by POV under orders authorizing such mode of transportation as more advantageous to the government, you will be reimbursed at the rate of so many cents per mile for POV or for motorcycle for the official distance in addition to the authorized per diem.

Per Diem. Per diem allowances are designed to cover the costs of room rentals, meals, and miscellaneous expenses incident to performing official government travel. Rates are specified in the Joint Federal Travel Regulations. For round-trips of ten hours or less within one calendar day, no per diem is authorized.

The military uses a "lodgings plus per diem" method of reimbursement for travel expenses. Under this method, the actual costs of lodging, ranging from a minimum of $40 per day up to a specified maximum for each area, is payable to the traveler. For example, the 2022 maximum lodging rate is now $365 per day for the New York City area. In addition to the costs for room rental, the traveler is entitled to a fixed reimbursement at rates dependent on the area, to cover expenses of meals and miscellaneous items. This allowance for meals and incidental expenses (M&IE) may be increased by special authorization for destinations where the meal expenses are known to exceed the stated maximum rate. On the day of departure and the day of return, the traveler receives reimbursement at 75 percent of the daily rate authorized for the TDY destination regardless of the time of departure or the time of return.

TRAVEL EXPENSES

Reimbursable Expenses. In the past, many officers paid travel expenses for which reimbursement might have been received. On a long trip, that loss can be material. The following are reimbursable:

Taxi Fares. Reimbursement is authorized for taxicab fares between places of abode or business and stations, wharves, airports, other carrier terminals, or local terminus of the mode of transportation used; between carrier terminals while en route when free transfer is not included in the price of the ticket or when necessitated by change in mode of travel; and from carrier terminals to lodgings and return in connection with unavoidable delays en route incident to the mode of travel. Itemization is required.

Allowed Tips. Tips incident to transportation, such as tips to baggage porters, redcaps, and so on, are reimbursable but are not to exceed customary local rates; tips for baggage handling at hotels are excluded; the number of pieces of baggage handled must be shown on the claim. Itemization is required.

Checking and Transfer of Baggage. Expenses incident to checking and transfer of baggage are reimbursable. The number of pieces of baggage checked must be shown on the claim. Itemization is required.

Excess Baggage. When excess baggage is authorized, actual costs for such excess baggage in addition to that carried free by the carrier are reimbursable. Receipt is required.

Registration Fees. Registration fees incident to attendance at meetings of technical, professional, scientific, or other nonfederal organizations are reimbursable when attendance is authorized or approved. Receipt is required. (See annual appropriation acts.)

Bachelor Officers' Quarters Fees. When government quarters are available and used, the cost of the lodging in the government quarters is reimbursable. A receipt from the billeting facility is required.

Government Auto. Cost of storage of government automobiles when necessary is reimbursable if government storage facilities are not available. Receipt is required.

Telephone, Internet Services, Facsimile, and Other Communication Services. Cost of official telephone, Internet, facsimile, and similar communication services is reimbursable when incident to the duty performed or in connection with items of transportation. Such services, when solely in connection with reserving a hotel room, and so on, are not considered official. Local and long-distance telephone calls are allowable when itemized.

Stenographic Services. Charges for necessary stenographic services or rental of typewriters or word processors in connection with the preparation of reports or official correspondence are reimbursable when authorized or approved by the headquarters directing the travel. This provision does not apply when stenographic services are performed by military personnel or government employees. Receipts are required.

Local Public Carrier Fares. Expenses incident to travel on public transportation or other usual means of local transportation are allowed. Itemization is required. Commercial or government (GSA) rental car expenses are reimbursable; authorizations for such rentals are normally made in advance and indicated on travel orders.

Toll Fares. Ferry fares and road, bridge, and tunnel tolls are reimbursable when travel is performed by government vehicle or by authorized hired conveyance, or when performed by privately owned conveyance within the surrounding area of a duty station.

Receipts. Receipts are required to support claims for all lodging expenses and for other reimbursable expenses greater than $75. Without a supporting receipt, your claim may be denied.

Advance Payment of Travel and Transportation Allowances. Travel and transportation allowances for an officer's travel are authorized to be paid in advance, except in connection with retirement and upon first entering active duty. However, it is recommended that you avoid using advance payments if possible. The added paperwork and the accounting necessary for reimbursement and/or repayment of funds advanced can lead to difficulties. If you must use advance travel pay, see the post finance officer.

Travel Expenses Not Payable by the Government. Travel expenses of the type listed in the examples below are not payable from government funds:

- Expenses incurred during periods of travel that are incident to other duties (such as traveling aboard a vessel in performance of temporary duty on such vessel).
- Travel from leave to official station for duty. Individuals departing from their official duty station on leave do so at their own risk. If ordered to return from leave, they must assume the expense involved.
- Travel under permissive orders to travel, in contrast to orders directing travel, requires the individual to pay the costs.
- Travel under orders but not on public business, for example, travel as a participant in an athletic contest. Such travel may be paid from unit or command welfare funds, which are generated through the operation of exchanges and motion picture theaters.
- Return from leave to duty abroad. Unless government transportation is available, such as space on an AMC flight, the individual on leave in the United States from an overseas command must defray his or her own return expenses.
- Attendance at public ceremonies or demonstrations whose expenses are borne by the sponsoring agency.

Frequent Flyer Miles and Promotional Items

On 28 December 2001, President Bush signed into law the National Defense Authorization Act for fiscal year 2002. Section 1116 permits federal military employees to accept promotional items such as frequent flyer miles earned when traveling in an official capacity. For DoD military personnel, the Joint Federal Regulation has been revised and can be found at http://141.116.74.201/regchgs.htm. Thus, provided that the entity that paid for the travel does not object, federal military personnel may retain frequent flyer miles that are derived from such travel.

Taxability of the Benefit. It is possible that such benefits may be considered to be additional compensation, and taxed accordingly. Until a ruling is received, it is recommended that personnel who redeem frequent flyer miles or other promotional benefits keep a record of such redemptions.

Supervisory Challenges. Under the new rules, it is unforeseeable that some personnel may attempt to schedule travel in order to acquire frequent flyer miles or other promotional items. Any such attempt, to the extent that it increases the costs of travel to the government, violates the Joint Ethics Regulation, and, in some cases, may violate criminal conflict of interest statutes.

Upgrade to First Class. Personnel on official travel may now use frequent flyer miles, because they belong to the individual, to upgrade to first class. However, each military department has issued guidance regarding the wearing of uniforms while traveling in first class accommodations.

TRAVEL OUTSIDE THE UNITED STATES

Travel expenses for travel outside the United States are furnished in advance or are reimbursable on essentially the same basis as temporary duty travel performed in the United States. That is, the traveler is entitled to the costs of transportation, per diem allowance, and costs of incidental necessary expenses. Maximum travel per diem allowances for foreign areas are available at www.travel.dod.mil/Travel-Transportation-Rates/Per-Diem/Per-Diem-Rate-Lookup/.

Per diem rates for overseas travel also vary according to quarters and mess availability and charges for the same. Consult your local finance officer for assistance before and after any overseas travel.

Certificates are required from the traveler and from the commanding officer or a designated representative of an installation at which a traveler performs temporary duty.

PASSPORTS

Passports issued by the Department of State are required for persons visiting foreign countries. There are four classes of passports: diplomatic (no-fee), official (no-fee), regular (no-fee), and regular (tourist).

Diplomatic passports are issued to officers accredited to any embassy or legation of the United States abroad and to members of the households of such officers. Field-grade officers and above assigned to military assistance program missions have been granted diplomatic passports.

Official passports are issued to officers proceeding abroad under orders in the discharge of their official duties. Family members accompanying or traveling to join bearers of official passports who are stationed abroad may apply for official passports.

Regular (no-fee) passports are issued to family members of a military member whose assignment does not warrant issuing the family members diplomatic or official passports.

Regular (tourist) passports are obtained and paid for by persons who are traveling abroad for personal reasons.

Each Army installation and activity in the United States, Guam, and Puerto Rico has a passport agent. Apply for passports from your station passport agent.

All servicemembers traveling overseas on official business to a country requiring a passport and all command-sponsored family members will obtain separate no-fee passports. This is true regardless of destination or the age of the family members. There are special rules for alien family members, for which you should see your passport agent.

Each passport application must be accompanied by two identical, clear, front-view, full-face photographs measuring two by two inches. The photographs may be either in color or black and white. Group photographs are unacceptable. Each photograph must be signed in the presence of the passport agent. A parent may sign for a minor child, as "Richard Doe by Jane Doe (mother)." It is recommended that the nonmilitary parent sign the passport applications and the photographs for minor children.

Each passport application must be accompanied by documentary evidence of citizenship.

A visa is an endorsement made on a passport by the proper authorities (usually embassy or consular officials) of a country to be visited, showing that the passport has been examined and that the bearer may proceed to that country.

PERMANENT CHANGE OF STATION TRAVEL

Travel of Family Members. Members of the uniformed services are entitled to transportation of family members upon a permanent change of station for travel performed from the old station to the new permanent station or between points otherwise authorized. As to officers, there are some important exceptions:

1. An officer assigned to a school or installation as a student, if the course of instruction is to be of less than twenty weeks' duration.
2. Separation from the service or relief from active duty under conditions other than honorable.
3. Call to active duty for training for less than one year.
4. Call to active duty for other than training duty for less than six months.
5. An officer who fails to receive revocation of PCS orders because he or she took advantage of leave of absence and the notice of revocation was received at the officer's old permanent station sufficiently in advance of the time that would have been required to proceed under the original orders.
6. When the family member is a member of the uniformed service on active duty on the effective date of the orders.

7. For any portion of travel performed by a foreign registered vessel or airplane, if American registered vessels or airplanes are available by the usually traveled route.
8. When the family members departed the old permanent station prior to the issuance of orders, and the voucher is not supported by a certificate of the commanding officer or a designated representative of the headquarters issuing the orders that the officer was advised prior to the issuance of change of station orders that such orders would be issued.
9. When dependency does not exist on the effective date of the order directing permanent change of station.
10. For family members receiving any other type of travel allowances from the government in their own right.

Mileage. Mileage is an allowance applicable to PCS travel under the following circumstances:

1. When travel is by privately owned conveyance.
2. When travel is by rail and available transportation requests were not used.
3. On relief from active duty.
4. On separation from the service.
5. On transfer to the temporary disability retired list.
6. On retirement.

Mileage is computed by finance officers from a table of official distances. These distances govern, regardless of the actual route followed by the traveler.

Reimbursement for Costs of Dependent Travel. An officer who transports lawful family members at personal expense from a location where transportation requests are not available may elect to be reimbursed for the actual costs of the transportation in lieu of the monetary allowances stated below.

An officer who elects to transport family members at personal expense may obtain reimbursement. Known as a monetary allowance in lieu of transportation (MALT), the amount is payable only after travel has been completed. The total entitlement is determined on the basis of a certain number of dollars per day (per diem) plus so many cents per mile for the servicemember; the mileage rate increases incrementally per mile if the member is accompanied by one dependent, two dependents, or if more than two dependents travel with the member. In addition, a flat rate of so many dollars per day is authorized for dependents aged twelve and over and a higher amount per day for dependents under twelve years of age. One day of travel time is allowed for each 350 miles of the official distance traveled. Payments also may be made for travel by family members in a second car during a PCS move. No special authorization is required.

It is also possible to obtain reimbursement for the movement of family members in three vehicles under certain specific circumstances. To obtain approval, the use of three cars must be deemed advantageous to the government, and one of the following sets of circumstances must apply:

- If more family members are traveling together than can reasonably be accommodated, with luggage, in two vehicles.

- If a dependent requires special accommodations because of age or physical condition that makes a third vehicle necessary.
- If dependents are prevented from accompanying the servicemember for acceptable reasons, such as school, the sale of property, settlement of business affairs, disposal of household goods, or inadequate housing at the new station.
- If dependents move before the servicemember because of acceptable reasons, such as enrollment in school.
- If dependents are involved in unaccompanied travel between authorized points, such as traveling to the new station while the servicemember is officially carried as TDY en route to the new assignment.
- If the travel is by two cars, or if travel by three cars is authorized, each car with a single driver is entitled to reimbursement of so many cents per mile. Additional dependents increase the entitlement as noted earlier for movement by a single vehicle.

Temporary Lodging Allowances. Special allowances, in addition to other moving and housing allowances, are provided to help defray the extra expenses involved in a PCS move. Members reporting to an overseas station are entitled to a temporary lodging allowance (TLA) of 65 percent of the local per diem, provided their expenses are that great. A member with one dependent is entitled to 100 percent of the local per diem rate, with an additional 25 percent for each additional family member under twelve years of age and 35 percent for each family member who is twelve or older. These are the maximum rates. Actual allowances are set based on the family size, the local per diem rate, actual cost of quarters, whether the quarters have cooking facilities, and other allowances the servicemember is receiving. The TLA normally is paid in ten-day increments, and is usually limited to sixty days for those arriving and ten days for those preparing to depart. The allowance ends if the member goes on leave.

A similar allowance, called the temporary lodging entitlement (TLE), is provided to members making PCS moves within CONUS. The TLE provides a certain number of dollars per day for ten days for temporary housing costs incurred before signing out from the old duty station or after arriving at the new duty station. The actual amount is determined by the same factors as are used for determining the TLA rates overseas. A member moving from CONUS to an overseas station is entitled to the TLE for up to five days prior to departure.

Dislocation Allowance (DLA). The purpose of DLA is to partially reimburse an officer, with or without dependents, for the expenses incurred in relocating the member's household on a PCS, housing moves ordered for the government's convenience, or incident to an evacuation. If the officer is single, or if the dependents do not move, the allowance is payable only if the officer is not assigned government quarters at the new station. DLA payment limitations are based on a fiscal year. Ordinarily officers are entitled to only one DLA payment per fiscal year. However, there are some exceptions: when officers are ordered to service schools or when approved by the Secretary of the Army. This allowance

PRIMARY DLA RATES

Grade	Without Dependent Rate	With Dependent Rate
O-10	$3,476.74	$4,279.86
O-9	$3,476.74	$4,279.86
O-8	$3,476.74	$4,279.86
O-7	$3,476.74	$4,279.86
O-6	$3,189.71	$3,853.65
O-5	$3,072.02	$3,714.50
O-4	$2,846.92	$3,274.40
O-3	$2,281.56	$2,709.05
O-2	$1,809.81	$2,313.19
O-1	$1,523.99	$2,067.85
O-3E	$2,463.71	$2,911.40
O-2E	$2,094.38	$2,626.84
O-IE	$1,801.00	$2,427.03
W-5	$2,892.42	$3,160.52
W-4	$2,568.67	$2,897.50
W-3	$2,158.90	$2,654.67
W-2	$1,917.33	$2,442.18

Effective 1 January, 2022-These rates are only payable when a second DLA is paid in accordance with JTR, par. 050507

is in addition to all other authorized allowances and may be paid in advance. The rate for a dislocation allowance is based on a servicemember's rank as well as dependency status on the effective date of the PCS. The current version of DLA rates may be accessed at www.travel.dod.mil/.

TRANSPORTATION OF HOUSEHOLD GOODS

Shipment of household goods consists of transportation, including packing, crating, drayage (at point of shipment and destination), temporary storage, uncrating, and unpacking at government expense. DA Pamphlet 55-2, *It's Your Move*, should be used as a basic reference by all officers preparing to move, and a copy should be in each officer's library of important official documents.

These services are performed or expenses paid for an officer ordered to active duty at a permanent station, or assigned to a new permanent station, or relieved from active duty, or retired.

As soon as you receive orders, consult the local installation transportation office (ITO) for the best possible guidance in preparing to ship household property so that it will proceed smoothly. Follow the guidance carefully. Special

SECONDARY DLA RATES

Grade	Without Dependent Rate	With Dependent Rate
O-10	$2,933.75	$3,611.43
O-9	$3,476.74	$4,279.86
O-8	$3,476.74	$4,279.86
O-7	$3,476.74	$4,279.86
O-6	$3,189.71	$3,853.65
O-5	$3,072.02	$3,714.50
O-4	$2,846.92	$3,274.40
O-3	$2,281.56	$2,709.05
O-2	$1,809.81	$2,313.19
O-1	$1,523.99	$2,067.85
O-3E	$2,463.71	$2,911.40
O-2E	$2,094.38	$2,626.84
O-1E	$1,801.00	$2,427.03
W-5	$2,892.42	$3,160.52
W-4	$2,568.67	$2,897.50
W-3	$2,158.90	$2,654.67
W-2	$1,917.33	$2,442.18

Effective 1 January, 2022-These rates are only payable when a second DLA is paid in accordance with JTR, par. 050507

attention should be given to the necessity for temporary storage and the allowable period for such storage up to ninety days. An additional ninety days may be authorized under exceptional circumstances.

Household Goods Shipped at Government Expense. The term *household goods* includes clothing, baggage, all other personal effects of similar character, professional books, papers, and equipment, as well as the items normally required to equip a home with furniture, appliances, and the like.

There are items that are excluded from government shipment, and the recommendation is to consult the ITO. There are provisions for shipping a mobile home or for handling the move yourself, as discussed below.

Authorized weight allowances are shown in the accompanying table. Reduced weight allowances are authorized in conjunction with TDY assignments. Consult your ITO for details.

Excess Costs. The transportation charges for unauthorized articles, excess weight, or excess mileage will be borne by the owner.

TABLE OF WEIGHT ALLOWANCES (POUNDS) ON CHANGE OF STATION

	Without Dependents	With Dependents
General and General of the Army	18,000	18,000
Lieutenant general	18,000	18,000
Major general	18,000	18,000
Brigadier general	18,000	18,000
Colonel	18,000	18,000
Lieutenant colonel and chief warrant officer (CW5)	16,000	17,500
Major and chief warrant officer (CW4)	14,000	17,000
Captain and chief warrant officer (CW3)	13,000	14,500
First lieutenant and chief warrant officer (CW2)	12,500	13,500
Second lieutenant and warrant officer (WO1)	10,000	12,000

Insurance. A claim against the government for damage or loss incident to a shipment may be relied on to obtain a fair and just settlement. However, personal property of high value should be covered by commercial insurance. Recommended is the coverage that can be obtained from the United Services Automobile Association, San Antonio, Texas, or from Armed Forces Insurance, Fort Leavenworth, Kansas. These two associations were established to serve the needs of officers and their families, and their rates probably are more favorable than you can obtain elsewhere. Consult your JAG claims officer if you have any questions.

Moving Companies. All reliable moving companies are in business to provide good service. These companies are doing all that is possible to ensure speedy, efficient moves, with the avoidance of confusion and disappointment. Still, not all meet the standard. It is essential that Army officials be able to identify the superior companies, the average ones, and the poor ones. Keeping current on this information requires the cooperation of Army families.

Fill out with care and accuracy the customer satisfaction report. Be accurate and prompt. Be fair to the company, but also be fair to yourself, the Army, and Army families who will make future moves. This report indicates to the transportation officials and to the transfer firm just what kind of service you received.

A good way to start is to confer with the ITO well in advance of the move. Get advice as to preparation, and learn precisely the service you are supposed to receive. In that way, most misunderstandings are prevented. You are supposed to receive fine service. Do your full part to get it. Report on the proper form just what you did receive.

What You Should Do to Assist in the Move

1. Contact your ITO as soon as possible after receipt of orders.
2. Advise the ITO that you have professional books and papers to be shipped so that they can be packed and weighed separately from your household goods.
3. Have sufficient copies of your change of station orders (usually six to nine for each shipment).
4. If you will proceed to your new duty station prior to the time you will want your household goods shipped, leave or send your spouse or agent sufficient copies of your change of station orders. Be sure that you or your duly authorized agent is on hand at the time of packing, loading, unpacking, and unloading of your household goods.
5. If you have silver, gold, or other valuables, it's a good idea to carry these items with you, or insure commercially as noted earlier.
6. Request storage at the point of origin whenever you are in doubt as to where you want your goods shipped. Be sure to check the allowable time limits for storage to match your plans for leave, house hunting, and the like.
7. If your household goods are moving by van, be sure to obtain a copy of the carrier's inventory from the driver; also, you will be requested to sign a DD Form 619, *Statement of Accessorial Services Performed*. The certificate contains an itemized list of the units of packing performed at your residence. Be sure to check the certificate carefully, and never sign before it has been filled in.
8. Notify your ITO immediately if your orders are canceled or modified or if you want to change the destination of the shipment.
9. Appliances are serviced for transporting at government expense. The ITO makes arrangements for you with the packing or moving firm. Similarly, after delivery, the appliances are to be "deserviced."
10. The refrigerator should be defrosted and well cleaned the day before the move, so that its interior will be dry at the time of loading. The shelves and trays will be removed by the packers and placed in suitable containers for safe movement.
11. Obtain from the ITO the approximate date of arrival of your household goods at your destination.
12. Be sure that you or your agent is at home on the day of the expected move, and make arrangements for receipt of the property at the destination.
13. Turn over all your household goods for the same destination at one time.
14. Clean china and cooking utensils before packers arrive.
15. Set aside and call to the attention of the movers extra-fragile items, such as chinaware and delicate glassware, and professional books that must be packed and weighed separately.
16. Keep nonperishable groceries and food supplies together in one area for proper packing.

17. Remove articles from drawers of the furniture intended for shipment. Let the packer determine which, if any, light, bulky articles may be shipped in the drawers.
18. Be sure to inventory your household goods with the van driver. Do not allow an entry of "marred and scratched" on the inventory form unless such entry is correct. This broad language may cover extensive damage. Insist upon accurate descriptions of the condition of the furniture, such as, "one-inch scratch, left leg," or "rubbed, right front corner."
19. Make arrangements to have telephone service and other utilities disconnected.
20. Dispose of opened but unused foods that might spill or spoil en route. They should never be stored or shipped.
21. Don't include plants, fresh fruits, or flowers in shipment, as this is prohibited in many states.
22. Separate and collect into one place all items that are not to be included in the shipment. Show the van operator the articles, if any, that have been set aside and are not to be included in the shipment.

Claims for Loss or Damage Incident to Shipment. Regardless of when your household goods are delivered, have the carrier or local agent unpack all boxes and deservice all appliances. Carefully note all damaged and lost items on all copies of the inventory and on DD Form 1840, *Joint Statement of Loss or Damage at Delivery*. Be as meticulous as the company was that packed you. Usually, the local agent will make immediate arrangements for repair or replacement. Once you have checked all your belongings, including the operation of appliances, consult with the local ITO, the claims officer, and the local agent. You may claim for the difference between what you believe fair and what the carrier or the insurance company will allow (see AR 27-20, *Claims*). All have deduction tables for depreciation in value according to how old the item is, and all have time limits for presenting claims.

Transportation of a Mobile Home. The ITO will arrange for a commercial hauler to move a mobile home in conjunction with a PCS move. The government will pay as much to move the home and contents as it would have paid to move the authorized weight allowance for the member. Any excess costs are borne by the member. Mobile home owners are entitled to a dislocation allowance and to in-transit mobile home storage of up to 180 days.

Do-It-Yourself (DITY) Moves. Servicemembers may elect to do their own packing, loading, and moving of household goods for a PCS move. Advance authorization is required. The government will reimburse expenses up to 100 percent of what it would have cost to make the move commercially. If the member can make the move for less than 95 percent of the estimated commercial move costs, the member pockets the difference between the cost incurred (verified with receipts) and 95 percent. The difference, though, is taxable as regular income. Costs in excess of 100 percent of commercial costs are borne by the member. It is possible to ship part of your household goods commercially and part by DITY. See your ITO for details.

Professional Assistance. An excellent source of information and assistance regarding PCS moves is available from Military on the Move (MOM), a relocation company devoted to assisting military families moving from one station to another within CONUS. Assistance also is available when departing to or returning from an overseas station. There are more than 650 MOM member agents located near military posts nationwide who are military spouses, former military members, or others who have affiliation with the military. They are experienced in handling the special problems encountered by military families making PCS moves, having made numerous such moves themselves. They understand the needs of a military family faced with a PCS move. Request information from MOM (Military on the Move), www.usmilitaryonthemove.com.

22

Uniforms and Appearance

This chapter contains essential information of special interest to officers about the Army's uniforms. Extracts and illustrations have been drawn from the following official publications: AR 670-1, Wear and Appearance of Army Uniforms and Insignia; DA PAM 670-1, Guide to the Wear and Appearance of Army Uniforms and Insignia; CTA 50-900, Clothing and Individual Equipment; and AR 700-84, Issue and Sale of Personal Clothing.

This chapter describes the wear and appearance for the Army Service Uniform (ASU) as the Class A uniform. The ASU is authorized for wear until 30 September 2026; however, the officer would be wise to transition to the Army Green Service Uniform (AGSU) soonest. In addition to the AR and DA PAM 670-1, as the AGSU is relatively new, modifications to its wear are being published as necessary. This chapter describes the wear and appearance of the ASU only, as to not confuse the reader. The next edition will only discuss the AGSU, as the ASU will be near the end of its wear-out period. While many measurements and alignments of insignia are similar on the AGSU and ASU (particularly for male officers), it is important to reference the correct regulations governing the uniform being prepared for wear. Always check with your unit S-1 for current information regarding changes to Army uniform policy or visit www.armyg1.army.mil/hR/uniform for new ALARACT messages.

GENERAL

There are three general categories of uniforms for both male and female Army officers: service, dress, and utility. Included are uniforms of appropriate weight and material for cold weather and warm weather, for wear in an office environment, for participation in social activities, or for performing duty in the field. Some are required, while the purchase of others is optional. Certain utility uniforms are issued by the organization if their wear is required.

The service uniforms are further classified as class A and class B; the utility, field, and organizational uniforms are referred to as class C. Each uniform, regardless of its class, is made up of prescribed components; the regulations describe how the uniform is to be worn. There are a number of accessory items and several types of grade and branch insignia, with associated rules as to how and when they may, or must, be worn. Decorations, service medals, and badges are authorized for wear on some uniforms, in prescribed locations and order of precedence, but are not authorized to be worn on other uniforms. As an Army officer, you must learn the details regarding the proper wear of uniforms. In the Army, discipline is judged, in part, by the manner in which individuals wear their uniforms. Unless you first set the example, you cannot expect your soldiers to present a neat, well-groomed appearance, which is fundamental to building the pride and esprit de corps of an effective military force.

This chapter contains general information regarding the wearing of uniforms and suggestions on how to select and care for uniforms, as well as discussion regarding personal appearance. It presents condensed descriptions and illustrations of the authorized uniforms with occasions for their wear. Insignia and ornamentation are discussed as they relate to proper wear on the uniform. Thoughtful study of this chapter and its companion, chapter 23, Military Awards, will enable you to dress appropriately for each occasion and to wear your uniform with the confidence and pride befitting an officer of the U.S. Army.

UNIFORM APPEARANCE AND FIT

Uniforms must be properly fitted, clean, and serviceable. Articles carried in pockets are not to protrude from the pockets; nor should they be so bulky as to cause the pockets to bulge. Uniforms must be worn buttoned, zippered, or snapped as appropriate; metallic devices must be kept in proper luster and free of scratches and corrosion; medals and ribbons must be clean and not frayed; shoes and boots must be clean and shined or brushed as appropriate.

At the discretion of the commander and when in performance of duties, the wearing of an electronic device on the belt, belt loops, or waistband of the uniform is permitted. Only one electronic device may be worn; it may be a pager, a Blackberry, or a cell phone. The body of the device may not exceed the size of a government-issued device.The device and the carrying case must be black; no other colors are authorized. If a security cord or chain is attached to the device, it must be concealed. No other electronic device is authorized for wear on the uniform. If the commander issues and requires the use of one in the performance of duties, it will be carried in the hand, pocket, briefcase, purse, bag, or some other carrying container. Wearing wireless and non-wireless ear pieces while in any Army uniform is prohibited except while operating a commercial or military vehicle (to include a motorcycle or bicycle.)

The wearing of headgear is not required to evening social events (after retreat) when wearing the Army Service Uniform or the mess and evening mess uniforms.

The proper fit of uniforms is prescribed in DA PAM 670-1 and TM 10-227. Refer to the regulations for details. The general fitting guidelines are as follows.

Male Officers. The sleeves of long-sleeved shirts extend to the center of the wrist bone; the sleeves of uniform coats and jackets extend 1 inch below the bottom of the wrist bone; the sleeve of the black all-weather coat should be 1/2 inch longer than the sleeve of the service coat. The bottom of the black all-weather coat extends 1½ inches below the midpoint of the knee (crease in the back of the knee).

Trousers are fitted and worn so that the bottom of the waistband is at the top of the hip bone, plus or minus ½ inch. The front crease should extend to the top of the instep, with the bottoms (no cuffs) finished on a diagonal so that the rear crease extends to a point about midway between the top of the heel and the top of a standard shoe. A slight break in the front crease is permissible (but not desirable).

Female Officers. Sleeve length of shirts, coats, and the black all-weather coat is the same as prescribed for male officers. The bottom of the black coat should reach a point 1 inch below the skirt hem, but not less than 1½ inches below the crease in the back of the knee. Slacks are fitted and worn so that the center of the waistband is at the natural waistline. The legs of slacks are finished as per the description for male trousers. Knee-length skirts and dresses extend to a point not more than 1 inch above or 2 inches below the crease in the back of the knee.

The coat front overlaps to form a straight line from the bottom of the lapel to the bottom of the coat. It is fitted to produce a military effect. It should not be buttoned so closely at the waist as to cause folds or wrinkles. The skirt should fit smoothly, so that it does not drape in folds under the coat. It should form a continuation of the lines of the coat and not flare at the sides; nor should it be pegged.

Hats and Caps. Hats and caps are part of the uniform and are worn with the uniform except as follows: (1) headgear need not be worn in private vehicles or commercial conveyances; (2) headgear may be removed if it would interfere with the proper operation of a military vehicle; (3) headgear is not worn indoors unless under arms in an official capacity or unless directed by the commander; and (4) headgear is not required with the mess and evening mess uniforms, or with the Army Service Uniform to an evening social event.

PERSONAL APPEARANCE

As an officer, you are responsible for maintaining the Army's professional image by knowing and enforcing standards of personal appearance and grooming. Unlike the Army Service and mess uniforms, which trace their lineage back over 100 years, hair length, grooming styles, and attitudes toward decorations such as tattoos, body piercing, and jewelry are constantly changing and directly reflect American culture. Refer to AR 670 1, chapter 3, and the Army G1 website to maintain awareness of current Army standards.

The best-fitted uniform is to no avail if it is worn by a soldier who does not have military bearing. Stand or sit erect, chest out, stomach in; don't slouch. Look the world and your associates squarely in the eye. You are an honorable person in an honorable profession. Do your best to look the part. This includes long-term attention to physical fitness and such self-control as is necessary to meet the Army's weight standards. Remember, it is much easier to put on a few pounds than it is to take them off.

All personnel are to maintain their fingernails clean and neatly trimmed, and of a length so as not to interfere with the performance of duty. Use of nail polish is discussed later.

Hair. ALARACT 040-2021 announced sweeping changes to the authorized hair lengths and styles for female soldiers. All leaders, male and female, would do well to read the ALARACT and AR 670-1 updates to familiarize themselves with these new standards.

Men. The hair on the top of the head must be neatly groomed. The length and bulk of the hair should not present a ragged, unkempt, or extreme appearance. The hair should present a tapered appearance when combed and not fall over the ears or eyebrows or touch the collar except for the closely cut hair at the back of the neck. In all cases, the bulk or length must not interfere with the normal wear of headgear or protective masks.

Sideburns must be neatly trimmed. The base cannot be flared, nor should it extend below the lowest part of the exterior ear opening. The bottom must be a clean-shaven horizontal line.

The face is to be clean-shaven, except that a mustache is permitted, provided it is kept neatly trimmed, tapered, and tidy. No portion of the mustache may cover the upper lip or extend sideways beyond the corners of the mouth, nor should it present a chopped-off appearance. Handlebar mustaches, goatees, and beards are expressly forbidden.

The wearing of a wig or hairpiece is forbidden except to cover natural baldness or physical disfiguration caused by an accident or medical procedure. When worn, it must conform to haircut criteria stated above. As an exception to this policy, Army National Guard (ARNG) and U.S. Army Reserve (USAR) personnel may wear a wig or hairpiece conforming to the haircut criteria during unit training assemblies or when serving on active duty for training or full-time training duty for periods of thirty days or less. When ordered to active duty for periods longer than thirty days, ARNG and USAR personnel fall under the same rules as other active-duty personnel.

Male soldiers' nails will be kept trimmed as to not extend beyond the finger tip. Males are authorized to wear clear nail polish.

Women. As stated above, ALARACT 040-2021 announced a multitude of regulation changes pertaining to female hairstyles. The changes are too extensive to briefly cover while doing justice to explaining the nuanced changes now allowed female soldiers. All leaders need to familiarize themselves with these changes. Hairstyles must not interfere with the proper wearing of headgear or protective masks. Hairnets are not authorized unless required for health or safety reasons. Wigs, weaves, and hairpieces may be worn so long as the hairpiece is a natural color and conforms to the criteria above. Hair-holding ornaments (barrettes, pins, clips), if used, must be transparent or similar in color to the hair and inconspicuously placed. Beads or similar ornamental items are not authorized.

Women may wear cosmetics applied conservatively and in good taste. Exaggerated or faddish styles are not authorized. Lipstick may be worn with all uniforms as long as the color is conservative and complements both the soldier's complexion and the uniform. Extreme shades, such as purple, bright pink, bright red, gold, blue, black, hot pink, green, yellow, ombre, and fluorescent/neon colors are not permitted. Natural colors, to include tinted glosses, are authorized. The optional wearing of lip liner is authorized, but colors must match the shade of lipstick being worn.

Women may wear solid-colored shades of nail polish with all uniforms as long as the color is not extreme. Extreme colors are defined as purple, bright pink, bright red, gold, blue, black, hot pink, green, yellow, ombre, and fluorescent/neon colors along with French manicure. Authorized colors include nude/natural shades, American manicure, and light pink. Nail shapes such as ballerina, stiletto, arrow, and coffin are not authorized. Square and rounded styles are authorized. Nail length will not exceed ¼ inch beyond the tip of the finger. Women will comply with cosmetics policy while in uniform or in civilian clothes when on on duty.

Jewelry and Eyeglasses. The wearing of a wristwatch, a wrist identification bracelet, and not more than two rings is authorized (a wedding set is considered one ring), provided the styles are conservative and in good taste, unless prohibited for health or safety reasons. In addition to the single item (wristwatch or identification bracelet) on each arm, soldiers may wear one activity tracker, pedometer, or heart rate monitor. No jewelry, watch chains, or pens and pencils should be exposed on the uniform. Authorized exceptions are a conservative tie tack or tie clasp, which may be worn with the black four-in-hand necktie; a pen or pencil may appear exposed on the hospital duty, combat vehicle crewman, and flight uniforms. Pens/pencils of any color also may be worn, exposed, in the pen/pencil slots on the ACU coat. Fad devices, vogue medallions, personal talismans, or similar items are not authorized for wear with the uniform or on duty. The wearing of religious articles and jewelry is authorized, provided they are not visible or apparent. Wearing of such articles may be prohibited when complete uniformity is desired, as for parades, color guards, or ceremonial units.

Conservative prescription civilian eyeglasses are authorized for wear. Conservative prescription and nonprescription sunglasses are authorized except when in formation. Eyeglasses or sunglasses that are faddish or have lenses or frames with initials or other adornments are not authorized. Wearing sunglasses for authorized medical reasons is allowed. Sunglasses are not authorized to be hung on uniforms or from restraints down the front of uniforms, attached to chains, bands or ribbons, while in a garrison environment. Tinted contact lenses, except for medical reasons, are not authorized.

Women are authorized to wear earrings in the ACU when not conducting physical training, deployment, or in areas where normal hygiene is not accessible. Stud earrings of screw-on, clip-on, or post in gold, silver, or clear diamond are authorized. Diamonds can be single or clustered. While pearls are not authorized for wear with the ACU, they are authorized with the service or dress uniform provided the earring does not exceed ¼ inch in diameter and is unadorned, spherical, or square. Hoop, two-sided, or drop earrings are not authorized.

Body Piercing. Soldiers may not attach, affix, or display objects, articles, jewelry, or ornamentation to or through the skin, tongue, or any other body part on or off duty—except for earrings for women. Women may wear earrings with the service, dress, mess, and evening mess uniforms, but not with Class C or utility uniforms. Earrings may be of the clip-on, screw-on, or post type; they must be unadorned and spherical in shape, not more than 6 millimeters in diameter, and of gold, silver, diamond, or white pearl. Earrings must be worn in matched pairs, not more than one earring per earlobe, and fit snugly against the ear. Women may wear any type of earrings off duty, on or off military installations, as long as they do not create or support ear gauging (enlarged holes in the lobe of the ear, greater than 1.6 millimeters).

Tattoos and Body Mutilation. Any tattoo or brand anywhere on the head or face is prohibited except for permanent makeup. Tattoos below the wrist bone and on hands are prohibited except for one ring tattoo on each hand. With the exception of earrings for women, soldiers are prohibited from any form of unauthorized mutilation of the body or any body parts in any manner. Examples include tongue bifurcation, unnatural shaping of teeth, ear pointing, scarification, or body modifications for the purpose of suspension. Refer to AR 670-1 and the latest G1 policies.

Identification Tags and Security Badges. Identification tags (dog tags) must be worn at all times while on duty in uniform unless otherwise directed by the commander. Security badges must be worn in restricted areas for identification as prescribed by the commander in accordance with applicable regulations.

Religious Accommodations. Modified uniform and grooming standards are available for soldiers who submit and receive approval for a religious accommodation request in accordance with AR 600-20. Religious accommodations are available for hijabs, beards, turbans and under-turbans, and leggings worn during physical fitness exercises. Consult AR 670-1 and AR 600-20 to understand the regulations and submit a request.

Personal Protective or Reflective Clothing. The wear of commercially designed protective headgear is authorized when riding a motorcycle, bicycle, or similar vehicle. Such headgear should be removed and authorized uniform headgear donned when travel is complete. Commanders may authorize the wear of protective or reflective outer garments with uniforms when safety considerations so dictate.

Civilian Clothing. Civilian clothing is authorized for wear when off duty unless prohibited by the installation commander in CONUS or by the MACOM commander overseas. Commanders down to unit level may restrict the wear of civilian clothes by those soldiers who have had their pass privileges revoked under the provisions of AR 600-8-10.

When on duty in civilian clothing, Army personnel must conform to the appearance standards stated previously unless specifically authorized by the commander for specific mission requirements. Many installations have "dress codes" for both soldiers and their family members. Entry to the base and its associated public areas implies agreement to conform to the installation dress code. Officers and their family members should take care to become familiar with the installation dress code in order to avoid embarrassing situations.

Civilian Gym Bags and Rucksacks. Civilian gym bags, rucksacks, and similar bags are authorized for use while in uniform. Authorized bags may be carried by hand, on one shoulder using a shoulder strap, or over both shoulders using both shoulder straps. If the choice is to carry a bag over one shoulder, the bag must be carried on the same side of the body as the shoulder strap, not with the bag slung across the body with the strap over the opposite shoulder. If a soldier is in uniform, shoulder bags must be black or match one of the colors in the authorized camouflage patterns. The contents of the bag must not be visible. There is no restriction on the color of civilian bags carried in the hand. These rules do not apply to purses, which are discussed separately. Commanders govern the wear of organizational issue rucksacks in garrison and field environments.

SELECTION, PURCHASE, AND CARE OF UNIFORMS

It is mandatory that all officers dress in accordance with their positions in the U.S. Army and with the traditions and customs of the service. Officers are responsible for procuring uniforms and equipment pertaining to their rank and duty and for maintaining them in neat and serviceable condition. The accompanying table lists the minimum quantities of uniforms normally prescribed by commanders. Officers are responsible for ensuring that their uniforms and insignia conform to the requirements of AR 670-1. In addition to the required uniforms, sufficient quantities of appropriate accessories, insignia, footwear, undergarments, headgear, and handgear are to be purchased and maintained.

Officers of all components and grades, of short service or long, are obliged to meet high uniform standards. There is no variation in official requirements. Still, as a practical matter, the standard will be established at the top, by general officers and field officers of a command. A high standard will be expected of career officers. Whatever your grade or component, do your part. Be meticulous in procurement of uniforms that fit well and are always maintained in a clean, well-pressed condition, so that you can wear them with pride.

The minimum quantities of uniforms shown in the table are just that. Depending upon your assigned duties and the availability of laundry and dry-cleaning facilities, more than the minimum number may be needed. The junior officer faces a dilemma of what to buy and how to dress well without becoming impoverished. The total expenditure for these uniform items is considerable.

The initial uniform allowance provided to reserve officers helps to defray the initial cost but does not cover all expenditures. A reserve officer on inactive-duty status is required to provide himself or herself with service uniforms and insignia of the branch in which commissioned for use when ordered to active duty. A proper minimum for such officers is a complete Army Service Uniform with extra trousers or skirts, and one or two extra shirts with the black all-weather coat. This minimum will permit the officer to report for duty and to perform duties in uniform until he or she can procure basic needs of the station and duty. The initial allowance provided for uniforms should cover these minimum purchases; it is intended to permit newly appointed officers to procure the uniforms needed for their start without financial sacrifice.

The wise officer will exercise care and judgment in selections to make certain that the articles bought meet service standards and fit well. To be an Army officer, you should strive not only to be a good officer but also to look like a good officer.

Where to Purchase Uniforms. The Army and Air Force Exchange Service handles the retail sale of official uniform items. These are uniform items purchased by the Defense Logistics Agency and are the same items as those issued to enlisted soldiers. They are stocked in the Military Clothing Sales Store (MCSS) or at www.aafes.com. The MCSS also stocks commercial and optional uniform items and accessories. The commercial items displayed along with the issue items afford an individual the opportunity to comparison shop. You will realize the maximum economy in price by purchasing your uniforms at the MCSS. Certainly junior officers, unless they are independently wealthy, will wish to make most of their purchases of required uniforms at the MCSS.

Civilian Tailoring Companies. There are a number of large and reliable companies in the United States that specialize in uniforms for officers. There are also many small custom-tailoring establishments well known as uniform specialists; some of these tailors have been in this field for years with well-established reputations. There is just one reason to choose a proven civilian source over the official outlets, and that is quality of individual tailoring. Officers who can afford the cost—career officers and senior officers especially—are justified in purchasing some of their uniforms from carefully chosen custom tailors. The added cost is the price of looking their military best.

The Commanding Officer's Responsibility. Commanders prescribe the uniform for wear in daily duty, formations, parades, honor guards, and color guards, and when in the field, deployed, and at war. Commanders will not establish seasonal wear dates for uniforms, nor may they require individuals to purchase optional uniform items. Likewise, they will not discourage soldiers from wearing optional authorized uniform items except in those instances where uniformity is required, such as parades or formations. The commanding officer is required to make periodic inspections of the uniforms worn by members of his or her command, including officers. Inspection includes checking as to the possession of the items required, that uniforms fit properly and are in serviceable condition, that they meet specifications for quality and design, and that only duly prescribed items of insignia and ornamentation are worn with the uniform.

Making Choices. Know what is required and what may soon change. Various kinds of cloth are authorized for the various uniforms, with a choice of weights to meet varying climatic conditions. Uniforms may not be mixed; that is, do not wear a serge coat with trousers or a skirt of gabardine. Reserve officers on limited tours of active duty may sensibly choose the wool serge material, because these uniforms can be obtained through the MCSS.

If you do buy uniforms from other than official sources, check for the warranty labels certifying the quality and the meeting of Army standards. Uniforms and accessories bought at bargain stores may have been rejected by government purchasers.

Apply the same kind of reasoning when choosing material for the Army blue dress uniform and for the various mess and evening mess uniforms if you decide to purchase these optional garments. These are all-year uniforms. You should seek a fine appearance in these uniforms, but be mindful of comfort in summer as well as winter.

There are numerous items of ornamentation that call for gold or gold-color materials. Here are the rules: Wherever gold lace or gold bullion ornamentation is prescribed for wear with the uniforms, gold-color nylon or rayon or synthetic metallic gold may be substituted subject to the following limitations:

- If trouser and sleeve ornamentation is gold bullion, cap decoration and shoulder strap insignia must be bullion.
- Ornamentation on the visors (male) or hatbands (female) of all Army service caps must be of gold bullion, synthetic metallic gold yarn, or anodized aluminum in 24-karat gold color.

Care of Uniforms. Good uniforms and appropriate accessories deserve the treatment that will ensure their maximum durability and good appearance. An old uniform of good quality that fits well and is clean, neat, and unfaded will look better than a new and costly one that is noticeably soiled or not pressed. The care that should be given to uniforms and equipment need not be burdensome, but it must be done regularly and correctly. This discussion, which includes points of common experience, should be helpful.

A modest amount of regular care of your uniforms is necessary. Upon removing uniform garments, brush them, inspect for spots or soil, and promptly use effective cleaning solvents, not forgetting soap and water; place garments on good wooden or plastic hangers and hang them where they can air and dry. Trousers and slacks are best hung at full length. Such care will result in restoration of the press and removal of small wrinkles and provide a uniform ready to wear when needed. It isn't necessary to have a uniform cleaned and pressed as frequently when these habits are followed.

Have a number of regulation neckties or neck tabs. Ties soil quickly and require dry cleaning. Replace them before they approach the point of unattractiveness.

Moths and mildew from excess humidity are the enemies of uniforms in storage. Be certain that such uniforms are clean and well brushed at the time of storage. Place them in a tight container with an adequate supply of moth preventive. Some dry-cleaning establishments provide mothproofing service.

Care of Ribbons, Decorations, and Service Medals. Ribbons for decorations and service medals must always be worn fresh and bright. Never wear frayed or soiled ribbons. Never use ribbons covered with transparent plastic or impregnated with a substance to increase their life. Dry cleaning will not injure them. Keep the metallic parts clean; some of them require shining. Wear them with pride.

Care of Brass Items. All Army "brass" items for sale in MCSS and through the AAFES website do not need to be shined. Vintage uniform items made of solid brass must still be hand shined.

Care of Gold Braid. Gold braid or an authorized substitute, now a part of several uniforms, is found on items such as caps, shoulder knots, insignia of grade on dress uniforms, sleeve ornamentation, and trouser stripes. Gold braid items are costly, and therefore it is unlikely that any officer will choose gold braid these days. However, should you elect to purchase items with gold braid, with correct care, you can keep them serviceable for a long time. Incorrect handling can ruin them quickly. If cleaning is attempted at all, it should be done only by somcone who has proven skill in doing it.

Tarnish is the enemy of gold braid. When not in use, these items should be stored where they will be dry, protected from light, and wrapped in tarnish-proof paper. Such paper may be obtained from jewelers. Ordinary paper contains sulfur and will cause tarnish, as will rubber. Dry cleaning will not injure trouser stripes or sleeve decorations of gold.

Care of Shoes. The most important point in preserving and prolonging the useful life of good shoes is to place them on properly fitted shoe trees as soon as they are removed. They will then dry in the correct shape, without wrinkles, and be comfortable when worn again. Have several pairs of shoes and rotate their use.

Clean shoes as required. Saddle soap works well. Castile or other mild soap is a good cleaning agent for leather. Applying coat upon coat of polish, without intermediate cleaning, merely piles polish on dirt. Leather that becomes dry and lifeless may be restored with leather dressings, or by a light application of neat's-foot oil on the flesh side. Use good polishes; the exchanges carry reliable brands.

Boots worn in field service must be strong, well fitted, comfortable, and treated to resist water penetration. The wise officer will keep one pair of such boots in top-notch condition ready for instant use.

UNIFORMS FOR MALE OFFICERS

Army Service Uniform (ASU)—male
Army Green Service Uniform (AGSU)—male
Army blue mess and evening mess uniforms—male
Army white mess and evening mess uniforms—male
Army utility uniforms, various (discussed in later section)

ARMY SERVICE UNIFORM (ASU)—MALE

The class A and class B Army Service Uniform (ASU) is authorized for year-round wear by all male personnel. All officers are required to own the ASU for wear on appropriate occasions. The class A service uniform consists of a dark blue coat, dark or light blue trousers, and a long-sleeved Army white shade 521 shirt with a turn-down collar. Black oxford shoes are worn with this uniform. When worn with the black four-in-hand necktie, it is considered an informal uniform. When worn with a black bow tie, it constitutes a formal uniform corresponding to a civilian tuxedo.

Army Service Uniform—Male

The class B service uniform consists of light blue trousers and a short- or long-sleeved Army white shade 521 shirt with a stand-up collar. The black necktie is always worn with the long-sleeved shirt. The short-sleeved shirt may be worn with or without a tie.

Occasions for Wear. The ASU, class A or class B, is the normal duty uniform, unless the nature of the duties requires one of the utility uniforms. Either version may be worn by all male personnel when on duty, off duty, and during travel. The ASU with bow tie is worn as a dress uniform for social functions of a general or official nature after retreat, unless other uniforms are prescribed by the host. When the uniform is not prescribed for formations or other occasions, the selection between class A and either of the class B versions is based on weather conditions, duties, and the formality of the occasion.

Authorized Materials. There is a choice of wool barathea, 14-ounce; wool gabardine, 11- or 14.5-ounce; wool elastique, 16-ounce; or wool tropical, 10.5-ounce, each in dark blue, Army shade 150; or polyester/wool blend in twill weave, 9.5-ounce, or in plain weave, 9.5-ounce, each in dark blue, Army shade 450. For general officers, both the coat and the trousers are of this material and

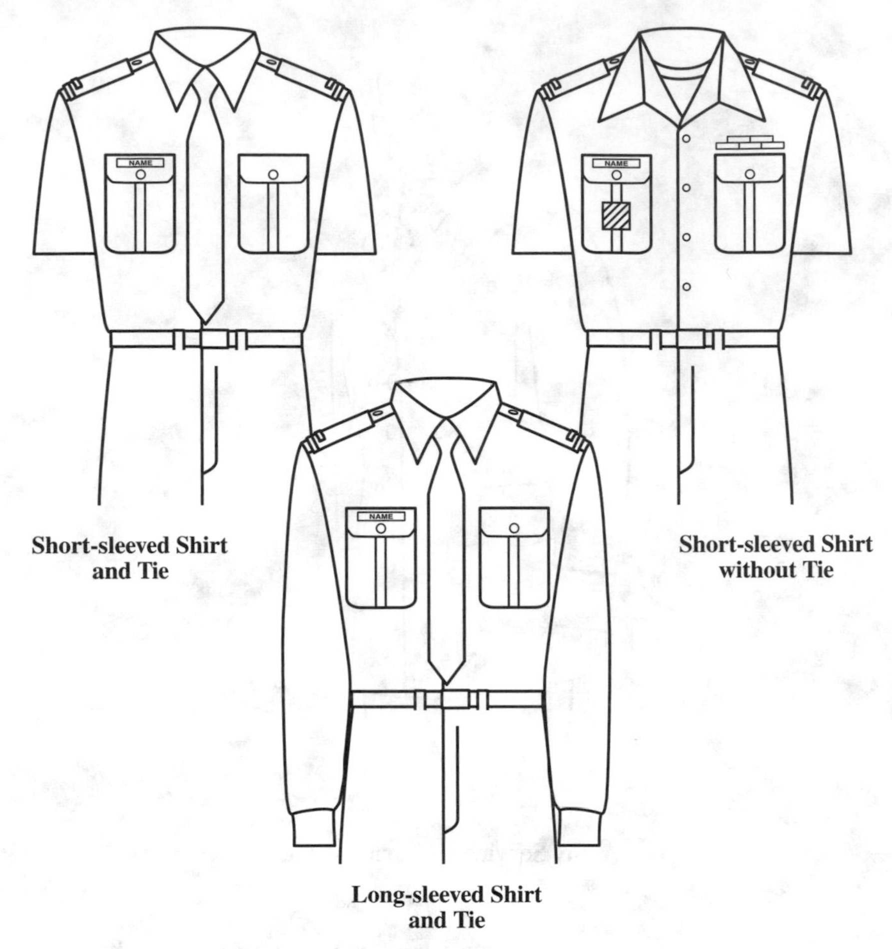

Class B Army Service Uniforms

shade. For other officers, the coat is of the material and shade noted above, but the trousers are light blue, Army shade 151 or 451, of the same material as the coat. The trousers may be high-waisted, suitable for wear also with the Army blue mess uniform; however, if high-waisted trousers are worn as part of the blue class A or class B uniform, they must be worn with the service coat, black windbreaker, black pullover, or black cardigan sweater.

Many people believe that the better-appearing Army blue uniforms are tailored from the heavier weights of cloth. However, since it is an all-year uniform, the lighter-weight fabrics provide the best year-round comfort. A well-fitted, carefully tailored uniform, clean and freshly pressed, will look good with any of the authorized weights of material. The lightweight fabrics are advised for reasons of economy and comfort.

Coat and Ornamentation. The Army service coat is a single-breasted, peak-lapel, four-button coat, extending below the crotch and fitting easily across the chest and shoulders, with a slight draped effect in front and back. Officers authorized to wear an aiguillette will attach a 20-ligne button on the left or right outside shoulder seam of the Army blue coat, depending on the position in which the aiguillette is worn. Officers authorized to wear a fourragère will attach a 20-ligne button on the left shoulder seam, $^1/_2$ inch outside the collar edge. Ornamental gold braid (bullion or one of the authorized substitutes) is worn on the sleeves of the blue coat. For general officers, a $1^1/_2$-inch gold braid is positioned with the bottom of the braid parallel to and 3 inches above the bottom of each sleeve. For all other officers, the braid consists of two $^1/_4$-inch gold stripes positioned $^1/_4$ inch apart over silk material of the first-named color of the officer's basic branch. The braid is positioned with the bottom parallel to and 3 inches above the bottom of each sleeve.

Trousers and Ornamentation. The low-waisted blue trousers are straight-legged without cuffs, with side and hip pockets. General officers wear dark blue trousers; all other soldiers wear light blue trousers. Trouser ornamentation consists of gold braid stripes (of the same material as used on the coat sleeves) sewn over the outer seams of the trouser legs. General officers wear two $^1/_2$-inch-wide stripes of gold braid, spaced $^1/_2$ inch apart. All other officers wear a $1^1/_2$-inch stripe of gold braid.

Short- and Long-Sleeved Shirt. The issue shirt is a dress type with a stand-up collar, tapered shoulder loops, and two plain pockets with button-down flaps. The shirt has three permanent creases on the back and one on each side of the front of the shirt running vertically through the pocket. The long-sleeved shirt has two button cuffs and is designed to be worn strictly with a tie. Wearing a tie with a short-sleeved shirt is optional when it is worn as an outer garment. A tie is also optional when wearing a short- or long-sleeved shirt with sweaters. If a tie is worn with either sweater, the collar of the shirt goes inside the sweater. If no tie is worn with the pullover sweater, the collar is worn outside. If no tie is worn with the cardigan sweater, the collar can be worn inside or outside the sweater. Cardigan sweaters may be worn buttoned or unbuttoned while indoors, but must be buttoned when outdoors. Note that either shirt is worn tucked into the trousers so that the shirt edge, the front fly opening of the trousers, and the outside (right) edge of the belt buckle form a straight "gig" line.

Headgear. The black beret is the standard headgear for wear with the ASU in any configuration; it is authorized for use with duty uniforms for special events only as prescribed by the command.

Organizational berets are authorized for wear with the class A and class B service uniforms by personnel assigned to ranger units (tan beret), special forces units (green beret), and airborne units (maroon beret). Except for color, they are the same, and all are organizational issue items.

The Army flash is the only flash authorized for wear on the black beret, unless authorization for another flash was granted before the implementation of the black beret as the standard Army headgear.

Wear of the beret, men and women

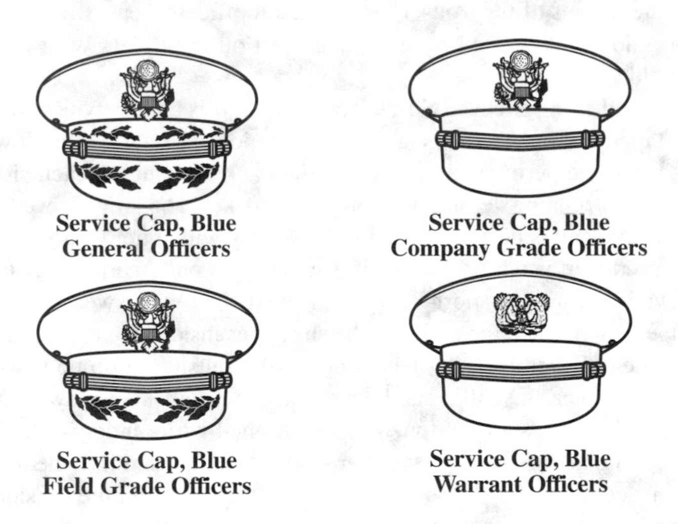

Army Service Uniform Headgear

The beret is worn with the headband straight across the forehead, 1 inch above the eyebrows, with the top draped over the right ear and the flash positioned over the left eye. The adjusting ribbon is cut off and the knot concealed inside the edge binding at the back of the beret.

Officers and warrant officers wear nonsubdued grade insignia centered on the beret flash, and chaplains wear their branch insignia. General officers may wear full, medium, or miniature size stars on the beret. Stars are worn point-to-point and may be mounted on a bar as an option.

Shoulder Straps, Army Service Uniform

The Army blue service cap of standard design and specification is worn with the ASU in its capacity as a formal dress uniform or with the class A version for units at the specific direction of the commander. The cap material is either the same material and shade as the coat, or officers may wear a cap of fur felt, 9-ounce, dark blue, Army shade 250. Officers also have an option to wear the cap frame with removable cover of the same material as the coat. The visor of the cap is of black leather or poromeric material with a leather finish. For general and field-grade officers, the top of the visor is of black cloth with two arcs of oak leaves in groups of two, embroidered in one of the authorized gold-color materials. General officers have similar ornamentation on the cap band, which is 1¾ inches in width, of blue-black velvet, extending around the entire outside rim of the cap. Other officers wear a similar hatband of grosgrain silk in the first-named color of the officer's basic branch with a ½-inch two-vellum gold (or authorized substitute) band at the top and bottom of the hatband. For company-grade officers and warrant officers, the top of the visor is of plain black shell cordovan finish leather.

Shoulder Straps. The insignia of grade for wear with the ASU are embroidered on shoulder straps 1⅝ inches wide by 4 inches long. The shoulder straps clip on to the shoulder of the blue coat near the edge of the shoulder seam. For general officers, the background of the shoulder straps is blue-black velvet. For all other officers, the background is a rayon grosgrain ribbon in the first-named color of the officer's basic branch. The strap has a ⅜-inch gold-color border surrounded on the inside and outside by a strand of gold jaceron. If the officer's basic branch has two colors, the second branch color is used as a ⅛-inch inside border instead of the gold jaceron. Grade insignia on the strap are rayon embroidery or bullion and jaceron. See the accompanying illustrations.

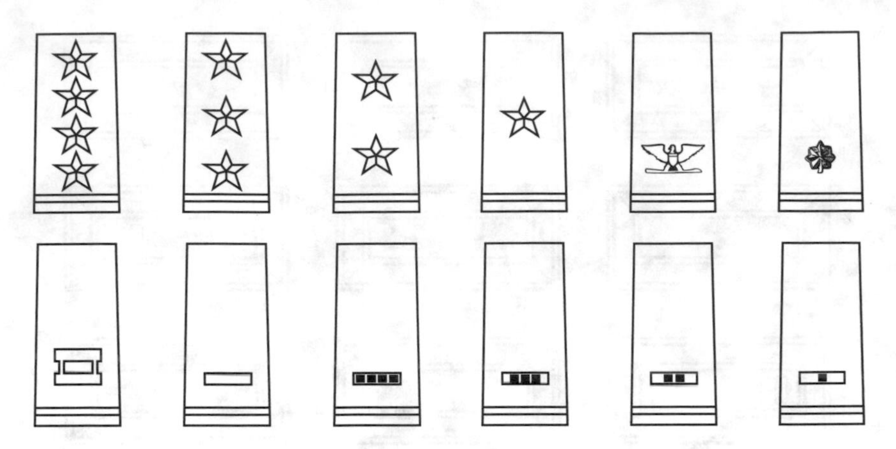

Marks for women are slightly smaller than those for men.

Shoulder Marks

Ornamentation and Insignia. Officers wear the following ornamentation and insignia on the Army Service Uniform:

Headgear insignia
U.S. insignia
Insignia of grade (shoulder straps)
Insignia of branch to which assigned or detailed
Authorized decoration ribbons and service medal ribbons—full-size or miniature decorations and service medals may be worn after retreat
Combat Service Identification Badge
U.S. badges (identification, marksmanship, combat, and special skill—full-size or miniature); dress miniature combat and special skill badges are worn when miniature medals are worn
Regimental distinctive insignia
Unit awards
Distinctive items for infantry personnel
Nameplate
Aiguillette (dress or service)
Fourragère/lanyards

The manner of attaching and wearing insignia and other accoutrements is discussed later.

For the ASU worn in a class A configuration, officers may wear overseas service bars on the wrist area of the jacket, a beret with rank, and shoulder boards, in addition to all the other insignia and rank prescribed for wear with the uniform in a formal dress capacity. Officers who wear green, tan, or maroon berets, are assigned to air assault-coded positions, or are military police on duty may blouse their trousers with the black leather combat boot.

For the ASU worn in a class B configuration, officers may wear a Combat Service Identification Badge (CSIB) on the white shirt, a beret with rank, shoulder boards, and a black necktie/tab with the long-sleeved white shirt. Officers who wear green, tan, or maroon berets, are assigned to air assault-coded positions, or are military police on duty may blouse their trousers with the black leather combat boot.

Accessory Items.

- belt with gold buckle (para 20-2)
- boots, combat, leather, black (optional for wear with class A and class B uniforms, only for those soldiers authorized to wear the tan, green, or maroon berets, those assigned to air assault-coded positions, and military police performing military police duties) (para 20-3)
- bow tie, black (worn after retreat) (para 20-18a)
- buttons (para 20-4)
- cape, black or blue (officer only) (para 20-5)
- chaplain's apparel (para 20-6)
- cufflinks and studs, gold (para 20-9)
- coat, black, all weather (para 20-7)
- gloves, black, leather, unisex, dress (worn with black all-weather coat or black windbreaker) (para 20-11b)
- gloves, white, dress (para 20-11c)
- handbag, black, fabric or leather (para 20-12b)
- handbag, black, shoulder (para 20-12d)
- handbag, black, clutch type, optional purchase (para 20-12a)
- hat, drill sergeant (authorized for wear with class A and class B uniforms) (para 20-13)
- judge's apparel (para 20-14)
- military police accessories (not authorized with the formal class A service uniform) (para 20-15)
- necktie, black, four-in-hand (worn on duty) (para 20-18c)
- neck tabs (para 20-17)
- scarf, black (only with black all-weather coat or black windbreaker) (para 20-21a)
- shirt, white, long-sleeved (para 20-22c)
- shirt, white, short-sleeved (para 20-22a)
- shoes, oxford, black (para 20-23a)
- shoes, pumps, black (para 20-23f and 23g)
- socks, black, cushion sole (worn with boots only) (para 20-24a)
- socks, black, dress (worn with trousers/slacks) (para 20-24b)
- stockings, sheer (para 20-24d)
- sweater, pullover, black (para 20-26c)
- sweater, unisex cardigan, black (para 20-26a)
- undergarments, white (para 20-28)

- umbrella, black (Soldiers may carry and use an umbrella, only during inclement weather, when wearing the service (class A and B), dress, and mess uniforms. Umbrellas are not authorized in formations or when wearing field or utility uniforms.) (para 20-27)
- windbreaker, black (only with class B uniform) (para 20-30)

Insignia, Awards, Badges and Accoutrements.

- aiguillettes, service (officers only; not authorized on the class B ASU) (para 21-26 and 21-27)
- airborne background trimming (para 21-32b)
- branch of service scarves (not authorized on the enlisted formal class A service uniform) (para 21-20)
- branch insignia (para 21-10 and 21-12a)
- brassards (not authorized on the dress blue ASU) (para 21-30)
- Combat Service Identification Badge (CSIB)

The CSIB will be worn when available in place of the Shoulder Sleeve Insignia-Former Wartime Service on the ASU. The CSIB will be worn centered on the wearer's right breast pocket of the ASU coat for male soldiers, and on the right side parallel to the waistline on the ASU coat for female soldiers. The CSIB is ranked fifth in order of precedence below the Presidential, Vice Presidential, Secretary of Defense, and Joint Chiefs of Staff Identification Badges. The CSIB can also be worn on the shirt when wearing the class B versions of the ASU (para 22-17a).

- decorations and service medal ribbons (para 22-6, 22-7 and 22-8)
- distinctive items authorized for infantry personnel (para 21-31)
- distinctive unit insignia (enlisted only) (para 21-22)
- foreign badges (para 22-18)
- fourragère/lanyards (para 22-10d)
- gold star lapel pin (para 22-6c)
- headgear insignia (para 21-3)
- insignia of grade (para 21-5, 21-6, 21-7, and 21-8)
- OCS/WOC insignia (para 21-14 and 21-15)
- nameplate (para 21-25c)
- organizational flash (para 21-32a)
- overseas service bars (optional) (para 21-29)
- regimental distinctive insignia (optional) (para 21-23)
- service stripes (enlisted personnel only) (para 21-28)
- unit awards (para 22-10)
- U.S. badges (identification, marksmanship, combat, and special skill) (para 22-17, 22-15, and 22-16)
- U.S. insignia (para 1-4)

ARMY GREEN SERVICE UNIFORM AND DRESS VARIATION—MALE

The AGSU is authorized for year-round wear by all personnel. The Class A and dress AGSU are authorized for wear both on and off duty, but they are not intended for wear as an all-purpose uniform when other uniforms are more appropriate. The

Class A AGSU includes the heritage green 564 coat, heritage taupe 565 trousers or skirt for females, heritage tan 566 short- or long-sleeved shirt, heritage green four-in-hand necktie, and a heritage walnut 567 belt. The Class B AGSU includes the heritage tan 566 short- or long-sleeved shirt and heritage taupe 565 trousers. When the long-sleeved shirt is worn without the Class A coat, soldiers will wear a four-in-hand heritage necktie. Authorized headgear for the AGSU are the Army green garrison cap, heritage green service cap, and the beret. Brown oxfords, pumps, or combat boots (when authorized) are worn with the AGSU. See DA PAM 670-1 for the full list of authorized accessories and insignia to be worn.

Ornamentation and Insignia. Officers wear the following ornamentation and insignia on the Army Service Uniform:

- Headgear insignia
- U.S. insignia
- Insignia of grade (shoulder straps)
- Insignia of branch to which assigned or detailed
- Authorized decoration ribbons and service medal ribbons—full-size
- Combat Service Identification Badge
- U.S. badges (identification, marksmanship, combat, and special skill—full-size or miniature); dress miniature combat and special skill badges are worn when miniature medals are worn
- Regimental distinctive insignia Unit awards
- Distinctive items for infantry personnel
- Aiguillette (dress or service) Fourragère/lanyards
- Special skill tabs, full color
- SSI, current organization
- SSI-MOHC
- Overseas service bars

ARMY BLUE MESS AND BLUE EVENING MESS UNIFORMS—MALE

The Army blue mess and blue evening mess uniforms are authorized for year-round wear by all male personnel.

Occasions for Wear. The blue mess and blue evening mess uniforms may be worn at social functions of a general or official nature after retreat and at private formal social functions after retreat. The blue mess uniform is for "black tie" functions and corresponds to the civilian tuxedo. The blue evening mess uniform is the most formal Army uniform and corresponds to the civilian "white tie and tails" tuxedo.

Composition. The Army blue jacket and dark or light blue high-waisted trousers constitute the basic uniform. Worn with a white, semiformal dress shirt with turn-down collar, black bow tie, and black cummerbund, it is the blue mess uniform. The blue evening mess uniform features a white formal dress shirt with wing collar, white bow tie, and white vest.

Authorized Materials. The authorized materials for the jacket are wool barathea, 14-ounce; wool gabardine, 11- or 14.5-ounce; wool elastique, 15-ounce;

Army Blue Mess Uniform

Army Blue Evening Mess Uniform

or wool tropical, 9-ounce, all in dark blue, Army shade 150; or polyester/wool blend in plain weave, 9.5-ounce, or in gabardine, 9.5-ounce, each in dark blue, Army shade 450. The trousers for general officers are of the same material and shade (dark blue) as the jacket. All other officers wear trousers of the same material as the jacket, but the color is light blue, Army shade 151 or 451.

Jacket and Ornamentation. The jacket is cut on the lines of an evening dress coat; it descends to the point of the hips and is slightly curved to a peak in the front and back. Two 25-ligne buttons, joined by a small gold or gold-color chain, about $1^{1}/_{2}$ inches long, may be worn in the upper buttonholes. Shoulder knots of gold bullion or substitute material are worn on the shoulders of the jacket by all officers.

Lapels. Lapels of the jacket are of rayon, acetate, or other synthetic fabric in colors as follows: general officers, except chaplains—dark blue; chaplains—black; all other officers—first-named color of the officer's basic branch.

Sleeves. General officers wear a cuff of blue-black velvet, 4 inches in width, positioned 1/8 inch from the bottom of each sleeve, with a band of gold oak leaves in groups of two, about 1 inch in width, placed 1 inch below the upper edge of the cuff. Insignia of grade, of embroidered silver bullion, is centered on the outside of the sleeve 1 inch above the upper edge of the sleeve cuff. Whenever insignia of branch is worn by a general officer, it is placed 1 inch above the top of the cuff, and the insignia of grade is placed 1 inch above the insignia of branch. Insignia of branch, if worn, is nonsubdued metal pin-on.

Other commissioned and warrant officers wear on each sleeve a band of two 1/4-inch two-vellum gold, synthetic metallic gold, or gold nylon or rayon braids, placed 1/4 inch apart over a silk stripe of the first-named color of their basic branch, the bottom of the sleeve band to be 3 inches above and parallel to the bottom of the sleeve. A trefoil consisting of a knot of three loops of 1/4-inch gold braid, one large upper loop and two small lower loops, interlaced at points of crossing, is placed on the outside of the sleeves with the ends resting on the sleeve band. Insignia of grade—nonsubdued, pin, or embroidered silver bullion—is worn vertically in the center of the space formed by the lower curves of the knot and the upper edge of the sleeve braid.

Trousers and Ornamentation. The trousers are cut on the lines of civilian dress trousers with a high waist and without pleats, cuffs, or hip pockets. Ornamentation is gold stripes as described for the Army Service Uniform trousers. Suspenders may be worn with these uniforms, but they may not be visible.

Headgear. The Army blue service cap is the authorized headgear with the blue mess and blue evening mess uniforms.

Accessories. Accessories are what differentiate the mess uniform and the evening mess uniform. A white, pleated, semiformal dress shirt, black bow tie, and black cummerbund are worn with the mess uniform for "black tie" affairs; a white, formal dress shirt (not pleated), white bow tie, and white vest are worn with the evening mess uniform for "white tie" affairs. Officers are authorized to wear a blue cape with these uniforms instead of the black all-weather coat. See the later section on accessories for a description of these items.

Ornamentation and Insignia. Insignia of grade and branch are worn as described for the jacket ornamentation. The only other authorized items of ornamentation are as follows:

Regimental distinctive insignia
Miniature decorations and service medals
Identification badges
Dress miniature combat and special skill badges
Dress aiguillette
See the later section for the manner of wear of these items.

Army White Mess Uniform

Army White Evening Mess Uniform

ARMY WHITE MESS AND WHITE EVENING MESS UNIFORMS—MALE

The Army white mess and white evening mess uniforms are authorized for optional wear by all male personnel. These uniforms normally are worn from April to October, except in clothing zones I and II (see CTA 50-900), where they may be worn year-round.

Occasions for Wear. The rules for wear are identical to the rules for the Army blue mess and blue evening mess uniforms.

Composition. The Army white jacket and black, high-waisted trousers make up the basic uniform. Worn with a white semiformal dress shirt with turn-down collar, black bow tie, and black cummerbund, it is the white mess uniform. The white evening mess uniform features a white formal dress shirt with wing collar, white vest, and white bow tie.

Authorized Materials. The jacket and vest are of white cotton twill, 8.2-ounce; white polyester/wool blend in plain weave, 9-ounce; white

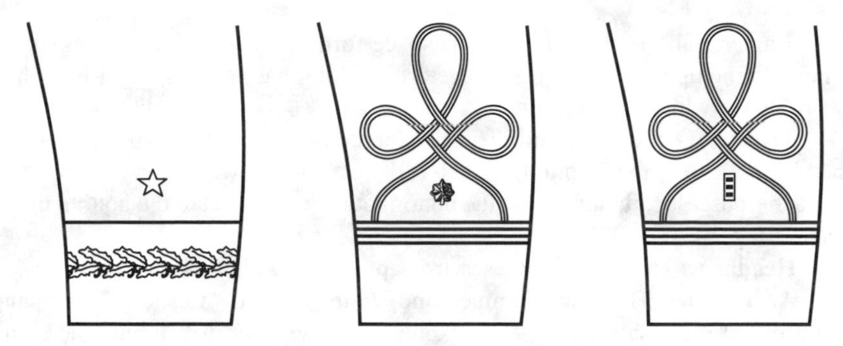

Sleeve Insignia, Blue Mess Jacket, Male and Female

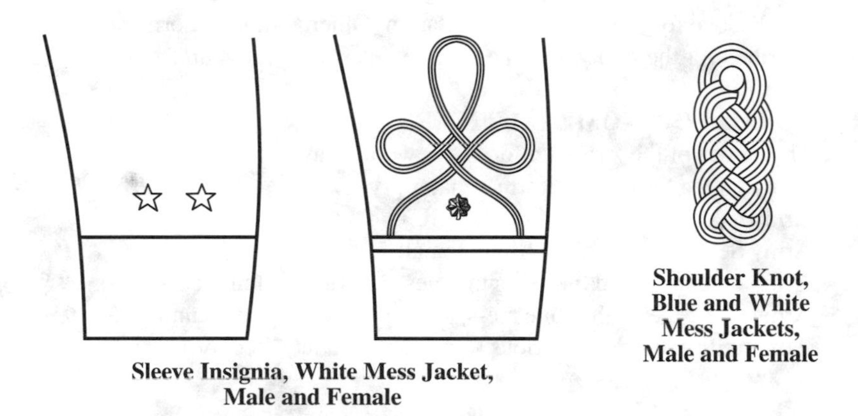

Sleeve Insignia, White Mess Jacket, Male and Female

Shoulder Knot, Blue and White Mess Jackets, Male and Female

polyester/wool blend in gabardine, 10.5-ounce; or white polyester texturized woven serge, 6.5-ounce. The trousers are black, commercial tuxedo design in a lightweight material.

Jacket and Ornamentation. The cut and fit of the jacket, the wearing of shoulder knots, and the use of an optional gold or gold-color chain joining the upper buttonholes are as described for the Army blue mess jacket.

General officers wear a white cuff of mohair or mercerized cotton braid, 4 inches in width, positioned 1/8 inch above the bottom of each sleeve. Insignia of grade is centered on the outside of the sleeves 1 inch above the upper edge of the cuff braids. If insignia of branch is worn by a general officer, it is nonsubdued metal pin-on, centered on the outside of the sleeve 1 inch above the upper edge of the cuff braid, and the insignia of grade is positioned 1 inch above the insignia of branch. The grade insignia is of embroidered white cloth or silver bullion.

All other commissioned and warrant officers wear a 1/2-inch band of white mohair or mercerized cotton braid with the lower edge parallel to and 3 inches above the bottom edge of the sleeve, with a trefoil consisting of a knot of three loops (one large upper loop and two small lower loops) of 1/4-inch white soutache braid, interlaced at points of crossing, with the ends resting on the sleeve bands. Insignia of grade, nonsubdued metal or embroidered white cloth, is worn vertically in the center of the space formed by the lower curves of the knot and the upper edge of the sleeve braid. Note that no insignia of branch is worn on the white mess jacket.

Trouser Ornamentation. The commercial-design, black tuxedo trousers have a black silk or satin braid, 3/4 inch to 1 inch wide, sewn on the outside seam of each trouser leg, running from the bottom of the waistband to the bottom of the trouser leg.

Headgear. The Army white service cap is worn with these uniforms.

Accessories. The black cummerbund, white vest, bow ties, and formal and semiformal dress shirts are the same ones worn with the Army blue mess and blue evening mess uniforms. They are described in a later section.

Ornamentation and Insignia. The authorized grade insignia are as described above for the jacket ornamentation. Other authorized ornamentation is as prescribed for the Army blue mess and blue evening mess uniforms.

UNIFORMS FOR WOMEN OFFICERS

The names of the Army uniforms for women are as follows:

Army Green Service Uniform—maternity
Army Green Service Uniform (AGSU)—female
Army Service Uniform (ASU)—female
Army blue mess and blue evening mess uniforms—female
Army white mess, all-white mess, and white evening mess uniforms—female
Army utility uniforms, various (discussed in a later section)

MATERNITY SERVICE UNIFORM

The maternity service uniform is authorized for year-round wear as a service or dress uniform by pregnant personnel when prescribed by CTA 50-900, AR 700-84, and the commander.

Composition. The class A AGSU-M maternity uniform consists of the heritage green 564 maternity tunic, maternity skirt or slacks in heritage taupe 565, a heritage tan 566 short- and long-sleeved maternity shirt , and the heritage green four-in-hand necktie. The class B uniform consists of the skirt or slacks and the short- or long-sleeved shirt. The necktie is always worn with the long-sleeved shirt; it is optional with the short-sleeved shirt.

Occasions for wear. The class A or class B maternity uniform may be worn by all female personnel on or off duty or during travel. These uniforms are acceptable for social functions of a formal or informal nature after retreat. Appropriate civilian maternity attire is also authorized in lieu of the uniform for social functions. Worn with the skirt, the class A uniform constitutes a dress uniform and is authorized for wear at social functions of a private or official nature either before or after retreat, and as designated by the host. Headgear, ornamentation, and insignia authorized for wear on the class A maternity service uniform are the same as for the class A and B uniforms. See DA PAM 670-1, chapter 15, for more details.

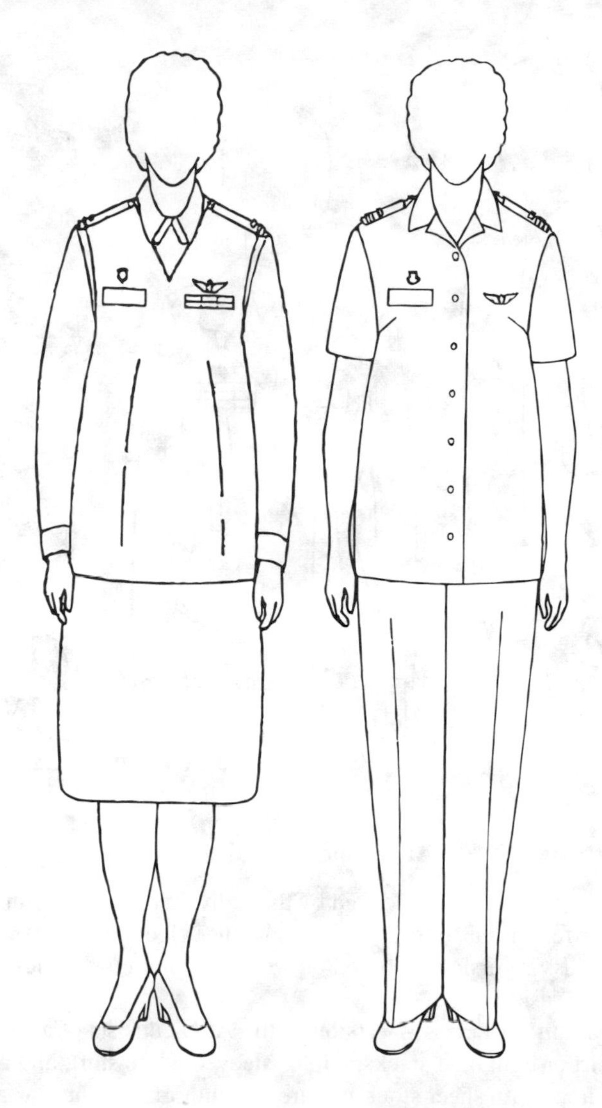

Maternity Service Uniforms

ARMY SERVICE UNIFORM (ASU)—FEMALE

The class A and class B Army Service Uniform is authorized for year-round wear by all female personnel. All active-duty female officers are required to own the uniform unless the period of active duty is for six months or less.

Occasions for Wear. The ASU, class A or class B, is the normal duty uniform, unless the nature of the duties requires one of the utility uniforms. Either the class A or class B version may be worn by all female personnel when on duty, off duty, and during travel. The class A version may be worn as a dress uniform for social functions of a general or official nature before or after retreat, and on

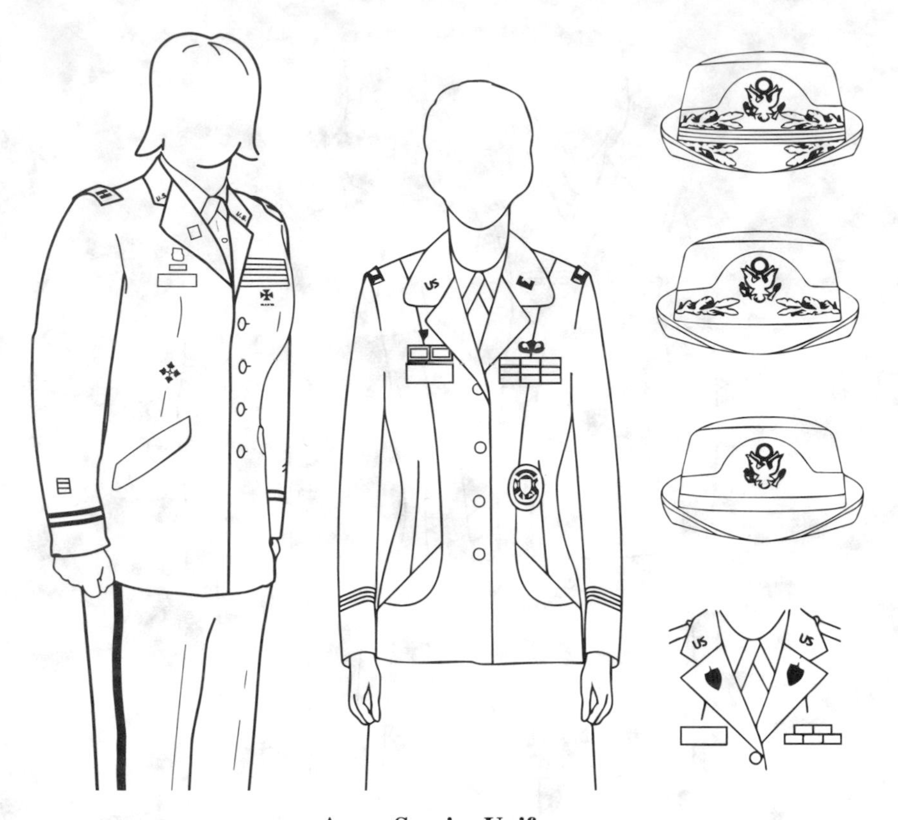

Army Service Uniform

other appropriate occasions as desired by the individual. When the uniform is not prescribed for formations or other occasions, the selection between class A and either of the class B versions is based on weather conditions, duties, and the formality of the occasion.

Composition. The class A version of the ASU consists of a dark blue coat, dark blue skirt or light blue slacks, a long-sleeved white shirt, and a black neck tab. Black pumps with sheer stockings are the authorized footwear along with a black leather dress handbag. The class B version consists of light blue trousers or a dark blue skirt with a long- or short-sleeved white Army shade 521 shirt. The black neck tab is always worn with the long-sleeved shirt but is optional for the short-sleeved shirt.

Authorized Materials. The dark blue coat, dark blue skirt, light blue slacks, and service hat are of the same material. Authorized materials are wool barathea, 12- or 14-ounce; wool gabardine, 11- or 14.5-ounce; wool elastique, 16-ounce; or wool tropical, 10.5-ounce, all Army blue shade 150; and polyester/wool blend in gabardine, 9.5-ounce, or in tropical weight, 9.5-ounce, Army blue shade 450.

Coat and Ornamentation. The coat is single-breasted, hip length, with two slanted flap front pockets, a notched collar, and side-body construction, with four

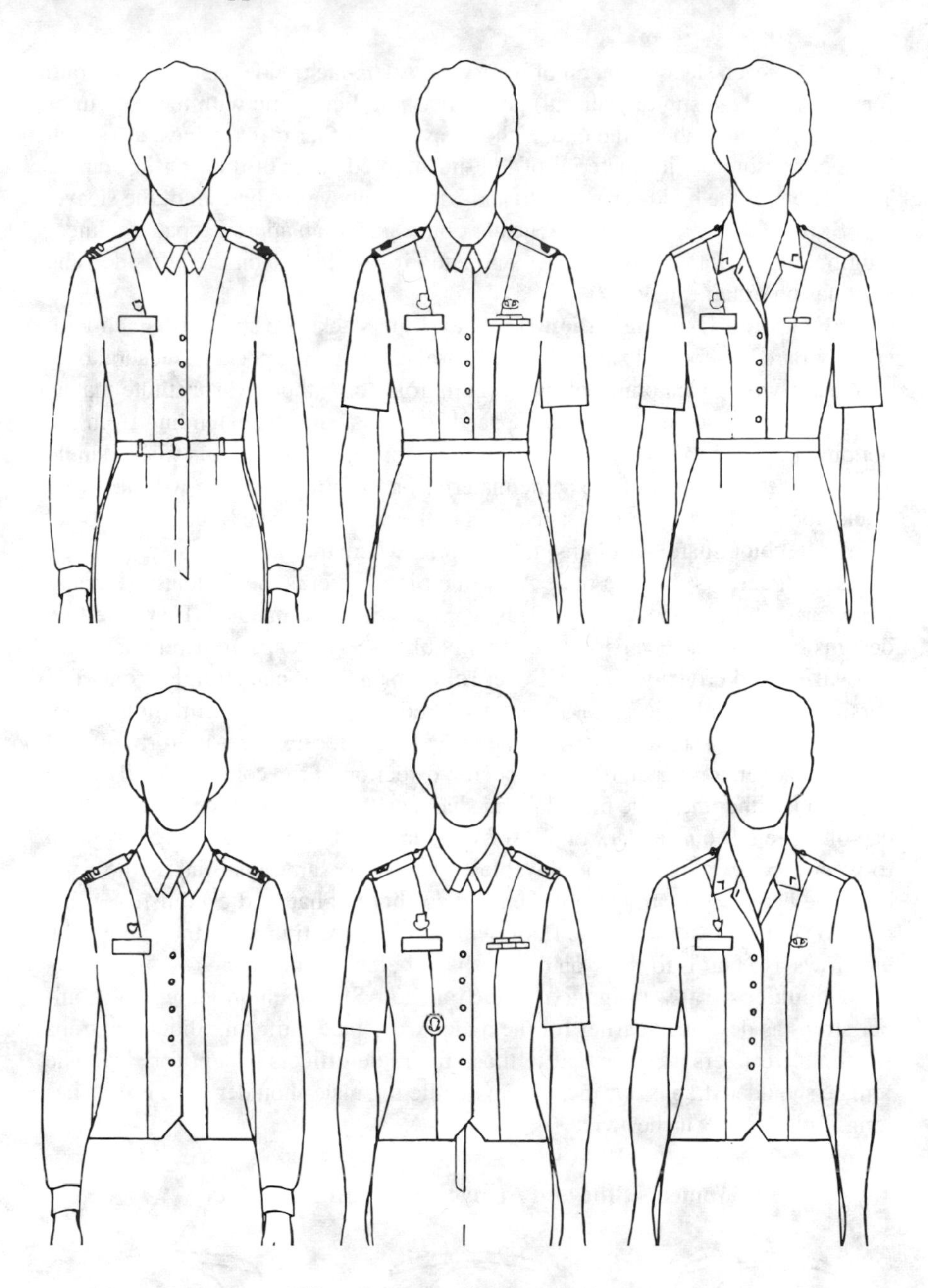

Class B Army Service Uniform

buttons. On each sleeve, general officers wear a $1\frac{1}{2}$-inch-wide band of gold braid (or authorized substitute material) positioned parallel to and with the bottom of the braid 3 inches above the end of the sleeve. All other officers have a $\frac{3}{4}$-inch braid consisting of silk material of the first-named color of their basic branch. The bottom of the braid is parallel to and 3 inches above the bottom of the sleeve.

Skirt and Ornamentation. The dark blue skirt is an approved pattern, knee-length, slightly flared with a waistband and zipper closure on the left side. The skirt has no ornamentation.

Slacks and Ornamentation. The dark blue slacks are straight-legged with slightly flared bottoms, a zipper front closure, and two side pockets. General officers wear two $\frac{1}{2}$-inch braids of two-vellum gold (or authorized substitute material) on the outer seam of each leg, spaced $\frac{1}{2}$ inch apart and extending from the bottom of the waistband to the bottom of the slack leg. Other officers wear a single 1-inch-wide gold braid of the same material on each leg. Women will not wear slacks for social functions unless required in the performance of their duties (band members, color guard, chaplains) in situations where the skirt is less appropriate.

Headgear. As with the male ASU, the black beret is the standard headgear for wear with the female ASU and is worn in the same manner. There are two designs for the female version of the Army blue service cap. Both versions have an oval-shaped crown with an all-over rolled brim back and straight front. The version for general officers has an embroidered brim, while the brim for all other soldiers is plain. The cap is worn with the ASU as a formal dress uniform or with the class A version for units at the specific direction of the commander. The cap material is either the same material and shade as the coat, or officers may wear a cap of fur felt, 9-ounce, dark blue, Army shade 250. Officers also have an option to wear the cap frame with removable cover of the same material as the coat.

The hat is worn straight on the head so that the hatband establishes a line around the head parallel to the floor, with the brim resting $\frac{1}{2}$–1 inch above the eyebrows. No hair is to show on the forehead below the brim.

Shoulder Straps. Insignia of grade for the ASU are embroidered on shoulder straps as described earlier for the male Army blue uniform. Shoulder straps for female officers are identical to those for male officers except for size. The female shoulder strap is $3\frac{1}{2}$ inches long, while the male shoulder strap is 4 inches long. Both are $1\frac{5}{8}$ inches wide.

Women's Blue and Army Service Uniform Hats

General and Field Grade Officer

Company Grade Officer

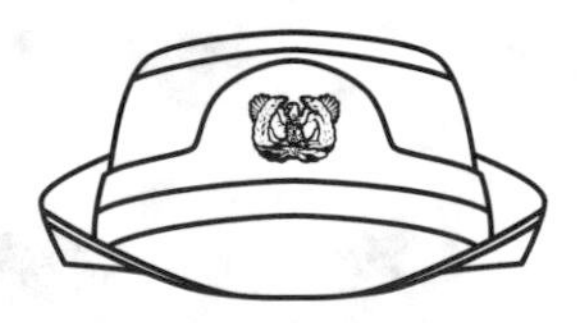

Warrant Officer

Army Blue Mess Uniform

Army Blue Evening Mess Uniform

Ornamentation and Insignia. As described earlier for the male ASU.
Accessory Items. Same as the accessory items listed for the male ASU.

ARMY BLUE MESS AND BLUE EVENING MESS UNIFORMS—FEMALE

The Army blue mess and blue evening mess uniforms are authorized for year-round wear by all female personnel.

Composition. The blue mess uniform consists of the Army blue jacket, a blue knee-length skirt, a white formal blouse with dress black neck tab, and a black cummerbund. The blue evening mess uniform substitutes a full-length skirt for the knee-length skirt. Black pumps with sheer stockings are worn with these uniforms, and a black fabric handbag (dress) is carried.

Occasions for Wear. The blue mess and blue evening mess uniforms are authorized for wear at social functions of a general or official nature after retreat, and at private formal social functions after retreat. The blue evening mess uniform is the most formal uniform worn by female personnel.

Authorized Materials. The authorized materials for the Army blue mess jacket and skirts are wool barathea, 14-ounce; wool elastique, 15-ounce; wool gabardine, 11- or 14.5-ounce; or wool tropical, 9-ounce, each in Army blue shade 150 or 450; and polyester/wool blend in gabardine, 9.5-ounce, or in tropical, 9.5-ounce, in Army blue shade 450.

Jacket and Ornamentation. The Army blue mess jacket is cut on the lines of an evening dress coat, descending to the point of the hips, and slightly curved to a peak in the front and the back. It is fully lined, with an inside vertical pocket on the right side. The coat front has six gold-color 20-ligne buttons. Two 20-ligne gold-color buttons may be joined by a small gold or gold-color chain in the upper buttonholes. Shoulder knots of gold bullion (or authorized substitute) are worn on the shoulders of the jacket by all officers.

Lapels. Jacket lapels are of rayon, acetate, or other synthetic fabric in colors as follows: general officers, except chaplains—dark blue; all chaplains—black; all other officers—first-named color of the officer's basic branch.

Sleeves. General officers wear a cuff of blue-black velvet braid, 4 inches wide, positioned $^1/_8$ inch from the bottom of each sleeve, with a band of oak leaves in groups of two, 1 inch in width, positioned with the top of the band 1 inch below the top of the cuff. The band of oak leaves is in gold bullion, synthetic metallic gold, or gold-colored nylon or rayon. Insignia of grade, of embroidered silver bullion, is centered on the outside of the sleeves, 1 inch above the upper edge of the cuff braid. If branch insignia is worn by a general officer, it is nonsubdued, metal pin-on, centered on the outside of the sleeve, 1 inch above the top of the cuff, and the insignia of grade is positioned 1 inch above the branch insignia. Other commissioned and warrant officers wear on each sleeve a $^3/_4$-inch braid consisting of two $^1/_4$-inch stripes of two-vellum gold (or authorized substitute) spaced $^1/_4$ inch apart over a silk stripe of the first-named color of their basic branch. The braid is positioned so that it is parallel to, with the bottom 3 inches above, the end of the sleeve. On each sleeve is a trefoil consisting of a knot of three loops of $^1/_4$-inch gold (or authorized substitute), one large upper loop and two small lower loops, interlaced at the points of crossing, with the ends resting on the sleeve band. Insignia of grade—nonsubdued, metal pin-on, or embroidered silver bullion—is worn vertically in the center of the space formed by the lower curves of the knot and the top of the braid.

Skirts and Ornamentation. The blue mess skirt is knee-length, and the blue evening mess skirt is full-length. Both have a one-piece front with waist darts on each side, a four-piece back, a zipper closure on the left side, and a sewn-on waistband closed with three hooks and eyes. Both skirts are fully lined. Neither has any special ornamentation.

Headgear. None.

Cape. Either the Army blue or the Army black cape is authorized for optional wear with the blue mess and blue evening mess uniforms. For a description of capes and other uniform accessories, see the later section on accessories.

Ornamentation and Insignia. Ornamentation and insignia are the same as for the male blue mess and blue evening mess uniforms, except female officers wear no headgear insignia.

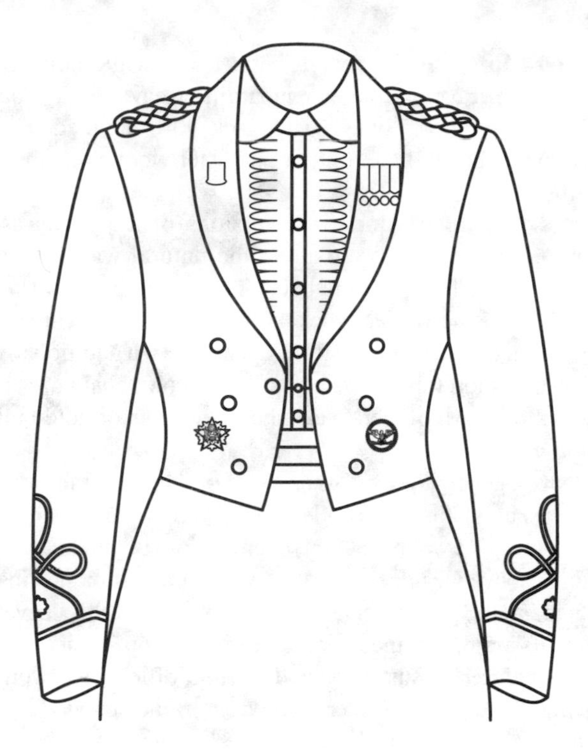

Army White Mess Uniform

ARMY WHITE MESS, ALL-WHITE MESS, AND WHITE EVENING MESS UNIFORMS—FEMALE

The Army white mess, all-white mess, and white evening mess uniforms are authorized for wear by all female personnel. These uniforms normally are worn from April to October, except in clothing zones I and II (CTA 50-900), where they may be worn year-round.

Occasions for Wear. These uniforms are worn to social functions of a general or official nature after retreat, and to private formal social functions after retreat.

Composition. The Army white mess uniform consists of the Army white jacket, a black knee-length skirt, the white formal blouse with dress black neck tab, and the black cummerbund. The all-white mess uniform is the same except a white knee-length skirt is substituted for the black skirt, and a white cummerbund is substituted for the black cummerbund. The white evening mess uniform uses a full-length black skirt instead of the knee-length skirt and is worn with the black cummerbund. Black pumps and sheer stockings are worn with the white mess and white evening mess uniforms, and a black fabric handbag may be carried. White pumps are worn with the all-white mess uniform, and a white fabric handbag may be carried. None of these uniforms have any headgear.

Authorized Materials. The white jacket and white skirt are made from either white polyester/rayon fabric in a gabardine weave, 6- or 8-ounce; or white texturized polyester serge, 6.5-ounce. The black skirts are made from wool tropical, 8.5-ounce, Army black shade 149; or polyester/wool blend in tropical weave, 10-ounce, Army black shade 332.

Jacket and Ornamentation. Two versions of the jacket are authorized.

New Version. The jacket is cut along the natural waistline and is slightly curved to a peak in the front and the back. The jacket has a shawl collar and is fully lined, with an inside vertical pocket on the right side. Three 20-ligne gold-color buttons are on each side of the front opening. Two additional 20-ligne gold-color buttons with a short (about 1 1/2 inches) gold-color chain may be used to join the upper buttonholes. All officers wear shoulder knots of gold bullion or authorized substitute material.

On the sleeves, general officers wear a cuff of white mohair or mercerized cotton braid, 4 inches in width, positioned 1/8 inch from the end of the sleeve. Insignia of grade, of embroidered white cloth or silver bullion, is centered on the outside of the sleeve, 1 inch above the upper edge of the cuff braid. If branch insignia is worn, it is nonsubdued, metal, pin-on centered on the outside of the sleeve, 1 inch above the top of the cuff braid, and the insignia of grade is positioned 1 inch above the branch insignia. Other commissioned and warrant officers wear on each sleeve a band of white mohair or white mercerized cotton, 1/2 inch in width, with the bottom of the band parallel to and 3 inches above the bottom edge of the braid on each sleeve. A trefoil adorns each sleeve, consisting of a knot of three loops (one large upper loop and two small lower loops) of 1/4-inch white soutache braid, interlaced at points of crossing, with the ends resting on the sleeve braid. Insignia of grade, pin-on metal or embroidered cloth, is worn vertically in the center of the space formed by the lower curves of the knot and the top of the sleeve band.

Old Version. The jacket is single-breasted with a natural waist length and a shawl-type collar. There are three 20-ligne gold-color buttons on each side of the front. General officers wear gold (or authorized substitute material) shoulder knots and sleeve ornamentation as described for the new version of the jacket. Other officers wear shoulder boards denoting grade and a band of 1/2-inch-wide white mohair or white mercerized cotton braid on each sleeve, with the bottom of the braid parallel to and 3 inches above the end of the sleeve. The old version of the white mess jacket may be worn until no longer serviceable.

Skirts and Ornamentation. Both the white mess skirt (black) and the all-white mess skirt (white) are knee-length with a one-panel front and a four-panel back of straight design, with a waistband and a zipper closure on the left side. The full-length black skirt for wear with the white evening mess uniform is of similar design but with an overlapped center back pleat. None of the skirts have any ornamentation.

Headgear. None.

Accessory Items. The black cummerbund, the white cummerbund, the formal blouse, and the Army blue or black capes (either of which may be worn with each of the uniforms) are described later. Black pumps with sheer stockings are worn with the white mess and white evening mess uniforms, and white pumps with sheer stockings are worn with the all-white mess uniform. The black dress fabric handbag is carried with the white mess and white evening mess uniforms, and the white dress fabric handbag is carried with the all-white mess uniform.

Ornamentation and Insignia. Ornamentation and insignia are the same as for the male white mess and white evening mess uniforms, except there is no headgear insignia for women.

FIELD AND UTILITY UNIFORMS

Field and utility uniforms for both male and female personnel and various organization uniforms—such as hospital duty and flight uniforms—are all categorized as class C uniforms. However, installation commanders in CONUS, MACOM commanders overseas, and the state adjutants general for ARNG personnel may publish exceptions to this policy. These uniforms are not intended to be worn when other uniforms are more appropriate.

ARMY COMBAT UNIFORM (ACU)

The ACU with its digitized gray/green pattern was replaced by the operational camouflage pattern fatigue uniform beginning in July 2015. The patrol cap is the normal headgear for the ACU, but the ACU sun hat, fleece cap, and beret may be worn when directed.

Occasions for Wear. The ACU is designed to be worn under body armor. It may be worn year-round by all personnel when prescribed by the commander. Personnel may wear the ACU off post unless prohibited by the commander. Like all uniforms, personnel will not wear the ACU in off-post establishments that primarily sell alcohol (bars and pubs). The ACU is a combat uniform; it is not normally considered appropriate for social or official functions off the installation, such as memorial services and funerals, when wear of class A, B, or dress uniforms is more appropriate. The commander may prescribe the addition of organizational and individual equipment items under provisions of CTA 50-900 when the ACU is worn at parades, reviews, and ceremonies. As of 1 June 2019, soldiers are authorized to wear the Improved Hot Weather Combat Uniform (IHWCU) but cannot mix outer elements of the IHWCU with the ACU.

Composition. Both combat uniform ensembles consist of a coat, trousers, T-shirt, rigger belt in either tan or coyote brown color, and boots in either tan or coyote brown as appropriate. The colors on the ACU are foliage green, desert sand brown, urban gray, and foliage green 504 in a digital pattern. The fabric of the ACU is 50-percent rayon and 50-percent cotton, while the fabric of the Fire Resistant (FR) Army combat uniform is 65-percent flame-resistant rayon, 25-percent para-aramid, and 10-percent nylon.

Accessories. The following items may be worn with the ACU: Army combat boots, temperate or hot weather (tan); socks (black/tan/green, cushion sole); moisture-wicking T-shirt (tan); silk-weight undergarments; patrol cap; beret with

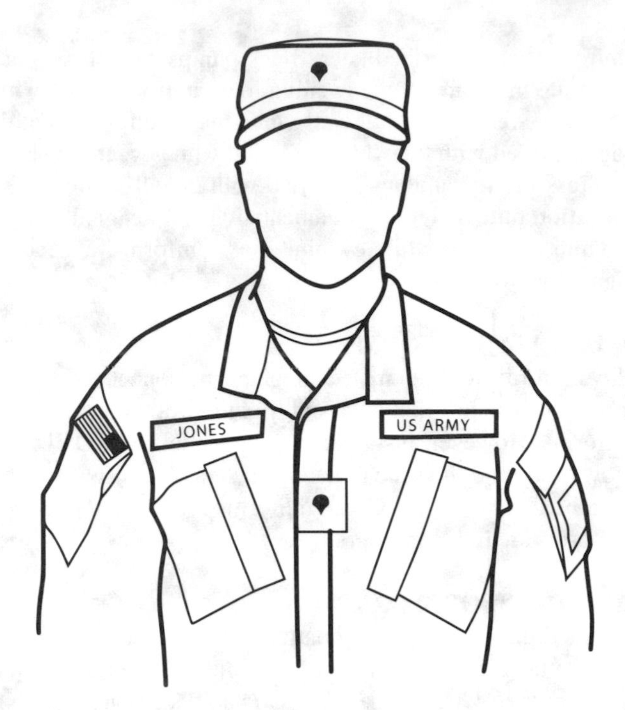

Army Combat Uniform

organizational flash; ACU sun ("boonie") hat; coat, cold-weather (field jacket); parka, cold-weather, camouflage (Gore-Tex jacket); trousers, cold-weather, camouflage (Gore-Tex pants); parka, wet-weather, camouflage (wet-weather jacket); trousers, wet-weather, camouflage; fleece cap; belt, web (rigger) with open-faced buckle; chaplain's apparel; gloves; handbags (women only); drill sergeant's hat; military police accessories; neck gaiter; scarves; organizational clothing and individual equipment (OCIE) as determined by the commander per CTA 50-900; medical items per CTA 8-100; and personal hydration systems as determined by the commander.

Insignia and Accoutrements. Only the following insignia and accoutrements are worn with the ACUs: hook-and-loop (Velcro) full-color or subdued U.S. flag insignia; hook-and-loop shoulder-sleeve insignia; embroidered grade insignia; subdued nametape and subdued U.S. Army tape; combat and special skill badges (pin-on or sew-on, subdued, optional, maximum of five); brassards; headgear insignia; subdued rank insignia; hook-and-loop subdued shoulder-sleeve insignia; former wartime service "combat" patch (optional); hook-and-loop skill tabs (optional); and drill sergeant and recruiter badges.

Soldiers may sew on the U.S. Army tape, nametape, insignia of rank, and all authorized badges as an option. The commander may authorize the wear of subdued pin-on combat, special skill, and identification badges in garrison but not in field or deployed units. The permanent infrared feedback squares affixed to each shoulder for nighttime identification will be covered when the insignia are not needed. The current unit shoulder-sleeve insignia, shoulder-sleeve insignia–former wartime service, and U.S. flag insignia are attached to the hook-and-loop pads located on each shoulder and may not be sewn on the uniform.

Manner of Wear. The ACU should be loose fitting and comfortable. Alterations to make them fit tightly are not authorized. The only alterations authorized are those listed in AR 700-84. Items should be fitted loosely enough to allow for some shrinkage without rendering the garment unusable.

The coat is worn hook-and-looped and zipped, with the coat outside the trousers, and the trousers worn with a belt. The coat may also be worn inside the trousers when directed by the commander (that is, when wearing the outer tactical vest, for example). The coat will not extend below the top of the cargo pocket on the pants and will not be higher than the bottom of the side pocket on the trousers. The mandarin collar is worn in the down position except when wearing the outer tactical vest or when weather conditions dictate the wear as prescribed by the commander; at these times, personnel wear the mandarin collar in the up position. The hook-and-loop sleeve cuff closure must be worn down and closed at all times, not rolled or cuffed. The elbow pouch with hook-and-loop closure for internal elbow pad inserts must be closed at all times. The integrated blouse billows for increased upper-body mobility. The T-shirt, either moisture-wicking or cotton, is worn underneath the coat and tucked inside the trousers at all times.

Personnel will wear the trousers tucked into the top of the boots or bloused using the drawstrings at the bottom of the trousers or commercial blousing devices. Personnel will not wrap the trouser leg around the leg tightly enough to present a pegged appearance or insert any items inside the trouser leg to create a round appearance at the bottom. When bloused, the trouser should not extend below the third eyelet from the top of the boot. The knee pouch with hook-and-loop closure for internal knee-pad inserts and the billowed calf storage pocket with hook-and-loop closure on the left and right legs will be worn closed at all times.

To maximize the service life and maintain optimum performance of the ACU, wash in cold water and tumble dry at low heat; do not use starch, sizing, or any process that involves dry cleaning or a steam press, as it will adversely affect the treatments and durability of the uniform. Place on a rust-proof hanger. Do not wring or twist.

Headgear. The ACU patrol cap is the primary headgear for all soldiers as the duty uniform headgear. Commanders retain the authority to prescribe the beret for special events such as parades or changes of command or responsibility. Soldiers may sew on the nametape and rank insignia as an option at their own expense. There is also an ACU sun hat, and a micro black fleece cap that may be worn at the commander's discretion. When not in use, the patrol cap may be stored neatly in the ACU cargo pocket.

The ACU patrol cap or sun ("boonie") hat is worn with the ACU in field environments when the Kevlar helmet is not worn. Personnel wear the ACU patrol cap straight on the head with no hair visible on the forehead beneath the hat. Sewn or pin-on subdued rank is worn on the hat. A nametape will be worn centered on the hook-and-loop pads or sewn on the back of the ACU patrol cap.

When authorized for wear, the drawstring on the sun ("boonie") hat can be worn under the chin, around the back of the head and neck, or tucked inside. The drawstring will not be worn over the top of the hat. The hat will not be worn rolled, formed, shaped, blocked, or with an upturned brim. Sewn or pin-on subdued rank is worn on the hat.

Personnel wear the black fleece cap pulled down snugly on the head, without rolling the edge of the cap. Standards of wear and appearance specified in paragraphs 1-7 and 1-8 of AR 670-1 apply at all times.

Sewing of the subdued rank insignia on the ACU patrol cap, ACU sun hat, and Kevlar camouflage cover is authorized.

MATERNITY WORK UNIFORM

The maternity work uniform is authorized for year-round wear by pregnant women when prescribed by the commander. In addition, there is a maternity service uniform; refer to DA PAM 670-1, chapter 15, for more information.

Occasions for Wear. The maternity work uniform is worn in lieu of the ACU and under the same restrictions for wear.

Composition. The maternity work uniform consists of the ACU cap, the cold-weather coat, and the maternity coat and maternity trousers, both in a camouflage pattern.

Accessories, Insignia, and Manner of Wear. The maternity work uniform is worn with the same accessories, insignia, and accoutrements and in the same manner as the ACU.

AIRCREW COMBAT UNIFORM

There are two types of uniform for aircrew personnel: the Army Aircrew Combat Uniform (A2CU) and the Fire Resistant Environmental Ensemble (FREE). The A2CU is a daily work, utility, and field uniform for aircrew members. The FREE is designed to be worn with the prescribed duty uniform to provide aviators and combat vehicle crewmembers with modular, flame-resistant protection. Both the A2CU and FREE are intended for wear in deployed environments only.

Composition. The A2CU is composed of a coat, trousers, boots, undershirt, undergarments, belt, and headgear. The FREE is composed of a jacket, trousers, undershirt, vest, parka, boxer briefs, and drawers. There are three types of jacket and trousers depending on the environment (light weather, intermediate weather, and extreme weather). The parka can be worn either with or without a parka liner. The foliage-green drawers are composed of three levels depending on weather conditions (an underlayer, mid-layer and a mid-weight fleece layer). The headgear for both types of uniform are the flight helmet when performing flight duties and the ACU patrol cap in all other situations.

Maternity Work Uniform

Accessories. Accessories for the A2CU include the FREE; gloves (Nomex, summer-weight flight gloves, maximum-grip NT gloves, and other gloves when not performing flight duties); neck gaiter; cold-weather coat; fleece cap; silk-weight undergarments; Army combat shirt; personal hydration systems; and OCIE items when prescribed by the commander. The accessories for the FREE include the A2CU; headgear; boots; rigger belt; fire-resistant socks; gloves (Nomex, summer-weight flight gloves, maximum-grip NT gloves, and other gloves when not performing flight duties); Army combat shirt; personal hydration systems; and OCIE items when prescribed by the commander.

Aircrew Combat Uniform

Manner of Wear. The A2CU coat is worn outside the trousers at all times. Sleeves of both uniforms are worn down at all times and are not authorized to be cuffed or rolled. The uniforms are loose fitting by design, and alterations to present a tailored look are not authorized. The proper wear of combat and skill badges, nametapes, U.S. Army tapes, insignia of grade, SSI-FWTS, and the U.S. flag insignia are worn in the same manner as the ACU. Like the ACU, if combat and skill badges are sewn on, the nametape, U.S. Army tape, and insignia of grade must be sewn on as well.

Hospital Duty Uniform

HOSPITAL DUTY UNIFORM—MALE

The male hospital duty uniform is authorized for year-round wear by all male officers in the Army Nurse Corps and the Army Medical Specialist Corps.

Occasions for Wear. The hospital duty uniform is worn on duty in medical care facilities as prescribed by the medical commander. It is not for travel and may not be worn off the military installation except when in transit between the station and the individual's quarters. Commanders may authorize exceptions to this policy when medical personnel are providing support to civilian activities, such as at parades and ceremonies.

Composition. There are three principal components of this uniform: a hip-length white smock with left breast pocket, front-button closure, and straight-cut bottom; white trousers of standard design with two slash pockets in front, two patch pockets in back, and front zipper closure, with belt loops; and a white, knee-length physician's smock with front-button closure and upper and lower pockets. The black web belt with open-face buckle is worn on the trousers.

Insignia and Accoutrements. Nonsubdued branch and grade insignia, headgear insignia, and a nameplate are the only authorized items worn with this uniform.

Manner of Wear. Officers assigned to the Army Nurse Corps wear the short white smock and white trousers with white oxford shoes and white socks. Officers assigned to the Medical, Dental, Veterinary, Medical Service, or Medical Specialist Corps may wear the physician's white smock over the service or utility uniforms while in the medical care facility or while on duty as directed by the facility commander.

HOSPITAL DUTY AND MATERNITY UNIFORMS—FEMALE

The female hospital duty and hospital duty maternity uniforms are authorized for year-round wear by all female officers in the Army Nurse Corps and the Army Medical Specialist Corps.

Occasions for Wear. Same as for the male hospital duty uniform.

Composition. Uniform materials are white cotton or white polyester. There are two principal variants of the hospital duty uniform. One is a knee-length, short-sleeved white dress with front-button closure, wing-tip collar, and belt. The second is a tunic and pants uniform; the tunic is an over-the-hip design with wing collars, short sleeves, and side pockets, and straight-legged pants. The physician's smock, as described previously, is part of the ensemble. A white unisex cardigan sweater may be worn.

The maternity variants of the uniform consist of a white maternity dress and white maternity slacks and tunic, both of any plain, unadorned, commercial design, with wing collars suitable for placement of insignia.

White oxfords and white stockings or socks (when wearing the pants) are worn with these uniforms by officers. The tunic is worn outside the pants.

Insignia and Accoutrements. Only nonsubdued grade and branch insignia, headgear insignia, and a nameplate are worn with this uniform.

COMBAT VEHICLE CREWMAN UNIFORM

The Combat Vehicle Crewman (CVC) Uniform is authorized for year-round wear by combat vehicle crewmen when issued in accordance with CTA 50-900 and prescribed by the commander.

Occasions for Wear. The CVC uniform is worn on duty when prescribed by the commander. It is not worn for travel or off the military installation except when in transit between the station and the individual's quarters. It is not intended as an all-purpose uniform when other uniforms are more appropriate.

Composition. The fabric is made of flame-resistant materials in the uniform camouflage pattern (UCP). The basic uniform consists of a one-piece coverall and a cold-weather jacket. The coveralls have a zippered front closure, drop seat, extraction strap located at the upper back, and pockets located on the left sleeve, right-left chest, right-left hips, right-left upper thighs, and right-left lower legs. All of the pockets have zipper closures. The cold-weather jacket is single-

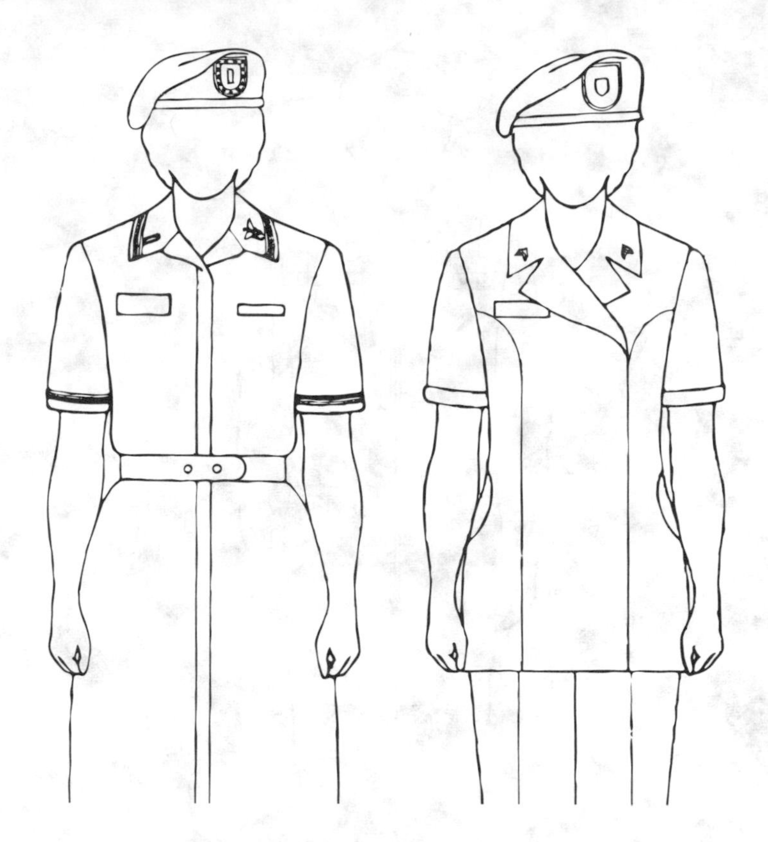

Hospital Duty and Maternity Uniforms

breasted with a zipper closure and an inside protective flap. The back has a yoke and retrieval strap opening. There are two slash pockets and a utility pocket on the left sleeve; the sleeves have elbow patches. The cuffs and waistband are rib knit. The uniform is worn with combat boots but not bloused. The ACU cap is the standard headgear when the CVC helmet is not being worn. The jacket may be worn only with the CVC uniform.

Accessories. FREE (see section on the aircrew uniform for further details); balaclava, hood; bib overalls; body armor, ballistic undergarment; CVC gloves, both cold-weather and summer; patrol cap; CVC helmet; organizational clothing and equipment as determined by the commander; and personal hydration system, as determined by the commander.

Insignia and Accoutrements. Only subdued grade insignia, headgear insignia, nametape and U.S. Army distinguishing tape, and shoulder-sleeve insignia (subdued, current organization) are authorized for wear with the CVC uniform.

The U.S. Army tape is worn approximately 1/2 inch up from the outside zipper seam on the left chest, horizontal to the ground. The nametape is worn on the right side in line with the U.S. Army tape. Subdued sew-on insignia of grade is

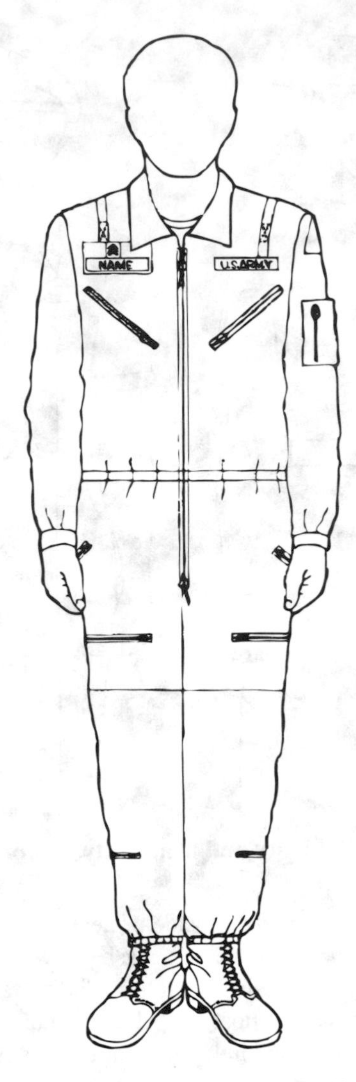

Combat Vehicle Crewman Uniform

worn centered and 1/4 inch above the nametape. On the cold-weather jacket, the nametape and U.S. Army tapes are positioned approximately 1 1/2 inches above the top of the pocket flaps. Otherwise, positioning of insignia on the coveralls and the jacket is the same.

PHYSICAL FITNESS UNIFORMS

The Army Physical Fitness Uniform (APFU) is a mandatory clothing item as of 1 October 2017. The APFU is authorized for year-round wear by all personnel when prescribed by the commander.

Occasions for Wear. The APFU may be worn on and off duty, on and off military installations, when authorized by the commander. Soldiers may wear all or part of the physical fitness uniform with civilian attire off the installation when authorized by the commander. The commander governs policy on the installation.

Composition. The APFU ensemble consists of a black and gold jacket; black pants with the Army logo; black, moisture-wicking trunks; a black, moisture-wicking, short-sleeved T-shirt; and a black, moisture-wicking, long-sleeved T-shirt. Only the black fleece cap is authorized for wear with the APFU. The cap may be folded but not rolled when worn.

Plain white or black socks, running shoes, gloves, long underwear, or other items appropriate to weather conditions may be worn with the uniform when authorized by the commander. Long underwear or other similar items must be concealed from view with the physical fitness uniform jacket and pants.

When wearing the physical fitness uniform as a complete uniform, soldiers must keep the sleeves down on the jackets and the legs down on the pants, as well as tuck the shirt inside the trunks.

Insignia and Accoutrements. No insignia are worn on the physical fitness uniform except for the Physical Fitness Training Badge, which may be worn on the upper-left front side of the T-shirt or jacket.

Wear during Pregnancy. Under HQDA policy, pregnant soldiers may wear the physical fitness uniform until it becomes too small or uncomfortable. Commanders may not require these soldiers to purchase larger-sized APFUs to accommodate the pregnancy. For comfort, pregnant soldiers may wear the T-shirt outside the shorts. When no longer able to wear the APFU, pregnant soldiers may wear equivalent civilian maternity physical fitness clothing.

ACCESSORY ITEMS

There are a number of articles used with all or a number of the uniforms. In addition, there are a number of optional items that may be worn or carried. See DA PAM 670-1 for detailed descriptions. Issue items may be purchased by officers through the MCSS at the post exchange. The PX also carries many of the optional items, purchased from commercial sources.

Bags. Soldiers are permitted to either carry or wear backpacks, gym bags, or similar articles while in uniform. Hand-carried bags must be conservative or professional in appearance. Bags worn over the shoulder must either be black or match one of the colors in the authorized camouflage patterns. They may be worn by using one strap over a single shoulder or by both shoulder straps over both shoulders, but are not authorized to be worn with a strap across the chest. The contents of the bag may not be seen; therefore clear or mesh bags are not authorized for wear.

Belts. There are three types of belts worn with class A, B, and most utility uniforms. The rigger belt is a 2-inch-wide web belt worn with utility uniforms that have belt loops (and all combat uniforms). It is worn with the tip passing

Army Physical Fitness Uniform

through the open-face buckle to the wearer's left. No more than 2 inches of webbing may protrude beyond the edge of the buckle.

The black web waist belt with black tip is also worn with utility uniforms (other than the combat uniform), with the black tip passing through the buckle to the wearer's left. As with the rigger belt, no more than 2 inches may extend beyond the buckle. The black web waist belt with brass tip is worn with the class A and B uniforms, with the tipped end passing through the brass buckle to the left for men and to the right for women. The tipped end passes through the buckle so that only the brass tip is visible.

Boots. Coyote "new buck" combat boots are issue items. The present-issue boot is made of coyote-colored cattlehide and nylon, with removable cushioned inserts, a closed-loop speed lace system, and drainage eyelets. In addition, jungle combat boots and boots of commercial design similar to the Army combat boot are authorized. The boots are diagonally laced, with the excess lace tucked into the boot top or under bloused (tucked in or by use of blousing bands) trousers or slacks. Organizational boots (safety boots) prescribed and issued by the commander may be worn instead of combat boots with field and utility uniforms.

Buttons. Buttons used on the service, dress, and mess uniforms are of prescribed design, gold-plated. Officers other than those of the Corps of Engineers wear the button bearing the U.S. Coat of Arms. Engineer officers wear the Essayons button. This is believed to have been designed by Col. Jonathan Williams, first chief engineer of the present Corps of Engineers, who was also the first superintendent of the U.S. Military Academy. The first authoritative reference to the special design is contained in General Orders No. 7, AGO, 18 February 1840.

Blouse, Formal. The white formal blouse worn by female officers with the mess and evening mess uniforms is an optional purchase item. It is tuck-in style, made of polyester and cotton, with a front closure having seven removable, dome-shaped buttons. Three rows of vertical ruffles are on each side of the front opening.

Cap, Cold-Weather. The cold-weather cap is an optional purchase item. It is of Army blue shade 450, 55/45 polyester/wool fabric, with a black synthetic fur visor and side flaps. It is mainly worn as a uniform item by ceremonial units such as the 3rd Infantry Regiment (The Old Guard). An eyelet is provided in the center of the visor to attach headgear insignia. Because of the thickness of the fur pile, headgear insignia worn on the cap must have a center post and screw; therefore, all soldiers will wear the male headgear insignia on this cap. The cap is worn straight on the head, with the headgear insignia centered and with no hair showing on the forehead. The side flaps are either fastened up or fastened under the chin if worn down. The cap is authorized for wear when wearing the black windbreaker with the ASU and when wearing the black all-weather coat with the service, dress, mess, and hospital duty uniforms. It may not be worn with the black pullover sweater.

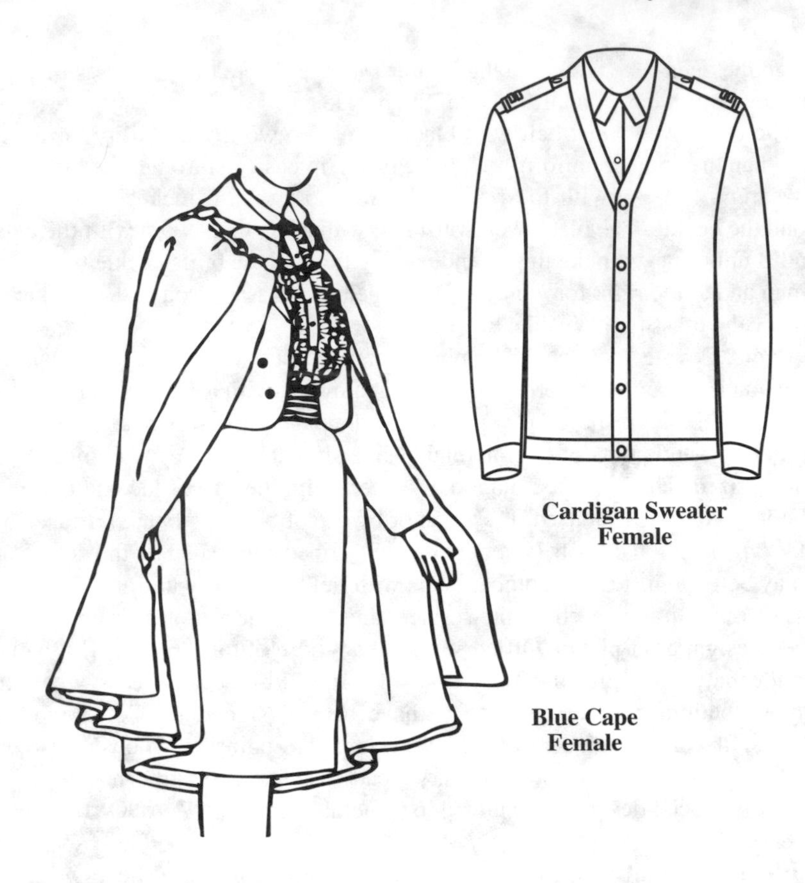

Cardigan Sweater
Female

Blue Cape
Female

Capes. Male officers may purchase a blue cape for wear with the Army blue dress, blue mess, and blue evening mess uniforms. The cape extends to the midpoint of the knee. The capes are fully lined: for general officers, the lining is dark blue, and for all other officers, the lining is the first-named color of the officer's basic branch.

Female officers may purchase either a blue cape or a black cape for wear with any of the mess and evening mess uniforms. The blue cape is fingertip length, with lining as described for the male cape. The black cape is knee-length with a white satin lining.

Chaplains' Apparel. Chaplains' scarves are organizational issue items. Black scarves for the Christian faith and black or white scarves for the Jewish and Muslim faiths have the U.S. Coat of Arms and the appropriate chaplain's insignia embroidered on each end. Chaplains are authorized to wear the military uniform, vestments, or other appropriate attire when conducting religious services.

Coat, All-Weather. The black all-weather coat is an issue item. The coat is worn by all personnel and may be worn with or without a zip-in liner with the service, dress, mess, hospital duty, and food service uniforms. Only nonsubdued pin-on insignia of grade is worn on the coat. Without insignia, the coat may be

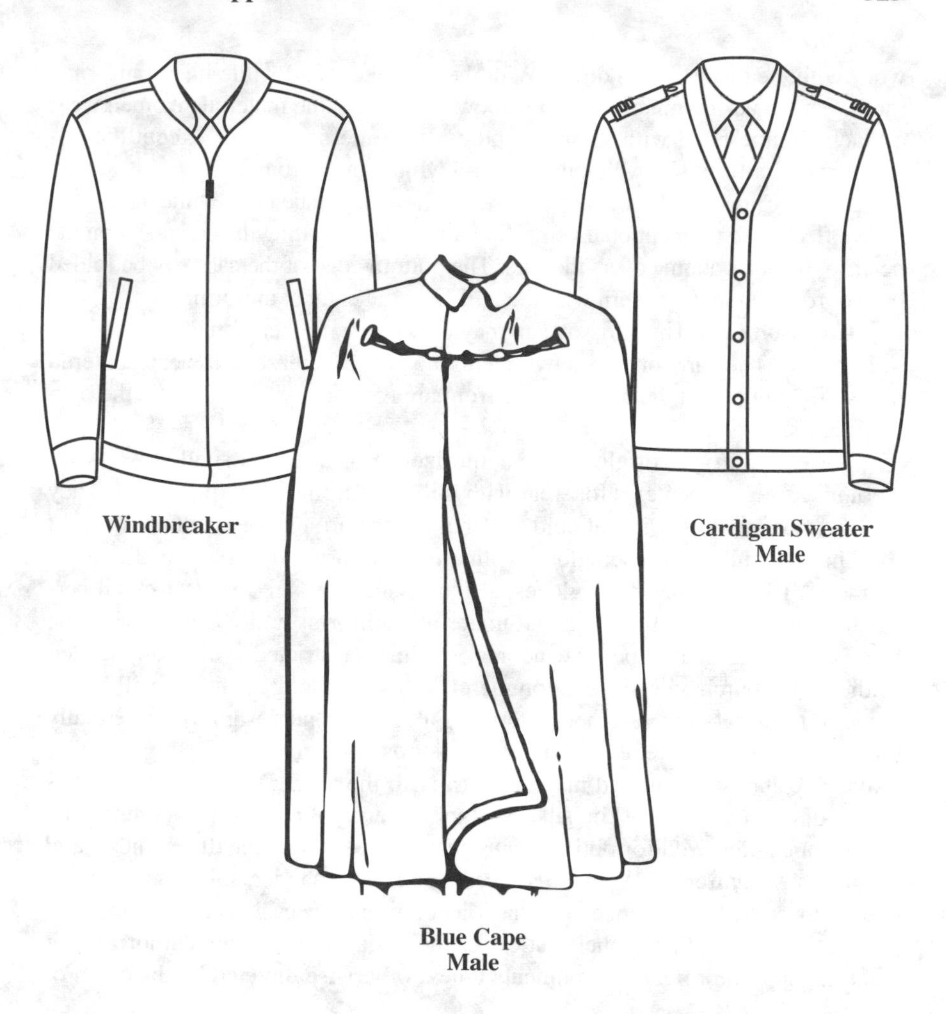

Windbreaker

Cardigan Sweater
Male

Blue Cape
Male

worn with civilian clothing. The current mandatory version is double-breasted with six buttons and a belt. The black scarf is authorized for wear with this coat.

Cover, Cap. A transparent plastic rain cap cover with a visor protector is authorized for optional purchase and wear by men when wearing the blue or white service caps.

Cufflinks and Studs. Cufflinks and studs are authorized for optional purchase and wear. Gold cufflinks—round, plain face, and $^1/_2$–$^3/_4$ inch in diameter—may be worn by male personnel with the Army blue uniform and with the Army blue mess and Army white mess uniforms. Studs for these mess uniforms also are gold-color, $^1/_4$–$^3/_8$ inch in diameter. Plain white (such as mother-of-pearl) cufflinks and studs, with or without platinum or white-gold rims, are worn with the blue evening mess uniform.

Cummerbunds. Cummerbunds are authorized for purchase and wear. Each is of commercial design with four or five pleats running the entire length and is worn with the pleats facing down. With the blue mess and white mess uniforms, men wear a black cummerbund with a bow tie of the same material. Women wear a black cummerbund with all mess and evening mess uniforms except the all-white mess uniform, for which they wear a white cummerbund.

Fleece Cap. The black fleece cap is a clothing bag issue item, while the coyote brown fleece cap is an optional item. Both are worn by pulling the cap down on the head so that the bottom covers the ears. The bottom edge of the cap may be folded but not rolled. It may be worn with the APFU or the combat uniform.

Garrison Cap. The garrison cap may be worn by all personnel with Class A or B AGSU. The garrison cap is worn with the vertical crease of the cap centered in a straight line with the nose and the front approximately 1 inch above the eyebrows.

Gloves. Six types of gloves are authorized. Black leather shell gloves with wool inserts are issue items for wear with utility uniforms by all personnel. They may be worn with or without cold-weather outer garments. Personnel may not wear inserts without the shell gloves with utility uniforms and cold-weather outer garments. Black leather dress gloves, also an issue item, are for wear by all personnel with the class A service and maternity uniforms and when wearing the black all-weather coat, cape, or windbreaker. Unless restricted by the commander, soldiers may purchase and wear commercial gloves as long as they are all black, made of leather, fabric, or other material of a commercial design. Commercially procured gloves must be plain and not have logos or other designs. Brown leather gloves are an issue item and may be worn with the AGSU. White dress gloves (made of cotton, kid, doeskin, silk, or other material of appropriate commercial design) are authorized for optional purchase for wear with the dress, mess, and evening mess uniforms. When prescribed by the commander, soldiers may wear white gloves with the service uniforms on ceremonial occasions or, like military police, in the execution of their duties. Flame-resistant gloves are authorized for wear in garrison or field environments unless otherwise directed by the chain of command.

Handbags. There are four types of handbags. A black service handbag, of polyurethane or leather, is an issue item and may be carried with the service, utility (only in a garrison environment), and Army blue uniforms, either carried in the hand or worn on the shoulder. It is not authorized to be worn with the strap across the body.

A black, clutch-type handbag of leather, polyurethane, or vinyl, with zipper, snap, or envelope-type closure, is authorized for optional purchase. It may be carried with the ASU and utility uniforms in a garrison environment. The handbag may have a wrist strap but not a shoulder strap.

A black dress handbag of fabric or leather is authorized for optional purchase. It is untrimmed, envelope or clutch style, of commercial design, with or without chain. The leather version may be carried with the ASU during and after duty hours. The fabric version may be carried with the black, white, and blue mess and evening mess uniforms.

A white dress handbag of fabric or leather is authorized for optional purchase. It is untrimmed, of commercial design, envelope or clutch style, with or without chain. The fabric handbag may be carried with the all-white mess uniform after duty hours.

Judges' Apparel. Judges' robes are organizational issue. They are of the type customarily worn in the U.S. Court of Military Appeals and are worn over the service uniform.

Neck Gaiters. The dark brown or tan neck gaiter is an optional item. It is a cylindrical tube 10–15 inches long, camouflage compatible, and can be worn with the ACU, maternity work uniform, and cold-weather uniforms. It may be worn as a neck warmer, hood, balaclava, ear band, or hat in cold, windy, or dusty environments. Commanders cannot require soldiers to purchase or wear neck gaiters on an individual basis; however, if the unit purchases them with available operating funds, the commander can require the unit to wear them.

Neck Tabs. The black service neck tab in Army shade 305 for women is an issue item. It is worn with the ASU uniform and when the long-sleeved shirt is worn without the ASU coat. The neck tab is required when the long-sleeved shirt is worn without the class A coat and when the long- and short-sleeved shirts are worn with the class A coat. The neck tab is optional for a shirt worn with the black pullover or cardigan sweaters.

Neckties. The heritage green necktie is an issue item. It may be tied in a four-in-hand, Windsor, or half-Windsor knot. It is worn with the Class A AGSU and with the short- and long-sleeved uniform shirts. The necktie is required when the long-sleeved shirt is worn without the Class A jacket and when either the short- or long-sleeved shirts are worn with the Class A jacket. It is optional when either shirt is worn with the green pullover sweater. The necktie is worn with the service uniform before retreat or on duty. Officers may wear the four-in-hand tie with the ASU after retreat, when the dress code is military informal.

The black necktie is worn in the same manner as the green necktie but with the Army Service Uniform.

A black bow tie is an optional purchase item for wear with the ASU and white mess uniforms after retreat. A white bow tie may be purchased for wear with the Army blue evening mess and white evening mess uniforms.

Overshoes. Black overshoes of rubber or synthetic material, commercial design, are authorized for optional purchase and wear with oxford shoes by men when not in formation during inclement weather. They may be worn with the service, dress, and mess uniforms.

Scarves. A black scarf of commercial design, about 12 x 52 inches, of wool, silk, or rayon, may be purchased and worn by all personnel with the black all-weather coat and the windbreaker. When worn, the scarf is folded in half lengthwise and crossed left over right at the neck, with the ends tucked neatly into the neckline of the outer garment. An organizational-issue green wool scarf may be worn with the field jacket or parka. The manner of wearing is the same as for the black scarf.

Shirts

See the section on the Army Green Service Uniform and Army Service Uniform for a description of the shirts worn with that uniform. This section describes shirts worn with the blue and white mess and evening mess uniforms.

Women. There are two types of formal shirt worn with the mess and evening mess uniforms. Both types have short sleeves, seven dome-shaped, removable pearl buttons, and a rounded collar. The main difference between the shirts is that the first type has three vertical rows of ruffles on each side of the front opening, while the second type has a pleated front.

Men. A white, semiformal dress shirt with long sleeves, pleats, a soft bosom, French cuffs, and a standard turn-down collar is worn by men with the blue mess and white mess uniforms. A white, formal dress shirt with long sleeves, an unpleated stiff bosom, French cuffs, and a wing-type collar is worn by men with the evening mess uniforms.

Shoes

Men. Black oxfords of leather (issue item), poromeric material, or patent leather without contrasting soles are worn with the service, dress, mess, evening mess, hospital duty, and food service uniforms. Brown oxfords are made of leather from an approved specification or pattern. When authorized by the commander, officers may wear jodhpur-style boots as long as they are black, without buckles and straps, and have noncontrasting heels of 2 inches or less in height. This option is not common practice with current uniforms. White oxfords are worn by Army Nursc Corps and Army Medical Specialist Corps officers with the hospital duty uniform.

Women. Black oxfords of leather (issue item), poromeric material, or patent leather with at least three eyelets, closed toe and heel, and heels no higher than 2 inches are worn with the service, hospital duty, and food service uniforms. Brown oxfords are made of leather from an approved specification or pattern. A jodhpur-type boot is also authorized when wearing slacks. Similar white oxfords are worn by officers with the hospital duty and hospital duty maternity uniforms.

Inclement-weather boots of black leather, rubber, or synthetic material may be purchased for optional wear by women. These are over-the-foot boots of commercial design, not more than knee-high, in a plain style with no trimming. They have inconspicuously placed zipper or snap closures and heels no higher than 2 inches. The boots may be worn during inclement weather with the service, dress, and mess uniforms, but must be exchanged for standard footwear when indoors.

Pumps, service, in black, brown, or white are of fine-grain calfskin, poromeric material, or patent leather, untrimmed, with closed toe and heel, heels from $^1/_2$ to 3 inches high, and soles not more than $^1/_2$ inch thick. The black service pumps are authorized for wear by all female personnel with the service dress and mess uniforms. White pumps are authorized for wear with the all-white mess uniforms.

Pumps, dress, in black or white fabric, are of commercial design, untrimmed, with closed toe and heel, heel height from 1/2 to 3 inches, and a sole not more than 1/2 inch thick. The black dress pumps are worn with the blue and white mess and all evening mess uniforms. The white pumps are worn with the all-white mess uniform after duty hours. Pumps and handbag must be of the same material.

Socks. Black socks are worn with the black oxfords by men. White socks are purchased by officers for wear with the white oxfords as part of the hospital duty uniform. Heritage green socks are worn with brown oxfords. Green, tan, and black cushion-sole socks are worn by all personnel when wearing combat or organizational-issue boots.

Women wear sheer or semi-sheer stockings without seams, and in flesh tones complementary to the wearer and the uniform, with the service, dress, and mess uniforms. White stockings, sheer or semi-sheer, are worn with the white oxfords with the hospital duty or maternity hospital duty uniform. Black socks may be worn when wearing the service uniform slacks, and white socks may be worn with the white oxfords when wearing the hospital duty uniform.

Suspenders. Suspenders may be purchased and worn by men with the dress and mess uniforms, but they must not be visible.

Sweaters. A black acrylic/wool cardigan sweater may be purchased and worn as an outer garment by officers with class B uniforms. The white unisex cardigan may be worn by hospital and food service soldiers. The black cardigan may not be worn with the white hospital duty and food service uniforms. The unisex sweater is a long-sleeved coat style and has epaulets and five buttons. The sweater must be worn fully buttoned outdoors, except for pregnant soldiers, and may be worn buttoned or unbuttoned indoors. If worn with the long- or short-sleeved shirt without necktie or neck tab, the shirt collar may be inside or outside the sweater. The collar of the hospital duty and food service uniforms will be worn outside the sweater. Officers will wear shoulder marks on the epaulets. Nameplate and distinctive unit insignia (DUI) will not be worn on the cardigan sweater. The sleeves may be worn cuffed or uncuffed, but may not be rolled or pushed up beyond the wrist. The cardigan sweater without rank insignia is authorized for wear with civilian clothes.

A black, 100-percent wool, V-neck pullover sweater is authorized for optional purchase and wear by officers with the class B uniform. The collar of the short- or long-sleeved AG 521 shirt is worn outside the sweater if no tie or neck tab is worn. The sweater may be worn under the black coat and the black windbreaker, but it must not show below the windbreaker. Officers wear shoulder marks indicating rank on the shoulders of the sweater. The nameplate is worn centered 1/4 inch above the bottom of the patch, with the DUI centered from left to right and from top to bottom above the nameplate. If an individual is not authorized a DUI, the regimental distinctive insignia is worn. Chaplains wear insignia of branch. The sweater sleeves may not be rolled or pushed up beyond the wrist. Without insignia, the sweater may be worn with civilian clothes.

The heritage green 564 pullover is authorized for optional purchase and wear by officers with the AGSU class B uniform. It is worn in the same manner as the black pullover sweater.

Umbrellas. Although still not common practice, soldiers may carry a black umbrella when wearing the service, dress, and mess uniforms. It may not be carried with field or utility uniforms.

Undergarments. Brassieres and panties of commercial design, in black, white, or neutral, are worn by women with all uniforms. Slips of commercial design in black, white, or neutral are worn by women with the service, dress, and mess skirts, and with the hospital duty and food service uniforms. Women are authorized to wear commercially purchased white, black, or neutral camisoles with all uniforms, except ACUs, with which the tan T-shirt is required. The camisole is not a substitute for the brassiere.

Drawers, either briefs or boxer style, are worn by men with all uniforms. Brown, boxer-style drawers are the issue item. Undershirts of commercial design, white with short sleeves and a V-neck or crew neck (T-shirts) are worn by men with the service, dress, mess, hospital duty, and food service uniforms.

A tan undershirt (issue item) is worn by all personnel with all utility uniforms except for hospital duty and food service uniforms.

Vest. A white vest, single-breasted, cut low with a rolling collar and pointed bottom and fastened with three detachable small white buttons is worn by men with the blue evening mess and white evening mess uniforms.

Windbreaker. A black windbreaker, in Army shade 458, is authorized for optional purchase and wear by all personnel with the class B service, hospital duty, and food service uniforms. The officer windbreaker has a knit collar, cuff, and waist. It is worn with nonsubdued pin-on insignia and must be worn zipped at least to the top of the second button from the top of the shirt. Without insignia, it may be worn with civilian clothing.

INSIGNIA

Insignia worn on the Army's uniforms identify the wearer as to status. Insignia denote grade, branch, organization, duty assignment, and prior Army service. Insignia are made of appropriate color metal or embroidery and are worn on prescribed uniforms, in precise locations.

U.S. Flag Insignia. U.S. flag cloth insignia will be worn by all soldiers throughout the force, regardless of deployment status. The reverse-side, full-color flag is to be worn on all utility uniforms (ACU, CVC, and maternity duty) unless deployed or in a field environment. When deployed or in a field environment, soldiers wear the subdued tactical flag insignia. The U.S. flag insignia (both full-color and subdued) is worn on the pocket flap on the right shoulder of the ACU and cold-weather jackets. The insignia is centered on the hook-and-loop pad attached to the pocket flap.

The "U.S." Insignia. Male and female officers wear the "U.S." insignia on both lapels of the ASU coat. On the male ASU coat, the insignia are positioned

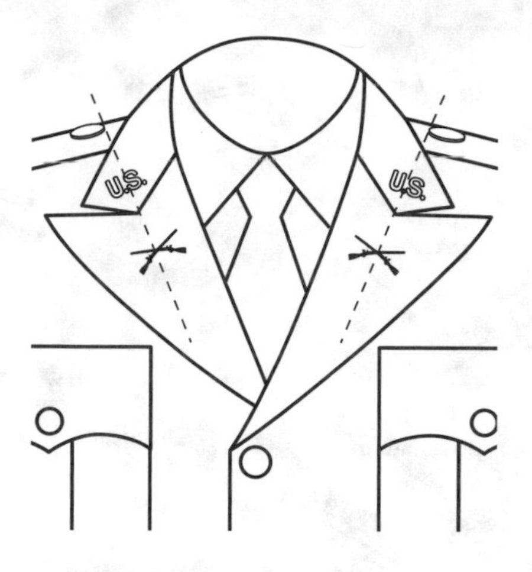

Male Officers Army Blue, White, and AGSU Uniforms

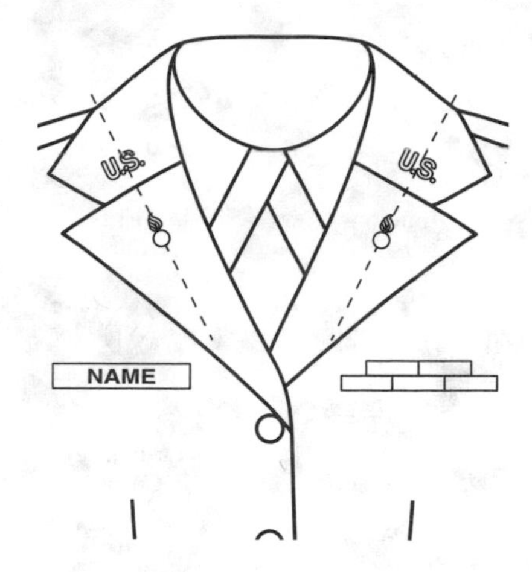

Female Officers Army Blue and White Uniforms

5/8 inch above the cut of the lapel, with the centerline of the insignia bisecting the notch and parallel to the inside edge of the lapel.

On the female ASU coat, the insignia are centered on the collars, 5/8 inch up from the collar and lapel seam, with the centerline of the insignia parallel to the inside edge of the lapel.

Branch Insignia. Male and female officers, except for most general officers, wear branch insignia on both lapels of the ASU and AGSU coat. The insignia are positioned 1 1/4 inches below the "U.S." insignia and with the centerline of the branch insignia coinciding with the centerline of the "U.S." insignia. Officers affiliated with a combat arms regiment wear regimental collar insignia instead of branch insignia. Regimental insignia is the branch insignia with numerals affixed indicating the number of the regiment.

Except for chaplains, branch insignia is not worn on class B and most class C uniforms. It is only worn on the ACU by chaplains. On the hospital duty uniform, both male and female officers wear the branch insignia on the left collar, centered between the inside edge and the outside edge, 1 inch from the lower edge of the collar and with the centerline parallel to the lower edge of the collar. Only nonsubdued, pin-on insignia of grade and branch are worn on the hospital duty uniform.

General officers may, at their option, wear branch insignia on the left collar. All other commissioned and warrant officers wear the insignia of their basic branch or the insignia of the branch to which detailed.

General Staff Corps Insignia. The General Staff Corps insignia is worn by those commissioned and warrant officers, other than general officers, whose assignments meet the following exact conditions:

Officer's Insignia of Branch

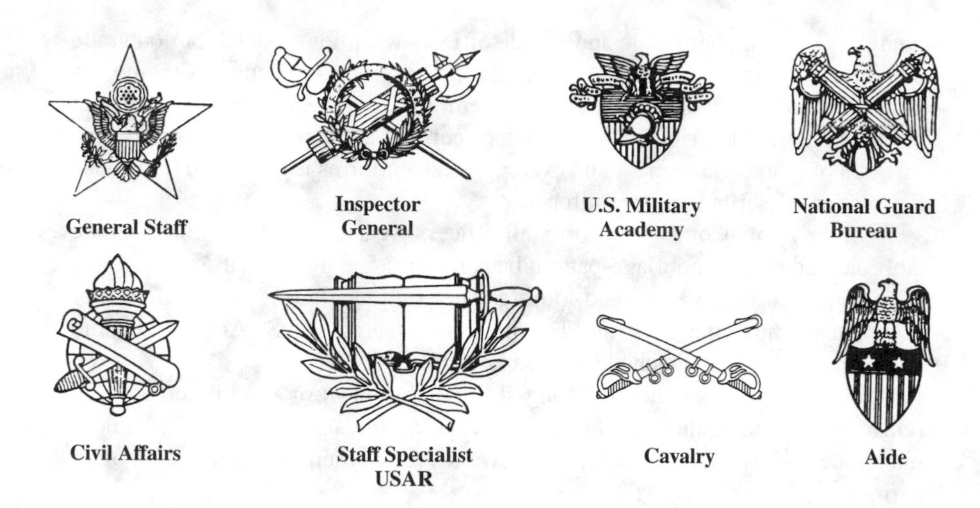

Officer's Insignia—Other than Basic Branch

1. Assigned to the offices of the Secretary of the Army, the Undersecretary of the Army, and the Assistant Secretary of the Army, who are authorized by the Secretary of the Army to wear this insignia during their tour of duty in these offices.
2. Detailed to duty on the Army General Staff.
3. Detailed to General Staff with troops (see AR 614-100).
4. As directed by the Chief of Staff.
5. Assigned to departmental or statutory tour Table of Distribution and Allowance (TDA) positions in the National Guard Bureau.

Inspector General Insignia. The Inspector General insignia is worn by the Inspector General and by those officers detailed as inspectors general under AR 614-100.

Judge Advocate General Corps Insignia. Officers detailed to the JAGC but not yet admitted to practice law before a federal court or the highest court of a state will wear the insignia of their basic branch. They may wear the JAGC insignia after they are admitted to practice.

Aide Insignia. Officers detailed as aides to general officers and other high government officials wear aide insignia appropriate to their position. See AR 670-1 for descriptions.

Other Insignia. Special insignia is provided for wear by officers assigned to the National Guard Bureau, to Civil Affairs (USAR), and to the Staff Specialist Reserve. See AR 670-1 for authorization for wear.

Insignia of Grade. Officers wear pin-on, nonsubdued insignia of grade on the shoulders of the black all-weather coat and the windbreaker. Insignia is centered on the shoulder loops, 5/8 inch from the outside shoulder seam. On the hospital duty uniform, both male and female officers wear nonsubdued, pin-on grade insignia, centered on the right collar, 1 inch from the lower edge of the collar.

All officers wear shoulder straps with embroidered grade insignia on the coat of the ASU, as described in the earlier sections concerning those uniforms. Insignia of grade on the mess and evening mess uniforms is discussed in the earlier sections describing those uniforms.

On the Army Combat Uniform, all officers wear subdued insignia of grade embroidered on a camouflage-pattern background. The insignia of grade may be either attached to the hook-and-loop fastener on the front of the ACU coat or sewn on in the same location. The insignia of grade, U.S. Army tape, and nametape must all be attached in the same manner.

When the AG 521 short- or long-sleeved shirt is worn as an outer garment (class B uniforms), shoulder marks are worn on the shoulder loops. Shoulder marks are also worn on the black pullover sweater when it is worn as an outer garment.

Cautions Regarding Affixing of Grade Insignia. Observe that there is an exact position for attaching each item of grade and other articles of insignia and ornamentation on the uniform. Here are some special ones which are easily overlooked.

A point of each of a general's stars points to the button of the shoulder loop and is placed point upward on the garrison cap, helmet, and sleeves of the mess uniform jackets. The beaks of each of a colonel's eagles are extended forward, never backward, and to the right or the front on headgear, as appropriate. An easy way to remember the rule is to picture an attacking eagle: beak forward and talons out. The stems of the silver leaf and the oak leaf, of lieutenant colonel and major, are on the outside or on the bottom, as affixed to shoulder loop, garrison cap, or sleeves of the mess uniform jackets.

Leader's Identification. The green leader's identification tab is worn by leaders, regardless of category (Active Army, Army National Guard, and Army Reserves). The specific leaders are as follows: commanders, deputy commanders, platoon leaders, command sergeants major, first sergeants, platoon sergeants, section leaders, squad leaders, tank commanders, team leaders, assistant SF detachment commanders, SF operational detachment "B" sergeants major, and SF operational detachment "A" senior sergeants. This identification is a green cloth loop, 1⅝ inches wide, worn on the center tab of the cold-weather parka under the insignia of grade. It ceases to be worn when an individual is reassigned from a command position or from a combat unit that had provided authority for its wear. *Caution:* Make certain of your right to wear this coveted device before you add it to your uniform.

Distinctive Unit Insignia. Subject to approval by the Institute of Heraldry, a distinctive unit insignia (DUI) is authorized for wear on the service uniforms by personnel designated in AR 670-1 as a means of promoting esprit de corps. When a DUI is authorized, it is worn by all assigned members except general officers. The DUI is authorized only in metal, or metal and enamel. For enlisted soldiers a

OFFICER INSIGNIA OF GRADE

AIR FORCE	ARMY	MARINES	NAVY
General of the Air Force	General of the Army	(None)	Fleet Admiral
General	General	General	Admiral
Lieutenant General	Lieutenant General	Lieutenant General	Vice Admiral
Major General	Major General	Major General	Rear Admiral (Upper Half)
Brigadier General	Brigadier General	Brigadier General	Rear Admiral (Lower Half)
Colonel	Colonel	Colonel	Captain
Lieutenant Colonel	Lieutenant Colonel	Lieutenant Colonel	Commander
Major	Major	Major	Lieutenant Commander

OFFICER INSIGNIA OF GRADE

AIR FORCE	ARMY	MARINES	NAVY
Captain	Captain	Captain	Lieutenant
First Lieutenant	First Lieutenant	First Lieutenant	Lieutenant Junior Grade
Second Lieutenant	Second Lieutenant	Second Lieutenant	Ensign
(None)	SILVER AND BLACK W-5 Chief Warrant Officer W-4 Chief Warrant Officer W-3 Chief Warrant Officer	SCARLET AND SILVER W-5 Chief Warrant Officer W-4 Chief Warrant Officer W-3 Chief Warrant Officer	W-4 Chief Warrant Officer W-3 Chief Warrant Officer
	SILVER AND BLACK W-2 Chief Warrant Officer W-1 Warrant Officer	SCARLET AND GOLD W-2 Chief Warrant Officer W-1 Warrant Officer	W-2 Chief Warrant Officer W-1 Warrant Officer

COAST GUARD

Coast Guard officers use the same rank insignia as Navy officers. Coast Guard enlisted rating badges are the same as the Navy's for grades E-1 through E-9, but they have silver specialty marks, eagles and stars, and gold chevrons. The badge of the Master Chief Petty Officer of the Coast Guard has a gold chevron and specialty mark, a silver eagle, and gold stars. For all ranks, the gold Coast Guard shield on the uniform sleeve replaces the Navy star.

ENLISTED INSIGNIA OF GRADE

AIR FORCE	ARMY	MARINES	NAVY
Chief Master Sergeant of the Air Force (CMSAF)	Sergeant Major of the Army (SMA)	Sergeant Major of the Marine Corps (SgtMajMC)	Master Chief Petty Officer of the Navy (MCPON)
Chief Master Sergeant (CMSgt) Command Chief Master Sergeant	Command Sergeant Major (CSM) Sergeant Major (SGM)	Sergeant Major (SgtMaj) Master Gunnery Sergeant (MGySgt)	Fleet/Command Master Chief Petty Officer Master Chief Petty Officer (MCPO)
Senior Master Sergeant (SMSgt) First Sergeant (E-8)	First Sergeant (1SG) Master Sergeant (MSG)	First Sergeant (1stSgt) Master Sergeant (MSgt)	Senior Chief Petty Officer (SCPO)
Master Sergeant (MSgt) First Sergeant (E-7)	Platoon Sergeant (PSG) or Sergeant First Class (SFC)	Gunnery Sergeant (GySgt)	Chief Petty Officer (CPO)
Technical Sergeant (TSgt)	Staff Sergeant (SSG)	Staff Sergeant (SSgt)	Petty Officer First Class (PO1)
Staff Sergeant (SSgt)	Sergeant (Sgt)	Sergeant (Sgt)	Petty Officer Second Class (PO2)
Senior Airman (SrA)	Corporal (CPL) Specialist (SPC)	Corporal (Cpl)	Petty Officer Third Class (PO3)
Airman First Class (A1C)	Private First Class (PFC)	Lance Corporal (LCpl)	Seaman (Seaman)
Airman (Amn)	Private E-2 (PV2)	Private First Class (PFC)	Seaman Apprentice (SA)
Airman Basic (AB) (no insignia)	Private E-1 (PV1) (no insignia)	Private (Pvt) (no insignia)	Seaman Recruit (SR)

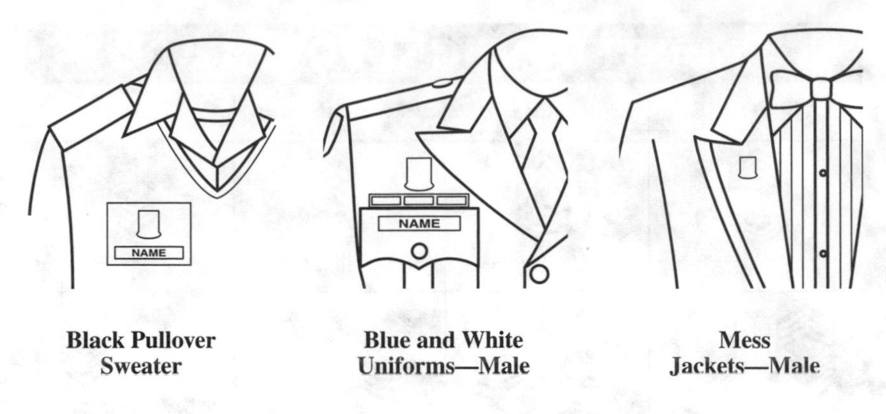

Black Pullover Sweater **Blue and White Uniforms—Male** **Mess Jackets—Male**

Wear of Regimental Crest

complete set of DUI consists of three pieces—one for each shoulder strap and one for the beret flash. Officers wear DUI centered above the nameplate on the black pullover sweater. Officers may also wear the DUI of their current unit or of a unit in which they previously served successfully in lieu of the regimental distinctive insignia.

Regimental Distinctive Insignia. Each of the Army's regiments has a regimental distinctive insignia (RDI). Personnel affiliated with the regiment are authorized to wear the RDI. Men wear the RDI centered and 1/8 inch above the right pocket flap of the ASU and the class B uniform shirt, or 1/4 inch above unit awards or foreign badges, if worn. On the blue and white mess jackets, the RDI is worn centered on the right lapel, 1/2 inch below the notch, with the vertical axis of the crest perpendicular to the ground.

Women wear the RDI centered and 1/2 inch above the nameplate or 1/4 inch above unit awards or foreign decorations, if worn, on the right side of the ASU, the class B shirt, and the maternity tunic. The RDI may be aligned to the right if it is obscured by the coat lapel.

On the blue mess jacket and the new version of the white mess jacket, the RDI is worn centered on the right lapel with the top of the crest aligned with the top row of miniature medals and the vertical axis of the crest perpendicular to the ground. On the old version of the white mess jacket, the RDI is worn centered on the right side (not lapel), aligned as above.

Airborne and Air Assault Background Trimming. Distinctive background trimming, oval-shaped, 1 3/8 inches high by 2 1/4 inches wide, is worn under the Airborne or Air Assault Badge. The trimming is subject to approval by the Institute of Heraldry when authorized by HQDA. When authorized, it is worn by all assigned personnel in the unit who have been awarded the Airborne or Air Assault Badge. Only one background trimming is worn. Background trimming is not authorized for wear by personnel who are not currently assigned to Airborne, Air Assault, or Special Operations units.

Distinctive Items for Infantry Personnel. Officers and enlisted personnel of the infantry who have been awarded the Combat Infantryman Badge or the Expert Infantryman Badge, or who have, as members of assigned infantry units, completed the basic unit phase of an Army training program or the equivalent thereof, wear the infantry shoulder cord of infantry blue. The cord is worn on the right shoulder of the ASU coats and the class B shirt. The blue cord passes under the arm, and the cord attaches to a 20-ligne button sewn on the shoulder seam 1/2 inch outside the collar edge.

Aiguillettes. Service and dress aiguillettes are provided to officers authorized to wear them (attachés and aides). Authorization for wear and manner of wear are as stated in AR 670-1.

Headgear Insignia

Men's Service Cap and Cold-Weather Cap. For male commissioned and warrant officers, the insignia is the coat of arms of the United States, 2 3/8 inches high, of gold-color metal.

Women's Service Hat and Cold-Weather Cap. For female commissioned and warrant officers, the insignia is the coat of arms of the United States, 1 5/8 inches high, of gold-color metal.

Organizational Beret. Officers wear nonsubdued insignia of grade centered on the organizational flash, which is sewn centered on the stiffener of the beret.

Patrol Cap. All officers wear subdued insignia on the patrol cap. Chaplains wear branch insignia instead of grade insignia. Insignia is worn centered on the front of the headgear.

Helmet Camouflage Cover. All officers wear subdued insignia of grade on the front, approximately 2 1/2 inches up from the bottom rim.

Baseball-Style Cap. When wear of this cap is authorized by the commander, officers wear nonsubdued insignia of grade centered on the front of the cap.

Insignia, Distinguishing, "U.S. Army." This insignia is a woven tape, 1 inch wide and either 4 1/2 inches long or the width of the pocket, olive green in color, with the inscription "U.S. Army" in black, block letters 3/4 inch high, either printed (issue) or embroidered (optional purchase). This tape is worn on the ACU coat and ACU field jacket and on organizational clothing when prescribed by the issuing commander. It is sewn on the uniform, immediately above and parallel to the top seam of the left breast pocket, or in a similar position on clothing with slanted or no pockets.

Nametapes and Nameplates. Nametapes are worn over the right breast pocket on the same uniforms and in the same manner as the U.S. Army distinguishing tape. The nametape and the U.S. Army distinguishing tape must be the same length, and each must be either printed or embroidered.

Nameplates are of black laminated plastic, either gloss or nongloss, 1 inch by 3 inches by 1/16 inch thick, with a white border not to exceed 1/32 inch in width. Lettering is block type, indented, 3/8 inch high, and centered on the nameplate. Only the last name is used.

Male personnel wear the nameplate on the flap of the right breast pocket, centered between the button and the top of the pocket, on the Army white shirts; on the ASU coat; and in a comparable position on the hospital duty uniform. The nameplate is worn centered on the patch of the black pullover sweater. If a DUI is also worn, the nameplate is worn 1/2 inch above the bottom of the patch with the DUI centered, left to right and top to bottom, above the nameplate.

Female personnel wear the nameplate between 1 and 2 inches above the top of the top button on the right side of the ASU coat, centered from side to side. It is worn in a comparable position on the white shirts, maternity tunic, and hospital duty uniform. On the black pullover sweater, the nameplate is worn as described for male personnel. The patch positioning may be adjusted to conform to individual figure differences.

Organization Shoulder-Sleeve Insignia. Approved designs of shoulder-sleeve insignia (SSI) are authorized for wear by personnel of units definitely assigned to an organization having Department of the Army authorization for its use (see AR 670-1).

Individuals entitled to the privilege may wear the SSI of their current unit on the left sleeve of the ACU coat and ACU field jacket. SSI is not worn on the black all-weather coat or windbreaker, on any class A uniform, or on organizational clothing, except for the Aircrew Combat and CVC Uniforms. The SSI is positioned 1/2 inch below the top of the shoulder seam. When the Ranger or Special Forces tab is worn, it is positioned 1/2 inch below the top of the shoulder seam, and the SSI is positioned 1/4 inch below the tab. Tabs that are an integral part of the SSI, such as Airborne or Mountain, are worn with no space between the tab and the rest of the SSI.

In the same manner, but on the right sleeve, SSI of a former wartime organization may be worn by individuals entitled to do so. AR 670-1 sets the entitlements.

Overseas Service Bars. The male version of this device is a golden-light-color rayon bar, 1 5/16 inches in length, 3/16 inch in width, on a blue background that forms a 3/32-inch border around the bar. The female version is the same except that the size is 1/8 inch wide by 7/8 inch long, with a 1/16-inch border.

One overseas bar is authorized for each period of six months of active federal service as a member of the Army or another U.S. service during periods of hostilities as designated in AR 670-1, when serving in hostile fire areas. Service of less than six months' duration that otherwise meets the requirements may be combined with additional service to determine the number of overseas bars authorized.

The overseas bar is worn centered on the outside bottom half of the right sleeve of the ASU, with the lower edge of the bar parallel to and 1/4 inch above the braid. Additional bars are worn parallel to and above the first bar with a 1/16-inch space between bars.

Colors of Branches. There are official colors of the branches of the Army. These colors appear as piping on uniform components, in facings, and elsewhere on the blue and blue mess uniforms. They are as follows:

Adjutant General Corps—dark blue and scarlet
Air Defense Artillery—scarlet
Armor—yellow
Army Medical Department—maroon and white
Aviation—ultramarine blue and golden orange
Cavalry—yellow
Chaplains—black
Chemical Corps—cobalt blue and golden yellow
Civil Affairs—purple and white
Corps of Engineers—scarlet and white
Cyber—steel gray and black
Dental Corps—maroon and white
Electronic Warfare—golden yellow and black
Field Artillery—scarlet
Finance Corps—silver gray and golden yellow
Infantry—light blue
Inspector General—dark blue and light blue
Judge Advocate General Corps—dark blue and white
Logistics Corps—soldier red
Military Intelligence—oriental blue and silver gray
Military Police Corps—green and yellow
National Guard Bureau—dark blue
Ordnance Corps—crimson and yellow
Psychological Operations—bottle green and silver gray
Quartermaster Corps—buff and blue
Signal Corps—orange and white
Special Forces—jungle green
Staff Specialist, USAR—green
Transportation Corps—brick red and golden yellow
Veterinary Corps—maroon and white
Warrant Officers—brown
Branch immaterial—teal blue and white

Branch of Service Scarves. Branch of service, bib-type scarves may be worn, when issued and prescribed for wear by the local commander, with the service and utility uniforms for ceremonial occasions only. The scarf for each branch is the first color listed in the colors of branches above. Camouflage scarves are also sometimes used when authorized by a local commander.

WEAR OF THE UNIFORM BY PERSONNEL OTHER THAN MEMBERS OF THE ACTIVE ARMY

Individuals whose current military status is other than a member of the Active Army are restricted as to when they may wear the uniform. AR 670-1 presents detailed regulations under which the categories of personnel listed below may wear the uniform.

Army National Guard personnel
Retired personnel
Persons who have been awarded the Medal of Honor
Separated personnel who have served honorably during war

Occasions of Ceremony. Occasions of ceremony are essentially military in character at which the uniform is more appropriate than civilian clothing (e.g., military balls, military parades, military weddings, military funerals, memorial services, and meetings or functions of associations formed for military purposes, the membership of which is composed largely or entirely of honorably discharged veterans of the Armed Forces or of Reserve personnel). Authorization includes wearing the uniform while traveling to and from the ceremony, provided such travel can be completed on the day of the ceremony.

Army National Guard and U.S. Army Reserve Officers, Inactive Status. Officers of the National Guard and of the Army Reserve, while on inactive status, are authorized to wear the uniform during periods of military instruction. They may also wear it on occasions of ceremony, as stated above. Members of the Army National Guard may also wear the uniform in the performance of state service when so authorized by their respective state adjutant general. Consult the commanding officer for authority to wear the uniform on other occasions.

Retired Personnel. On occasions of ceremony as discussed above, retired personnel may wear, at their option, either the uniform for persons of their grade and branch at the time of retirement, or the uniform of persons on the active list. However, the two uniforms cannot be mixed. A U.S. Army Retired shoulder patch has been developed for wear by retired personnel.

Persons Who Have Been Awarded the Medal of Honor. Persons who have been awarded the Medal of Honor are authorized to wear the uniform at any time except in connection with the promotion of a political or commercial interest or when engaged in civilian employment; when participating in public speeches, interviews, picket lines, marches, rallies, or public demonstrations, except as authorized by competent authority; when attending any meeting or event that is a function of or sponsored by an extremist organization; when wearing the uniform would bring discredit on the Army; and when specifically prohibited by regulation.

23

Military Awards

The purpose of military awards is to promote mission accomplishment by recognizing valor, achievement, service, special skills or qualifications, and heroism not involving actual combat. Individual military awards include decorations, campaign and service medals, service ribbons, badges and tabs, and certificates and letters. Unit awards also exist to recognize outstanding contributions by military organizations in peacetime, combat, or other operational deployments.

The U.S. Armed Forces adopted the idea of individual awards and decorations very slowly. In the major European armies during the eighteenth, nineteenth, and even early twentieth centuries, individual decorations were often bestowed by monarchs; many carried other honorifics and titles concomitantly, and some even came with the promise of land in conquered territories. All this was alien to American culture and undoubtedly stifled the development of the sort of elaborate and extensive military awards systems that evolved elsewhere.

The first American *decoration* was the original Purple Heart, a cloth device for "singularly meritorious action" to be worn on the left breast of the uniform. Only three are known to have been awarded by the originator of the decoration, Gen. George Washington, although rumors exist about there being several more. (Washington's initiation of this first U.S. Army decoration is commemorated by the inclusion of his likeness on the modern Purple Heart Medal.) During the Mexican War (1846–48), the Army established the "Certificate of Merit," which brought with it a $2 per month bonus as long as the awardee remained on active duty. No ribbon or medal was associated with this certificate, however.

The first *medal* authorized for award to American soldiers was the Medal of Honor, created by act of Congress in July 1862. Beginning in 1904, efforts were made to raise the standards for the Medal of Honor through the creation of campaign medals and ribbons, all of which were retroactively awarded to veterans of wars as far back as the Civil War. By the time hundreds of thousands of American soldiers were destined for combat in Europe in 1918, the Distinguished Service Medal and Distinguished Service Cross were authorized for award to soldiers who committed acts or performed service of note, but less than the extraordinary gallantry fitting for recognition by the Medal of Honor. The variety of decoration increased between the world wars, with the creation of the Silver Star Medal (replacing the citation star affixed to the Victory Medal) and the Distinguished Flying Cross and the reinstitution of the Purple Heart in 1932 in its modern role as recognition for wounds sustained in combat (replacing the wound chevrons formerly worn on the lower right sleeve of the class A uniform). The variety of awards grew even further during the Second World War as well as the expansion of the Purple Heart's awarding criteria, now including posthumous bestowment to a soldier's next of kin.

The burgeoning range of special skills concomitant to war in the Industrial Age resulted in unprecedented differentiation of military occupational specialties. During World War II, those placing soldiers in particular danger (such as flight or parachute duty) were especially recognized, and a wide variety of badges was created. Also, by the end of this war, the basics of the current system of theater ribbons with campaign stars and unit citations was in place.

The post–World War II era witnessed a new phase for the Army awards program. Increasingly, deeds not involving heroism or the performance of duties of "great responsibility" were eligible for recognition with the award of medals. In December 1945, the Army Commendation Ribbon (which later became the Army Commendation Medal) was authorized for recognition of heroism, meritorious achievement, or meritorious service in war or peace. The "ARCOM" was joined in ensuing years by several more medals that could be used for recognition of noncombat achievements or service, as well as an entire hierarchy of "Defense" or joint awards, intended to recognize outstanding performance of duty while assigned to a joint billet.

During the half century following World War II, the number of badges and service medals or ribbons also proliferated to the point that individual accomplishments and service are openly celebrated today far more than in previous years. Thus, it is not really fair to compare the medals earned by an officer who served from 1906 to 1926 with those of an officer serving from 1981 to 2001. The philosophies and systems were too different to allow meaningful comparison, although it is interesting and meaningful to consider what the bases of the different philosophies were, and the relative effectiveness of each.

This chapter is based on official publications and is intended to be sufficiently comprehensive for the personal needs of the officer. For the administration of the program, or for official action regarding the program, reference should be made to the official publications, with their changes. The basic publication for the awards program is AR 600-8-22. Wearing of decorations, medals, and badges is discussed in AR 670-1. Other documents of reference are cited as they apply in the discussions in this chapter. The Army G-1's website offers a collection of pertinent references (https://www.armyg1.army.mil/hR/uniform/), as does the Institute of Heraldry (https://tioh.army.mil/default.aspx).

The broad categories of awards include the following:

Individual Awards

- Decorations for valor or achievement
- Good Conduct Medal (for enlisted personnel only)

Service medals
Combat and special skill badges and tabs
Foreign individual awards
Certificates and letters
Unit Awards
Unit decorations
Infantry and medical streamers
Campaign streamers, war service streamers, and campaign silver bands
Foreign unit decorations

Important Definitions. Definitions of words or terms that are in common use in recommending individuals for awards, or in making awards, must be understood. The following are important.

Above and Beyond the Call of Duty. Exercise of a voluntary course of action the omission of which would not justly subject the individual to censure for failure in the performance of duty. It usually includes the acceptance of existing danger or extraordinary responsibilities with praiseworthy fortitude and exemplary courage. In its highest degree, it involves the voluntary acceptance of additional danger and risk of life. (This definition is the most important.)

Active Federal Military Service. All periods of active duty, excluding periods of active duty for training. Service as a cadet at the USMA is active duty. For the award of the Armed Forces Reserve Medal, active duty for training counts in determining eligibility.

Citation. A written, narrative statement of an act, deed, or meritorious performance of duty or service for which an award is made.

Combat Heroism. An act or acts of heroism by an individual engaged in actual conflict with an armed enemy or in military operations that involve exposure to personal hazards due to direct enemy action or the imminence of such action.

Combat Zone. The region where fighting is going on; the forward area of the theater of operations where combat troops are actively engaged. It extends from the front line to the front of the communications zone.

Distinguished Himself (Herself) By. A person who has distinguished himself or herself must, by praiseworthy accomplishment, be set apart from other persons in the same or similar circumstances. Determination of this distinction requires careful consideration of exactly what is or was expected as the ordinary, routine, or customary behavior and accomplishment for individuals of like rank and experience in the circumstances involved.

Duty of Great Responsibility. Duty that, by virtue of the position held, carries the ultimate responsibility for the successful operation of a major command, activity, agency, installation, or project. The discharge of such duty must involve the acceptance and fulfillment of the obligation so as to greatly benefit the interests of the United States.

Duty of Responsibility. Duty that, by virtue of the position held, carries a high degree of the responsibility for the successful operation of a major command, activity, agency, installation, or project, or that requires the exercise of judgment and decision affecting plans, policies, operations, or the lives and well-being of others.

Extraordinary Heroism. An act or acts of gallantry in combat involving the risk of life. The minimum level of valorous performance is that consistent with a recommendation for award of the Distinguished Service Cross.

Gallantry in Action. Spirited and conspicuous acts of heroism and courage in combat. The minimum level of valorous performance is that consistent with a recommendation for award of the Silver Star.

Heroism. Extreme courage in combat demonstrated in attaining a noble end. Varying levels of documented heroic actions are necessary to substantiate recommendations for award of the Bronze Star with V device, the Air Medal with V device, or the Army Commendation Medal with V device.

Impact Awards. Unofficial term for decorations presented in impromptu circumstances by a commander authorized to approve them, usually immediately upon the commander's realization of the heroism or meritorious achievement being recognized (thus the "impact" of awarding the decoration soon after the deed). Officers must ensure that all necessary documentation is completed and forwarded as appropriate as soon as possible after the award presentation.

In Connection with Military Operations Against an Armed Enemy. This phrase covers all military operations, including combat, support, and supply, that have a direct bearing on the outcome of an engagement or engagements against armed opposition. To perform duty or to accomplish an act of achievement in connection with military operations against an armed enemy, the individual must have been subjected to either personal hazard as a result of direct enemy action or the imminence of such action, or must have had the conditions under which his or her duty or accomplishment took place complicated by enemy action or the imminence of enemy action.

Key Individual. A person who is occupying a position that is indispensable to an organization, activity, or project.

Meritorious Achievement. An act that is well above the expected performance of duty. The act should be an exceptional accomplishment with a definite beginning and ending date. The length of time involved is not a primary consideration, but speed of accomplishment may be a factor in determining the value of an act.

Meritorious Service. Service that is distinguished by a succession of outstanding acts of achievement over a sustained period of time.

Officer. Except where stated otherwise, the word *officer* means commissioned or warrant officer. And *he, his, him* includes *she, hers, her* as appropriate.

Valor. Heroism performed under combat conditions.

Army Personal Decorations. A decoration is awarded in recognition of performance of duty involving heroism, or high achievement. There are degrees of heroism and of achievement, and for that reason there are awards to recognize these varying standards. All soldiers are expected to do their duty, and to accept the normal hazards of duty, *for which there is no special individual award.* Consider again the definition of "above and beyond the call of duty." This definition is the key to understanding. In the field of combat heroism, the Medal of Honor is our highest award; other awards for heroism in descending scale are the Distinguished Service Cross, Silver Star, Distinguished Flying Cross, Bronze Star Medal, Air Medal, Joint Service Commendation Medal, and Army Commendation Medal. In the area of achievement, the Defense Distinguished Service Medal is the highest award. Those that follow are the Distinguished Service Medal, Defense Superior Service Medal, Legion of Merit, Distinguished Flying Cross, Bronze Star Medal, Defense Meritorious Service Medal, Meritorious Service Medal, Joint Service Achievement Medal, and Army Achievement Medal. Some awards may be awarded both for heroism and for achievement. The Purple Heart, the oldest decoration in our service, is awarded only for wounds or death resulting from wounds.

The accompanying table, based on AR 600-8-22, lists the awards to Army individuals in the order of precedence, and the order in which they are worn on the uniform.

Approval Authority for Decorations. The regulations are specific and detailed as to the approval authority for award of the various decorations. The criteria vary depending on whether the award is made during peacetime or wartime. Consult the regulation, AR 600-8-22, for details.

UNITED STATES ARMY AND DEPARTMENT OF DEFENSE DECORATIONS

Medal of Honor. The Medal of Honor, established by act of Congress in 1862, is the highest and most rarely awarded decoration conferred by the United States. The deed for which the Medal of Honor is awarded must have been one of personal bravery or self-sacrifice so conspicuous as to clearly distinguish the individual for gallantry and intrepidity above his or her comrades and must have involved risk of life. Incontestable proof of the performance of the service is exacted, and each recommendation for the award of this decoration is considered on the standard of extraordinary merit.

Presentation of the Medal of Honor is made only by the President.

No special personal privileges or exemptions from military obligations accompany the award. Medal of Honor winners may receive free air transportation (AMC) on a space-available basis, however. Sons and daughters of Medal of Honor recipients, otherwise qualified for admission to the United States Military Academy, are not subject to quota requirements.

Army personnel holding the Medal of Honor may apply to Commander, HRC (TAPC-PDA), Alexandria, Virginia, 22332-0471, to have their names entered on the Medal of Honor Roll. Persons on the roll and otherwise eligible may, upon application, qualify for a special lifetime pension of $400 per month. *Note:* Although awarded by the President in the name of Congress, this decoration is *never* properly called the "Congressional Medal of Honor," but rather the "Medal of Honor."

Distinguished Service Cross. Established by act of Congress on 9 July 1918 and amended by act of 25 July 1963, this medal is awarded to a person who,

United States Military Decorations

Decorations (in order of precedence)	Awarded for		Awarded to			
			United States Personnel		Foreign Personnel	
	Heroism	Achievement or Service	Military	Civilian	Military	Civilian
Medal of Honor	Combat		War[1]			
Distinguished Service Cross	Combat		War	War[2]	War[1]	War[2]
Defense Distinguished Service Cross		War Peace	War Peace			
Distinguished Service Medal		War Peace	War Peace	War[2]	War Peace	War[2]
Silver Star	Combat		War	War[2]	War	War[2]
Defense Superior Service Medal		War Peace	War Peace			
Legion of Merit		War Peace	War Peace		War[4] Peace[4]	
Distinguished Flying Cross	Combat Noncombat	War Peace[7]	War Peace		War	
Soldier's Medal	Noncombat		War Peace[7]		War Peace[7]	
Bronze Star Medal	Combat[3]	War Peace	War Peace	War Peace	War Peace	War Peace[2]
Purple Heart	Wounds Received in Combat		War Peace[6,7]	War[6,7] Peace[6,7]		
Defense Meritorious Service Medal		Peace	Peace			
Meritorious Service Medal		Peace	Peace		Peace	
Air Medal	Combat[3] Noncombat	War Peace[7]	War Peace[7]	War	War	War
Joint Service Commendation Medal	Combat[3] Noncombat	War Peace	War Peace			
Army Commendation Medal	Combat[3] Noncombat	War Peace	War[5] Peace[5]	War	War[5] Peace[5]	
Joint Service Achievement Medal		Peace	Peace[5]			
Army Achievement Medal		Peace	Peace[5]		Peace[5]	

Notes:

1. The Medal of Honor is awarded only to United States military personnel
2. Only rarely awarded to these personnel
3. Awarded with bronze V device for valor in combat
4. Awarded to foreign military personnel in one of four degrees
5. Not awarded to general officers
6. Awarded to military and civilian personnel wounded by terrorists or while members of a peacekeeping force
7. Approval authority for peacetime award is Headquarters, PERSCOM

while serving in any capacity with the Army, distinguishes himself or herself by extraordinary heroism not justifying the Medal of Honor while engaged in an action against an enemy of the United States, while engaged in military operations involving conflict with opposing foreign forces, or while serving with friendly foreign forces engaged in an armed conflict against an opposing armed force in which the United States is not a belligerent party.

Defense Distinguished Service Medal. Established by Executive Order 11545 of 9 July 1970, this medal is awarded by the Secretary of Defense to any officer of the armed forces of the United States whose exceptional performance of duty and contributions to national security or defense have been at the highest level. It ranks between the Distinguished Service Cross and the Distinguished Service Medal in order of precedence. It will not be awarded to any individual for a period of service for which a Distinguished Service Medal or similar decoration is awarded.

Distinguished Service Medal. Established by Congress on 9 July 1918, this medal is awarded to any person who, while serving in any capacity with the Army, distinguishes himself or herself by exceptionally meritorious service to the government in a duty of great responsibility. For service not related to actual war, the term *duty of great responsibility* applies to a narrower range of positions than in wartime and requires evidence of conspicuously significant achievement. Awards may be made to persons other than members of the armed forces of the United States for wartime service only, under exceptional circumstances, and with the approval of the President.

Silver Star. Established by act of Congress 9 July 1918 and amended by act of 25 July 1963, the Silver Star is awarded to a person under the same circumstances as described above for the Distinguished Service Cross but when the gallantry is of a lesser degree but performed with marked distinction.

Defense Superior Service Medal. Established by Executive Order 11904 of 6 February 1976, this medal is awarded by the Secretary of Defense to any member of the armed forces of the United States who, after 6 February 1976, renders superior meritorious service while in a position of significant responsibility. It ranks between the Silver Star and the Legion of Merit in order of precedence. It will not be awarded to any individual for a period of service for which a Legion of Merit or similar decoration is awarded.

Legion of Merit. Established by Congress on 20 July 1942, this medal is awarded to any member (usually key individuals) of the armed forces of the United States or of a friendly foreign nation who has distinguished himself or herself by exceptionally meritorious conduct in the performance of outstanding services and achievement. For service not related to war, the term *key individuals* applies to a narrower range than in wartime. Awards are made to U.S. nationals without reference to degree, and, for each award, the Legion of Merit (Legionnaire) is issued. The award may be made to foreigners, under conditions prescribed in AR 627-7, in one of four degrees—Chief Commander, Commander, Officer, or Legionnaire.

Distinguished Flying Cross. Established by Congress on 2 July 1926, this medal is awarded to any person who, while serving in any capacity with the

Army, distinguishes himself or herself by heroism or extraordinary achievement while participating in aerial flight.

Soldier's Medal. Established by Congress on 2 July 1926, this medal is awarded to any member of the armed forces of the United States or of a friendly foreign nation who, while serving in any capacity with the Army of the United States, distinguishes himself or herself by heroism not involving actual conflict with an armed enemy. The same degree of heroism is required as for a Distinguished Flying Cross. The award is not made solely on the basis of having saved a life.

Bronze Star Medal. The Bronze Star Medal was established by Executive Order 9419 in 1944, which was superseded by Executive Order 11046, 24 August 1962. It is awarded to any person who, while serving in any capacity in or with the Army of the United States after 6 December 1941, distinguishes himself or herself by heroism or meritorious achievement or service, not involving participation in aerial flight, in connection with military operations against an armed enemy or while engaged in military operations involving conflict with an opposing armed force in which the United States is not a belligerent party.

Awards may be made for acts of heroism that are of lesser degree than required for award of the Silver Star.

Awards may be made for achievement or meritorious service that, while of lesser degree than that required for award of the Legion of Merit, must nevertheless have been meritorious and accomplished with distinction.

Purple Heart. The Purple Heart, established by Gen. George Washington at Newburgh, New York, on 7 August 1782 and revived by the President as announced in War Department General Orders 3, 22 February 1932, as amended by Executive Order 11016, 25 April 1962, as further amended by Executive Order 12464, 23 February 1984, and Public Law 98-525, 19 October 1984, is awarded in the name of the President of the United States to any member of an armed force or civilian national of the United States who, while serving under competent authority in any capacity with one of the United States armed services after 5 April 1917, has been wounded or killed or has died or may die after being wounded in the following circumstances:

1. In any action against an enemy of the United States.
2. In any action with an opposing armed force of a foreign country in which the armed forces of the United States are or have been engaged.
3. While serving with friendly foreign forces engaged in an armed conflict against an opposing armed force in which the United States is not a belligerent party.
4. As the result of an act of any such enemy or opposing armed force.
5. As the result of an act of any hostile foreign force.
6. After 28 March 1973, as a result of an international terrorist attack against the United States or a foreign nation friendly to the United States.
7. After 28 March 1973, as a result of military operations while serving outside the territory of the United States as part of a peacekeeping force.

A Purple Heart is authorized for the first wound suffered under conditions indicated above, but for each subsequent award an oak-leaf cluster shall be awarded to be worn on the medal or ribbon.

A Purple Heart will be issued to the next of kin of each person entitled to a posthumous award. Issue will be made automatically by the commanding general, PERSCOM upon receiving a report of death indicating entitlement.

Defense Meritorious Service Medal. Established by Executive Order 12019, 3 November 1977, it is awarded in the name of the Secretary of Defense to any member of the armed forces of the United States who distinguishes himself or herself by noncombat meritorious achievement or service. It ranks between the Bronze Star Medal and the Meritorious Service Medal. It will not be awarded to any individual for a period of service for which any similar decoration has been awarded.

Meritorious Service Medal. Established by Executive Order 11448, 16 January 1969, as amended by Executive Order 12312, 2 July 1981, the Meritorious Service Medal is awarded to a member of the armed forces of the United States or to any member of the armed forces of a friendly foreign nation who, after 16 January 1969, distinguishes himself or herself by outstanding meritorious achievement or service in a noncombat situation. It ranks between the Defense Meritorious Service Medal and the Joint Service Commendation Medal as a noncombat award.

Air Medal. Established by Executive Order 9158 on 11 May 1942, and amended by Executive Order 9242-A, 11 September 1942, this medal may be awarded to any person who, while serving in any capacity in or with the Army, distinguishes himself or herself by meritorious achievement while participating in aerial flight. The medal may be awarded for heroism in combat, for single acts of meritorious service involving superior airmanship, and for meritorious service involving sustained distinction in the performance of duties that require regular and frequent participation in aerial flight for a period of at least six months in combat.

Joint Service Commendation Medal. Authorized by the Secretary of Defense on 25 June 1963, and implemented by DoD 1348.33-M, this decoration is awarded in the name of the Secretary of Defense and takes precedence with, but before the Army Commendation Medal. It is awarded to any member of the armed forces who distinguishes himself or herself by meritorious achievement or service. Awards may include the V device if the citation is approved for valor in a designated combat area.

Army Commendation Medal. The Army Commendation Medal, established by War Department Circular 377, 18 December 1945, and amended in Department of the Army General Orders 10, 31 March 1960, is awarded to any member of the armed forces of the United States who, while serving in any capacity with the Army after 6 December 1941, distinguishes himself or herself by heroism, meritorious achievement, or meritorious service. Award may also be made to a member of the armed forces of a friendly foreign nation for an act of heroism, extraordinary achievement, or meritorious service that has been of mutual benefit to a friendly foreign nation and the United States.

Awards may be made for acts of valor performed under circumstances described above that are of lesser degree than required for award of the Bronze Star Medal. These acts may involve aerial flight.

An award may be made for acts of noncombatant-related heroism that do not meet the requirements for an award of the Soldier's Medal.

The Army Commendation Medal is not awarded to general officers.

Joint Service Achievement Medal. Authorized by the Secretary of Defense on 3 August 1983 and implemented by DoD 1348.33-M, this medal may be awarded to any member of the U.S. Armed Forces below the grade of colonel (O-6) who distinguishes himself or herself by outstanding performance of duty and meritorious achievement or service after 3 August 1983.

Army Achievement Medal. Established by the Secretary of the Army on 10 April 1981, this medal is awarded to any member of the armed forces of the United States or to any member of the armed forces of a friendly foreign nation who, while serving in any capacity with the Army in a noncombat area on or after 1 August 1981, distinguishes himself or herself by meritorious service or achievement of a lesser degree than required for award of the Army Commendation Medal. This medal may not be awarded to general officers.

U.S. ARMY AND DEPARTMENT OF DEFENSE UNIT AWARDS

Unit awards are authorized in recognition of group heroism or meritorious service, usually during a war, as a means of promoting esprit de corps. They are of the following categories: unit decorations, infantry and medical streamers, campaign streamers, war service streamers, and campaign silver bands. Only personnel who were assigned to the unit during the period for which the unit award is made are entitled to wear an emblem signifying receipt of the decoration as a permanent part of the uniform. Streamers and silver bands accrue to the unit only and are displayed on the guidon or color.

United States Unit Decorations. United States unit decorations, in order of precedence shown below, have been established to recognize outstanding heroism or exceptionally meritorious conduct in the performance of outstanding services:

Presidential Unit Citation (Army and Air Force)
Presidential Unit Citation (Navy)
Joint Meritorious Unit Award
Valorous Unit Award
Meritorious Unit Commendation (Army)
Navy Unit Commendation
Meritorious Unit Commendation (Navy)
Air Force Outstanding Unit Award
Air Force Organizational Excellence Award
Army Superior Unit Award

These awards may be worn permanently by those who served with the unit during the cited period. The Presidential Unit Citation (Army), the Valorous Unit Award, the Meritorious Unit Commendation, and the Army Superior Unit Award may be worn temporarily by those serving with the unit subsequent to the cited period. The Presidential Unit Citation (Air Force) is awarded under the same criteria as established for the Presidential Unit Citation (Army). The Air Force award was derived from the Army award, and the two awards are equal in precedence.

Presidential Unit Citation. The Presidential Unit Citation (until 1966, officially called the "Distinguished Unit Citation") is awarded to units of the armed forces of the United States and cobelligerent nations for extraordinary heroism in action against an armed enemy occurring on or after 7 December 1941. The unit must display such gallantry, determination, and esprit de corps in accomplishing its mission under extremely difficult and hazardous conditions as to set it apart from and above other units participating in the same campaign. The degree of heroism required is the same as that which would warrant award of a Distinguished Service Cross to an individual. Extended periods of combat duty or participation in a large number of operational missions, either ground or air, is not sufficient. Only on rare occasions will a unit larger than a battalion qualify for award of the decoration.

The Presidential Unit Emblem (Army) is a blue ribbon set in a gold-colored metal frame of laurel leaves. It is authorized for purchase and wear as a permanent part of the uniform by those individuals who served with the unit during the cited period.

Joint Meritorious Unit Award. Awarded in the name of the Secretary of Defense to joint activities of the DoD for meritorious achievement or service, superior to that normally expected, during combat with an armed enemy of the United States, during a declared national emergency, or under extraordinary circumstances that involved the national interest.

Valorous Unit Award. Criteria are the same as those for the Presidential Unit Citation except that the degree of valor required is that which would merit award of the Silver Star to an individual. The initial eligibility date is 3 August 1963. The emblem is a scarlet ribbon with the Silver Star color design superimposed in the center, set in a gold-colored metal frame with laurel leaves.

Meritorious Unit Commendation. Awarded to service and support units for at least six months of exceptionally meritorious conduct in performance of outstanding services during periods of military operations against an armed enemy after 1 January 1944. Service is not required to be in the combat zone, but it must be directly related to the combat effort. CONUS-based units and units outside the area of operation are excluded from this award. The degree of achievement is that which would merit the award of the Legion of Merit to an individual. The emblem is a scarlet ribbon set in a gold-colored metal frame with laurel leaves.

Army Superior Unit Award. Awarded for outstanding meritorious performance of a difficult and challenging mission under extraordinary circumstances by a unit during peacetime. Peacetime is defined as any period during which wartime or combat awards are not authorized in the geographical area in which the mission was executed. The emblem is a scarlet ribbon with a vertical green stripe in the center on each side of which is a narrow yellow stripe, set in a gold-colored metal frame with laurel leaves.

Infantry Streamers. Infantry streamers are awarded to U.S. infantry units that have participated in combat or that have been designated as expert infantry units. When 65 percent or more of the TOE strength of an infantry unit, brigade or smaller, has been awarded the combat infantryman badge, the unit is awarded the *combat infantry streamer.* It consists of a white streamer with the words "Combat Infantry (Brigade) (Battalion) (Company)" embroidered in blue. Effective 20 December 1989, Special Forces units meeting the above criteria are awarded the combat infantry streamer. The *expert infantry streamer* is awarded when 65 percent of the TOE strength has been awarded either the combat infantryman badge or the expert infantryman badge. It consists of a white streamer with the words "Expert Infantry (Brigade) (Battalion) (Company)" embroidered thereon.

Medical Streamers. The *combat medical streamer* is awarded when 65 percent of the TOE strength of a medical unit authorized a color, distinguishing flag, or guidon has been awarded the combat medical badge. It is a maroon streamer with a 1/16-inch white stripe on each edge and the words "Combat Medical Unit" embroidered in white. The *expert medical streamer* is awarded when 65 percent or more of the assigned strength of a medical unit authorized a color, distinguishing flag, or guidon has been awarded the combat medical badge or the expert field medical badge.

Campaign Streamers and War Service Streamers. These streamers are awarded to organizations that have been authorized an organizational color or standard. Campaign streamers are awarded for active federal military service to recognize award of campaign participation credit. War service streamers are awarded to recognize active federal military service in a theater of operations under circumstances when a campaign streamer is not authorized for the same service. Appendix B to AR 600-8-22 lists the campaigns, service requirements, and inscriptions prescribed for streamers.

Campaign Silver Bands. Campaign silver bands are awarded to units authorized a guidon to recognize combat participation credits. They are awarded only if the unit is not an organized element of a separate battalion, brigade, regiment, or larger unit that is authorized a streamer for the same campaign.

OTHER U.S. ARMY AWARDS

Prisoner of War Medal. This award is authorized for all U.S. military personnel who were taken prisoner of war after 5 April 1917 during an armed conflict and who served honorably during the period of captivity.

Good Conduct Medal (Army). This is a medal awarded only to enlisted personnel in recognition of exemplary behavior, efficiency, and fidelity under prescribed conditions as to time and ratings. A distinctive clasp is awarded for each successive period of three years' service that meets the requirements.

Army Reserve Components Achievement Medal. Established by the Secretary of the Army on 3 March 1971 this medal may be awarded upon recommendation of the unit commander each four years since 3 March 1972, for exemplary behavior, efficiency, and fidelity. Service must have been consecutive, in the grade of colonel or below, and in accordance with the standards of conduct, courage, and duty required by law and customs of the service for an active-duty member of the same grade. After 28 March 1985, the period of service was reduced to three years. Second and succeeding awards are recognized by an oak-leaf cluster. The reverse of this medal is struck in two designs for award to personnel whose service has been in the Army Reserve or in the National Guard.

UNITED STATES MILITARY SERVICE MEDALS AND RIBBONS

Service (campaign) medals denote the honorable performance of military duty within specified limiting dates in specified geographical areas. They may also denote military duty anywhere within specified time periods.

They are worn as described later in this chapter.

U.S. service medals are those awarded by the Army, Navy, and Air Force. Other service medals of the federal government or awarded by a state or other inferior jurisdiction are civilian service medals and are not worn on the uniform.

Service Medals Authorized Prior to Vietnam War. Service medals, appurtenances, and devices that pertain only to wars or campaigns prior to the Vietnam War are not included, since there are no officers still on active duty who might wear such items. An exception is the Army of Occupation Medal, which was issued to soldiers serving in the occupying forces in Germany, Austria, Italy, Japan, and Korea during specified periods following World War II. This service medal was awarded to individuals serving in Army units stationed in Berlin until 2 October 1990. See AR 600-8-22 for details on these awards.

National Defense Service Medal. This medal has been authorized for honorable active service for any period between 27 June 1950 and 27 July 1954, between 1 January 1961 and 14 August 1974, from 2 August 1990 to 30 November 1995, and most recently from 11 September 2001 to 31 December 2022. A second award of this medal is designated by a service star. Persons on active duty for purposes other than extended active duty are not eligible for this award.

Antarctica Service Medal. The Antarctica Service Medal is awarded to persons serving with U.S. expeditions to the Antarctic from 1 January 1946 to a date to be determined by the Secretary of Defense.

Armed Forces Expeditionary Medal. This medal, established by executive order on 4 December 1961, is authorized for specified U.S. military operations, U.S. operations in direct support of the United Nations, and U.S. operations of assistance for friendly foreign nations. The second and subsequent awards are denoted by service stars. Generally, thirty days or the full period of the operation are required.

The following U.S. military operations, areas, and dates have been designated for this award:

Quemoy and Matsu Islands—23 August 1956 to 1 June 1963
Lebanon—1 July 1958 to 1 November 1958
Taiwan Straits—23 August 1958 to 1 January 1959
Berlin—14 August 1961 to 1 June 1963
Cuba—24 October 1962 to 1 June 1963
Congo—23 to 27 November 1964
Dominican Republic—28 April 1965 to 21 September 1966

Korea—1 October 1966 to 30 June 1974
Cambodia (Operation Eagle Pull)—11 to 13 April 1975
Vietnam (Operation Frequent Wind)—29 to 30 April 1975
Mayaguez —15 May 1975
Grenada (Operation Urgent Fury)—23 October 1983 to 21 November 1983
Libya (Operation El Dorado Canyon)—12 April 1986 to 17 April 1986
Panama (Operation Just Cause)—20 December 1989 to 31 January 1990
Haiti (Operation Uphold Democracy)—16 September 1994 to 31 March 1995

The following designated U.S. operations in direct support of United Nations have been also designated for this award:

Congo—14 July 1960 to 1 September 1962
Somalia (Operations Restore Hope and United Shield)—5 December 1992 to 31 March 1995
Former Republic of Yugoslavia (Operations Joint Endeavor and Joint Guard)—1 June 1992 to 20 June 1998, only for participants deployed in Bosnia-Herzegovina and Croatia
Former Republic of Yugoslavia (Operation Joint Forge)—21 June 1998 to a date to be determined

The following designated U.S. operations of assistance to a friendly foreign nation have been designated for this award:

Vietnam—1 July 1958 to 3 July 1965
Laos—19 April 1961 to 7 October 1962
Cambodia—29 March 1973 to 15 August 1973
Thailand (only those in direct support of Cambodia operations)—29 March 1973 to 15 August 1973
El Salvador—1 January 1981 to 1 February 1992
Lebanon—1 June 1983 to 1 December 1987
Persian Gulf (Operation Earnest Will)—24 July 1987 to 1 August 1990
Southwest Asia (Operation Southern Watch)—1 December 1995 to a date to be determined
Southwest Asia (Maritime Intercept Operation)—1 December 1995 to a date to be determined
Southwest Asia (Vigilant Sentinel)—1 December 1995 to 15 February 1997
Southwest Asia (Operation Northern Watch)—1 January 1997 to a date to be determined
Southwest Asia (Operation Desert Thunder) —11 November 1998 to 22 December 1998
Southwest Asia (Operation Desert Fox)—16 to 22 December 1998
Southwest Asia (Operation Desert Spring)—31 December 1998 to 18 March 2003

Vietnam Service Medal. The Vietnam Service Medal was established by Executive Order 11231 on 8 July 1965. It was awarded to servicemembers serving in Vietnam or contiguous waters or air space from 3 July 1965 through 28 March 1973.

Individuals who served in Vietnam between 1 July 1958 and 3 July 1965 and are qualified to wear the Armed Forces Expeditionary Medal, discussed above, are qualified to receive the newer award in lieu of the Expeditionary Medal. No person may be awarded both medals for Vietnam service, however.

Southwest Asia Service Medal. Established by Executive Order 12754, 12 March 1991, the Southwest Asia Service Medal is authorized for award to individuals who served in southeast Asia on or after 2 August 1990 to 30 November 1995.

Kosovo Campaign Medal. The Kosovo Campaign Medal was established by Executive Order 13154, 3 May 2000, to recognize the accomplishments of military servicemembers who participated in or were in direct support of the Kosovo operations within the Kosovo Air Campaign (24 March 1999 to 10 June 1999) or the Kosovo Defense Campaign (11 June 1999 to 31 December 2013.

Eligible personnel must be *bona fide* members of a unit participating or engaged in direct support of the operation for 30 consecutive days in the AOE or for 60 nonconsecutive days as long as this support involves entering the operation's AOE or meets one or more of the criteria stated in AR 600-8-22.

Afghanistan Campaign Medal. On 29 November 2004, President Bush signed an executive order establishing the Afghanistan Campaign Medal with suitable appurtenances for soldiers who served in direct support of Operation Enduring Freedom on or after 24 October 2001 until 31 August 2021. The area of eligibility includes all the land area of Afghanistan and all the air space above the land. Soldiers must serve 30 consecutive or 60 nonconsecutive days in either area, or be engaged in combat during their service; be wounded or injured requiring evacuation from the area.

The medal replaces the former Global War on Terrorism Expeditionary Medal (GWOTEM). Soldiers who already have the GWOTEM, for service in Afghanistan, may apply for the new medal in lieu of the earlier one. Once a soldier elects to receive a campaign medal in lieu of the GWOTEM, then the GWOTEM will be removed from the soldier's records and he or she will only be authorized to wear the new campaign medal. However, it is possible for a soldier to wear both the GWOTEM and one of the campaign medals simultaneously. For this to occur, the soldier must have been deployed to theater on two different occasions to earn the two awards independently.

The medal may be awarded posthumously to any person covered by and under regulations prescribed in accordance with this order.

Iraq Campaign Medal. On 29 November 2004, President Bush signed an executive order establishing the Iraq Campaign Medal with suitable appurtenances for soldiers who served in direct support of Operation Iraqi Freedom on or after 19 March 2003, to 31 December 2011. The area of eligibility encompasses all land area of the country of Iraq, the contiguous water area out to 12 nautical miles and all air spaces above them. Soldiers must serve thirty consecutive or sixty nonconsecutive days in either area; or be engaged in combat during their service; or be wounded or injured requiring evacuation from the area.

The medal replaces the former Global War on Terrorism Expeditionary Medal. Medal exchange rules are the same as for the Afghanistan Campaign Medal. It may also be awarded posthumously.

Inherent Resolve Campaign Medal. Established on 30 March 2016 by executive order, this medal is awarded to soldiers serving in operations within the approved AOE on or after 15 June 2014 to a date to be determined. Service in the area of eligibility must be for thirty consecutive days or sixty nonconsecutive days. Soldiers engaged in combat, killed, or evacuated due to wounds immediately qualify for the award. Approved campaign phases are (1) Abeyance: 15 June 2014 to 24 November 2015; (2) Intensification: 25 November 2015 to 14 April 2017; (3) Defeat: 15 April 2017 to 1 July 2020; and (4) Normalize: 2 July 2020 to a date to be determined.

Global War on Terrorism Expeditionary Medal (GWOTEM). Established by executive order on 12 March 2003, this medal is awarded to soldiers who have deployed abroad for service in Global War on Terrorism (GWOT) operations on or after 11 September 2001, to a date to be determined. The initial award is limited to soldiers deployed abroad in Operations Enduring Freedom and Iraqi Freedom in specific geographic areas of eligibility (AOE) listed in AR 600-8-22. To be eligible for the award, a soldier must be assigned, attached, or mobilized to a unit participating in these operations in the AOE for thirty consecutive days or sixty nonconsecutive days, or meet one of the following criteria: (1) be engaged in actual combat against the enemy and under circumstances involving grave danger of death or serious bodily injury from enemy action, regardless of time in the AOE; (2) while participating in the designated operation in the AOE, who, regardless of time, is killed, wounded, or injured and required medical evacuation; or (3) soldiers participating as a regularly assigned air crew member flying sorties into, out of, within, or over the AOE in direct support of these operations, with each day that one or more sorties are flown counting as one day toward the thirty consecutive or sixty nonconsecutive day requirement. Only one award of the GWOTEM may be authorized for any individual. Battle and service stars will be addressed at a later time.

As of early 2005, a soldier can elect to receive the GWOTEM or the Afghanistan or Iraq Campaign medal, but not both, for an individual deployment.

Global War on Terrorism Service Medal (GWOTSM). This medal was authorized by executive order on 12 March 2003 for soldiers who participated in, or served in support of, the GWOT outside the AOE designated for the GWOTEM, on or after 11 September 2001, to 10 September 2022, for thirty consecutive or sixty nonconsecutive days. Initial award of the GWOTSM is limited to airport security operations from 27 September 2001 through 31 May 2002 and to soldiers who supported Operations Noble Eagle, Enduring Freedom, and Iraqi Freedom. Only one award of the GWOTSM may be authorized. Effective 11 September 2022, the GWOTSM will be awarded only to personnel serving in approved AOEs in support of Global War on Terror campaigns.

Korea Defense Service Medal (KDSM). Congress authorized this medal on 2 December 2002 for members of the armed forces who have served on active duty for thirty consecutive or sixty nonconsecutive days in support of the defense of the Republic of Korea from 28 July 1954 to a date to be determined. The AOE encompasses all land area of the Republic of Korea, the contiguous water out to 12 nautical miles, and the air spaces above the land and water areas. Servicemembers must have been assigned or attached to units operating in the AOE or have been engaged in combat during an armed encounter, regardless of the time in the AOE; be wounded or injured in the line of duty and required medical evacuation from the AOE; or participated as a regularly assigned air crewmember flying sorties into, out of, or within the AOE in direct support of military operations, with each day that one or more sorties are flown counting as one day toward the thirty- to sixty-day requirement. The nonconsecutive service period for eligibility remains cumulative throughout the entire period. The KDSM may be awarded posthumously without regard to length of service. Only one award of the KDSM is authorized for any individual.

Armed Forces Service Medal. Established by Executive Order 12985, 11 January 1996, this medal is awarded to U.S. military personnel who, subsequent to 1 June 1992, participate as members of U.S. military units in a U.S. military operation that is deemed to be a significant activity, and encounter no foreign armed opposition or imminent threat of hostile action. These activities generally include peacekeeping operations and prolonged humanitarian operations and must not have been recognized by the award of another U.S. campaign or service medal.

Humanitarian Service Medal. Established by Executive Order 11965 on 19 January 1977, the Humanitarian Service Medal is awarded by the Secretary of Defense or the Secretary of Transportation to members of the Coast Guard when that service is not operating as a military service in the Navy. It is awarded to members of the armed forces of the United States who meritoriously participate in a significant military act or operation of a humanitarian nature. Award of this medal does not preclude or conflict with other medals awarded on the basis of valor, achievement, or meritorious service. Only one Humanitarian Service Medal may be awarded for participation in a given military act or operation, however. Meritorious participation in subsequent acts or operations is recognized by award of service stars. Appendix C of AR 600-8-22 provides a listing of humanitarian service operations qualifying for award of this medal.

Military Outstanding Volunteer Service Medal. Established by Executive Order 12830 on 9 January 1993, this medal is awarded to members of the armed forces of the United States and the reserve components who, subsequent to 31 December 1992, perform outstanding volunteer community service of a sustained, direct, and consequential nature. The award is intended to recognize exceptional support to either the civilian or military communities over time and not a single act or achievement.

Armed Forces Reserve Medal. Honorable and satisfactory service is required in one or more of the reserve components of the armed forces for a period of ten years, not necessarily consecutive, provided such service was performed within a period of twelve consecutive years. Periods of service as a member of a regular component are excluded from consideration. Required also is the earning of a minimum of 50 retirement points per year (AR 135-180). Individuals are advised to consult the unit instructor of their organization as to individual eligibility.

Ten-year Device. One ten-year device is authorized to be worn on the suspension and service ribbon to denote service for each ten-year period in addition to and under the same conditions as prescribed for the award of the medal. It is a bronze hourglass with a Roman numeral X superimposed thereon, 5/16 inch in height.

U.S. ARMY SERVICE AND TRAINING RIBBONS

Army Sea Duty Ribbon. Established by the Principal Deputy Assistant Secretary of the Army (M&RA) on 17 April 2006, the Army Sea Duty Ribbon is awarded to soldiers for completion of designated periods of sea duty as defined in AR 56-9, Table 1-1. The ribbon is also authorized for duty aboard qualifying vessels in accordance with AR 600-88, para. 1-7.

Army Service Ribbon. Established by the Secretary of the Army on 10 April 1981, the Army Service Ribbon is awarded to officers upon successful completion of their basic/orientation or higher-level course, or upon completion of four months of honorable service. This ribbon is awarded only once.

Overseas Service Ribbon. Established by the Secretary of the Army on 10 April 1981, the Overseas Service Ribbon is awarded for completion of a creditable overseas tour that is not recognized by award of another service medal. Subsequent awards are designated by numerals. Effective 3 February 2004, the Overseas Service Ribbon is no longer authorized for overseas tours in the Republic of Korea. All soldiers who received the OSR for service in Korea prior to 3 February 2004, are not currently required to relinquish previous awards of their OSRs. They may add the Korea Defense Service Medal to their ribbon rack and retain any awards of the OSR.

Army Reserve Components Overseas Training Ribbon. Established by the Secretary of the Army on 11 July 1984, the ribbon is awarded to members of the reserve components of the Army (Army National Guard and U.S. Army Reserve) for successful completion of annual training or active duty for training for a period of not less than ten days on foreign soil. Numerals are used to denote second and subsequent awards.

NON-U.S. SERVICE MEDALS

United Nations Service Medal. Established by the United Nations General Assembly Resolution 483(V) on 12 December 1950 to honor soldiers participating in the Korean War. Presidential acceptance for the United States Armed Forces was announced by DoD on 27 November 1951. Period of qualifying service for this medal is from 27 June 1950 to 27 July 1954.

Inter-American Defense Board Medal. Established by the Inter-American Defense Board on 11 December 1945 and authorized by Executive Order 11446 on 18 January 1969, as amended by Executive Order on 28 February 2003. U.S. military personnel who have served on the Inter-American Defense Board for at least one year as Chairman of the Board, delegates, advisors, officers of the staff, officers of the secretariat, or officers of the Inter-American Defense College may wear this ribbon permanently.

United Nations Medal. Established by the United Nations Secretary-General, on 30 July 1959, and accepted by the President in Executive Order 11139 on 7 January 1964, the United Nations Medal is awarded for service of not less than six months with United Nations teams and forces that are in or have been in Lebanon, Palestine, India and Pakistan, Hollandia, and the former Yugoslavia.

NATO Medal. Awarded by NATO to U.S. military personnel who served under direct NATO command or in direct support of NATO operations in the former Yugoslavia for thirty days of continuous or accumulated service between 1 July 1992 and a date to be announced.

Multinational Force and Observers Medal. Established by the Director General, Multinational Force and Observers, 24 March 1982, and accepted by DoD Memorandum of 28 July 1982, this medal may be awarded to personnel who have served with the international peacekeeping force, the MFO, in the Sinai Peninsula for a period of at least ninety days after 3 August 1981. Effective 15 March 1985, the period of service must be six months (170 days minimum). Subsequent awards are designated by numerals.

Republic of Vietnam Campaign Medal. Authorized for award to individuals by DoD Instruction 1348.33-M for six months' service in Vietnam or in direct combat support of operations in Vietnam during the period 1 March 1961 to 28 March 1973; during this period, the recipient must have met the criteria established for the Vietnam Service Medal or the Armed Forces Expeditionary Medal (Vietnam). Lesser periods of service are acceptable if the individual was wounded, captured, or killed by hostile forces.

Kuwait Liberation Medal—Kingdom of Saudi Arabia. Awarded by the Saudi Arabian government to U.S. military personnel who participated in Operation Desert Storm during the period 17 January 1991 to 28 February 1991.

Kuwait Liberation Medal—Government of Kuwait. Awarded by the government of Kuwait to U.S. military personnel who were assigned to one of several designated areas in and around Kuwait during the period 2 August 1990 through 31 August 1993. To be eligible, personnel must have been attached to, or regularly served for one day or more, with an organization participating in ground and/or shore operations; with a naval vessel directly supporting military operations; or as a crew member in one or more aerial flights that directly supported military operations in the designated areas. Temporary duty for thirty consecutive days or sixty nonconsecutive days supporting such operations during the designated period also qualifies for award of the medal. The time requirement may be waived for TDY soldiers who actually participated in combat operations.

Republic of Korea War Service Medal. Originally offered to the United States armed forces by the Republic of Korea on 15 November 1951. Approved on 20 August 1999 by the Assistant Secretary of Defense (Force Management Policy) for acceptance and wear by veterans of the Korean War.

U.S. ARMY BADGES AND TABS

Badges and tabs are appurtenances of the uniform. In the eyes of their wearers, several badges have a significance equal to or greater than all but the highest decorations. There is no established precedence with badges as there is with decorations and service medals or ribbons. The badges are of three types: combat and special skill badges, marksmanship badges and tabs, and identification badges. Badges are awarded in recognition of attaining a high standard of proficiency in certain military skills. Subdued combat and special skill badges and the Ranger and Special Forces tabs are authorized on field uniforms.

Combat and Special Skill Badges. These include the combat infantryman badge; expert infantryman badge; combat medical badge; expert field medical badge; combat action badge; Army astronaut badges; aviator, flight surgeon, and aircraft crewman badges; glider badge; parachutist badges; combat parachutist badges; pathfinder badge; air assault badge; Ranger tab; Special Forces tab; diver badges; driver and mechanic badge; explosive ordnance disposal badges; nuclear reactor operator badges; and the parachute rigger badge.

These badges are awarded to denote excellence in performance of duties under hazardous conditions and circumstances of extraordinary hardship as well as for special qualifications and successful completion of prescribed courses of training. (See AR 600-8-22 for details.)

Combat Infantryman Badge. This badge was created at the behest of Lt. Gen. Leslie McNair, CG, Army Ground Forces, during World War II. It was created for formal recognition of the unique dangers and conditions of infantry duty in combat. The contributions made and hardships sustained by members of other branches were considered but were deemed to be sufficiently recognizable by existing awards. Then, as now, the combat infantryman badge is a singular award presented to those who have withstood the worst conditions of combat by virtue of their assignment to units specifically designed to take part in them.

The combat infantryman badge is awarded for participation in ground infantry combat subsequent to 6 December 1941 by infantry and special forces personnel in the rank of colonel or below who satisfactorily perform duty while assigned or attached as members of an infantry, ranger, or special forces unit of brigade size or smaller during any approved period. Personnel other than infantrymen or Special Forces soldiers are not eligible, regardless of the circumstances.

The periods for which award of the combat infantryman badge is authorized are listed in AR 600-8-22, with changes and updates for the several periods since 1995 listed in various MILPERCEN messages. Changes and updates are posted at https://www.hrc.army.mil/site/active/TAGD/awards/index.htm.

Expert Infantryman Badge. Awarded to infantry and Special Forces personnel of the active Army, ARNG, and USAR who satisfactorily complete prescribed proficiency tests.

Combat Medical Badge. Awarded to members of the Army Medical Department, the Naval Medical Department, or the Air Force Medical Service in the grade of O-6 or below who have satisfactorily performed medical duties in ground combat while assigned or attached to an infantry unit at brigade level or below. Specifically, this does *not* include medical personnel assigned to the medical companies of forward support battalions or other noninfantry units, unless they actually accompany infantrymen into combat as replacements for, or augmentees to, the medics assigned to the battalion medical platoon. The other criteria are exactly the same that apply to the combat infantryman badge. In 1991, the Chief of Staff, Army, authorized a limited expansion of combat medical badge eligibility to include medical personnel assigned or attached to armor and ground cavalry units, provided they meet all other qualifying criteria.

Combat Action Badge

Expert Field Medical Badge. Awarded to Army Medical Department (AMEDD) personnel who satisfactorily complete prescribed proficiency tests.

Combat Action Badge. On 4 May 2005, the Army approved a new badge for any soldier, in any branch or military occupational specialty who performs his or her assigned duties in an area where hostile fire pay and imminent danger pay are authorized and who is personally present and actively engaged (under fire) by the enemy. Soldiers must be performing satisfactorily, within the guidelines of the rules of engagement authorized for the combat area.

Expert Soldiers Badge. Approved on 14 June 2019, the ESB is awarded to soldiers who are not infantry, special forces, or medics who are active Army, ARNG, and USAR who satisfactorily complete prescribed proficiency tests.

Stars for Combat Infantryman, Combat Medical, and Combat Action Badges. The second and succeeding awards of the combat infantryman, combat medical, and combat action badges, made to recognize participation and qualification in additional wars, are indicated by the addition of stars to the basic badges.

Only one award of the CIB is authorized for service during the so-called Vietnam "era," which terminated on 10 March 1995 and included combat operations in Vietnam, Laos, the Dominican Republic, Korea on the DMZ (4 January 1969 to 31 March 1994), El Salvador, Grenada, Panama, the Persian Gulf War, and Somalia, regardless of whether an individual had served one or more tours in one or more of these areas. Effective 5 December 2001, the Army authorized subsequent awards of the CIB and CMB (indicated by the addition of a star) for combat service in selected geographical areas under the same circumstances prescribed for award of these badges in the previous era. Operation Enduring Freedom began the fourth conflict period qualifying for award of the badges, the previous eras having been World War II, Korea, and Vietnam.

Note that either, but not both, the combat infantryman badge or the combat medical badge may be awarded for the same period of service in these areas.

Army Aviator Badges. There are nine badges relating to army aviation—three each for Army aviators, flight surgeons, and aircraft crewmen—in the degrees of basic, senior, and master. The following publications provide information on these badges:

AR 600-8-22, *Military Awards.*

AR 600-105, *Aviation Service of Rated Army Officers.*

The master Army aviator badge, the senior Army aviator badge, and the Army aviator badge are awarded upon satisfactory completion of prescribed training and proficiency tests as outlined in AR 600-105.

The master flight surgeon badge, the senior flight surgeon badge, and the flight surgeon badge are awarded to Army Medical Corps officers who complete the training and other requirements prescribed by AR 600-105.

The master aircraft crewman badge, the senior aircraft crewman badge, and the aircraft crewman badge are authorized for award to enlisted personnel who meet the prescribed requirements. (AR 672-5-1).

Army Astronaut Device. The Army astronaut device, a gold-colored device 7/16 inch in length consisting of a star emitting three contrails encircled by an elliptical orbit, may be awarded to personnel who complete one operational mission in space (50 miles above the earth). The device is affixed to the appropriate Army aviator badge, flight surgeon badge, or aircraft crewman badge to which the recipient is entitled. An individual who has not been awarded one of these badges, but who otherwise meets the astronaut criteria, will be awarded the basic aircraft crewman badge with the Army astronaut device.

Parachutist Badges. Both static-line and free-fall parachute skills and service are recognized by Army parachutist badges. The parachutist badge is awarded for satisfactory completion of the airborne course conducted by the Infantry School, or while assigned or attached to an airborne unit or for participation in at least one combat jump.

The senior parachutist badge is awarded to soldiers who have been rated excellent in character and efficiency with at least 30 jumps, of which at least 15 must have been with normal TOE equipment. Additionally, at least two jumps must have been at night (one as a jumpmaster), and at least two must have been as part of tactical operations. Qualifying soldiers must also be graduates of an Army-approved jumpmaster course, must have served as a jumpmaster on at least one combat jump, or must have served as a jumpmaster on 15 noncombat jumps. Finally, the recipient must have served at least 24 total months on jump status. (See AR 600-8-22 for definitions of "TOE equipment," "tactical operations," and "approved jumpmaster course" for the purposes of this qualification.)

The master parachutist badge is awarded to soldiers who have been rated excellent in character and efficiency with at least 65 jumps, of which at least 25 must have been with normal TOE equipment. Additionally, at least four jumps must have been at night (one as a jumpmaster), and at least five must have been as part of tactical operations. Qualifying soldiers must also be graduates of an

Army-approved jumpmaster course, must have served as a jumpmaster on at least one combat jump, or must have served as a jumpmaster on 33 noncombat jumps. Finally, the recipient must have served at least 36 total months on jump status. (See AR 600-8-22 for definitions of "TOE equipment," "tactical operations," and "approved jumpmaster course" for the purposes of this qualification.)

All three parachutist badges may be revoked for several reasons. Essentially, such reasons include UCMJ punishment for jump refusals; voluntary, self-initiated termination of jump status with fewer than 36 cumulative months on jump status; withdrawal of certain Special Forces MOS or branch qualification; or dismissal from the service, dishonorable discharge, or conviction by court-martial for desertion in wartime. See AR 600-8-22 for details.

The basic military free-fall parachutist badge is awarded to special operations forces personnel who have satisfactorily completed a prescribed program of instruction in military free-fall approved by the Army John F. Kennedy Special Warfare Center and School or who have executed at least one military free-fall combat jump.

The military free-fall parachutist badge-jumpsmaster is awarded to special operations forces personnel who have satisfactorily completed prescribed military free-fall basic and jumpmaster programs of instruction approved by the Army's John F. Kennedy Special Warfare Center and School.

The free-fall parachutist badges may also be revoked for several reasons. Essentially, such reasons include UCMJ punishment for jump refusals; voluntary, self-initiated termination of jump or military free-fall basic or jumpmaster status; or dismissal from the service, dishonorable discharge, or conviction by court-martial for desertion in wartime. See AR 600-8-22 for details.

Combat Parachutist Badges. One bronze star affixed to the appropriate parachutist badge is authorized for each combat jump in which the recipient participates.

Pathfinder Badge. Awarded upon successful completion of the Pathfinder Course conducted at the Infantry School.

Air Assault Badge. Awarded to personnel who have satisfactorily completed either the Training and Doctrine Command (TRADOC) prescribed training course or the standard air assault course while assigned or attached to the 101st Airborne (Air Assault) Division since 1 April 1974.

Ranger Tab. Awarded to any person who successfully completes a Ranger Course conducted by the Infantry School.

Sapper Tab. Approved by the Chief of Staff, Army on 28 June 2004, the Sapper Tab is awarded to U.S. and foreign military and civilian personnel who successfully complete the Sapper Leader's Course conducted by the U.S. Army Engineer School on or after 14 June 1985.

Special Forces Tab. Awarded to any person who successfully completes the Special Forces Qualification Course and officers who complete the Special Forces Officer Course conducted by the John F. Kennedy Special Warfare Center. This tab also may be awarded for former wartime service. See the regulation (AR 600-8-22) for details.

Diver Badges. Awarded after satisfactory completion of prescribed proficiency tests (AR 611-75). Five badges are authorized for enlisted personnel.

Driver and Mechanic Badge. Awarded only to enlisted personnel to denote a high degree of skill in the operation and maintenance of motor vehicles.

Explosive Ordnance Disposal Badges. There are three badges under this heading, any of which may be awarded to officers: master explosive ordnance disposal badge, senior explosive ordnance disposal badge, and explosive ordnance disposal badge. They are awarded to individuals assigned to duties involving the removal and disposition of explosive ammunition under hazardous conditions.

Parachute Rigger Badge. Officers who successfully complete the Aerial Delivery and Materiel Officer Course, the Parachute Maintenance and Aerial Supply Officer Course, the Parachute Maintenance and Airdrop Course, or the Parachute Rigger Course, all conducted by the U.S. Army Quartermaster School, qualify for award of the parachute rigger badge.

Physical Fitness Training Badge. Established by the Secretary of the Army on 25 June 1986, this badge is awarded to soldiers who obtain a minimum score of 290 on the Army Physical Fitness Test (APFT) and who meet the weight control requirements of AR 600-9. Once awarded, the badge may be retained as long as the individual achieves a minimum passing score on subsequent APFTs and continues to meet the weight requirements. This badge is worn only on the left breast of the physical training uniform.

Marksmanship Badges and Tabs. These include basic marksmanship qualification badges, excellence in competition badges, distinguished designation badges, the United States distinguished international shooter badge, and the President's Hundred tab. Members of the armed forces of the United States, civilian citizens of the United States, and foreign military personnel who qualify as prescribed are eligible for award of basic qualification badges. Only U.S. military and civilian personnel are eligible for the other badges and the tab.

Qualification badges for marksmanship are of three types: basic qualification, excellence in competition, and distinguished designation. Basic qualification badges (including expert, sharpshooter, and marksman badges) are awarded to those individuals who attain the qualification score prescribed in the appropriate field manual for the weapon concerned. Excellence in competition badges are awarded to individuals in recognition of an eminent degree of achievement in firing the rifle or pistol. Distinguished designation badges are awarded to individuals in recognition of a preeminent degree of achievement in target practice firing with the military service rifle or pistol.

The distinguished international shooter badge is awarded to military or civilian personnel in recognition of an outstanding degree of achievement in international competition.

A President's Hundred tab is awarded each person who qualifies among the top 100 contestants in the President's Match held annually at the National Rifle Matches.

Foreign Badges. Commanders serving in the rank of brigadier general or higher (or colonels exercising general court-martial authority) are authorized to approve the acceptance, retention, and permanent wear of many foreign badges, which are prescribed in appendix D of AR 600-8-22.

Permission to accept, retain, and wear all other foreign badges can be granted only by PERSCOM. The basic criteria for approval are the criteria established by the foreign government for award of the badge. Badges not meeting these criteria (i.e., "honorary" badges) may be approved for acceptance but not for wear. Forward requests and justification, including all documentation, to HQ, PERSCOM, ATTN: TAPC-PDA, Alexandria, Virginia, 22332-0471.

Identification Badges. These include the presidential service identification badge; vice-presidential service identification badge; Secretary of Defense identification badge; Joint Chiefs of Staff identification badge; Unified Combatant Command identification badge; National Defense University identification badge; Army staff identification badge; Guard, Tomb of the Unknown Soldier identification badge; Army ROTC Nurse Cadet Program identification badge; drill sergeant identification badge; U.S. Army recruiter badge; the U.S. Army Reserve recruiter badge; U.S. Army National Guard recruiter badge; and career counselor badge.

Presidential Service Identification Badge. The presidential service certificate and the presidential service badge were established by Executive Order 11174, 1 September 1964, as amended by Executive Order 11407, 23 April 1968; Executive Order 11520, 25 March 1970; and Executive Order 12793, 20 March 1992.

The presidential service certificate is awarded in the name of the President of the United States to members of the armed forces who have been assigned to duty in the White House or to other positions in direct support of the Executive Office of the President for at least one year subsequent to 21 January 1989.

The presidential service badge is issued to members of the armed forces who have been awarded the presidential service certificate. Once this badge is awarded, it may be worn as a permanent part of the uniform.

Vice-Presidential Service Identification Badge. The vice-presidential service badge was established by Executive Order 11926, 19 July 1976. It may be awarded, upon recommendation of the military assistant to the Vice President, to personnel who have been assigned to duty in the Office of the Vice President for at least one year subsequent to 19 December 1974. Once this badge is awarded, it may be worn as a permanent part of the uniform.

Secretary of Defense Identification Badge. Military personnel who have been assigned to duty and have served not less than one year after 31 January 1961 in the Office of the Secretary of Defense are eligible for this badge. Once awarded, it may be worn as a permanent part of the uniform. It also is authorized for temporary wear by personnel assigned to specified offices of the Secretary of Defense. The governing directive is DoD 1348.33-M.

Joint Chiefs of Staff Identification Badge. This badge may be awarded to military personnel who have been assigned to duty and who have served not less than one year after 14 January 1961 in a position of responsibility under the direct cognizance of the Joint Chiefs of Staff. Once awarded, the badge may be worn as a permanent part of the uniform.

Unified Combatant Command Identification Badge. This badge is authorized for personnel to wear while assigned to the staff of a combatant command and while assigned to subordinate unified commands and direct reporting units to the unified command, as determined by the unified combatant commander. The design of the badge is unique to each command. The badge is worn on the left side on class A and B, Army blue, mess, and evening mess uniforms.

National Defense University (NDU) Identification Badge. Personnel assigned to the faculty and staff of NDU, National War College, Industrial College of the Armed Forces, and Armed Forces Staff College are authorized to wear the badge during their assignment. The badge is worn on the right side.

Army Staff Identification Badge. This badge has been awarded by the Army since 1920 and is the oldest of the five types of identification badges now authorized for officers. It was instituted to give a permanent means of identification to those commissioned officers who had been selected for duty on the War Department General Staff, with recommendation for award based on performance of duty. It has been continued under the present departmental organization.

The requirements for award of this badge include service of not less than one year between 1 August 1977 and 28 May 1985 as a commissioned officer, or between 22 August 1980 and 28 May 1985 as a warrant officer, while detailed to duty on the Army General Staff and assigned to permanent duty in a TDA position on the Army General Staff, to the Office of the Secretary of the Army, the National Guard Bureau, or the Office, Chief Army Reserve. Between 30 September 1979 and 28 May 1985, the badge could also be awarded to the Sergeant Major of the Army and to other senior staff NCOs (SGM E9) assigned to duty with the same staff units. Effective 29 May 1985, qualifying service must be on the Army General Staff or assigned to the Office of the Secretary of the Army. Once awarded, this badge may be worn as a permanent part of the uniform.

Other Badges. For information on the following badges, see AR 600-8-22: Guard, Tomb of the Unknown Soldier identification badge, Army ROTC Nurse Cadet Program identification badge, drill sergeant identification badge, U.S. Army recruiter identification badge, Army National Guard Recruiter identification badge, U.S. Army Reserve recruiter identification badge, career counselor badge.

APPURTENANCES TO DECORATIONS AND SERVICE MEDALS

Appurtenances are authorized as indicated in the following paragraphs.

Oak-Leaf Cluster. A bronze (or silver) twig of four oak leaves with three acorns on the stem is issued in lieu of a decoration for the second or succeeding awards of United States military decorations (other than the Air Medal), the Army Reserve Components Achievement Medal, and unit awards. A silver oak-leaf cluster is worn in lieu of five bronze oak-leaf clusters for the same decoration. Oak-leaf clusters are worn with the stem of the oak leaves pointing down toward the wearer's right and are attached to the ribbons of the decorations to which they pertain.

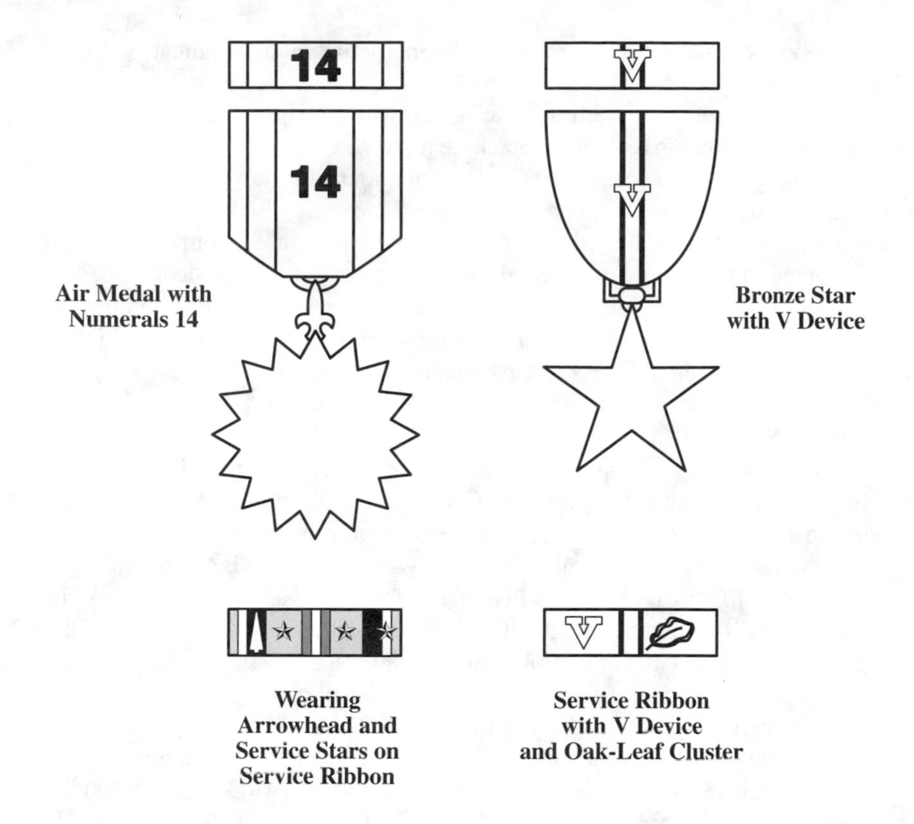

Numerals. Arabic numerals 3/16 inch in height are issued in lieu of a decoration for second and succeeding awards of the Air Medal, the Multinational Force and Observers Medal, the Overseas Service Ribbon, and the Army Reserve Components Overseas Training Ribbon. The numerals are centered on the ribbon of the medal and on the ribbon bar. The ribbon denotes the first award, and numerals, starting with the numeral 2, denote the second and succeeding awards.

Letter V Device. A bronze block letter V is worn on the suspension and service ribbons of the Bronze Star Medal, the Air Medal, the Joint Service Commendation Medal, and the Army Commendation Medal to denote an award made for acts of heroism involving conflict with an armed enemy. Not more than one V will be worn. When worn with oak-leaf clusters or numerals on the same ribbon, the V is to the wearer's right.

Arrowhead Device. A bronze arrowhead device is awarded for wear on the appropriate service medal ribbon to signify that the wearer participated in a combat parachute jump, a helicopter assault landing, a glider landing, or an amphibious assault as a member of an organized force carrying out an assigned tactical mission. The device is worn with the point upward. It is placed to the right of all service stars.

Service Stars. Service stars, signifying participation in a combat campaign, are worn on service medal ribbons, point of star upward. A silver star is worn in lieu of five bronze stars. A bronze star is also worn on the parachutist badges to denote participation in a combat parachute jump.

Miniatures. Miniature decorations and appurtenances are replicas of the corresponding decorations and appurtenances on the scale of one-half. Miniatures are not presented or sold by the Army but may be purchased from civilian dealers. There is no miniature of the Medal of Honor. Except for the Medal of Honor, only miniature decorations and service medals, dress miniature combat and special skill badges, and dress miniature versions of Ranger and Special Forces tabs may be worn on the mess and evening mess uniforms.

FOREIGN DECORATIONS AND GIFTS

The Constitution requires the consent of Congress for an individual holding a federal office or position of trust to accept a foreign decoration. Individuals may participate in ceremonies and receive the tender of a foreign award or gift, however. Foreign awards may be accepted and worn after receiving approval of the Commander, PERSCOM (TAPC-PDA). The foreign award with accompanying documents will be retained by the individual until the individual is informed of the final DA action. Gifts of minimal value (retail value of less than $200) may be accepted and retained by the individual. The burden of proof as to the value is the responsibility of the individual accepting the gift, however. For gifts of more than minimal value, receipt of the gift should be immediately reported through command channels to the Commander, PERSCOM, ATTN: TAPC-PDO-IP, 200 Stoval Street, Alexandria, Virginia, 22332-0474, and within sixty days the gift must be forwarded to the same address.

CERTIFICATES AND LETTERS FOR SERVICE

Several types of certificates may be awarded to individuals, as follows:

Certificates for Decorations. Each individual who has been awarded a decoration is entitled to a certificate on a standard Department of the Army form, bearing a reproduction of the decoration.

Certificate of Honorable Service. The certificate (DA Form 1563) is issued to the next of kin of those who die in the line of duty while on active service in time of peace. In time of war, an accolade with facsimile signature of the President has been used.

Certificate of Achievement. Commanding officers may issue such a certificate in recognition of faithful service, acts, or achievements. DA Form 2442, Certificate of Achievement, or a similar form of local design may be used.

Letters of Commendation and Appreciation. Acts or services that do not meet the criteria for decorations or the various certificates may be recognized by letters of commendation or appreciation. Such letters, typed on letterhead stationery, may be issued to military personnel and, as specified in AR 672-20, to civilians or civilian groups.

ORDER OF PRECEDENCE OF DECORATIONS AND AWARDS

There is a definite ranking among categories of decorations and awards, and within each category there is also an order of precedence. This section provides the rules. Consult it before you assemble authorized decorations and awards on your uniform to be sure you have them in the right sequence.

Categories of Medals. The listing below indicates the order of precedence by category when medals of two or more categories are worn simultaneously. The same order of precedence applies when service ribbons are worn in lieu of decorations and service medals.

U.S. military decorations
U.S. unit awards
U.S. nonmilitary decorations
Good Conduct Medal
Army Reserve Components Achievement Medal
U.S. campaign and service medals
U.S. service and training ribbons
U.S. Merchant Marine awards
Foreign military decorations
Foreign unit awards
Non-U.S. service awards

United States Military Decorations. United States military decorations of the Army, Navy, Air Force, and Coast Guard are worn by Army personnel in the following order:

Medal of Honor (Army, Navy, Air Force)
Distinguished Service Cross
Navy Cross
Air Force Cross
Defense Distinguished Service Medal
Distinguished Service Medal (Army, Navy, Air Force, Coast Guard)
Silver Star
Defense Superior Service Medal
Legion of Merit
Distinguished Flying Cross
Soldier's Medal
Navy and Marine Corps Medal
Airman's Medal
Coast Guard Medal
Bronze Star Medal
Purple Heart
Defense Meritorious Service Medal
Meritorious Service Medal
Air Medal
Joint Service Commendation Medal
Army Commendation Medal

Navy Commendation Medal
Air Force Commendation Medal
Coast Guard Commendation Medal
Joint Service Achievement Medal
Army Achievement Medal
Navy Achievement Medal
Air Force Achievement Medal
Coast Guard Achievement Medal
Prisoner of War Medal
Combat Action Ribbon

United States Unit Awards. These awards may be worn as a permanent part of the uniform by personnel who were assigned to the unit during any part of the period during which the unit earned the award. In addition, as regards the four Army unit awards, personnel who were not present with the unit at the time the award was earned may wear the award temporarily while they are assigned to the unit at some later date. For elements of regiments organized under the New Manning System or the Combat Arms Regimental System, the emblem may be worn temporarily only by personnel assigned to the earning unit. The awards are worn in the following order:

Presidential Unit Citation (Army and Air Force)
Presidential Unit Citation (Navy)
Joint Meritorious Unit Award
Valorous Unit Award (Army)
Meritorious Unit Commendation (Army)
Navy Unit Commendation
Air Force Outstanding Unit Award
Coast Guard Unit Commendation
Army Superior Unit Award
Meritorious Unit Commendation (Navy)
Navy "E" Ribbon
Air Force Organizational Excellence Award
Coast Guard Meritorious Unit Commendation

United States Nonmilitary Decorations. These awards may be worn on the uniform only with one or more military decorations or service medals. They are worn after the Combat Action Ribbon and before service medals in this order:

Presidential Medal of Freedom
Presidential Citizen's Medal
President's Award for Distinguished Federal Civilian Service
Department of Defense Distinguished Civilian Service Award
Secretary of Defense Medal for the Defense of Freedom
Secretary of Defense Meritorious Civilian Service Award
Office of the Secretary of Defense Exceptional Civilian Service Award
Surgeon General's Exemplary Service Medal
NASA Space Flight Medal

Public Health Service Commendation Medal
Public Health Service Achievement Medal
Department of State Superior Honor Award
Decoration for Exceptional Civilian Service
Meritorious Civilian Service Award
Superior Civilian Service Award
Commander's Award for Civilian Service
Achievement Medal for Civilian Service

The Good Conduct Medal Worn with Other Decorations. The Good Conduct Medal takes precedence immediately after all authorized United States military and nonmilitary decorations. Good Conduct Medals from the other services follow the Army Good Conduct Medal.

Army Reserve Components Achievement Medal. The Army Reserve Components Achievement Medal takes precedence immediately following the Army Good Conduct Medal and/or the Good Conduct Medals of the other services.

Service Medals. United States service medals are worn following the military and nonmilitary decorations, the Good Conduct Medal, and the Army Reserve Components Achievement Medal, if worn. Except for the Army of Occupation Medal, service medals for actions or time periods prior to the Vietnam War are not included here. The order or precedence is as follows:

Army of Occupation Medal
National Defense Service Medal
Antarctica Service Medal
Armed Forces Expeditionary Medal
Vietnam Service Medal
Southwest Asia Service Medal
Kosovo Campaign Medal
Afghanistan Campaign Medal
Iraq Campaign Medal
Global War on Terrorism Expeditionary Medal
Global War on Terrorism Service Medal
Korea Defense Service Medal
Armed Forces Service Medal
Humanitarian Service Medal
Military Outstanding Volunteer Service Medal
Armed Forces Reserve Medal
NCO Professional Development Ribbon
Army Service Ribbon
Overseas Service Ribbon
Army Reserve Components Overseas Training Ribbon

Service medals of the U.S. Navy, Air Force, Marine Corps, Coast Guard, and Merchant Marine may be worn on the uniform. Others issued by state and local governments, fraternal societies, U.S. Maritime Service, and professional groups and organizations are prohibited for wear on the uniform.

Foreign Decorations. The wearing of foreign decorations on the uniform is governed by both law and regulations. The reader is referred to AR 600-8-22 for a check of complete information regarding authorized wearing of foreign decorations. Foreign decorations or unit awards will not be worn under any circumstances unless at least one U.S. decoration or service medal or ribbon is worn at the same time.

Non-U.S. Service Medals and Ribbons.

United Nations Medal
NATO Medal
Multinational Force and Observers Medal
Republic of Vietnam Campaign Medal
Kuwait Liberation Medal, Kingdom of Saudi Arabia
Kuwait Liberation Medal, Government of Kuwait

Combat and Special Skill Badges. Combat and special skill badges are grouped into categories with an order of group precedence by category as follows:

Group 1 Combat infantryman badge (three awards), combat action badge (three awards), and expert infantryman badge.

Group 2 Combat medical badge (three awards) and expert field medical badge.

Group 3 Army astronaut device (three degrees), Army aviator badge (three degrees), flight surgeon badges (three degrees), and aircraft crewman badge (three degrees).

Group 4 Parachutist badge (three degrees), combat parachutist badges, free-fall parachutist badge (two degrees, only for personnel assigned to a component of the U.S. Special Operations Command), pathfinder badge, and air assault badge. *Note:* For purposes of classification and wear policy, the Ranger and Special Forces metal tab replicas are considered Group 4 badges.

Group 5 Diver badge (five badges), driver and mechanic badge (six clasps), explosive ordnance disposal badges (three degrees), nuclear reactor operator badge (four badges), and parachute rigger badge.

Group 6 Physical fitness training badge.

Marksmanship Badges. Marksmanship badges authorized for wear on the Army uniform are worn in the following order of precedence:

U.S. distinguished international shooter badge
Distinguished rifleman badge
Distinguished pistol shot badge
National trophy match badge
Interservice competition badge
U.S. Army excellence in competition rifleman badge
U.S. Army excellence in competition pistol shot badge
Marksmanship qualification badge (expert, sharpshooter, and marksman)

In addition, the President's Hundred tab is a marksmanship award, but it must compete with other tabs (see below) for wear on the uniform.

Special Skill Tabs. The authorized tabs are the President's Hundred tab, the ranger tab, and the special forces tab. Both the special forces tab and the ranger tab may be worn on the uniform. When worn together, the special forces tab takes precedence.

GUIDE FOR WEARING AWARDS

When an officer wears decorations, service medals, badges, and other uniform accoutrements, he or she must be certain to wear them correctly. This means wearing them on proper occasions, on the correct garments, and in prescribed order or arrangement. There is a prescribed place and position for each item authorized in relation to others that may be worn at the same time. Metallic portions must be clean and bright, ribbons clean. Wear them correctly—or don't wear them at all. You can check correct order on any number of military Web apps, including www.uniformribbons.com or www.ezrackbuilder.com.

Military tailors assemble the ribbons of decorations and service medals in rows, when a wearer is entitled to several, and attach them to the garments in correct position. The wearer must indicate the order of arrangement and be certain of its accuracy upon completion. This is the simplest and best practice. They can also be assembled on pins for use on any appropriate garment.

Occasions for Wearing Decorations, Service Medals, or Their Ribbons. Commanding officers may prescribe the wearing of these items at parades, reviews, inspections, funerals, and on ceremonial and social occasions. They may be worn at the option of the wearer on the service and dress uniforms on normal duty (when not prohibited) and when off duty. Also, miniature medals may be worn on the left lapel of formal civilian attire at formal social functions when it is inappropriate or unauthorized to wear the Army uniform.

Prohibited Wearing. The items are not worn on any uniform other than as authorized by AR 670-1 (as detailed in this chapter), by officers while suspended from either rank or command, by enlisted personnel while serving a sentence of confinement, or when wearing civilian clothing except on formal civilian attire when wear of the uniform is inappropriate. A lapel button in the form of a miniature service ribbon may be worn.

Penalty for Unauthorized Wearing. It is a violation of law to wear decorations other than those to which an individual is entitled, or to wear their ribbons or other decoration substitutes. Serious penalties may be invoked.

The preceding section lists the categories of awards and itemizes the relative order of precedence both among and within categories. This section describes when, how, and where the authorized awards may be worn on the uniforms. The rules are specific. If in doubt after studying these two sections, consult AR 670-1.

Though the rules for wear may seem quite involved, no one individual needs to know all the rules pertaining to all the awards. Just be sure that you understand the rules that pertain to the awards you are authorized to wear so that

Ribbons—Men

Full Size Medals—Men & Women

Miniature Medals on Mess Uniform—Men

Ribbons—Women

Full Size Medals—Women

Miniature Medals on Mess Uniforms—Women

Wear of Ribbons and Medals

you may assemble them correctly and wear them on your uniform with confidence and pride.

The descriptions and the accompanying illustrations should answer most questions about the correct placement of decorations and awards. For all decorations and awards, the place of honor is on the wearer's right and at the top. The order of precedence then is from right to left and from top to bottom.

Service Ribbons. Decoration and service medal ribbons may be worn on the coat of the Army green service uniforms, on the Army blue and Army white uniform coats, and on the shirts and the maternity tunic. The ribbons are worn in one or more rows in order of precedence, with either no space or a 1/8-inch space between rows. No more than four ribbons may be worn in any row, but a second row is not started unless four or more ribbons are worn. The first two rows must contain the same number of ribbons (either three or four) before starting a third row. The third and succeeding rows contain the same number as or fewer ribbons than the first two rows. The top row is centered on the row beneath or aligned to the left edge of the row underneath, whichever presents the best appearance.

Men wear the ribbons centered 1/8 inch above the left breast pocket in as many rows as necessary. Women wear the ribbons on the left side with the bottom of the bottom row on the same horizontal line as the bottom of the nameplate. Placement may be adjusted to conform to individual figure differences.

Full-Size Decorations and Service Medals. Full-size decorations and service medals may be worn on the Army blue and Army white uniforms after retreat and by enlisted personnel on the Army blue uniform when worn for social functions. The medals are worn in order of precedence, in one or more rows, with 1/8-inch spacing between rows. The second and subsequent rows should not contain more medals than the row below. The number of medals worn in a row depends on the size of the coat; they must not overlap within a row. Service and training ribbons are not worn with the full-size medals. When full-size medals are worn, up to three full-size or miniature combat and special skill badges may be worn above the medals in order of precedence.

Men wear the first row of medals attached immediately above the seam of the left breast pocket. Women wear the medals so that the bottom of the bottom row of pendants is on the same horizontal line as the bottom of the nameplate.

Miniature Decorations and Service Medals. Miniature medals are half-scale replicas of the full-size decorations and service medals. They are available for all medals except the Medal of Honor, which is worn only in full size. Except for the Medal of Honor, only miniature medals are authorized for wear on the various mess and evening mess uniform jackets. They may also be worn on the Army blue and Army white dress uniforms after retreat when these uniforms are worn as formal attire (bow ties). Only the dress miniature versions of the combat and special skill badges may be worn with miniature medals. Service and training ribbons are not worn in combination with miniature medals. Miniatures may also be worn on the left lapel of formal civilian attire on occasions when wear of the uniform is inappropriate or unauthorized.

Miniature medals are mounted on holding bars in order of precedence from the wearer's right to left, in more than one row if necessary. They are worn side by side when there are four or fewer medals. They may be overlapped up to 50 percent when five, six, or seven medals are in the row. The overlap is equal for all medals, with the right medal showing in full. When more than one row of medals is worn, the pendants on the lower row must be fully visible.

Men. On the Army blue and Army white dress uniforms, the medals are worn as described for the full-size medals. Up to three of the dress miniature combat and special skill badges may be worn above the medals in order of group precedence.

On the blue and white mess jackets, the medals are worn centered on the left lapel, about 1/2 inch below the notch. They do not extend beyond the edge of the lapel. Up to four of the dress miniature combat and special skill badges may be worn with the miniature medals on the mess jackets. They are worn in order of group precedence above the miniature medals. When two are worn, they are placed side by side immediately above the medals. When a third is worn, it is centered 1/4 inch above the first two. If a fourth badge is worn, it is placed 1/4 inch above the third badge.

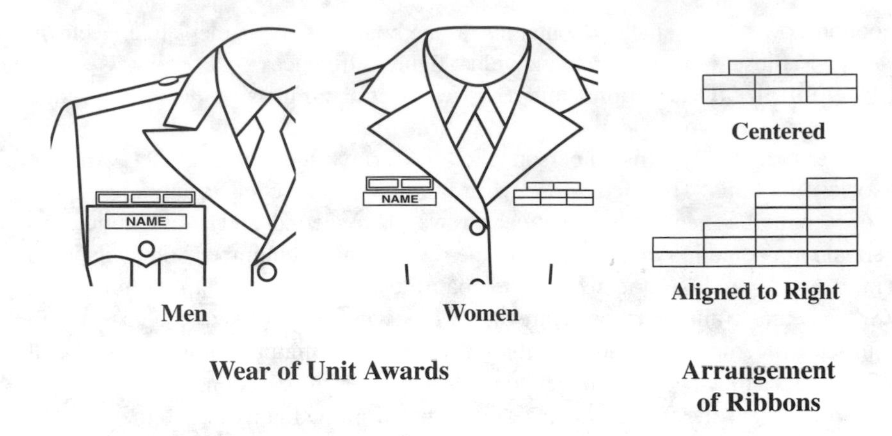

Wear of Unit Awards

Arrangement of Ribbons

Women. Miniature medals are worn centered on the left lapel of the blue mess and the new version of the white mess uniform jackets. They are worn centered on the left side (not the lapel) of the old version of the white mess uniform jacket. The bottom of the bottom row of pendants is positioned approximately in line with the top edge of the top button on the Army blue and the Army white coats, and in a comparable position on the blue and white mess jackets. Dress miniature combat and special skill badges are worn as described for the men.

U.S. and Foreign Unit Awards. Unit award emblems are worn in order of precedence, not more than three per row, with no space between emblems, and up to 1/8-inch space between rows. A permanent U.S. unit award has a gold-colored metal frame with laurel leaves, and a temporary award (authorized for wear only while assigned to the unit that earned the award) does not have the frame.

The framed awards are worn with the laurel leaves pointing upward. The unit award emblems may be worn with service ribbons or full-size medals but are not worn with miniature medals. Foreign unit awards are worn after U.S. unit awards in order of date of receipt of the award. The foreign unit awards from the Vietnam War era, authorized only for permanent wear, are the Vietnam Presidential Unit Citation badge, the Republic of Vietnam Gallantry Cross Unit Citation badge, and the Republic of Vietnam Civil Actions Unit Citation badge.

Men wear the unit awards on the coats of the Army blue and Army white uniforms, and the shirt. The awards are worn centered and with the bottom row 1/8 inch above the seam of the right breast pocket flap.

Women wear the unit awards on the coats of the Army blue and Army white uniforms, and on the shirt and the maternity tunic. The awards are centered on the right side with a 1/2-inch space between the bottom of the bottom row and the top of the nameplate.

Marksmanship Badges and Tabs. Not more than three marksmanship badges may be worn at one time, and only on the coats of the Army blue and Army white uniforms, and on the shirts and the maternity tunic. These badges are worn by men on the flap of the left breast pocket, under the service ribbons, and by women in a comparable position. There must be no more than three clasps per badge, and the total number of marksmanship badges and special skill badges worn on the pocket flap or beneath the ribbons cannot exceed three. At least one marksmanship badge is normally worn by all personnel except for those individuals who are exempted by regulation.

Men wear a single marksmanship or special skill badge centered on the left breast pocket flap, with the top of the badge 1/8 inch below the seam. When two such badges are worn, they are placed side by side about 1/8 inch below the seam with the special skill badge to the wearer's right and the badges equally spaced on the flap from right to left. Three such badges in any combination are worn in similar fashion except when one special skill and two marksmanship badges are worn. In this case the special skill badge may be worn centered on the pocket flap 1/8 inch below the seam, with the marksmanship badges centered between the button and the right side of the pocket flap, and with the bottom of the badge coinciding with the bottom of the flap.

Women wear marksmanship and special skill badges on the left side, approximately 1/4 inch below the bottom row of ribbons, but otherwise as prescribed for men.

The President's Hundred tab is worn 1/2 inch below the shoulder seam on the left sleeve of the Army green uniform coat. It competes in this position with the ranger and special forces tabs. It may not be worn in conjunction with the special forces or ranger tab.

Combat and Special Skill Badges and Tabs. As stated earlier, combat and special skill badges and tabs are categorized into six groups and an order of group precedence is established. The only badge in group 6 is the physical fitness badge, which is worn only on the physical training uniform. Of the remaining badges, a maximum of four may be worn at one time on the coats of the Army blue and Army white uniforms, and on the shirts and the maternity tunic. This total does not include the ranger or special forces tabs, which may be worn on the left shoulder of the green uniform coat. Only one badge each from groups 1, 2, 3, and 5 may be worn at a time. Two badges from group 4 may be worn if no badge from group 5 is worn.

Within a group, a combat badge takes precedence over a skill badge. For example, the combat infantryman badge would be worn instead of the expert infantryman badge if both are authorized. Up to three badges from groups 1 through 3 (one from each group) may be worn one above another, 1/2 inch apart, and with the bottom badge 1/4 inch above the ribbons or the pocket flap if no ribbons are worn. If no badges from groups 1 through 3 are worn, a total of two special skill badges from groups 4 and 5 (two from group 4 or one from group 4 and one from group 5) may be worn above the pocket or the ribbons as described for badges in groups 1 through 3. When badges from groups 1 through 3 are worn, badges from groups 4 and 5 are worn on the pocket flap or underneath the ribbons as described earlier under Marksmanship Badges and Tabs.

Wear of full-size or miniature (but not mixed sizes) combat and special skill badges is authorized on the blue and white dress uniforms. When the number of

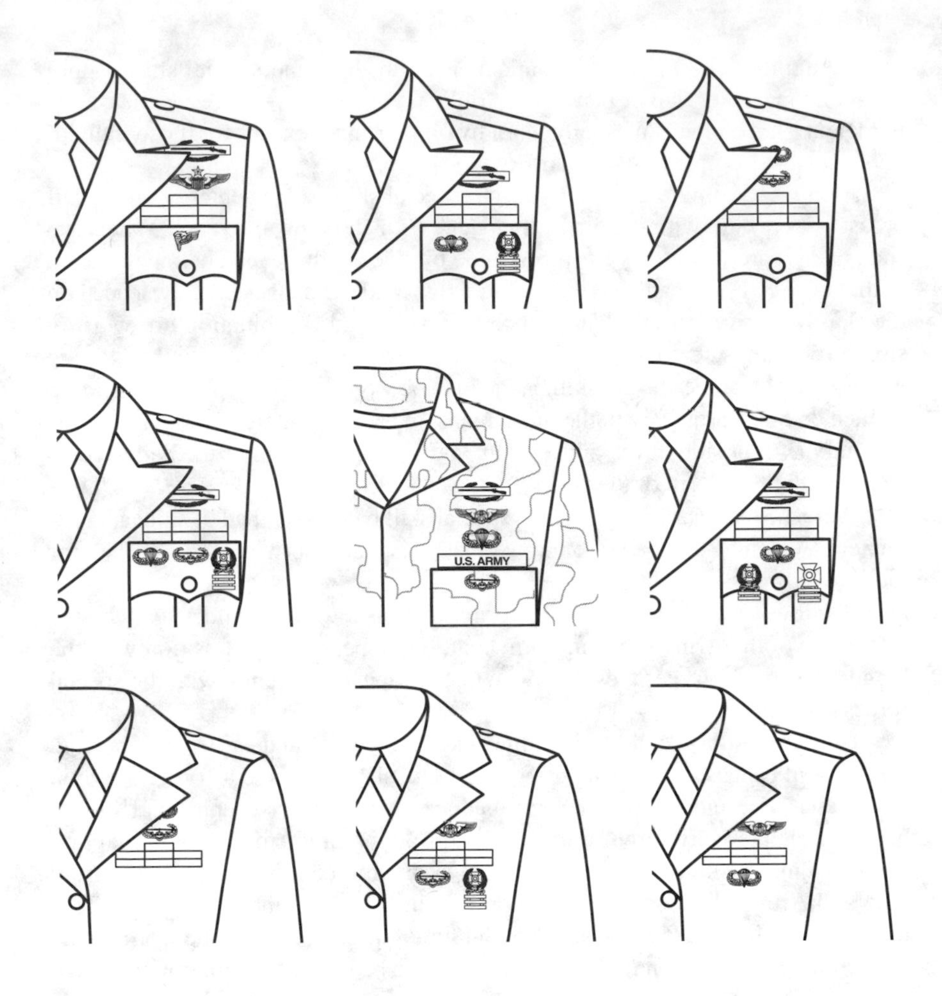

Wear Position for Various Combinations of Combat, Special Skill, and Marksmanship Badges

ribbons or medals worn causes the badges to be obscured by the lapel, the badges may be worn aligned with the left edge of the ribbons or medals.

On the Army blue and Army white dress uniforms, when wearing miniature medals, only dress miniature combat and special skill badges are worn. A maximum of three of these badges from groups 1 through 5 may be worn one above another, above the medals. Special skill and marksmanship badges are not worn beneath the miniature medals.

On the mess uniform jackets, up to four dress miniature combat and special skill badges may be worn. Two medals are worn side by side, centered above the miniature medals. When three badges are worn, the third is centered 1/4 inch above the first two badges, and a fourth badge, if worn, is centered 1/4 inch above the third badge.

Dress miniature combat and special skill badges may be worn on the shirt, but they cannot be mixed with full-size badges.

Subdued pin-on or embroidered sew-on badges may be worn on the cold-weather uniform shirt. When one badge is worn, it is centered over the U.S. Army tape, with the bottom approximately 1/4 inch above the tape. If two badges are worn, they are worn one above the other, centered over the U.S. Army tape with 1/2 inch between badges. If three badges are worn, two are as described above, and the third is worn centered on the pocket flap or in a comparable position on uniforms without pockets. If four badges are worn, three are worn in a vertical line centered above the U.S. Army tape, and one is worn centered on the pocket flap or in a comparable position.

Ranger, special forces, and sapper tabs are worn on the left shoulder of the Army green uniform coat, sewn 1/2 inch below the top of the shoulder seam. Either or both of these tabs or the President's Hundred tab may be worn at one time. When both the special forces tab and the ranger tab are worn, the special forces tab is positioned 1/2 inch below the top of the shoulder seam and the ranger tab is 1/8 inch below the special forces tab. Full-size metal replicas of the special forces and ranger tabs may be worn with ribbons or full-size medals on the Army blue and Army white dress uniforms and the shirts. In this case, they are considered group 4 badges, and placement is as discussed earlier. Dress miniature versions of the ranger and special forces tabs may be worn with miniature medals. In this case they are also considered and handled as group 4 badges. If no badge from groups 1 to 3 is worn, the metal tab replicas may be worn above the ribbons in the same precedence as for wear on the shoulder of the green coat. Subdued special forces and ranger tabs may be worn on the left shoulder of the ACU coats or the field jacket, sewn 1/8 inch below the shoulder seam.

Identification Badges. Identification badges are worn on the pockets or in comparable positions on the uniform coats. No more than two badges may be worn on one pocket or on one side of the coat at a time. When two badges are worn on one pocket or on one side of the coat, the order of precedence is from the wearer's right to left. The Presidential Service identification badge, the Vice-Presidential Service identification badge, the Army staff identification badge, and the Guard, Tomb of the Unknown Soldier identification badge are worn on the right side of the uniform. The Secretary of Defense identification badge and the Joint Chiefs of Staff identification badge are worn on the left side of the uniform. See above for men of the Unified Command and NDU identification badges.

Men. On the blue and white dress uniforms and the shirt, the ID badge is worn centered on the breast pocket, midway between the bottom of the flap and the bottom of the pocket. If two badges are worn, they are equally spaced from right to left on the pocket. On the blue and white mess jackets, the ID badge is worn centered between the upper two buttons on the appropriate side of the jacket.

Blue and White Uniforms—Men

Mess Jackets—Men

Blue and White Uniforms—Women

Mess Jackets—Women

Wear Position for Identification Badges

Women. On the blue and white dress uniforms, the shirt, and the maternity tunic, the ID badge is worn at waist level and centered on either the right or left side as appropriate for the badge. If no other awards or decorations are worn on the shirt, the ID badge may be placed at the same level as the nameplate or approximately 1 inch above the nameplate, depending on which side the badge is worn.

Foreign Badges. Foreign badges, if authorized, may be worn only one at a time, and then only if worn in conjunction with at least one U.S. medal or service ribbon, and only on the blue and white dress uniform coats.

Men wear the foreign badge 1/8 inch above the seam of the right pocket flap or 1/2 inch above any unit awards. Women wear the foreign badge 1/2 inch above the nameplate or 1/2 inch above any unit awards.

EFFECTIVE USE OF AWARDS

The great variety of military awards currently authorized by the Army provide a means of promoting excellence through recognition of outstanding duty performance or the possession of special skills. The current Army system allows unparalleled flexibility through its combination of medals, ribbons, badges, and certificates. Thus, today's officers have at their disposal more tools than ever to motivate their subordinates to achieve the increasingly high expectations of the Army.

Pitfalls multiply with choices, however. Precisely because of this flexibility, the awards that officers accumulate—especially achievement and service medals—generally tend to vary greatly according to the philosophies of their commanders. Some commanders are lavish, while others may be stingy. Some unduly restrict the number that may be awarded, while others may literally force the award of medals by designating percentages of their soldiers who must be recognized periodically. Fortunately, most use their authority wisely and fall somewhere in between. As a result of this wide variance, though, it is difficult to judge an officer by the ribbons he or she wears, unless the viewer has an intimate knowledge of the policies in effect in each unit in which that officer has served. This is one of the reasons why officer promotions and school or command selections are rarely based on the noncombat medals indicated on an ORB or in the official DA photo (although they can influence the elimination or retention of some officers). The old adage "You can't judge a book by its cover" certainly pertains to officers, and all would do well to evaluate fellow officers by their duty performance and professional acumen for exactly this reason.

A Word About Badge Hunting. It is possible for junior officers (and enlisted soldiers as well) to quickly amass a chest full of medals and badges, and their attractive appearance on a uniform can become a siren song for misplaced ambition and misbegotten deeds. Although it is admirable to aspire to the achievements, service, and skills often documented by rows of ribbons and multiple badges and tabs, their accumulation should never become the *main* motivation for an officer. "Badgefinders" are generally unpopular among their peers and mistrusted by their soldiers, because they appear to confuse their duty concept with their vanity. Some commanders, unquestionably, expect their officers to demonstrate certain abilities or to attend certain schools to enhance their professionalism, but it is a good idea to avoid substituting of appearance for substance.

PROMOTING EXCELLENCE BY RECOGNIZING OUTSTANDING PERFORMANCE OF DUTY

A balanced, comprehensive awards program can be a cornerstone in the foundation of a great Army unit. A paucity of awards can frustrate soldiers and limit their confidence in the chain of command. It can also actually hurt their careers, since the promotion system for junior NCOs is partly based on a points system that awards a certain number of points for each medal, badge, or certificate earned by the soldier being considered for promotion. Take the time to write up deserving members of your unit, and insist that the entire chain of command do the same. There is no valid reason why outstanding acts should go unrecognized by the Army. The adaptation of the following techniques can help ensure that soldiers are being properly recognized:

1. Take advantage of the full range of Army awards, including driver's badges, marksmanship badges, and certificates of appreciation and commendation. With the computers and software available today, high-quality certificates are relatively easy to create, and DA Certificates of Achievement are available, too. Any certificate of commendation signed by a battalion commander or higher counts for enlisted promotion points. More importantly, they are a unique memento from the unit and, in later years, will be especially important to recipients. They are also flexible and can legitimately be awarded to family members, helpful post agencies, local businesses, and organizations, as well as soldiers. Finally, they are the appropriate award for excellence in unit competitions, contests, excellence in PT, and so forth and preserve the value and integrity of decorations.
2. Write personal memorandums of appreciation or commendation to deserving soldiers. This takes time, but that is exactly why the recipient will appreciate it. Such a format also allows you to go into detail not normally possible on the certificates that accompany decorations. This is an especially useful technique with senior NCOs and subordinate officers, for whom Army Achievement and Commendation medals are less important professionally. Such correspondence is particularly appropriate when you are leaving the assignment and wish to express your appreciation for the support and hard work rendered "on your watch."
3. Periodically review each soldier's performance, and consider them for achievement awards.
4. Whenever the unit receives special accolades—or whenever *you* do—immediately seek to recognize those who were instrumental in making it possible. After especially successful major tactical exercises, command inspections, and operational deployments, consider those whose performance stood out and initiate achievement awards appropriately. Consider making such a review part of your unit recovery SOP.
5. Track upcoming changes of assignment, transfers, discharges, and retirements, and consider each soldier for a service award. As noted in AR 600-8-22, merely changing station is not a sufficient reason for awarding a decoration, although it may warrant a certificate of appreciation or commendation for the soldier and his or her spouse. Be sure to initiate awards so that they will be ready for presentation before the soldier departs the unit; few things reflect as poorly on unit efficiency or speak more about the command climate than an award that must be forwarded to the soldier's new command. Would the same thing ever happen with an Article 15?

Too many medals can be just as dangerous as too few. In fact, it is actually a more pernicious situation, because it is both an abuse of the system and difficult to undo. Too many medals can skew soldier expectations to the point that unrewarded routine actions can be a source of disgruntlement and can actually *retard* duty performance when soldiers begin to slack off if they are not receiving awards. There are several ways of avoiding this counterproductive situation:

1. Never set quotas for military decorations. This cheapens the awards for all concerned, including those who receive them under these circumstances.
2. Never use decorations as prizes in contests, or set preconditions for awards. These practices are specifically forbidden by AR 600-8-22 (and several subsequent MILPER messages), they cheapen the value of decorations, and they actually limit your flexibility for motivating your soldiers. If you award an AAM to anyone who maxes the PT test, what are you going to award a specialist for three years of outstanding service in the unit? Use trophies (purchased with authorized appropriated funds or perhaps donated by the regimental unit association or the local AUSA chapter), certificates of appreciation or commendation, or memorandums instead. Getting the commanding general, other appropriate commander, or the command sergeant major (CSM) to award coins will work, too, and the awarding officers or CSMs will probably appreciate the opportunity to meet good soldiers.
3. Avoid awarding decorations for routine achievements. Although it sounds strange to most, there are many recent documented cases of soldiers being awarded decorations for digging a good fighting position, maintaining his or her assigned vehicle well, and the like. Such actions are far more appropriately rewarded with a sincere, public verbal compliment than with a decoration. Such actions warrant a certificate at most.
4. Beware of the "automatic award" syndrome. Sometimes, in an effort to ensure that soldiers are recognized for meritorious service in a timely fashion (a good thing), awards can be generated for the undeserving (a very bad thing indeed). Pay attention to that roster of 30-60-90 day losses that gets passed out at the command and staff meeting, and consult the chain of command regarding the merits of an award. Sometimes, the answers will be surprising.

PREPARING AWARD RECOMMENDATIONS AND CITATIONS

With the availability of computers and automated forms, there is less excuse than ever for late or poorly prepared award recommendations or certificates. Nevertheless, the Information Age has been accompanied by a documented decline in literacy across the entire American cultural spectrum. Even though the recommendation forms for many awards have become "bulletized," it may be difficult for many NCOs and junior officers to prepare lucid, grammatically correct, personalized citations that will reflect well on the Army and the recipient. Take special care to mentor your subordinates to create citations with a "professional and dignified" appearance, as required by AR 600-8-22. After all, even if the certificate gets tossed in a footlocker, still in the green padded folder, today, it will evoke pride in a parent's eye tomorrow and may someday be looked upon with awe and envy by the recipient's grandchildren. If you have trouble writing an appropriate citation, seek help from a qualified mentor, and then practice, practice, practice. It is a basic competency you owe to your soldiers, to their families, and to the Army.

DERIVING MAXIMUM BENEFIT FROM MILITARY AWARDS PROGRAMS

1. As with all issues in the Army, be fair. Don't let the soldiers whose platoon or squad leader majored in English in college garner a disproportionate share of decorations. Similarly, make sure that the good soldiers serving under an unconscientious or semiliterate leader still get recognized for their achievements and service, and take action to correct the sloth or communications problems among the leadership.
2. Do not base awards on the grade of the intended recipient. This is a gross violation of the intent, as well as the letter, of the regulation. AR 600-8-22 states that awards "should reflect both the individual's level of responsibility and his or her manner of performance. The degree to which an individual's achievement or service enhanced the readiness or effectiveness of his or her organization will be the predominant factor." There is no reason why a specialist who performs brilliantly and has a major positive impact on unit effectiveness should not get a Meritorious Service Medal (MSM) as an achievement or service award. Similarly, nothing says that a major has to get an MSM. Beyond the clear mandate of the regulation, there is also a practical reason: such artificial constraints limit the extent to which a soldier may be rewarded for outstanding performance. If you want limited performance from your subordinates, then limit their rewards.

   Similarly, undeserved awards, based solely on rank, can contribute to setting standards that are too low. When an officer or NCO receives an award for "punching his ticket" in a job, and another one who makes notable contributions receives the same award, a clear message is perceived by some. Remember, efficiency reports are personal and confidential, but awards are presented in front of the world.
3. Never base an award on the recipient's past awards. This (il)logic, which has been used in some units in recent years, is so alien to the U.S. Army culture that it is difficult to determine the exact origin of this obnoxious practice. In other armies—particularly those of monarchies or formerly monarchical states, where awards once equated with privilege or caste—awards were or are cumulative. This is absolutely not the case in the U.S. Army, where our meritocratic traditions dictate that multiple similarly outstanding acts be recognized in a comparable fashion. This is why we have oak leaf clusters, while some other armies have a cumulative hierarchy of awards.

4. Whenever possible, present awards in a dignified setting, with "an air of formality and with fitting ceremony," as directed by AR 600-8-22. *You* may already have three AAMs, four ARCOMs, and an MSM or two, but for the specialist who is being discharged after four years of outstanding service, that AAM or ARCOM is likely to be the ultimate recognition of his or her military contributions. Make it meaningful.
   — Do it in front of the unit, with the citation read by someone who reads flawlessly and has a commanding voice.
   — Invite family members to be present.
   — Ensure that the moment of presentation is photographically recorded, and get the pictures or tapes to the recipients ASAP.
   — Consider holding the formation or ceremony at a place with special meaning for the unit and the individual recipients. Monuments commemorating famous unit or individual acts, the unit museum or trophy room, or even a site with a dramatic, dignified background are all appropriate locations.
   — Make sure that the presenter—whether it is you or a higher-ranking officer—has some sincere, personal, pertinent remarks to make. It is better to say nothing at all than to utter some phony, artificial, or, worse, inaccurate comments about the recipient. Such a gaffe will ruin the moment for all concerned and justly earn the derision of every soldier in the audience.
5. Every U.S. military decoration has a space on the reverse of the medal for engraving the name of the awardee. Most units have engraving capability; see whoever engraves the unit's coins. Ensure that all medals are engraved in advance of presentation; if impracticable, inform the recipient that he or she may mail the decoration (or Good Conduct Medal) to the Commander, U.S. Army Support Activity—Philadelphia, 2800 South 20th Street, Philadelphia, Pennsylvania, 19101-3460, for engraving at government expense.
6. If, for whatever reason, an award is not ready when a soldier leaves the assignment in which it was earned, follow this procedure:
   — Determine the recipient's preferences regarding the forwarding address. When the decoration is finally ready, does he or she want it to go to the gaining commander, or to him or her personally?
   — Forward the award, citation, orders, and a personal letter of apology, if appropriate (and it usually is), to the awardee via registered mail, signed receipt requested, or some other means facilitating positive control. Follow up to ensure that it gets to the soldier. Probably the only thing that reflects more poorly on a unit than failing to recognize outstanding service or achievement before a soldier departs is the "ARCOM's in the mail" promise that never comes true.

24

Retirement

The laws governing retirement from the armed forces are important to the government in order to maintain capably led armed forces, and they are important to the individual who elects to follow a military career. The purpose of retirement, voluntary and involuntary, is to assure that the Army has an officer corps of the highest caliber. To achieve this standard, officers are retired by statutory provisions, involving physical limitations, age, and time in grade, and by board action when performance of duty or conduct is below standard. The officer who serves well need not fear early retirement. The officer who desires a second career may retire with full and deserved honor (but at reduced pay) at twenty years, creating a promotion vacancy and stimulus for those who choose to serve a longer career. Retirement laws that are fair and equitable, both to the government and to the individual officer, are essential to military personnel administration.

This chapter will serve to orient the officer on this important subject; however, serious study of retirement should include the following references: AR 600-8-24, *Officer Transfers and Discharges*; AR 635-40, *Disability Evaluation for Retention, Retirement, or Separation*; and AR 600-8-7, *Retirement Services Program*. For additional information about retirement, visit the Army Retirement Services Office website, https://myarmybenefits.us.army.mil.

STATUS OF RETIRED OFFICERS

It is noteworthy that an officer who is retired retains status as an officer with many of its rights and privileges. He or she remains on the official roles of the Army and may be returned to active duty under applicable laws and departmental regulations. By tradition and law, retired Army personnel are considered to be in a real sense members of the Army. Therefore, close affinity does exist between the active Army and its retired personnel. The retired list is not a roster of former officers; it is a designation of personnel who, by age, length of service, or disability, should be regarded as having been transferred from one Army category to another.

Upon retirement from the Army, members are placed on one of the following retired lists: U.S. Army Retired List, for Regular Army commissioned officers, warrant officers, and enlisted personnel retired for any reason, who are granted retired pay; Army of the United States Retired List, for officers other than Regular Army officers, who are members and former members of the reserve components, and personnel who served in the Army of the United States without component, who are granted retired pay, and retired warrant officers and enlisted personnel of the Regular Army who by reason of service in temporary commissioned grade are entitled to receive retired pay of the commissioned grade; and Temporary Disability Retired List, for officers and enlisted personnel placed on the list in accordance with law for physical disability that may be of a permanent nature.

It is true that the beginning of military retirement calls for a major readjustment in the lives of most officers and their families. There is a real need for the officer to gain a clear understanding of his or her own post-retirement situation and the probable wishes of family members as they can be foreseen. A reasonable degree of planning for the period of retirement is advisable, and it should start long before retirement is requested or ordered.

Medical Service for Retired Individuals and Their Family Members. Retired personnel and their family members are authorized to receive medical care and hospitalization at facilities of the uniformed services, when available, and at civilian facilities if military facilities are unavailable. Care in civilian facilities is under the provisions of Tricare until age 65; thereafter coverage is under the provisions of Tricare for Life. See chapter 19, Personal and Financial Planning.

Use of Post and Station Facilities. Retired members, their family members, and unremarried widowed spouses are authorized the use of various facilities on military installations when facilities are available. Proper identification is required. This privilege includes authority to use and patronize the following (DA Pamphlet 600-5):

- Commissaries.
- Post exchanges.
- Clothing sales stores.
- Laundry and dry-cleaning plants.
- Military theaters.
- Army recreation service facilities.
- Officers' clubs (upon application to and approval of the club concerned).

As interservice custom, retired members are welcomed into club and other facilities of the sister services.

Travel to Home. At any time within one year after retirement, an officer is entitled to be moved, at government expense, to any authorized location in the world where he or she intends to establish a bona fide home. For details, consult the installation transportation office (ITO).

Use of Military Title after Retirement. See chapter 1, Heritage, Customs, and Courtesies of the Army.

Retirement Services Offices. Retirement services offices have been established at Headquarters, Department of the Army, in Army areas throughout the United States, and in some overseas commands. These are the offices from which retired individuals may obtain guidance and assistance regarding their rights, benefits, and privileges. Retired members are invited to write or visit these units whenever assistance is needed in their personal affairs.

Retired Grade. In accordance with the provisions of DOPMA, officers promoted to lieutenant colonel and above after 15 September 1981 must serve in the new grade for at least three years to be eligible to retire in that grade. Unless specifically exempted by the President, officers who elect to retire prior to completion of three years' service in grade will be retired at the next lower grade in which they have served for a minimum of six months.

Retired Status. A Regular Army officer placed on the retired list is still an officer of the United States Army (31 Ct. Cl. 35), subject to return to active duty under provisions of our laws. In time of war or emergency, many officers are recalled, although few (if any) are restored to active duty without their consent.

Certificate of Retirement. Each member of the Army, upon retirement, will be furnished a certificate of retirement, DD Form 363A, by the Adjutant General.

VOLUNTARY RETIREMENT

Voluntary retirement includes all types of retirement that the officer initiates by his or her own application. Approval of an officer's voluntary retirement may be mandatory or discretionary, dependent on specific provisions of the law under which retirement is sought.

After Twenty Years' Service. A commissioned officer of the Army who has at least twenty years of active federal service, at least ten years of which have been as a commissioned officer, may upon personal application and in the discretion of the Secretary of the Army be retired.

After Thirty Years' Service. An officer of the Army who has at least thirty years of service may upon his or her own application and in the discretion of the President be retired. After fifty years of service, an officer shall be retired upon personal request without need of approval.

Policy on Acceptance of Retirement Applications. When an officer has completed all service obligations prescribed in AR 600-8-24, his or her retirement application normally is approved. During periods of emergency, the Department of the Army may announce restrictions on the approval of retirement applications. Individuals who possess critical skills or hold assignments in key positions may be denied retirement (or resignation). Officers should consult their unit personnel officer for information that may influence their action.

MANDATORY RETIREMENT

Mandatory retirements are those retirements required by law and must be effected regardless of the desire of the individual or of the preferences of the Department of the Army. Computation of service is based on AR 600-8-24.

Retirement for Age. Unless retired or separated at an earlier date, each commissioned officer whose regular grade is below major general, other than a professor or the registrar of the USMA, shall be retired on becoming sixty-two years of age. Unless retired or separated earlier, each commissioned officer whose regular grade is major general, and whose retirement for length of service has been deferred, shall be retired at age sixty-two or, if not deferred, will be retired at age sixty-four.

Retirement for Length of Service. Unless retired or separated at an earlier date, officers shall be retired for length of service as follows:

Major Generals. Each officer in the grade of major general shall be retired on the fifth anniversary of the date of appointment in that grade or on the thirtieth day after completion of thirty-five years of service, whichever is later.

Brigadier Generals. Each officer in the grade of brigadier general, other than a professor of the USMA, shall be retired on the fifth anniversary of the date of his or her appointment in that grade or on the thirtieth day after completion of thirty-five years' service, whichever is later.

Colonels. Each officer in the grade of colonel shall be retired on the fifth anniversary of the date of his or her appointment in that grade or on the thirtieth day after completion of thirty years of service, whichever is later. Also, upon the determination of the Secretary of the Army that there are too many colonels on the active list, the Secretary may convene a board of five general officers to recommend such officers for early retirement.

Lieutenant Colonels. Each officer in the grade of lieutenant colonel shall be retired on the thirtieth day after completion of twenty-eight years of service.

In each of the above cases, retirement takes place on the first day of the month following the month in which service requirements are met. If the officer is on a recommended list for promotion to the next higher grade, the officer will be retained and will fall under the criteria for the next grade when promoted.

Selective Retirement. In accordance with the Defense Officer Personnel Management Act (DOPMA), lieutenant colonels and colonels may be selected for early retirement by board action, when rapid reduction of the numbers of officers in these grades is necessary.

Deferred Officers Not Recommended for Promotion—Majors, Captains, and First Lieutenants. Under the provisions of the DOPMA, officers in the grades of first lieutenant, captain, and major who twice fail to be selected for promotion will be involuntarily separated from the Army. However, DOPMA also contains provisions that allow the Secretary of the Army to selectively continue in grade certain majors to fill Army specialty needs. In such cases, majors may be continued in grade to a maximum of twenty-four years' service.

Exceptions. There are exceptions for officers of the Army Nurse Corps and the Army Medical Specialist Corps. The Secretary of the Army may defer the retirement of a lieutenant colonel in either of those corps until the last day of the month in which the officer completes thirty years and thirty days of service.

Removal from Active List. If an officer is removed from the active list of the Regular Army under the provisions of Title 10, USC, Chapter 359, and, if on the date of removal he or she is eligible for voluntary retirement under any provision of law then in effect, the officer shall be retired in the grade and with the retired pay to which entitled had the retirement been upon his or her own application.

RETIREMENT FOR PHYSICAL DISABILITY

The first stage in any proceeding for separation for physical reasons is a finding by the service that the person, by reason of a disability, is not qualified to perform his or her duties. If a person is kept on duty, there are, of course, no separation proceedings. But if a finding is made that the person cannot be retained in service, the proceedings enter a second stage. If the disability was due to intentional misconduct or willful neglect or incurred during unauthorized absence, the government gives the member nothing and merely separates him or her.

If the disability was not due to misconduct or neglect, the next question is: Is the disability 30 percent or more under the VA standard rating? (Loss of an eye or loss of use of a limb, and chronic, severe, high blood pressure are disabilities of 30 percent or more; loss of one or two fingers or one or two toes, loss of hearing in one ear, or defects or scars that do not seriously interfere with functions are not.) If the disability is less than 30 percent, no retirement is given (unless the individual qualifies for retirement by length of service). Instead, the person is given *disability severance pay*, which is two months' basic active-duty pay for each year of active service, to a maximum of two full years' active pay. Half or more of a year counts as a full year.

Officers with less than eight years' service whose disability is not the proximate result of the performance of active duty are entitled only to severance pay.

An officer of the Regular Army retired for physical disability incurred while serving under a temporary appointment in a higher grade shall have the rank and receive retired pay computed as otherwise provided by law for officers of such higher grade.

The laws are quite favorable as to the protection afforded officers who are recalled from retirement and incur physical disability or additional physical disability.

A final point about retirements involving physical disabilities. Annual physical screenings are required to determine the ongoing nature of the disability. Disabled retirees can stay on the Temporary Disability Retired List (TDRL) for at least five years by law. If a subsequent medical screening determines that the disability has improved, the retiree would be moved to either the U.S. Army Retired List, the Army of the United States Retired List, or could even be returned to active duty if still eligible. If, however, a reevaluation determines that the disability is permanent and has no chance of improving, the retiree would then be moved to the Permanent Disability Retired List (PDRL).

AFTER ARMY RETIREMENT—WHAT NEXT?

It is prudent for Army officers to anticipate the life they wish to lead after military retirement, and to plan with long-range foresight the vocation they may choose to follow as a second career. Under our current laws, officers who begin their Army careers at the normal ages of twenty-one to twenty-four will retire for length of service in their early fifties; or if they elect to retire after twenty years' service, in their early forties. Unless disability is a factor, many officers will wish to enter a new vocation. This is a critical decision because salability in the commercial job market begins to drop dramatically after about age forty. Give thorough consideration to the relative advantage of an additional ten years of base pay plus the increased retired pay versus the likelihood of obtaining acceptable civilian employment after age fifty. Officers may require additional income during these years to meet family obligations or to realize special desires. They may follow a new vocation to achieve results in a different field. All the reasons will be personal, but, it is prudent for everyone to thoroughly consider the possibilities, starting long before retirement.

COMPUTATION OF RETIRED PAY

Computation of retired pay used to be a relatively straightforward matter. It was set as a percentage of active-duty pay and increased with increases in active-duty pay. However, the Military Pay Act of 1963 established the principle that retired pay would be subject to cost-of-living adjustments without regard to active-duty pay scales. Further changes to the law in 1980 and 1986 complicated the calculation and resulted in the military now operating on three distinct, although related, retirement systems, depending on when the retired member entered military service. Your understanding of how these laws affect you is essential to your long-range planning.

Service Prior to 8 September 1980. For soldiers whose date of initial entry into military service (DIEMS) is prior to 8 September 1980, retired pay is calculated on the basis of 2.5 percent × the years of service × the final basic pay to which entitled prior to retirement. Thus, a soldier with a final base pay of $4,000 per month ($48,000 per year) retiring after 20 years' service would receive retired pay of 2.5 percent × 20 × $48,000 = 50 percent × $48,000 = $24,000 per year. The maximum retired pay of 75 percent of basic pay is reached at 30 years' service.

Soldiers Entering After 8 September 1980 and Prior to 1 August 1986. For soldiers whose DIEMS is on or after 8 September 1980, but before 1 August 1986, retired pay is calculated using the same multiplier for years of service, but using the average of the three highest years of basic pay. Thus, if the soldier in the previous example had a final base pay of $4,000 per month but had received this pay for only one year, with the previous two years being at the rate of $3,600 per month, the average base pay for the three-year period would be $44,800. Retired pay would then be calculated as 50 percent × $44,800 = $22,400 per year.

Soldiers Entering After 1 August 1986. For soldiers whose DIEMS is on or after 1 August 1986, the same high-three average pay years are used, and retirement is still calculated on the basis of 2.5 percent for each year of service, but with a penalty of 1 percent applied for each year of service less than 30 years. Using the previous example, for retirement at 20 years' service, the multiplier becomes 40 percent instead of 50 percent, calculated as 2.5 percent × 20 – 1 percent × 10. The retired pay based on the high-three years of base pay thus would be 40 percent × $44,800 = $17,920 per year. At the member's age sixty-two, the penalty would be removed and the retired pay recalculated as 50 percent of the high-three average base pay. For soldiers who entered the Army on or after 1 August 1986, any cost-of-living adjustment (COLA) increase in retired pay will be at a reduced rate based on the increase in the consumer price index minus 1 percent (CPI – 1). At the member's age sixty-two, there would be a one-time adjustment to the full COLA level, after which any further adjustments will be at CPI – 1 for life.

Beginning in 2001, an alternative was offered to the straight high-three program. Officers whose DIEMS is on or after 1 August 1986 may now be eligible to elect a $30,000 career status bonus (CSB) in their fifteenth year of active duty if they agree to receive reduced retired pay and cost-of-living adjustments under the CSB/REDUX retired pay plan. The CSB provides current cash for investing, major purchases, or setting up a business after retirement, and the REDUX portion determines retirement income. Thus, the longer one's career, the higher that income.

CBS/REDUX and high-three systems differ in their multiplier calculation. Also, CBS/REDUX has a catch-up increase at age sixty-two that brings the REDUX retired pay back to the same amount paid under the high-three system. CBS/REDUX is the only military retirement system with this feature.

For example, under the CBS/REDUX program, each of the first 20 years of service is worth 2.0 percent toward the retirement multiplier. But each year after the 20th is worth 3.5 percent. Hence, 2.0 percent × 20 years = 40 percent. But a 30-year career is computed by 2.0 percent times the first 20 years plus 3.5 percent for the 10 years beyond 20, resulting in 75 percent. This means that the retirement multiplier for a 30-year career is the same for both REDUX and high-three.

REDUX and high-three systems also differ in the annual cost-of-living adjustments (COLAs). Under the REDUX system, the COLA is 1 percent less than under the high-three system. Also, REDUX has a catch-up increase at age sixty-two that brings the REDUX retired pay back to the same amount paid under the high-three system. REDUX is the only military retirement system with this feature.

The $30,000 CSB can be taken in one lump sum or over a period of up to five years. In addition, you can defer your tax liability on the money received under the CBS/REDUX plan by putting it in a Thrift Savings Plan, discussed below.

Which program is right for you depends on your personal circumstances. Factors which should be considered in choosing your retirement plan include length of military career, age, use of bonus money, inflation, and return on investment. To fully understand this complex decision which affects the rest of your life, you should talk to your family, a retirement services coordinator, and your accountant.

RETIREMENT: AFTER TWENTY OR AFTER THIRTY YEARS' SERVICE?

An officer may apply for retirement after completing twenty years of active service. Such retirements retain the right to retiree pay and other benefits that are discussed in chapter 19, Personal and Financial Planning. The genesis of the present system was the Officer Personnel Act of 1947.

This law was enacted by Congress after thorough consideration. At the time of its enactment, vacancies in the higher grades of the Army had been so limited as to create a history of severe promotion stagnation. It was recognized that to provide attractive Army careers there must be sufficient vacancies to permit capable individuals to reach grades above lieutenant colonel at ages permitting adequate utilization by the government. Retirement of officers who wished to terminate their active service at twenty years permitted other officers who wished to remain to be promoted and to hold the positions of high responsibility to which their grades pertained. The fact that many officers have chosen to retire with less than thirty years' service has been of considerable concern to some members of Congress. The changes to the law in 1980 and 1986 regarding computation of retired pay were obvious attempts to induce officers to serve longer careers to avoid the cost penalty to the government of early retirement. However, there is still an opportunity for retirement between twenty and thirty years' service if desired.

Although many officers wish to remain in active service until the date of compulsory retirement, that is not the case with all officers. Situations develop during long years of service, which could not possibly have been foreseen at appointment, that make it necessary or desirable for an officer to terminate a military career, often with great regret. Such reasons are usually very personal. They may be caused by a cruel health situation of a family member that precludes the change-of-station situation of the Army officer. They may result from financial requirements, coupled with opportunity to secure a higher income. The officer may contemplate his or her military record, conclude that it offers little chance of high success during the remaining years, and prefer to leave the Army to start an entirely new career. There are other personal reasons, all understandable but not necessarily overriding those for not retiring.

The Wisdom in Thorough Analysis. The officer should approach this decision with care. There is need for thorough analysis of his or her present and future prospects in the Army, the forces suggesting retirement, and the civilian expectancy for employment and living. Retirement is final. There have been tragic mistakes by some officers who found after retirement that their decision was unwise. Others have been highly successful and entered rewarding lives. It is because of these diverse developments that this discussion is included. It is not intended to advise officers to retire or not to retire, as they attain eligibility to choose. The purpose is to assist in understanding and evaluating.

The Change in Living Conditions. The officer who loves the service life, and his or her family who finds it equally attractive, will undergo a serious adjustment if the tie is severed. Some of the customs and benefits officers and their families take for granted have real values. These intangibles merit thought.

The Rewards of the Final Ten Years. During the final ten years of service, your assignments will be among those of heaviest responsibility and the greatest interest. You will have the opportunity for selection to high position. You will receive some of the benefits of seniority, such as assignment of more desirable quarters. The officer with an outstanding record who terminates active service forfeits prospects of great value and of high rewards.

Consider the Pay Shrinkage. The need for prudent analysis is illustrated by comparing the retirement pay of a lieutenant colonel with twenty years' service who might have retired as a colonel with thirty years' service. Other examples may be computed from the pay tables.

In the case at hand, however, the colonel's monthly retired pay will be nearly twice that of the lieutenant colonel, the lieutenant colonel's retired pay being computed at only 50 percent of basic active-duty pay, but the colonel's based on 75 percent. This disparity increases to a factor of nearly three using the retirement rules for officers entering the service after 1 August 1986.

In candor, there is another side to this coin. Current promotion selection rates require many good officers to retire as majors after twenty years of service. Also, fewer than half the lieutenant colonels can attain the grade of colonel under present law, because of the lack of vacancies. A similar situation exists for warrant officers contemplating promotion from CW3 to CW4 and CW5. Lieutenant colonels may anticipate mandatory retirement upon completing twenty-eight years' service. Suppose a commissioned officer approaching the end of his or her twentieth year contemplates the future and estimates the chances of reaching the grade of colonel as low, an estimate perhaps gained by consulting the career manager of OPMD in Washington. Now the concern involves the difference in retired pay computed at 50 percent (twenty years' service) versus 70 percent (twenty-eight years' service), both computations based on the pay of a lieutenant colonel. Comparable numbers for an officer who entered the Army on or after 1 August 1986 would be 40 percent (twenty years' service) and 68 percent (twenty-eight years' service).

This comparison discloses a different (and difficult) situation. The officer who wishes to make the most of remaining employable years may find these facts to be convincing reasons for twenty-year retirement to start a new career. He or she may have financial obligations that cannot be satisfied by the prospective retired income in the grade of lieutenant colonel. This discussion states the factual comparison. It advocates neither retirement at the earliest opportunity nor remaining on active duty without regard to the probabilities. The decision is wholly personal and should be based on personal preferences, financial considerations, opportunities in another vocation, and the many likes and dislikes of the individuals concerned.

Consider the Reduction in Security. As discussed in chapter 19, Personal and Financial Planning, important benefits provided active-duty personnel are terminated or curtailed upon retirement. These are subjects to be identified and evaluated. They must be faced by officers who retire after thirty years' service, too, but the officer retiring at an earlier year accepts them at a younger age.

In Your Own Case, What About Civilian Employment? It is certainly true that many officers of long service who have resigned or retired have obtained civilian employment with above-average earnings, and a large number hold positions of the highest importance in civilian fields. We may be proud of our fellow officers who have won such outstanding recognition. One might assume it to be a group expectancy, but it isn't universal, and no certainty exists in changing times and in nongovernmental, highly competitive, and variable occupations. It is an individual matter entirely, worth most careful examination before forwarding the retirement application.

What is your education? What has been your service training and experience fitting you for specific civilian employment? Have you planned to undertake training to prepare yourself for a profession or employment?

In any case, you will have to do your homework. Recent experience has shown that it takes at least one year for the average officer to find a good job in the civilian economy. The good news is that the job search can be started while you are still on active duty.

Retiree Recalls. Given the stress on Army manpower during the Global War on Terrorism, the Army used this program to recall large numbers of highly trained and qualified retirees who volunteer to be called back to active duty for finite terms, usually for 365 days. Given the end of the missions in Iraq and Afghanistan, this program has been cut back significantly; however, opportunities for recall do still exist. Retired officers interested in this opportunity must meet Army medical and height/weight standards, obtain a memorandum of justification signed by the first general officer in the gaining unit's chain of command, and be slotted against a valid, vacant Active Component duty position. More information can be obtained from your local servicing Retirement Services Office.

Summary. Early voluntary retirement may be wise for some officers, heartbreakingly unwise for others. Decisions should be reached only after an evaluation such as you have been trained to make in military service: get all the facts, analyze them carefully, then decide which is best for you and for your family.

THRIFT SAVINGS PLAN

Part of your retirement planning should be your earlier decisions about investing in the Thrift Savings Plan (TSP) instituted in the DoD in 2001.

The TSP is a federal government-sponsored retirement savings and investment plan, offering the same type of savings and tax benefits that many private corporations offer their employees under so-called "401K" plans.

Created as part of the National Defense Authorization Act for Fiscal Year 2001, the TSP is an optional plan, which allows a uniformed servicemember to set aside funds for retirement, which is separate from retired pay.

The contribution to the TSP is collected and deducted before taxes are calculated, and the funds put into the TSP which is tax-deferred, thereby reducing your tax bill now.

Under TSP, you can contribute from 1 to 7 percent of your basic pay each pay period, as well as from 1 to 100 percent of any incentive pay or special pay (including bonus pay) you receive, up to the limits established by the Internal Revenue Code.

These funds are yours even if you do not serve twenty or more years in the Army.

The Federal Retirement Thrift Investment Board administers the TSP and is required by law to manage the TSP prudently and solely in the interests of the participants and their beneficiaries. Money in the TSP and earnings on that money cannot be used for any purpose other than paying benefits to participants and their beneficiaries and paying TSP administrative expenses.

The TSP offers a choice of five investment funds: Government Securities Investment (G) Fund; Fixed Income Index Investment (F) Fund; Common Stock Index Investment (C) Fund; Small Capitalization Stock Index Investment (S) Fund; and International Stock Index Investment (I) Fund.

If you have invested in the TSP while on active duty, you would have to roll over your TSP earnings into an IRA or other account upon retirement, as contributions to TSP cannot be made after retirement. Military retirees who are subsequently employed by the federal government could then enroll in the Federal Employee TSP. For more information on the TSP Program, visit www.tsp.gov.

RETIREMENT OF NATIONAL GUARD AND RESERVE OFFICERS

Retirement for Physical Disability. The laws governing retirement for physical disability apply equally to all officers on active duty whether of the Regular Army or the reserve components. This principle was first established by enactment of the Act of 3 April 1939 and is affirmed in the Career Compensation Act of 1949 and codified in Title 10, USC, Chapter 61.

Retirement for Age and Length of Service. National Guard and Reserve officers become eligible for retirement based on age and length of service in accordance with the provisions of Title 10, USC, Section 1331. They must meet the following criteria to be eligible:

- Have attained age sixty.
- Have completed a minimum of twenty years of qualifying service.
- Have served at least the last eight years of qualifying service as a member of a reserve component. The last eight years of qualifying service need not be the last eight years of military service, nor do they have to be continuous.

In addition, to qualify for receipt of retired pay under the provisions of Title 10, the officer must meet the following requirements:

- Must not be entitled to retired pay from the armed forces under any other provision of law.
- Must not have elected to receive disability severance pay in lieu of retired pay at age sixty.
- Must not fall within the purview of the Hiss Act (i.e., must not have been convicted of a national security-type offense, or must not have refused to testify before a duly constituted judicial or congressional proceeding on a matter related to the national security).

There are many forms and categories of qualifying service. Officers are advised to consult AR 135-180 and their personnel officer for information as to the creditability of their own service.

PROFESSIONAL ASSOCIATIONS

Military Officers Association of America. The purpose of the Military Officers Association of America (MOAA) is to aid retired personnel in every proper and legitimate manner. Included is the presentation of subjects important to retired officers before the appropriate members of Congress. It provides assistance in securing employment and offers many other helpful services. It is located at 201 North Washington Street, Alexandria, Virginia, 22314; phone (703) 549-2311, (800) 234-6622. Their website is http://www.moaa.org.

United States Army Warrant Officer Association. The United States Army Warrant Officer Association is a professional association of Army warrant officers of all components, both active and retired, dedicated to recommending programs for the improvement of the Army and the Warrant Officer Corps and to disseminating professional information to warrant officers in the field. Services include a bimonthly newspaper to members and liaison with government agencies. Membership is open to any Army warrant officer who holds or has held a warrant issued by the Secretary of the Army. The address of the association is 462 Herndon Parkway, Suite 207, Herndon, Virginia, 20170; phone (703) 742-7727. The website is www.usawoa.org.

25

Professional Military Web Forums

The cutting-edge knowledge of the Army resides in the minds of leaders at the tip of the spear—leaders in the experience right now. Connecting those leaders in conversation with each other as well as with those who will follow in their footsteps brings together the Army's greatest knowledge resources. Professional forums (aka online communities) give leaders access to each other and to the knowledge that they create together—improving our effectiveness and advancing the profession.

The Center for Junior Officers (https://juniorofficer.army.mil)) is one example. It is a one-stop shop to find sites and resources for platoon leaders and company commanders to learn from and interact with peers from across the Army. Additional professional forum examples include:

Company Command: CC *is* company commanders. It is where we connect. It is our HMMWV hood gathering place. Join the conversation . . . become more effective. (http://CC.army.mil)

S1 NET: Part of an Army-wide knowledge management system supporting the flow of knowledge throughout the Human Resources (HR) community. (https://s1net.army.mil)

S3-XO Net: Connects geographically distributed S3s and XOs through collaboration, information exchange, and sharing of best practices across all boundaries. (https://s3-xonet.army.mil)

LOG NET: A professional forum where logisticians share their latest thoughts, ideas, tactics, techniques, and procedures (TTPs), regardless of rank or duty position. (https://lognet.army.mil)

Warrant Officer Net: A place for all warrant officers, current, past and future to connect, share ideas and experiences. (http://bcks.army.mil)

FRG Leader: A community of past, present, and future Family Readiness Group leaders committed to unlocking the potential of their FRGs. (http://frgleader.army.mil)

• • •

All of the forums use AKO Single Sign On (SSO). Participants log in using their AKO ID or Common Access Card (CAC) and their password, and request membership to the communities they wish to join.

26

Professional Movie Guide

The following motion pictures, listed alphabetically in subject order, will be of great interest to officers of all ranks who seek a better understanding of their profession. Most of them could form the basis of great professional development classes, too. Most are available as DVDs through Netflix or on Amazon Prime.

EIGHTEENTH AND EARLY NINETEENTH-CENTURY WARFARE

***Master and Commander: The Far Side of the World* (2003)**

Although dealing with the British Navy of the nineteenth century, this epic film richly portrays the role of an officer who is torn between friendship and his duty. It is an object lesson in the loneliness of command when mission requires one to choose the harder right over the easier and more popular wrong. It addresses the themes of maintaining discipline and the internal struggle of sending men into situations where they may die. An excellent study in leadership under pressure and the embodiment of the adage "Mission First, Men Always."

***Mutiny on the Bounty* (1935)**

Another nautical saga with universal leadership implications. Important as a study in the trade-offs involved in balancing "mission" and "soldiers" (or in this case, sailors), this film provides no clear-cut solution. Nevertheless, it offers a great depiction of the dangers inherent in allowing this yin and yang relationship to become unhinged. Carefully watch the scenes in the officers' mess and learn the origin of the term "eating cheese." You can watch the later versions (the 1962 film of the same name, or 1984's *The Bounty*), but you won't find out about the cheese.

THE CIVIL WAR

***Gettysburg* (1993)**

This film adaptation of Michael Shaara's influential novel *The Killer Angels* is a powerful study in leadership and command during the American Civil War's most important battle. The sections of the film presenting Colonel Joshua

Lawrence Chamberlain's crucial defense of Little Round Top represent a stirring tribute to America's citizen soldiers.

***Glory* (1989)**
In addition to the exposition of the story of the segregated 54th Massachusetts Infantry Regiment during the Civil War, this film is important as a study in leadership, both good and bad. The regimental commander's dealings with uncooperative higher staff officers, unscrupulous peers, and subordinates with widely varying attitudes provide great food for thought—and some, but not all, of the answers. Also a great film for understanding the "family" nature of a fine unit.

***The Red Badge of Courage* (1951)**
A Civil War epic, based on the novella by Stephen Crane, that provides a serious examination of the meaning of courage. (Hint: it's not the absence of fear.) Against a backdrop of some spectacular Civil War battle scenes, the film also provokes thought about the theme of redemption within the military context. World War II icons Audie Murphy and cartoonist Bill Mauldin turn in superb performances as conflicted Civil War soldiers.

THE INDIAN WARS

***Fort Apache* (1948)**
Great study of the "Old Army," set on the southwestern plains frontier in the late nineteenth century. Want to learn about the origins of the Army social system? Need to know more about the evolution of the roles of officers and NCOs? Ever wonder where the expression "Yo!" started? Watch this one. This film is also a fine examination of the relationships between strong-minded officers and the effects of those on the unit and their subordinates.

***She Wore a Yellow Ribbon* (1949)**
Another John Ford–directed film of the Fort Apache genre. Allows additional insights into the Army traditions of duty, loyalty, and honor, all set against the backdrop of hard soldiering on the late nineteenth-century western frontier. Consider the simplicity—and austerity—of Army life in the days before stress-reduction classes and tobacco-cessation policies, but with all the challenges of the full spectrum of "operations other than war."

***Ulzana's Raid* (1972)**
This story of a small cavalry patrol and its civilian scout chasing renegade Apaches in the American Southwest is the most authentically brutal depiction of Indian Wars combat ever filmed. During the second half of the nineteenth century nearly the entire U.S. Army was deployed in America's Frontier West, and that decades' long effort constituted one of the Army's most defining eras. This film rejects the all-too-common "good guys versus bad guys" stereotypes to honestly depict the protagonists on both sides. The problems of leadership and command

faced by the cavalry patrol leader—a young lieutenant leading his first combat mission—present timeless lessons for all officers.

LATE NINETEENTH/EARLY TWENTIETH-CENTURY WARFARE

Breaker Morant (1979)

Set in South Africa during the Second Boer War (1899–1902), this film provides chilling examinations of what happens when officers lose control of their units, lose control of themselves, and stop being soldiers and start being politicians. A riveting examination of the importance of integrity, as well as a provocative look at war crimes, real and fancied.

The Four Feathers (1939)

A tale of loyalty and honor, set in the British Army in the Sudan in the late nineteenth century. Especially useful for examining the concepts of redemption and, even more importantly, friendship as they might apply in a military environment. A more modern version of this film was made in 2002, which provides a more accurate and less romantic vision of the story.

Khartoum (1966)

Although it addresses the issues and incidents surrounding the failures of British foreign policy in the Sudan in the late nineteenth century, this film also provokes thought about some issues of great import for modern American officers. In fact, beneath the lavish, DeMille-style mass action scenes, the movie is really about the tension between obedience to the law and to one's conscience, as well as the dangers of using military force to accomplish convenient, purely domestic political goals. This film is timelier than ever for leaders of an Army increasingly committed to the missions expected of the world's only superpower.

Zulu (1964)

Want to know why rear-area security is so important? Take a sobering look at this (mostly true) depiction of the incredibly heroic action of the British infantry company that was left behind to guard the supply trains during the British invasion of Zululand in 1879. Along the way, absorb some things about the nature of professionalism, duty, discipline, and coolness under stress. Paying attention to the actions of both sides will also yield some insights into the importance of bravery and tenacity. Finally, anyone who does not feel the little hairs on the back of the neck stand up in the scene before the final great Zulu charge does not possess the warrior's soul. (This scene did not occur in the real battle, by the way, but it should have.)

WORLD WAR I

1917 (2019)

A firsthand look at the horrors of World War I as two British soldiers travel through no-man's-land to deliver a message to call off an attack destined to fail.

Realistic and gritty, the film takes an honest and personal look at what drives soldiers and leaders.

***Gallipoli* (1981)**
Superb account of Australian soldiers caught up in the ultimately failed Allied campaign intended to take Turkey out of the war by invading the Gallipoli Peninsula on the Mediterranean Sea's eastern shore in 1915. Told through the experiences of two Aussie soldiers, the story reveals the bonds of friendship and camaraderie, the callousness of senior commanders to the suffering of their subordinates, and the courage and sacrifice of common soldiers fighting for each other.

***Paths of Glory* (1957)**
One of the finest war films ever made, this brutally realistic depiction of the horrors of World War I trench warfare follows a French Army infantry regiment as it attempts to carry out a suicidal attack ordered by a ruthlessly ambitious general. When the attack inevitably fails, the general callously puts his own soldiers on trial for "cowardice" to cover his own blunder. The acclaimed film is a classic study of leadership—good and bad!

***Sergeant York* (1941)**
The true story of a conscientious objector who went on to become the U.S. Army's most revered soldier in World War I. A first-rate think piece about the issues surrounding some of the more basic requirements of the military profession—like killing other human beings. If you haven't sorted these things out yet, watch this one ASAP.

WORLD WAR II

***Attack!* (1956)**
Powerful indictment of cowardly leadership by an infantry company commander during the December 1944 Battle of the Bulge that threatens to destroy the trust and confidence that is so vital to maintaining unit cohesion. The moral courage demonstrated by the company executive officer and a platoon leader are timeless examples of character and integrity for all officers to emulate.

***Band of Brothers* (2001)**
This excellent miniseries follows an airborne infantry rifle company of the 101st Airborne Division from initial formation in the United States through the end of World War II. Each of the chronological episodes builds on the previous ones as characters become more developed and experience more combat. With excellent portrayals of the family nature of a tightly knit unit, this series explores themes of leadership, loyalty, duty, and the hard choices leaders must make.

***Battleground* (1949)**
Portrayal of an infantry platoon of the 101st Airborne Division at Bastogne during the Ardennes Offensive in December 1944. Memorable for any number of reasons, but perhaps most importantly for its timeless depiction of the platoon sergeant who holds the unit together through some of the toughest combat of the war in western Europe.

***The Big Red One* (1980)**
Depiction of diverse episodes in combat with a squad of the 1st Infantry Division in World War II, directed by a veteran of that unit. Taken together, they build a sobering and evocative picture of the challenges of combat leadership. Give special attention to the scene on Omaha Beach, where the squad has to employ a bangalore torpedo; it defines the essence of discipline and the hair-raising choices implicit to combat leadership.

***Bridge on the River Kwai* (1957)**
Although the film departed significantly from the book on which it was based (in the blacklisted screenwriter's attempt to turn it into an "antiwar" story), this movie is a study in the dynamics of willpower. Useful for its stark depiction of the brutal conditions of a Japanese prison camp in Southeast Asia in World War II, more importantly it examines the issues of courage, obstinacy, and tenacity from several different perspectives.

***A Bridge Too Far* (1977)**
Some terrible casting almost cost this film a place on this list. Suffice it to say that several of the major commanders were utterly unlike the actors who portrayed them. On the positive side, this film teaches great lessons about duty, whether telling the boss the truth—even if he doesn't want to hear it—or defending an objective to the bitter end. There is also a scene (which, like everything else in this film, is based closely on actual events) that defines loyalty—in this case, an NCO's loyalty to his commander. Watch this scene closely; rerun it and think about what it takes to inspire this kind of love. Then go out and live it.

***The Caine Mutiny* (1954)**
A naval classic that transcends the service in which it is set, this film poses hard questions about integrity and even more difficult questions about the meaning of loyalty—above, below, to peers, the unit, and the service. It brilliantly conveys the notion that leaders do not get smarter in emergencies—only "scareder," and possibly "deader."

***Cross of Iron* (1977)**
A stark look at the merciless warfare waged on the Eastern Front in World War II. Useful not only for its combat scenes but also for the various leadership dilemmas faced by the chain of command. Highly recommended for anyone who,

through honest misunderstanding, idolizes the German armed forces of the Second World War. After seeing this one, you'll be glad of your U.S. Army heritage.

***Flags of our Fathers* (2006)**
The Battle of Iwo Jima, the U.S. Marine Corps' costliest invasion of the Pacific War, produced the most iconic image of World War II—the 23 February 1945 flag raising atop Mount Suribachi as Marines fought desperate combat against the island's Japanese defenders. This is the poignant story of the five Marines and one Navy corpsman who raised the flag—how it happened and what became of the flag raisers. The film's depiction of the horrific combat to invade and finally secure Iwo Jima is brutally authentic.

***From Here to Eternity* (1953)**
Although it departs from James Jones's classic epic in several important ways, this is the story of relationships in the Army, as timeless as the profession of arms itself. The challenges of marriage, dating, drinking, discipline, friendship, and a host of others are addressed in this film, which does not always provide the answers or the ending we would like to see. It is a bittersweet retrospective of service in peacetime and during the abrupt transition to war, but watch it most of all for the light it sheds on what it means to be a "thirty-year man in the Reg'lar Army."

***Go For Broke* (1951)**
Somewhat corny but solid story of training and combat with the segregated (Japanese-American) 442nd Regimental Combat Team in World War II. The larger, more important themes include the fact that every assignment can be what you make it, and that leaders are as good as the soldiers they lead.

***Hell is for Heroes* (1962)**
Highly realistic picture of the dynamics within a squad in combat. In the jargon of the military sociologist, it provides a unique examination of a situation in which a subordinate with considerable referent power challenges formal group leaders. In other words, it's about some of the dangers of operating in an information and leadership vacuum. As an added bonus, the film is refreshingly set in something other than a famous unit (in this case, the 95th Infantry Division) in one of World War II's less examined campaigns—the attack on the Siegfried Line.

***Letters from Iwo Jima* (2006)**
Clint Eastwood's sequel to his *Flags of Our Fathers* presents the story of the Battle of Iwo Jima from the Japanese defenders' perspective. The film puts a human face on an enemy that was typically demonized during World War II, and thereby provides officers important and valuable insight into the psychology of warfare. Although officers in combat should not be expected to empathize with the enemy, successful combat leadership requires that they *understand* their opponents.

***The Longest Day* (1962)**
Panoramic epic about the first day of the Normandy Invasion, 6 June 1944, based on actual accounts by veterans of the operation. Notable for its all-star cast and accurate depiction of the massive scale of the operation, which, like any military effort, was really a mosaic of countless individual and small-unit efforts.

***MacArthur* (1977)**
General of the Army Douglas MacArthur was America's most distinguished soldier of the twentieth century. He was also one of the country's most controversial leaders, particularly when he clashed with President Harry S Truman during the Korean War. This epic story presents MacArthur's career from the dark days opposing the Japanese invasion of the Philippines in 1941 through his relief from command in 1951 by Truman. Particularly noteworthy is the film's presentation of MacArthur's famed "Duty, Honor, Country" speech to West Point cadets in 1962, which frames the film's beginning and end.

***Mister Roberts* (1955)**
Although set aboard a naval vessel, this is a study in military life that transcends service or epoch. It provides a kaleidoscopic view of leadership styles and the dynamics of the relationships between and among officers, and between officers and enlisted personnel. Truly one of the great film examinations of "doing the right thing" and its consequences. Warning: even the most hardened viewer will choke up near the end.

***Once an Eagle* (1976)**
Anton Myrer's classic 1968 novel that had such an important impact on generations of U.S. military officers was produced as a nine-hour TV miniseries. This fictional story of an officer devoted to selfless service and his arch nemesis, a ruthless careerist, from World War I to the post–World War II era is a timeless study of leadership and character. The miniseries was released on DVD in 2010.

***The Pacific* (2010)**
This Emmy–winning, ten-part TV miniseries focuses on the true experience of three U.S. Marines (Eugene Sledge, Robert Leckie, and Medal of Honor recipient John Basilone) as they fight in Pacific War battles including Guadalcanal, Iwo Jima, and Okinawa. Created by the same team that made *Band of Brothers, The Pacific* realistically presents the "Marines' eye view" of the war that defeated Imperial Japan.

***Patton* (1970)**
A thoughtful study of egotism and ambition, this film is probably most important for its especially accurate portrayal of the U.S. Army general staff corps during the Second World War. Compare the methods of commanders and staffs who won a major war with those of today, and draw your own conclusions.

***Red Tails* (2012)**

This is the inspiring story of the Tuskegee Airmen, America's first African-American combat pilots, who forged an enviable record in the skies over World War II Europe. They not only destroyed German planes, but also shattered egregious racial stereotypes and helped pave the way for integration of the U.S. Armed Forces in the post–World War II era.

***Run Silent, Run Deep* (1958)**

World War II in the Pacific as experienced by a U.S. Navy submarine crew authentically captures the claustrophobic environment and, often, sheer terror of waging the undersea war. The plot is propelled by the dramatic confrontation between the sub's commander and his executive officer as they clash over mission, tactics, and leadership.

***Saving Private Ryan* (1998)**

Not only an example of the new "state of the art" in combat cinematic special effects, but also a thought-provoking examination of the "reasons why"—why soldiers fight, why the individual can sometimes be more important than the group, and why leaders are only as good as the soldiers they motivate and inspire. The film's first thirty minutes immerse viewers in the horrors of the D-Day invasion at Omaha Beach in one of the most realistic combat segments ever filmed.

***A Soldier's Story* (1984)**

Probably the best portrayal of the segregated Army, in this case as it existed during the Second World War. Examines the troubling, complex issues that surrounded the segregationist policies of the pre-1948 Army and strongly hints at the qualities possessed by Americans in general that allowed the Army to lead the nation out of the era of Jim Crow and "separate but equal."

***The Thin Red Line* (1988)**

Based on the James Jones novel, this is the story of soldiers fighting the battle of Guadalcanal. It addresses issues such as the conflict of whether or not to fight, victory in battle as a vehicle for personal gain and promotion, and the dissolution of a marriage while the soldier is away from home. Actor Nick Nolte's portrayal of an infantry battalion commander desperate for a tactical victory is one of the most realistic and convincing performances on film. An earlier adaptation of the same novel was made in 1964.

***To Hell and Back* (1955)**

An inspirational look at the wartime career of the man who was, quite possibly, the greatest American soldier ever. This true story of the exploits of the 3rd Infantry Division's Audie Murphy illustrates too many important themes to list here. Perhaps the two most important are never judge a book by its cover, and recognize the complex difference between being a friend to your soldiers and being friends with your soldiers.

***Twelve O'Clock High* (1949)**
A searing study of military leadership, applicable in peace or war. Keenly explores relationships between leaders and the led. Especially notable for examination of the inevitable balance between "mission" and "soldiers." Future adjutants should watch Major Stovall's attitudes and actions carefully; future commanders should pray for one like him.

***A Walk in the Sun* (1945)**
Want to know what your soldiers are thinking and talking about in the field or in combat? Watch this realistic portrayal of a platoon of the 36th Infantry Division in the invasion of Italy at Salerno in 1943. Stick with the whole thing, because the last scene is a stunning illustration of the "bottom line" in the profession of arms.

THE KOREAN WAR

***The Bridges at Toko-Ri* (1954)**
U.S. Navy carrier pilots during the Korean War face a dangerous and determined enemy while dealing with their own doubts about service, sacrifice, and who should ultimately bear the burden of war and combat. The film emphasizes courage—both physical and moral—and the challenges of leadership.

***Fixed Bayonets!* (1951)**
Gritty portrayal of a unit of U.S. soldiers charged with being the rear guard to protect the withdrawal of American and South Korean units after the massive Chinese intervention in the Korean War in November 1950. Directed by Samuel Fuller (*Big Red One* director and World War II combat veteran), the film superbly captures the dynamics of the interaction of a hard-pressed infantry squad and what it really takes to kill an enemy.

THE VIETNAM WAR

***Forrest Gump* (1994)**
Although this Academy Award–winning film covers the title character's entire life, the portions of the movie recounting his adventures in the Vietnam War offer some valuable insight into the bonds of friendship forged in combat. Particularly noteworthy is the film's depiction of the tremendous problems of readjustment suffered by "Lt. Dan" after he is seriously crippled in combat—it is a powerful reminder to all officers that for our Wounded Warriors, the war does not end after the last bullet is fired.

***Full Metal Jacket* (1987)**
One of the best film depictions of the Vietnam War experience, from boot camp through combat during the 1968 Tet Offensive. This powerful film follows the story of several U. S. Marines as they train under a brutal drill sergeant who must transform them from civilians into "killers," and then follows their experiences as they face the horrific realities of combat against the North Vietnamese Army and the Viet Cong.

***Go Tell the Spartans* (1978)**
A rare look at the Vietnam conflict in the early days of American advisers' involvement. Beyond its unique subject, this film is useful for examining the pitfalls of exaggerated personal ambition and other unprofessional attitudes. There is much to be learned about the sublimeness of duty in this one.

***The Great Santini* (1979)**
A vivid portrait of military family life. Set in the Marine Corps in the early 1960s, it could just as easily have been an Army family in the 2000s, with the challenges of frequent moves, more frequent separations, and all the good and bad that can come from both. Although there are some great laughs along the way, this film should make any military viewer uncomfortable; there is probably at least a little bit of the protagonist, Bull Meechum, in just about all of us, whether as parents or as officers. Watch and figure out which parts you want to keep.

***Hamburger Hill* (1987)**
Combat with the 101st Airborne Division in the A Shau Valley in 1968. Many professionals who were there claim that this is the best of the Vietnam movies, although most of the critics—practically none of whom had been there—did not agree. First-class exposition of combat leadership dynamics and the racial tensions that marred that period of U.S. Army history.

***Platoon* (1986)**
Director Oliver Stone's semi-autobiographical story of an infantry platoon in Vietnam combat and the struggle of two sergeants (played by Willem Dafoe and Tom Berenger) to win over the allegiance of a young soldier. The climactic night combat scene of a North Vietnamese attack on the U.S. infantry company is one of the best depictions of Vietnam War infantry combat.

CURRENT OPERATIONS

***American Sniper* (2014)**
Director Clint Eastwood shows the effect of intense combat on American warriors—and their families—in this acclaimed film based on the life of Navy SEAL Chris Kyle. This is a "must-see" film to understand that combat is only the beginning of what it means to serve, and that post-traumatic stress disorder is a real and tangible issue our troops—and those who lead them—must be expected to deal with and confront.

***Black Hawk Down* (2001)**
This film depicts the events surrounding the October 1993 battle fought by Task Force Rangers in Somalia when one hundred Rangers participated in a mission to capture warlords in Mogadishu. As the unstable situation spirals out of control and two Black Hawk helicopters are shot down, the importance of contingency planning and on-the-spot decision-making are highlighted. Pay close attention to

the pre-combat checks and inspections, and the consequences of deviating from training.

***The Hurt Locker* (2008)**
This film, which won multiple Oscars, recounts the actions of an explosive ordnance disposal (EOD) team during the Iraq War. The tension and daily—even hourly—suspense of the team's nerve-shattering job is palpable as the action plays out in this riveting film. "Tip of the spear" infantry and Special Forces troops seem to get the lion's share of coverage and praise, but this film reveals the true heroism and sacrifice of those who "police up the battlefield" in the wake of the maneuver forces.

***Restrepo* (2010)**
Sebastian Junger's riveting and visually compelling documentary film of the Afghanistan War (2001–present) is a yearlong examination of an American airborne infantry outpost manned by soldiers of the 173rd Infantry Brigade. The defense of this obscure outpost—named after PFC Juan Restrepo, a combat medic killed earlier in the campaign—is documented in a no-holds-barred presentation that brings the Afghan intervention "up close and personal."

***The Outpost* (2020)**
Based on the book *The Outpost: An Untold Story of American Valor* by Jake Tapper, this film brings to the screen a 2009 fight at COP in Afghanistan in which two soldiers would eventually be awarded the Medal of Honor. The desperate battle at COP Keating, in which the base and those inside are surrounded and attacked by the Taliban, is a story that highlights leadership and the personal courage needed to overcome overwhelming odds.

27

Professional Reading List

The Army Chief of Staff releases a reading list after assuming his position as the Army's most senior uniformed leader; a compilation of several CSA's lists follows below. The second list, compiled by a group of Army historians, addresses general military history by topic, allowing for additional insight into your profession.

THE U.S. ARMY CHIEF OF STAFF'S PROFESSIONAL READING LIST

The U.S. Army Chief of Staff's Professional Reading List is divided into six categories: "Strategic Environment," "Regional Studies," "History and Military History," "Leadership," "Army Profession," and "Fiction." Taken together, these readings will help soldiers or Army civilians sharpen their critical faculties and broaden their understanding of the military art. These books also complement materials currently used in the Army educational system and can help bridge the intervals between periods of formal instruction at Army schools. It is imperative for members of the Army profession to be well-read in all aspects of our honorable and selfless calling.

Any professional reading list is, of course, only a brief introduction to the many books worth reading on Army history, heritage, leadership, and world events. The list is just a starting point on a journey of discovery and development. This selection of books also does not imply that the Chief of Staff endorses the authors' views or interpretations. Nevertheless, these books contain thought-provoking ideas and information relevant to our dynamic Army today and into the future. https://www.army.mil/leaders/csa/readinglist/.

Armies at War: Battles and Campaigns

Anderson, Fred. *Crucible of War: The Seven Years' War and the Fate of Empire in British North America, 1754–1766*. New York: Knopf Publishers, 2000.

From the forest of frontier Ohio to the streets of London, Anderson provides rich detail and insightful analysis to explain how Great Britain, France, Spain, the Colonies, and the Native Americans battled for supremacy in North America. The Seven Years' War, often called the "first world war," featured a cast of state and non-state actors who competed in complex ways as commercial, diplomatic, and military interests maneuvered to seek advantages. The resultant security challenges of the eighteenth century are remarkably similar to those of today's global environment. As soldiers and policymakers seek to define national interests and work with allies, the lessons of the eighteenth century—during which the choices made by strategic leaders ultimately led to the American Revolution and the birth of the United States—bears reexamination.

Cowper-Coles, Sherard. *Cables from Kabul: The Inside Story of the West's Afghanistan Campaign*. London: Harper Press, 2011.

The author, who served as the British ambassador to Afghanistan from 2007 to 2010, wrote this pessimistic view of the war in Afghanistan at the conclusion of his tenure. In his opinion, the military continued to give overly optimistic assessments. In addition to providing a frank account of diplomacy in action, he also looks at the transatlantic alliance and the ambitious U.S. military strategy and offers conclusions on the war. Broader than just Afghanistan, this book furnishes insights on Western diplomacy and the way the North Atlantic Treaty Organization operates, particularly in hostile territories.

Fischer, David Hackett. *Washington's Crossing*. New York: Oxford University Press, 2006.

This meticulously researched and superbly written study of military operations and leadership recounts a time during the Revolutionary War when America nearly lost, but then saved, its recently declared independence. The title alludes not just to Gen. George Washington's famous crossing of the Delaware River in December 1776, but also to his maturation as a leader and to the Continental Army's transformation into a competent military force. In analyzing the 1776 campaign at the strategic, operational, and tactical levels, Fischer reflects upon the development of a uniquely American way of war and on the values that continue to guide our Army.

Daddis, Gregory A. *Westmoreland's War, Reassessing American Strategy in Vietnam*. New York: Oxford University Press, 2014.

Nearly fifty years after the United States committed ground combat troops to Vietnam, Americans still have a lot to learn—or unlearn—about the nation's first lost war. In this stimulating reappraisal of the conflict, Daddis argues that America's failure owed less to the much-maligned "strategy of attrition" than to broader flaws in national policy, including the belief that America could transform South Vietnam. In taking a fresh look at U.S. strategy during the Vietnam War, this book enhances our understanding of both the war itself and the challenges that continue to face soldiers and policymakers when intervening in the internal conflicts of foreign countries.

Gordon, Michael. *The Endgame: The Inside Story of the Struggle for Iraq, from George W. Bush to Barack Obama*. New York: Pantheon Books, 2012.

Clausewitz concluded that war is a continuation of politics by other means. War also requires an understanding of the human domain. Michael Gordon's *The Endgame* is an exhaustively researched account of the war in Iraq that seeks to explain the political dynamics that underlay the conflict, the motivations of the human actors who took part, and the social fabric against which a violent struggle for power took place. Gordon analyzes in detail the way our national security decision-making process operated during the Iraq war and the relationship between policy and strategy. This is an essential work for any Army leader who wishes to understand the complex character not just of the Iraq war, but of any war in the contemporary age.

Grant, Ulysses S. *Personal Memoirs: Ulysses S. Grant*. 1885; reprinted, New York: Modern Library, 1999.

A classic and honest study by one of America's greatest generals, this is one of the finest autobiographies of a military commander ever written. It has valuable insights into leadership and command that apply to all levels and in all times. Grant's resiliency under almost unimaginable stress during critical junctures of America's most bloody war makes him a fascinating and human case study of the "epitomized" soldier.

Grotelueschen, Mark E. *The AEF Way of War: The American Army and Combat in World War I*. New York: Cambridge University Press, 2007.

This exemplary case study of doctrinal and tactical innovation under fire shows how four divisions of the American Expeditionary Forces adapted, or failed to adapt, to conditions on the Western Front during World War I. The 1st and 2nd Divisions perfected artillery infantry liaison so that by November 1918 they had achieved "state-of-the-art" tactical skills as practiced by the Allied armies. Both the 26th and 77th Divisions failed to achieve this level of skill—the 26th because its commander failed to maintain control of his subordinate units, and the 77th because its commander remained wedded to prewar doctrine.

Hastings, Max. *Retribution: The Battle for Japan, 1944–45*. New York: Knopf, 2008.

A sweeping account of the closing years of World War II in the Pacific, when the fate of Japan was largely sealed yet massive armed forces continued to fight across the Pacific, inflicting horrific casualties. How and why the killing proceeded—and what it meant to people on all sides caught in the war's fury—makes a compelling and illustrative story of the sheer violence of war and the ways that rational military planning can lead to uncontrollable destruction. Hastings creates a riveting portrayal of combat largely overlooked until now that serves as both a warning and a lesson for operational and strategic leaders.

James, D. Clayton, with Anne Sharp Wells. *Refighting the Last War: Command and Crisis in Korea, 1950–1953*. New York: The Free Press, 1993.

This readable study diverges from the usual chronological narrative of the Korean War, instead devoting individual chapters to assessing key American senior leaders and their important command decisions during the war. Author D. Clayton James, the foremost authority on Gen. Douglas MacArthur, teamed up with his longtime research assistant to produce a work that will give the reader insight into the role of personalities during the war as well as the complexities of some of the decisions with which they wrestled.

Keegan, John. *The Face of Battle: A Study of Agincourt, Waterloo, and the Somme*. New York: Penguin Books, 1983.

One of the classics of modern military history, *The Face of Battle* brings to life three major battles: Agincourt (1415), Waterloo (1815), and the first battle of the Somme (1916). The author describes the sights, sounds, and smells of battle, providing a compelling look at what it means to be a soldier and how hard it is to describe realistically the dynamics of combat.

Lacey, Sharon Tosi. *Pacific Blitzkrieg: World War II in the Central Pacific*. Denton: University of North Texas Press, 2013.

Lacey closely examines the planning, preparation, and execution of ground operations at the corps and division level for five major invasions in the Central Pacific: Guadalcanal, Tarawa, the Marshalls, Saipan, and Okinawa. The commanders had to integrate the U.S. Army and Marine Corps into a single striking force, something that would have been difficult enough in peacetime but was a monumental task in the midst of a great global war. Yet, ultimate success in the Pacific rested on this crucial, if somewhat strained, partnership and its accomplishments. This book also details the collection of lessons learned after each battle and the incorporation—or not—of these lessons in the training, planning, and equipping for each subsequent operation, which provided the roots of all future joint operations.

McManus, John C. *Grunts: Inside the American Infantry Combat Experience, World War II Through Iraq*. New York: NAL Caliber, 2010.

Historian John C. McManus covers six decades of warfare, from the fight on the island of Guam in 1944 to today's counterinsurgency in Iraq. He demonstrates that the foot soldier has been the most indispensable and much overlooked factor in wartime victory despite the advancement of weaponry. McManus stresses that the importance of the human element in protecting the United States is too often forgotten and advances a passionate plea for fundamental change in our understanding of war.

McPherson, James M. *Battle Cry of Freedom: The Civil War Era*. New York: Oxford University Press, 1988.

McPherson has written a brilliant account of the American Civil War—the war that made the country what it is today. He discusses in clear, incisive detail the causes of the war, the military operations, the soldiers and leaders, and the political, economic, and social aspects of life in the Union and the Confederacy before and during the war. With many experts judging it to be the best one-volume history of the Civil War, it provides an excellent introduction to the most significant war fought by the American Army.

Mullaney, Craig M. *The Unforgiving Minute: A Soldier's Education*. New York: Penguin, 2009.

U.S. Army captain Craig Mullaney recounts the hard lessons that only war can teach while fighting al Qaeda in Afghanistan. This is a portrait of a junior officer grappling with the weight of war and coming to terms with what it means to lead others in combat.

Parker, Geoffrey, ed. *Cambridge Illustrated History of Warfare: The Triumph of the West*. New York: Cambridge University Press, 2000 (revised and updated 2008).

Written in a digestible, compelling manner, Parker's authors cover the gamut of Western warfare from antiquity to the present, including the development of warfare on land, sea, and air; weapons and technology; strategy, operations, and tactics; and logistics and intelligence. Throughout, there is an emphasis on the socioeconomic aspects of war, the rise of the West to global dominance, and the nature of the aggressive military culture that has been its hallmark.

Porch, Douglas. *Counterinsurgency: Exposing the Myths of the New Way of War*. New York: Cambridge University Press, 2013.

This book reviews American, British, and French experiences with counterinsurgency operations since the nineteenth century and argues that the strategy of winning "hearts and minds" has never worked as advertised. Porch demonstrates that force, rather than benevolent social engineering and nation building, has historically been the key to successful counterinsurgent operations. This book is a stimulating and thought-provoking counterpoint to existing American national security thinking and U.S. Army doctrine that, in Porch's opinion, is simply wishful thinking wrapped in myth.

Stewart, Richard W., ed. *American Military History, Vol. 2: The United States Army in a Global Era, 1917–2008*. Washington, D.C.: U.S. Army Center of Military History, 2010.

Created initially as a Reserve Officers' Training Corps textbook, this second book in a two-volume overview of the Army's story covers the period from World War I to the early days of the wars in Afghanistan and Iraq. Written in an engaging style and enhanced by sophisticated graphics and recommended readings, the work is an excellent source of general service history in the modern world.

Van Creveld, Martin. *Supplying War: Logistics from Wallenstein to Patton*. New York: Cambridge University Press, 1977.

Surveying four centuries of military history, noted historian Martin Van Creveld clearly points out the reasons "amateurs study tactics; professionals study logistics." Most battlefield results would not have been possible without the careful organization and allocation of logistical resources. Leaders who fail to consider logistics in all of their plans and operations will do so at their peril.

The Army Profession

Bouvard, Marguerite Guzmán. *The Invisible Wounds of War: Coming Home From Iraq and Afghanistan*. Amherst, NY: Prometheus, 2012.

The war does not simply go away when a soldier returns home. The demands of the recent wars in Iraq and Afghanistan have been particularly high because the Army's small size required multiple deployments by many soldiers, causing greater stress and damage for soldiers as well as their families. Through interviews with many veterans and their families, the author examines the causes of post-traumatic stress disorder and addresses the sometimes lamentable care veterans have received at home.

Brown, John Sloan. *Kevlar Legions: The Transformation of the U.S. Army, 1989–2005*. Washington, D.C.: U.S. Army Center of Military History, 2011.

Former Chief of Military History and retired Brig. Gen. John S. Brown brings his formidable analytical skills to bear in a detailed study of how the senior Army leadership formulated, managed, and executed a multiyear transformational effort after the end of the Cold War. A consistent vision from the senior leadership allowed the Army to cope with the changed international environment after the collapse of the Soviet Union, with high operational tempo contingency operations, and with the opening salvos of the Global War on Terror and still craft a flexible, adaptable Army that is the premier ground force in the world today.

Cox, Edward. *Grey Eminence: Fox Conner and the Art of Mentorship*. Stillwater, OK: New Forums Press, 2011

Because Fox Conner's name was synonymous with mentorship, he was nicknamed the "Grey Eminence" within the Army. His influence and mentorship helped shape the careers of George S. Patton, George C. Marshall, and, most notably, President Dwight D. Eisenhower. While little is known about Conner himself, the author uses stories about Conner's relationship with Eisenhower. Fox Conner's four-step model for developing strategic leaders still holds true today. First, be a master of your craft. Second, recognize and recruit talented subordinates. Third, encourage and challenge protégés to develop their strengths and overcome their weaknesses. Fourth, do not be afraid to break the rules of the organization to do it. Everyone can learn something to apply to his or her life from Fox Conner.

Grossman, Dave. *On Killing: The Psychological Cost of Learning to Kill in War and Society*. Revised edition, New York: Back Bay Books, 2009.

The book investigates the psychology of killing in combat and stresses that human beings have a powerful, innate resistance to the taking of life. The author examines the techniques developed by the military to overcome that aversion during the Vietnam War, revealing how American soldiers were more lethal during this conflict than at any other time in history. Grossman argues that the combination of the breakdown of American society, the pervasive violence in the media, and interactive video games is conditioning our children to kill in a manner similar to the Army's conditioning of soldiers.

Hubbard, Elbert. *A Message to Garcia*. Lexington, KY: Seven Treasures Publications, 2009.

This classic essay from 1899, based on the true story of Lt. Andrew Rowan, is a notable testament of initiative and responsibility. This work provides commonsense advice on the importance of personal responsibility, loyalty, hard work, and enterprise.

Kotter, John P. *Leading Change*. Boston: Harvard Business School Press, 1996.

In this now-classic book on leadership, Kotter describes a proven eight-step change process: establish a sense of urgency, create the guiding coalition, develop a vision and strategy, communicate the change vision, empower others to act, create short-term wins, consolidate gains and produce even more change, and institutionalize new approaches in the future. Leaders across the Army and at all levels should study and use Kotter's change process to assist in leading complex Army organizations through uncertainty and difficult circumstances.

Linn, Brian McAllister. *The Echo of Battle: The U.S. Army's Way of War*. Boston: Harvard University Press, 2007.

The Army develops weapons, doctrine, and commanders for future wars in the "echo of battle." Linn argues that since its early years the Army has had three enduring, and often antagonistic, intellectual traditions—that is, three different ways of war—with assumptions and concepts that have remained remarkably consistent. Linn surveys the past assumptions and errors of each throughout the Army's history and also notes the consistent tendency to discourage critical thinking, thereby enforcing complacency and a view of war comfortable to officers in each tradition. The result has been an Army often ill-prepared for the wars it was called upon to fight, as opposed to the wars each tradition envisioned.

Myrer, Anton. *Once an Eagle*. New York: HarperTorch, 2001.

A historical novel, *Once an Eagle* traces the career of a fictitious soldier from World War I to Vietnam. The book realistically portrays the confusion of combat, the bonds that form between fighting men, the tensions between line and staff officers, and the heavy responsibility of command. This is a great work for young leaders contemplating a career in the profession of arms and looking for a deeper understanding of Army culture.

Perry, Mark. *Partners in Command: George Marshall and Dwight Eisenhower in War and Peace*. New York: Penguin, 2007.

This book is a balanced biographical view of the relationship between Generals George C. Marshall and Dwight D. Eisenhower and provides an illustrative glimpse at the connection between the two men as they developed a grand alliance and forged the strategies that led to victory in Europe in World War II. It is a good read for strategic leaders who wish to better understand the complexities of coalition, joint, and civil-military relations.

Puryear, Jr., Edgar F. *19 Stars: A Study in Military Character and Leadership*. New York: Presidio Press, 2003.

This valuable work studies the lives and careers of Generals Dwight D. Eisenhower, Douglas MacArthur, George C. Marshall, and George S. Patton through their own eyes as well as the recollections of hundreds of others who worked with and knew them personally. Puryear examines elements common to their success, including obvious attributes such as their thorough preparation and capacity for work as well as their more subtle qualities of character and, of course, luck. This is a great work for up-and-coming officers to better understand the fundamentals of leadership, preparation, and the need for luck.

Shadley, Robert D. *The GAMe: Unraveling a Military Sex Scandal*. Edina, MN: Beaver's Pond Press, 2013.

Shadley, a retired Army major general, writes about a scandal in which a number of Army drill instructors colluded and competed to sexually assault female trainees. This gross violation of Army ethics and values was uncovered in 1996 at the Aberdeen Proving Grounds—the post where the author commanded the U.S. Army Ordnance Centers and Schools.

Williams, T. Harry. *Lincoln and His Generals*. New York: Vintage Books, 2011.

First published in 1952, *Lincoln and His Generals* remains one of the definitive accounts of President Abraham Lincoln's wartime leadership. Williams dramatizes Lincoln's long and frustrating search for an effective leader of the Union Army and traces his transformation from a politician with little military knowledge into a master strategist. Explored in-depth are Lincoln's often fraught relationships with generals such as George B. McClellan, John Pope, Ambrose Burnside, Joseph Hooker, John C. Fremont, and of course, Ulysses S. Grant.

Strategy and the Strategic Environment

Cirincione, Joseph, Jon B. Wolfsthal, and Miriam Rajkumar. *Deadly Arsenals: Nuclear, Biological and Chemical Threats*. 2nd Ed., New York: Carnegie Endowment for International Peace, 2005.

This volume provides an informative assessment of global nuclear, biological, and chemical arsenals, and the risks associated with their potential proliferation and employment. Presented in clear terms, the technical information about the growing dangers of weapons of mass destruction is timely and compelling. Future war will likely involve increased use of these terrible weapons, a subject deserving much more attention from soldiers and policymakers alike.

Ferguson, Niall. *The Ascent of Money: A Financial History of the World*. New York: Penguin Press, 2008.

Economic historian and Harvard professor Niall Ferguson provides a historical narrative on the ascent of money and how it has both contributed to the expansion of the global marketplace and caused conflict. This is a cautionary tale that relates to the current fiscal crisis in Europe and the United States, interdependence between the United States and Chinese economies, the financial crash of 2008, and the subsequent Occupy Wall Street Movement and slow economic growth around the world. It is an important read for those who want to understand how the global fiscal and economic crisis could impact our defense strategy and defense force structure.

Freedman, Lawrence. *A Choice of Enemies: America Confronts the Middle East.* New York: PublicAffairs, 2008.

A leading professor from the United Kingdom provides an objective study of U.S. engagements in the Middle East. This volume looks at U.S. strategies from President Jimmy Carter to President George W. Bush, evaluating the evolution of U.S. policy toward the region and offering thoughtful analysis of how and why the United States has confronted regional issues with global impact during five presidencies.

Gaddis, John Lewis. *The Cold War: A New History*. New York: Penguin Books, 2005.

Preeminent historian John Lewis Gaddis captures a lifetime of scholarship in this penetrating history of the Cold War, a long contest that shaped the affairs of all the world's nations for half of the twentieth century by pitting great powers against each other around the globe. This critical period challenged U.S. foreign policy by demanding that leaders devise new strategies to accommodate all forms of national power, including the potential use of nuclear weapons. How American foreign policy evolved and how the Army changed its doctrine and organization to meet the changing threats—including hot wars in Korea and Vietnam—are central to the story and applicable to the development of military thinking today.

Habeck, Mary. *Knowing the Enemy: Jihadist Ideology and the War on Terror.* New Haven: Yale University Press, 2007.

In this primer on a small splinter group of Islam, Habeck traces current jihadism from an early fourteenth-century scholar and the eighteenth-century founder of the harshly restrictive Islam (predominantly in Saudi Arabia) to four twentieth-century figures who inspired a host of radical reactionary organizations, including Hamas and al Qaeda. The author's purpose is to reveal the origins of jihadism. In so doing, she contributes in considerable detail and with admirable clarity one of the most valuable books on the ongoing Middle East—and global—conflicts. This is an important volume for all leaders to understand what drives people to attack democracies.

Kaplan, Robert D. *Monsoon: The Indian Ocean and the Future of American Power*. New York: Random House, 2010.

Kaplan provides a sweeping examination of the countries that compose "Monsoon Asia": China, India, Pakistan, Indonesia, Burma, Oman, Sri Lanka, Bangladesh, and Tanzania. This book is a must-read for Army professionals, especially those whose focus has been weighted toward the challenges in Iraq and Afghanistan over the past decade. As the United States pivots strategically to the Pacific, Kaplan's unique perspective on the implications of rising powers, the shifting global balance, and the potential contingencies that might arise is timely and important.

Kissinger, Henry A. *On China*. New York: Penguin Press, 2011.

This is a sweeping and insightful history of China by a distinguished international scholar, diplomat, and statesman. It provides a fascinating historical view of China and its relations with its neighbors and the United States over the last forty years. It is useful for readers to understand the world's second largest economy and the rising global player, which will impact U.S. national and economic security for decades to come.

Kilcullen, David. *Out of the Mountains: The Coming Age of the Urban Guerrilla*. New York: Oxford University Press, 2013.

This new book takes a fresh look at what could be the Army's most challenging combat environment: large urban centers. The sprawling cities of both the developing and developed worlds pose increasing and complex security challenges that governments are struggling to accommodate. At the same time, communications technologies are assisting those who seek to commit violence amid such weakly governed spaces. The result is a formidable range of threats that the Army must better understand and prepare to meet in the coming years.

Meredith, Martin. *The Fate of Africa: A History of the Continent Since Independence*. Revised edition, New York: PublicAffairs, 2011.

This classic history has been revised to incorporate recent developments such as Darfur, Sudan, Robert Mugabe's longevity, aid, development, the influence of China, and the North African/Arab Spring. Africa is remarkably diverse, complex, and challenging to understand, and even more difficult to influence. To date, the potential for the peoples and states of the continent to assert their international influence has not been met. This will likely change as Africans take their place on the world's stage in the coming decades. A lucid read with clear implications for U.S. involvement (or lack of) in Africa.

Morozov, Evgeny. *The Net Delusion: The Dark Side of Internet Freedom*. New York: PublicAffairs, 2011.

Morozov argues against the conventional wisdom that the Internet and social media will automatically promote and expand freedom around the world. While social networking may enable popular mobilization, authoritarian regimes can use the same tools to suppress free speech, monitor opposition, and disseminate propaganda. The author asserts that while "digital diplomacy" and "internet freedom" sound good rhetorically, foreign policy based on these terms may actually hinder the promotion of democracy, cause dictators to become more repressive, and harm the reform efforts of dissidents.

Myers, B. R. *The Cleanest Race: How the North Koreans See Themselves—and Why it Matters*. New York: Melville Press, 2010.

In this candid and compelling work, B. R. Myers provides an incisive and important interpretation of North Korea's regime. He explains how a racially centric ideology operates in Pyongyang to inform regime decision-making that leads to successive provocations on the Korean Peninsula. As American foreign policy turns to the Asia-Pacific, U.S. forces in Korea are at the center of an increasingly dangerous confrontation, now including the prospect of weapons of mass destruction. Myers's book provides necessary understanding for Army leaders to better comprehend this potential adversary.

Naim, Moises. *The End of Power: From Boardrooms to Battlefields and Churches to States, Why Being In Charge Isn't What It Used to Be*. New York: Basic Books, 2013.

An intriguing and provocative analysis of power as it is wielded by key institutions around the globe—whether government, military, religion, or business—and how these bastions of authority are now being tested, and in many cases, decisively eroded. The author identifies a three-way revolution of "more, mobility, and mentality" to demonstrate how changes in population, wealth, migration, and information are fundamentally challenging traditional hierarchies and causing ferment in societies around the world. The security implications are tremendous, and soldiers who will operate in such a mutating environment must consider how their roles must change as well.

Paret, Peter, ed. *Makers of Modern Strategy: From Machiavelli to the Nuclear Age*. Princeton: Princeton University Press, 1986.

A wonderful anthology on the evolution of strategic thought. Moving from Machiavelli to the present in twenty-eight insightful essays, the authors examine topics such as the role of doctrine, the genius of Napoleon, the limits of airpower, and nuclear strategy. A primer for all military leaders who must think strategically on a variety of issues, *Makers of Modern Strategy* summarizes the classic military thinkers, underlining the enduring lessons that remain relevant today.

Ramo, Joshua Cooper. *The Age of the Unthinkable: Why the New World Disorder Constantly Surprises Us and What We Can Do About It*. New York: Back Bay Books, 2010.

This book challenges conventional assumptions, world views, and thinking in an increasingly complex world. The author proposes controversial ways of considering global challenges, such as studying why Hezbollah is the most efficiently run Islamic militant group. Ramo uses economics, history, complexity theory, and network science to describe an ambiguous reality that has many innovative possibilities.

Rashid, Ahmed. *Descent into Chaos: The United States and the Failure of Nation Building in Pakistan, Afghanistan, and Central Asia*. New York: Viking Penguin, 2008.

Rashid, a Pakistani writer, describes how the war against Islamic extremism was being lost in Pakistan, Afghanistan, and central Asia. He examines the region and the corridors of power in Washington and Europe to see how the promised nation-building in these countries has progressed. His conclusions are devastating: an unstable and nuclear-armed Pakistan, a renewed al Qaeda profiting from a booming opium trade, and a Taliban resurgence and reconquest. He argues that failing states pose a graver threat to global security than the Middle East.

Reid, Michael. *Forgotten Continent: The Battle for Latin America's Soul*. New Haven: Yale University Press, 2009.

In examining a vast continent often overlooked by the West, Reid argues that Latin America's efforts to build more equitable and more prosperous societies make that region one of the most dynamic places in the world. In that area, a series of democratic leaders are attempting to lay the foundations for faster economic growth and more inclusive politics while addressing the region's seemingly intractable problems of poverty, inequality, and social injustice. Failure will not only increase the flow of drugs and illegal immigrants, but also jeopardize the stability of a region rich in oil and other strategic commodities and threaten some of the world's most majestic natural environments. The study provides a vivid, current, and informed account of a dynamic continent and its struggle to compete in a globalized world.

Singer, P. W. *Wired for War: The Robotics Revolution and Conflict in the 21st Century*. New York: Penguin Press, 2009.

P. W. Singer explores robotic warfare as a revolution in military affairs on a par with the atom bomb or gunpowder. The author discusses the impact of new military technology on the ways war is fought and the influence robots will have on the future battlefield, as well as the ramifications robotic warfare will have on ethics, law, politics, and economics.

Singer, P. W., and Allan Friedman. *Cybersecurity and Cyberwar: What Everyone Needs to Know*. New York: Oxford University Press, 2014.

A fascinating and well-written examination of the technology that supports our society and much of the world. The authors discuss how cyberspace and security work, how they affect us, and what every individual—and soldier—should know. If you have a computer, smartphone, or digital account, you should read this book.

Smith, Rupert. *The Utility of Force: The Art of War in the Modern World.* New York: Vintage Books, 2008.

Drawing on his experience as a commander during Operations Desert Shield and Desert Storm and in Bosnia, Kosovo, and Northern Ireland, British general Rupert Smith gives us a probing analysis of modern war and questions why we try to use military force to solve our political problems, and why, when our forces win military battles, we still fail to solve the problems. He demonstrates why today's conflicts must be understood as intertwined political and military events and makes clear why the current model of total war has failed in Iraq, Afghanistan, and other recent campaigns.

Yergin, Daniel. *The Quest: Energy, Security, and the Remaking of the Modern World.* New York: Penguin Press, 2011.

Widely recognized as the most authoritative voice on the geopolitics of energy and resource competition, Daniel Yergin has produced the most comprehensive and balanced work now available on the topic. He presents his concerns about the security, financial, technological, and ecological problems stemming from the role of fossil fuels in global energy production. This volume not only explains the deeper history of energy production and the dramatic changes of the past decade, but also provides valuable policy advice for minimizing the potential for catastrophic disruptions in energy security and the conflicts this could produce.

Zegart, Amy. *Flawed by Design: The Evolution of the CIA, JCS, and NSC.* Stanford: Stanford University Press, 1999.

Zegart looks at the interagency process and challenges the traditional opinion that the key national security institutions—the Central Intelligence Agency, the Joint Chiefs of Staff, and the National Security Council—operate as the National Security Act of 1947 intended. The reality has been that politics and bureaucratic and budgetary turf wars effectively impair the original purpose of these organizations and have caused systematic failures. However, proposals for reform of the entire interagency process must be carefully considered. This is an important read for those who wish to learn how key institutions were intended to work and how they actually function today in an era of increasing uncertainty.

ARMY HERITAGE

American Battle Monuments Commission. *American Armies and Battlefields in Europe.* Washington, D.C.: Government Printing Office, 1938.

Frazer, Robert W. *Forts of the West.* Norman: University of Oklahoma Press, 1972.

Johnson, Douglas W. *Battlefields of the World War.* New York: Oxford University Press, 1921.

Lawliss, Chuck. *The Civil War Sourcebook: A Traveler's Guide.* New York: Harmony Books, 1991.
Linenthal, Edward T., and Robert M. Utley. *Sacred Ground: Americans and Their Battlefields.* Champaign: University of Illinois Press, 1993.
Osborne, Richard E. *World War II Sites in the United States: A Tour Guide and Directory.* Florence, OR: Sagamon, 1997.
Stevens, Joseph E. *America's National Battlefield Parks: A Guide.* Norman: University of Oklahoma Press, 1991.

War of Independence

Bill, Alfred. *Valley Forge: The Making of an Army.* New York: Harper, 1952.
Forneaux, Rupert. *The Battle of Saratoga.* New York: Stein & Day, 1971.
Galvin, John R. *The Minute Men. The First Fight: Myths and Realities of the American Revolution.* New York: Pergamon/Brassey's, 1989.
Higginbotham, Don. *The War of American Independence: Military Attitudes, Policies, and Practices.* New York: Macmillan, 1971.
Jacobs, James R. *The Beginnings of the U.S. Army.* Princeton, NJ: Princeton University Press, 1947.
Ketchum, Richard. *Saratoga: Turning Point of America's Revolutionary War.* New York: Henry Holt, 1997.
Montross, Lynn. *Rag, Tag, and Bobtail: The Story of the Continental Army, 1775–1783.* New York: Harper, 1952.
Shy, John. *A People Numerous and Armed.* New York: Oxford University Press, 1976.

War of 1812

Elting, John R. *Amateurs to Arms! A Military History of the War of 1812.* Chapel Hill, NC: Algonquin Books, 1991.
Hickey, Donald R. *The War of Eighteen Twelve: A Forgotten Conflict.* Champaign: University of Illinois Press, 1990.

Mexican War

Bauer, K. Jack. *The Mexican War 1846–48.* New York: Macmillan, 1974.
Eisenhower, John S. D. *So Far from God: The U.S. War with Mexico, 1846–1848.* New York: Random House, 1989.
Winders, Richard B. *Mr. Polk's Army: The American Military Experience in the Mexican War.* College Station: Texas A&M University Press, 1997.

Civil War

Catton, Bruce. *Army of the Potomac: Glory Road.* Garden City, NY: Doubleday, 1952.
———. *Army of the Potomac: Mr. Lincoln's Army.* Garden City, NY: Doubleday, 1951.

———. *Army of the Potomac: A Stillness at Appomattox.* Garden City, NY: Doubleday, 1953.

Hagerman, Edward. *The American Civil War and the Origins of Modern Warfare: Ideas, Organization, and Field Command.* Bloomington: Indiana University Press, 1992.

McMurry, Richard. *Two Great Rebel Armies.* Chapel Hill: University of North Carolina Press, 1989.

McPherson, James M. *The Battle Cry of Freedom.* New York: Oxford University Press, 1988.

The Era of Manifest Destiny

Ambrose, Stephen. *Upton and the Army.* Baton Rouge: Louisiana State University, 1993.

Gates, John M. *Schoolbooks and Krags: The U.S. Army in the Philippines, 1898–1902.* Westport, CT: Greenwood Press, 1973.

Linn, Brian McAllister. *The Philippine War, 1899–1902.* Lawrence: University Press of Kansas, 2000.

Utley, Robert M. *Frontier Regulars: The United States Army and the Indian, 1866–1891.* Lincoln: University of Nebraska Press, 1984.

Spanish-American War

Cosmas, Graham A. *An Army for Empire: The United States Army in the Spanish-American War.* College Station: Texas A&M University Press, 1998.

Trask, David. *The War with Spain in 1898.* New York: Macmillan, 1981.

Walker, Dale L. *The Boys of '98: Theodore Roosevelt and the Rough Riders.* New York: Forge, 1998.

World War I

Asprey, Robert B. *At Belleau Wood.* Denton: University of North Texas Press, 1996.

Congdon, Don. *Combat: World War I.* New York: Dell, 1964.

Dalessandro, Robert J. and Michael G. Knapp. *Organization and Insignia of the American Expeditionary Force 1917–1923.* Atglen, PA: Schiffer Publishing, 2008.

Pershing, John J. *My Experiences in the World War,* vols. 1–2. New York: Stokes, 1931.

Stallings, Laurence. *The Doughboys: The Story of the AEF, 1917–1918.* New York: Harper & Row, 1963.

World War II

Blumenson, Martin. *Mark Clark: The Last of the Great World War II Commanders.* New York: Congdon & Weed, 1984.

Bradley, John H., Jack W. Dice, and Thomas E. Greiss. *The Second World War,* vols. 1–3. Wayne, NJ: Avery Publishing Group, 1989.

Bradley, Omar N. *Bradley: A Soldier's Story.* New York: Henry Holt, 1951.

Doubler, Michael D. *Closing with the Enemy: How GIs Fought the War in Europe, 1944–45*. Lawrence: University Press of Kansas, 1994.
Eisenhower, Dwight D. *Crusade in Europe.* Garden City, NY: Doubleday, 1948.
MacDonald, Charles B. *A Time for Trumpets: The Untold Story of the Battle of the Bulge.* New York: William Morrow, 1985.
Manchester, William. *American Caesar: Douglas MacArthur, 1880–1964.* Boston: Little, Brown, 1978.
Mosley, Leonard. *Marshall: Hero for Our Times.* New York: Hearst, 1982.
Ryan, Cornelius. *The Longest Day: June 6, 1944.* New York: Simon & Schuster, 1959.
Truscott, Lucian K., Jr. *Command Missions.* New York: Dutton, 1954.
Weigley, Russell. *Eisenhower's Lieutenants,* vols. 1 and 2. Bloomington: Indiana University Press, 1981.

Korea

Appleman, Roy E. *South to the Naktong, North to the Yalu: U.S. Army in the Korean War.* Washington, D.C.: Government Printing Office, 1961.
Blair, Clay. *The Forgotten War: America in Korea, 1950–1953.* New York: Times Books, 1987.
Fehrenbach, T. R. *This Kind of War: A Study in Unpreparedness.* New York: Bantam Books, 1963.
Whelan, Richard. *Drawing the Line: The Korean War, 1950–1953.* Boston: Little, Brown, 1990.

Vietnam

Braestrup, Peter. *Big Story: How the American Press and Television Reported and Interpreted the Crisis of Tet 1968 in Vietnam and Washington.* vols. 1–3, Boulder, CO: Westview, 1976.
Bunting, Josiah, *The Lionheads.* New York: Braziller, 1972.
Krepinevich, Andrew F., Jr. *The Army and Vietnam.* Baltimore: Johns Hopkins University Press, 1986.
Moore, Harold G. *We Were Soldiers Once . . . and Young: Ia Drang: The Battle that Changed the War in Vietnam.* New York: Harper Perennial Library, 1993.
Phillips, William R., and William C. Westmoreland. *Night of the Silver Stars: The Battle of Lang Vei.* Special Warfare Series. Annapolis: U.S. Naval Institute, 1997.

Post–Vietnam

DeLong, Kent, and Steven Tuckey. *Mogadishu! Heroism and Tragedy.* Westport, CT: Praeger, 1994.
Pagonis, LtG William. *Moving Mountains: Lessons in Leadership and Logistics from the Gulf War.* Boston: Harvard Business School Press, 1992.
Scales, Robert H., Jr. *Certain Victory: The U.S. Army in the Gulf War.* Dulles, VA: Brassey's, 1998.

THE AMERICAN SOLDIER THROUGH THREE CENTURIES

Fisher, Ernest F. *Guardians of the Republic: A History of the Noncommissioned Officer Corps of the U.S. Army.* Mechanicsburg, PA: Stackpole Books, 2001.

Garland, Albert N., ed. *A Distant Challenge: The U.S. Infantryman in Vietnam, 1967–1972.* Nashville: Battery Press, 1983.

Hoyt, Edwin P. *The GI's War: The Story of American Soldiers in Europe in World War II.* New York: McGraw-Hill, 1988.

Jones, James. *From Here to Eternity.* New York: Scribner's, 1951.

Kautz, August V. *The 1865 Customs of Service for Noncommissioned Officers and Soldiers.* Mechanicsburg, PA: Stackpole Books, republished 2001.

———. *The 1865 Customs of Service for Officers of the Army.* Mechanicsburg, PA: Stackpole Books, republished 2002.

Leinbaugh, Harold P., and John D. Campbell. *The Men of Company K: The Autobiography of a World War II Rifle Company.* New York: William Morrow, 1985.

Marshall, S. L. A. *Pork Chop Hill: The American Fighting Man in Action, Korea, Spring, 1953.* Nashville: Battery Press, 1986.

Mason, F. van Wyck. *The Fighting American.* New York: Reynal & Hitchcock, 1943.

McManus, John C. *The Deadly Brotherhood: The American Combat Soldier in World War II.* Novato, CA: Presidio, 1998.

Mitchell, Reid. *Civil War Soldiers.* New York: Penguin, 1997.

Nichols, David. *Ernie's War: The Best of Ernie Pyle's World War II Dispatches.* New York: Random House, 1986.

Peterson, Harold L. *The Book of the Continental Soldier.* Harrisburg, PA: Stackpole Books, 1968.

Rickey, Don, Jr. *Forty Miles a Day on Beans and Hay: The Enlisted Soldier Fighting the Indian Wars.* Norman: University of Oklahoma Press, 1973.

Simons, Anna J. *The Company They Keep: Life Inside the U.S. Army Special Forces.* New York, 1997.

Utley, Robert M. *Frontier Regulars: The United States Army and the Indian, 1866–1891.* Lincoln: University of Nebraska Press, 1984.

Wilson, George C. *Mud Soldiers: Life Inside the New American Army.* New York: Scribner's, 1989.

LEADERSHIP

Collins, Arthur S., and Daniel P. Bolger. *Common Sense Training: A Working Philosophy for Leaders.* Novato, CA: Presidio, 1998.

MacDonald, Charles B. *Company Commander.* New York: Bantam, 1984.

Malone, Dandridge M. *Small Unit Leadership: A Commonsense Approach.* Novato, CA: Presidio, 1995.

Newman, Aubrey S. *Follow Me I: The Human Element in Leadership.* Novato, CA: Presidio, 1997.

———. *Follow Me II: More on the Human Element in Leadership.* Novato, CA: Presidio, 1997.

———. *Follow Me III: Lessons on the Art and Science of High Command.* Novato, CA: Presidio, 1997.

Nye, Robert H. *The Challenge of Command: Reading for Military Excellence.* Wayne, NJ: Avery Publishing Group, 1986.

Mitchell, George C. *Matthew B. Ridgway: Soldier, Statesman, Scholar, Citizen.* Mechanicsburg, PA: Stackpole Books, 2001.

Puryear, Jr., Edgar F. *19 Stars: A Study in Military Character and Leadership.* Novato, CA: Presidio, 1997.

CONTEMPORARY INTEREST AND PROFESSIONAL TOPICS

Antal, John F. *Armor Attacks: The Tank Platoon: An Interactive Exercise in Small-Unit Tactics and Leadership.* Novato, CA: Presidio, 1992.

———. *Combat Team: The Captains' War; An Interactive Exercise in Company-Level Command in Battle.* Novato, CA: Presidio, 1998.

———. *Infantry Combat: The Rifle Platoon: An Interactive Exercise in Small-Unit Tactics and Leadership.* Novato, CA: Presidio, 1995.

Bolger, Daniel P. *Dragons at War: Land Battle in the Desert.* New York: Ivy, 1991.

———. *Savage Peace: Americans at War in the 1990s.* Novato, CA: Presidio, 1995.

Finney, Nathan K., and Tyrell O. Mayfield. *Redefining the Modern Military: The Intersection of Profession and Ethics.* Annapolis, MD: Naval Institute Press, 2018.

Stoneberger, Brett A. *Combat Leader's Field Guide.* Mechanicsburg, PA: Stackpole Books, 1999.

MILITARY THOUGHT

Clausewitz, Carl von. *On War.* Trans. and ed. by Michael Howard and Peter Paret. Princeton, NJ: Princeton University Press, 1984.

Earle, Edward Meade. *Makers of Modern Strategy: Military Thought from Machiavelli to Hitler.* Princeton, NJ: Princeton University Press, 1971.

Freedman, Lawrence. *The Future of War.* New York: Public Affairs, 2017.

Jomini, Baron de. *The Art of War.* Westport, CT: Greenwood Press, 1971.

Leonhard, Robert R. *Fighting by Minutes: Time and the Art of War.* Westport, CT: Praeger, 1994.

Nolan, Cathal J. *The Allure of Battle: A History of How Wars Have Been Won and Lost.* New York: Oxford University Press, 2017.

Paret, Peter, and Gordon A. Craig. *Makers of Modern Strategy: From Machiavelli to the Nuclear Age.* Princeton, NJ: Princeton University Press, 1986.

Peters, Ralph. *Fighting for the Future: Will America Triumph?* Mechanicsburg, PA: Stackpole Books, 1999.

Phillips, Thomas R. *Roots of Strategy: A Collection of Military Classics.* Mechanicsburg, PA: Stackpole Books, 1985.

Weigley, Russell F. *The American Way of War: A History of United States Military Strategy and Policy.* New York: Macmillan, 1973.

THE ARMY AND SOCIETY

Huntington, Samuel P. *Soldier and the State: The Theory of Politics of Civil-Military Relations.* Cambridge, MA: Belknap, 1981.

Janowitz, Morris. *The Professional Soldier: A Social and Political Portrait.* Glencoe, IL: Free Press, 1964.

Moskos, Charles C. *All That We Can Be: Black Leadership and Racial Integration the Army Way.* New York: Basic Books, 1997.

Skelton, William B. *An American Profession of Arms: The Army Officer Corps, 1784–1861.* Lawrence: University Press of Kansas, 1992.

GENERAL

Emerson, William K. *Encyclopedia of United States Army Insignia and Uniforms.* Norman: University of Oklahoma Press, 1996.

Ganoe, William A. *History of the United States Army,* rev. ed. New York: Appleton-Century, 1942.

Hastings, Max, and Simon Jenkins. *Battle for the Falklands.* New York: Norton, 1983.

Huston, James A. *The Sinews of War: Army Logistics, 1775–1953.* Army Historical Series, Washington, D.C.: Government Printing Office, 1966.

King, Benjamin, Richard C. Biggs, and Eric R. Criner. *Spearhead of Logistics: A History of the United States Army Transportation Corps.* Fort Eustis, VA: U.S. Army Transportation Center, 1994.

Risch, Erna. *Quartermaster Support of the Army: A History of the Corps, 1776–1918.* Princeton, NJ: Princeton University Press, 1962.

GENERAL MILITARY HISTORY

Asprey, Robert B. *War in the Shadows: The Guerilla in History,* vols. 1 and 2. Garden City, NY: Doubleday, 1975.

Cooper, Matthew. *The German Army, 1933–1945.* Chelsea, MI: Scarborough House, 1978.

Odom, William E. *The Collapse of the Soviet Military.* New Haven, CT: Yale University Press, 1998.

Taylor, Telford. *Sword and Swastika: Generals and Nazis in the Third Reich.* New York: Simon & Schuster, 1952.

28

Selected Acronyms

AAC	Army Acquisition Corps
AAFES	Army and Air Force Exchange Service
ABDU	Aircraft Battle Dress Uniform
ACDUTRA	Active Duty for Training
ACS	Army Community Service
ACU	Army Combat Uniform
ADA	Air Defense Artillery
AER	Army Emergency Relief
AERS	Army Education Requirements System
AFCS	Active Federal Commissioned Service
AFN	Armed Forces Network
AFS	Active Federal Service
AFSC	Armed Forces Staff College
AG	Adjutant General; Army Green
AGC	Adjutant General Corps
AHEC	Army Heritage and Education Center
AHM	Army Heritage Museum
AKO	Army Knowledge Online
ALO	Authorized Level of Organization
AMC	Air Mobility Command; U.S. Army Materiel Command
AME	Average Monthly Earnings
AMEDD	Army Medical Department
AMP	Advanced Management Program
AOC	Area of Concentration
AOE	Army of Excellence
APC	Armored Personnel Carrier
APFT	Army Physical Fitness Test
APO	Army Post Office
AR	Army Regulation
ARNG	Army National Guard
AUS	Army of the U.S. (Total Active Army Force)
AUSA	Association of the U.S. Army
AWC	Army War College
AWCCSC	Army War College Corresponding Studies Course
AWOL	Absent without Leave

BAH	Basic Allowance for Housing
BAS	Basic Allowance for Subsistence
BCT	Brigade Combat Team
BDU	Battle Dress Uniform
BOQ	Bachelor Officer Quarters
CAB	Combat Action Badge
CAC	Combined Arms Center
CASCOM	Combined Arms Support Command
CBRNE	Chemical, Biological, Radiological, Nuclear and Explosive
CFSSC-K	Community, Family, and Soldier Support Command, Korea
CGSC	Command and General Staff College
CIB	Combat Infantryman Badge
CID	Criminal Investigation Division
COIN	Counterinsurgency
COLA	Cost-of-Living Allowance
CONUS	Continental United States
CONUSA	Continental U.S. Army
COP	Common Operating Picture
CPI	Consumer Price Index
CPX	Command Post Exercise
CQ	Charge of Quarters
CS	Chief of Staff
CSA	Chief of Staff, Army
CTA	Common Table of Allowance
CVC	Combat Vehicle Crewman
DA	Department of the Army
DCP	Degree Completion Program
DCS	Defense Communications System
DCSI	Deputy Chief of Staff for Intelligence
DCSLOG	Deputy Chief of Staff for Logistics
DCSOPS	Deputy Chief of Staff for Operations and Plans
DCSPER	Deputy Chief of Staff for Personnel
DDE	Department of Distance Education
DEERS	Defense Enrollment Eligibility Reporting System
DIC	Dependency and Indemnity Compensation
DITY	Do-It-Yourself (Move)
DoD	Department of Defense
DoDEA	Department of Defense Educational Activity
DOPMA	Defense Officer Personnel Management Act
DOTMLPF	Doctrine, Organization, Training, Materiel, Leader Development, Personnel and Facilities
DPSC	Defense Personnel Support Center
DUI	Distinctive Unit Insignia
DVQ	Distinguished Visitor Quarters

ECWCS	Extended Cold-Weather Clothing System
EML	Environmental and Morale Leave
EOD	Explosive Ordnance Disposal
FA	Functional Area
FHA	Federal Housing Authority
FM	Field Manual
FOB	Forward Operating Base
FORSCOM	Forces Command
FTX	Field Training Exercise
GCM	General Court-Martial
GED	General Educational Development
GSA	General Services Administration
GTA	Graphical Training Aids
HHC	Headquarters and Headquarters Company
HHG	Household Goods
HQ	Headquarters
HQDA	Headquarters, Department of the Army
HRO	Housing Referral Office, Germany
IADC	Inter-American Defense College
ICAF	Industrial College of the Armed Forces
ID	Identification
IG	Inspector General
INSCOM	U.S. Army Intelligence and Security Command
IRA	Individual Retirement Account
IRS	Internal Revenue Service
ISB	Intermediate Staging Base
ITO	Installation Transportation Office
JAG	Judge Advocate General
JDA	Joint Duty Assignment
JFTR	Joint Federal Travel Regulations
JPME	Joint Professional Military Education
JRTC	Joint Readiness Training Center
JSO	Joint Specialty Officer
JUSMAG	Joint U.S. Military Advisory Group
KATUSA	Korean Augmentation to the U.S. Army
KP	Kitchen Police
LES	Leave and Earnings Statement
LFA	Lead Federal Agency
MACOM	Major Army Command
MALT	Monetary Allowance in Lieu of Transportation
MCM	Manual for Courts-Martial

MCSS	Military Clothing Sales Store
MDO	Multi Domain Operations
MDW	U.S. Army Military District of Washington
MECH	Mechanized
MEL	Military Education Level
MFO	Multinational Force and Observers
MI	Military Intelligence
MIHA	Move-in Housing Allowance
MOOTW	Military Operations Other Than War
MOS	Military Occupational Specialty
MP	Military Police
MS or MSC	Medical Service Corps
MSA	Morale Support Activity
MSC	Military Sealift Command
MTMC	Military Traffic Management Command
MTOE	Modified Tables of Organization and Equipment
MWR	Morale, Welfare, and Recreation
NASA	National Aeronautics and Space Administration
NATO	North Atlantic Treaty Organization
NAUS	National Association of Uniformed Services
NCO	Noncommissioned Officer
NDU	National Defense University
NTC	National Training Center
NWC	National War College
OCDETF	Organized Crime Drug Enforcement Task Force
OCO	Office, Chief of Ordnance
OCONUS	Outside the Continental United States
OCS	Officer Candidate School
OER	Officer Evaluation Report
OHA	Overseas Housing Allowance
OIC	Officer-in-Charge
OJT	On-the-Job Training
OMPF	Official Military Personnel File
OOD	Officer of the Deck
OPMD	Officer Personnel Management Directorate
OPMS	Officer Personnel Management System
ORB	Officer Record Brief
OTRA	Other than Regular Army
PAC	Personnel and Administration Center
PAM	Pamphlet
PAO	Public Affairs Officer
PCS	Permanent Change of Station
PERSCOM	U.S. Total Army Personnel Command

PIO	Public Information Office
PLL	Prescribed Load List
PM	Preventive Maintenance
POE	Port of Embarkation
POV	Privately Owned Vehicle
PROV	Provisional
PX	Post Exchange
QM	Quartermaster
QMC	Quartermaster Corps
RA	Regular Army
RASL	Reserve Active Status List
R&D	Research and Development
RDI	Regimental Distinctive Insignia
REINF	Reinforced
REMBASS	Remotely-Monitored Battlefield Sensor System
RHIP	Rank Has Its Privileges
ROK	Republic of Korea
RON	Remain Overnight
ROPMA	Reserve Officer Personnel Management Act
ROTC	Reserve Officers Training Corps
R&R	Rest and Recuperation
RSO	Reconnaissance Staff Officer
SBCT	Stryker Brigade Combat Team
SBP	Survivor Benefit Plan
SGLI	Servicemember's Group Life Insurance
SOP	Standard Operating Procedure
SSB	Special Separation Benefit
SSC	Senior Service College
SSI	Shoulder Sleeve Insignia
TA	Table of Allowances; Training Aids
TAD	Temporary Additional Duty
TAG	The Adjutant General
TC	Transportation Corps
TCS	Total Commissioned Service
TD	Table of Distribution
TDA	Tables of Distribution and Allowance
TDY	Temporary Duty
TEXCOM	Test and Experimentation Command
TIG	Time in Grade
TIS	Time in Service
TLA	Temporary Lodging Allowance
TLE	Temporary Lodging Entitlement

TM	Technical Manual
TOE	Tables of Organization and Equipment
TR	Transportation Request
TRAC	TRADOC Analysis Command
TRADOC	U.S. Army Training and Doctrine Command
TRANSCOM	U.S. Transportation Command
TWI	Training with Industry
TWOS	Total Warrant Officer System
UCMJ	Uniform Code of Military Justice
USA	U.S. Army (Regular Army)
USACE	U.S. Army Corps of Engineers
USAF	U.S. Air Force
USAISC	U.S. Army Information Systems Command
USAR	U.S. Army Reserve
USARAL	U.S. Army, Alaska
USARC	U.S. Army Reserve Command
USAREC	U.S. Army Recruiting Command
USAREUR	U.S. Army, Europe
USARJ	U.S. Army, Japan
USARPAC	U.S. Army, Pacific
USARSO	U.S. Army, South
USASOC	U.S. Army Special Operations Command
USC	United States Code
USMA	U.S. Military Academy
USO	United Services Organization
USSOCOM	U.S. Special Operations Command
VA	Department of Veterans Affairs
VEAP	Veterans Educational Assistance Act of 1984
VGLI	Veterans' Group Life Insurance
VIP	Very Important Person
VOQ	Visiting Officer Quarters
VSI	Voluntary Separation Incentive
WO	Warrant Officer
WOBC	Warrant Officer Basic Course
WOCC	Warrant Officer Candidate Course
WOMA	Warrant Officer Management Act
WOS	Warrant Officer Service
WOSC	Warrant Officer Senior Course
WOSSC	Warrant Officer Senior Staff Course
YG	Year Group

Index

About the Author

LTC Eric Hiu, USA (Ret.), was commissioned in the U.S. Army after graduating from Officers Candidate School at Fort Benning, Georgia, in 1995 following his graduation from St. Mary's College of Maryland in 1994. He obtained a Bachelor's of Arts in History from SMCM and Master's degree in international relations from Troy University in 2005.

After beginning his career in 1995 as a second lieutenant of infantry, he served in a wide variety of leadership and staff assignments, including platoon leader, company executive officer, two infantry company commands, infantry captain's career course small group instructor, and lead planner for the U.S. Second Fleet. As an army strategist, he served as a planner in numerous positions at the operational level and higher. LTC Hiu retired as the chief of joint concepts at the Army Futures Command's Futures and Concepts Center. He and his wife Katherine have four great Army kids, the oldest of whom is a captain in the United States Army.

LTC Hiu's numerous awards and decorations include the Legion of Merit, Bronze Star, Defense Meritorious Service Medal, Meritorious Service Medal with four Oak Leaf Clusters, Army Commendation Medal, Joint Service Achievement Medal, Army Achievement Medal with Oak Leaf Cluster, Joint Meritorious Award, Humanitarian Service Medal, National Defense Service Medal (second award), Global War on Terrorism Service Medal, Korea Defense Service Medal, Afghanistan Campaign Medal w/Campaign Star, Iraq Campaign Medal with Campaign Star (second award), Army Service Ribbon, Overseas Service Ribbon (sixth award), NATO Medal, Expert Infantryman Badge, Parachutist Badge, and Air Assault Badge.